Inverse Problems and Applications

Gunther Uhlmann

CONTEMPORARY MATHEMATICS

615

Inverse Problems and Applications

Conference in Honor of Gunther Uhlmann
on Inverse Problems
June 18–22, 2012
University of California, Irvine, CA

International Conference in Honor
of Gunther Uhlmann's 60th Birthday
on Inverse Problems and Applications
September 17–21, 2012
Yuquan Campus, Zhejiang University, Hangzhou, China

Plamen Stefanov
András Vasy
Maciej Zworski
Editors

American Mathematical Society
Providence, Rhode Island

2010 *Mathematics Subject Classification.* Primary 35R30.

Library of Congress Cataloging-in-Publication Data

Conference on Inverse Problems (2012 : Irvine, Calif.)

Inverse problems and applications : Conference on Inverse Problems, in honor of Gunther Uhlmann, June 18-22, 2012, University of California, Irvine, CA : International Conference on Inverse Problems, in honor of Gunther Uhlmann's 60th birthday, September 17-21, 2012, Zhejiang University, Hangzhou, China / Plamen Stefanov, Andras Vasy, Maciej Zworski, editors.

pages cm. – (Contemporary mathematics ; volume 615)

Includes bibliographical references.

ISBN 978-1-4704-1079-7 (alk. paper)

1. Inverse problems (Differential equations)–Congresses. I. Uhlmann, Gunther, 1952– honouree. II. Stefanov, Plamen, 1959- editor of compilation. III. Vasy, András, editor of compilation. IV. Zworski, Maciej, editor of compilation. V. International Conference on Inverse Problems (2012 : Hangzhou, China). VI. Title.

QA378.5.C665 2012
515′.353–dc23 2013039861

Contemporary Mathematics ISSN: 0271-4132 (print); ISSN: 1098-3627 (online)

DOI: http://dx.doi.org/10.1090/conm/615

∞ The paper used in this book is acid-free and falls within the guidelines
established to ensure permanence and durability.
Visit the AMS home page at http://www.ams.org/

10 9 8 7 6 5 4 3 2 1 19 18 17 16 15 14

Contents

Preface

This volume presents original research and expository papers by speakers at two conferences on Inverse Problems and Applications, held in 2012, to celebrate the work of Gunther Uhlmann. The first conference was at the University of California, Irvine, CA, in June 2012; the second one was in Hangzhou, China, in September 2012.

The topics covered include inverse problems appearing in medical imaging, scattering theory, geometry and image processing, and the mathematical theory of cloaking, as well as methods related to inverse problems.

This broad range of topics gives a cross-section of the vibrant current research in inverse problems.

Contemporary Mathematics
Volume **615**, 2014
http://dx.doi.org/10.1090/conm/615/12244

Spectral theory of a Neumann-Poincaré-type operator and analysis of cloaking by anomalous localized resonance II

Habib Ammari, Giulio Ciraolo, Hyeonbae Kang, Hyundae Lee, and Graeme W. Milton

ABSTRACT. If a body of dielectric material is coated by a plasmonic structure of negative dielectric constant with nonzero loss parameter, then cloaking by anomalous localized resonance (CALR) may occur as the loss parameter tends to zero. The aim of this paper is to investigate this phenomenon in two and three dimensions when the coated structure is radial, and the core, shell and matrix are isotropic materials. In two dimensions, we show that if the real part of the permittivity of the shell is -1 (under the assumption that the permittivity of the background is 1), then CALR takes place. If it is different from -1, then CALR does not occur. In three dimensions, we show that CALR does not occur. The analysis of this paper reveals that occurrence of CALR is determined by the eigenvalue distribution of the Neumann-Poincaré-type operator associated with the structure.

1. Introduction

If a body of dielectric material is coated by a plasmonic structure of negative dielectric constant (with nonzero loss parameter), then anomalous localized resonance may occur as the loss parameter tends to zero. This phenomena, first discovered by Nicorovici, McPhedran and Milton [**35**] (see also [**32**]), is responsible for the subwavelength focussing properties of superlenses [**37**], and also occurs in magnetoelectric and thermoelectric systems [**32**]. The fields blow-up in a localized region, which moves as the position of the source is moved, which is why it is termed anomalous localized resonance. Remarkably, as found by Milton and Nicorovici [**29**] the localized resonant fields created by a source can act back on the source and cloak it. This invisibility cloaking has attracted much attention [**2, 7, 8, 11, 20, 27, 29–31, 33, 34, 36, 39**].

To state the problem and results in a precise way, let Ω be a bounded domain in $\mathbb{R}^d$, $d = 2, 3$, and D be a domain whose closure is contained in Ω. For a given

2010 *Mathematics Subject Classification.* Primary 35R30; Secondary 35B30.

Key words and phrases. Anomalous localized resonance, plasmonic materials, quasi-static cloaking, Neumann-Poincaré operator.

This work was supported by the ERC Advanced Grant Project MULTIMOD–267184 and NRF grants No. 2010-0004091, and 2010-0017532, and by the NSF through grants DMS-0707978 and DMS-1211359.

loss parameter $\delta > 0$, the permittivity distribution in $\mathbb{R}^d$ is given by

$$\epsilon_\delta = \begin{cases} 1 & \text{in } \mathbb{R}^d \setminus \overline{\Omega}, \\ \epsilon_s + i\delta & \text{in } \Omega \setminus \overline{D}, \\ \epsilon_c & \text{in } D, \end{cases} \tag{1.1}$$

where $-\epsilon_s$ and ϵ_c are positive. We may consider the configuration as a core with permittivity ϵ_c coated by the shell $\Omega \setminus \overline{D}$ with permittivity $\epsilon_s + i\delta$. For a given function f compactly supported in $\mathbb{R}^d \setminus \overline{\Omega}$ satisfying

$$\int_{\mathbb{R}^2} f\, dx = 0 \tag{1.2}$$

(which is required by conservation of charge), we consider the following dielectric problem:

$$\nabla \cdot \epsilon_\delta \nabla V_\delta = f \quad \text{in } \mathbb{R}^d, \tag{1.3}$$

with the decay condition $V_\delta(x) \to 0$ as $|x| \to \infty$. The problem of cloaking by anomalous localized resonance (CALR) can be formulated as the problem of identifying the sources f such that first

$$E_\delta := \int_{\Omega\setminus D} \delta |\nabla V_\delta|^2\, dx \to \infty \quad \text{as } \delta \to 0, \tag{1.4}$$

and second, $V_\delta/\sqrt{E_\delta}$ goes to zero outside some radius a, as $\delta \to 0$:

$$|V_\delta(x)/\sqrt{E_\delta}| \to 0 \quad \text{as } \delta \to 0 \quad \text{when } |x| > a. \tag{1.5}$$

Physically the quantity E_δ is proportional to the electromagnetic power dissipated into heat by the time harmonic electrical field averaged over time. Using integration by parts we have the identity

$$E_\delta = \Im \int_{\mathbb{R}^d} (\epsilon_\delta \nabla V_\delta) \cdot \nabla \overline{V_\delta}\, dx = -\Im \int_{\mathbb{R}^d} f \overline{V_\delta}\, dx \tag{1.6}$$

which equates the power dissipated into heat with the electromagnetic power produced by the source, where $\overline{V_\delta}$ is the complex conjugate of V_δ. Hence (1.4) implies an infinite amount of energy dissipated per unit time in the limit $\delta \to 0$ which is unphysical. If we rescale the source f by a factor of $1/\sqrt{E_\delta}$ then the source will produce the same power independent of δ and the new associated potential $V_\delta/\sqrt{E_\delta}$ will, by (1.5), approach zero outside the radius a: cloaking due to anomalous localized resonance (CALR) occurs.

In the recent paper [**2**] the authors develop a spectral approach to analyze the CALR phenomenon. In particular, they show that if D and Ω are concentric disks in $\mathbb{R}^2$ and $\epsilon_c = -\epsilon_s = 1$, then there is a critical radius r_* such that for any source f supported outside r_* CALR does not occur, and for sources f satisfying a mild condition CALR takes place. The critical radius r_* is given by

$$r_* = \sqrt{r_e^3/r_i}, \tag{1.7}$$

where r_e and r_i are the radii of Ω and D, respectively. It is worth mentioning that these results were extended in [**20**] to the case when the core D is not radial by a different method based on a variational approach.

The purpose of this paper is to extend some of the results in [**2**] in two directions. We consider the two dimensional case when ϵ_c and $-\epsilon_s$ are not both 1 and CALR in three dimensions. The results of this paper are threefold: Let Ω and D be concentric disks or balls in $\mathbb{R}^d$ of radii r_e and r_i, respectively. Then, the following results hold:

- If $d = 2$ and $\epsilon_s = -1$, then CALR occurs. When $\epsilon_c = 1$ the critical radius r_* is given by (1.7) and when $\epsilon_c \neq 1$ the critical radius is

$$r_* = \frac{r_e^2}{r_i}. \tag{1.8}$$

  That is, for almost any source f supported inside r_* (satisfying the gap condition specified in section 4) CALR occurs and for any source f supported outside r_* CALR does not occur. When $\epsilon_c \neq 1$ the cloaking radius r_e^2/r_i matches that found in [**29**] for a single dipolar source (see figure 5 there and accompanying text).
- If $\epsilon_s \neq -1$, then CALR does not occur.
- If $d = 3$, then CALR does not occur whatever ϵ_s and ϵ_c are.

We emphasize that the result on non-occurrence of CALR in three dimensions holds only when the dielectric constant ϵ_s is constant. In the recent work [**3**] we show that CALR does occur in three dimensions if we use a shell with non-constant (anisotropic) dielectric constant.

It turns out that the occurrence of CALR depends on the distribution of eigenvalues of the Neumann-Poincaré (NP) operator associated with the structure (see the next section for the definition of the NP operator). The NP operator is compact with its eigenvalues accumulating towards 0. It is proved in [**2**] that in two dimensions the NP operator associated with the circular structure has the eigenvalues $\pm\rho^n$ for $n = 1, 2, \ldots$, where $\rho = r_i/r_e$. We show that in three dimensions the NP operator associated with the spherical structure has the eigenvalues

$$\pm\frac{1}{2(2n+1)}\sqrt{1 + 4n(n+1)\rho^{2n+1}}, \quad n = 0, 1, \ldots. \tag{1.9}$$

The exponential convergence of the eigenvalues in two dimensions is responsible for the occurrence of CALR and the slow convergence (at the rate $1/n$) in three dimensions is responsible for the non-occurrence.

2. Layer potential formulation

Let G be the fundamental solution to the Laplacian in $\mathbb{R}^d$ which is given by

$$G(x) = \begin{cases} \dfrac{1}{2\pi}\ln|x|, & d = 2, \\ -\dfrac{1}{4\pi}\dfrac{1}{|x|}, & d = 3. \end{cases}$$

Let $\Gamma_i := \partial D$ and $\Gamma_e := \partial\Omega$. For $\Gamma = \Gamma_i$ or Γ_e, we denote the single layer potential of a function $\varphi \in L^2(\Gamma)$ as $\mathcal{S}_\Gamma[\varphi]$, where

$$\mathcal{S}_\Gamma[\varphi](x) := \int_\Gamma G(x - y)\varphi(y)\, d\sigma(y), \quad x \in \mathbb{R}^d.$$

We also define the boundary integral operator $\mathcal{K}_\Gamma$ on $L^2(\Gamma)$ by

$$\mathcal{K}_\Gamma[\varphi](x) := \int_\Gamma \frac{\partial G(x - y)}{\partial\nu(y)}\varphi(y)\, d\sigma(y), \quad x \in \Gamma,$$

and let $\mathcal{K}_\Gamma^*$ be the L^2-adjoint of $\mathcal{K}_\Gamma$. Hence, the operator $\mathcal{K}_\Gamma^*$ is given by

$$\mathcal{K}_\Gamma^*[\varphi](x) = \int_\Gamma \frac{\partial G(x-y)}{\partial \nu(x)} \varphi(y)\, d\sigma(y), \quad \varphi \in L^2(\Gamma).$$

The operators $\mathcal{K}_\Gamma$ and $\mathcal{K}_\Gamma^*$ are called Neumann-Poincaré operators. These operators are compact in $L^2(\Gamma)$ if Γ is $\mathcal{C}^{1,\alpha}$ for some $\alpha > 0$.

The following notation will be used throughout this paper. For a function u defined on $\mathbb{R}^d \setminus \Gamma$, we denote

$$u|_\pm(x) := \lim_{t\to 0^+} u(x \pm t\nu(x)), \quad x \in \Gamma,$$

and

$$\frac{\partial u}{\partial \nu}\Big|_\pm(x) := \lim_{t\to 0^+} \langle \nabla u(x \pm t\nu(x)), \nu(x)\rangle\,, \quad x \in \Gamma,$$

if the limits exist. Here and throughout this paper, ν is the outward unit normal vector field to Γ and $\langle\ ,\ \rangle$ denotes the scalar product on $\mathbb{R}^d$.

The following jump formula relates the traces of the normal derivative of the single layer potential to the operator $\mathcal{K}_\Gamma^*$. We have

$$\frac{\partial}{\partial \nu} \mathcal{S}_\Gamma[\varphi]\Big|_\pm(x) = \left(\pm\frac{1}{2}I + \mathcal{K}_\Gamma^*\right)[\varphi](x), \quad x \in \Gamma. \tag{2.1}$$

See, for example, [**5**, **12**].

Let F be the Newtonian potential of f, *i.e.*,

$$F(x) = \int_{\mathbb{R}^d} G(x-y) f(y) dy, \quad x \in \mathbb{R}^d. \tag{2.2}$$

Then F satisfies $\Delta F = f$ in $\mathbb{R}^d$, and the solution V_δ to (1.3) can be represented as

$$V_\delta(x) = F(x) + \mathcal{S}_{\Gamma_i}[\varphi_i](x) + \mathcal{S}_{\Gamma_e}[\varphi_e](x) \tag{2.3}$$

for some functions $\varphi_i \in L^2_0(\Gamma_i)$ and $\varphi_e \in L^2_0(\Gamma_e)$ (L^2_0 is the collection of square integrable functions with zero mean-value). The transmission conditions along the interfaces Γ_e and Γ_i satisfied by V_δ read

$$\begin{aligned} (\epsilon_s + i\delta)\frac{\partial V_\delta}{\partial \nu}\Big|_+ &= \epsilon_c \frac{\partial V_\delta}{\partial \nu}\Big|_- \quad \text{on } \Gamma_i, \\ \frac{\partial V_\delta}{\partial \nu}\Big|_+ &= (\epsilon_s + i\delta)\frac{\partial V_\delta}{\partial \nu}\Big|_- \quad \text{on } \Gamma_e. \end{aligned}$$

Hence the pair of potentials (φ_i, φ_e) appearing in (2.3) is the solution to the following system of integral equations:

$$\begin{cases} (\epsilon_s + i\delta)\dfrac{\partial \mathcal{S}_{\Gamma_i}[\varphi_i]}{\partial \nu_i}\Big|_+ - \epsilon_c \dfrac{\partial \mathcal{S}_{\Gamma_i}[\varphi_i]}{\partial \nu_i}\Big|_- + (\epsilon_s - \epsilon_c + i\delta)\dfrac{\partial \mathcal{S}_{\Gamma_e}[\varphi_e]}{\partial \nu_i} \\ \qquad = (-\epsilon_s + \epsilon_c - i\delta)\dfrac{\partial F}{\partial \nu_i} \quad \text{on } \Gamma_i, \\ (-1 + \epsilon_s + i\delta)\dfrac{\partial \mathcal{S}_{\Gamma_i}[\varphi_i]}{\partial \nu_e} - \dfrac{\partial \mathcal{S}_{\Gamma_e}[\varphi_e]}{\partial \nu_e}\Big|_+ + (\epsilon_s + i\delta)\dfrac{\partial \mathcal{S}_{\Gamma_e}[\varphi_e]}{\partial \nu_e}\Big|_- \\ \qquad = (1 - \epsilon_s - i\delta)\dfrac{\partial F}{\partial \nu_e} \quad \text{on } \Gamma_e. \end{cases}$$

Note that we have used the notation ν_i and ν_e to indicate the outward normal on Γ_i and Γ_e, respectively. Using the jump formula (2.1) for the normal derivative of the single layer potentials, the above equations can be rewritten as

$$\begin{bmatrix} z_i^\delta I - \mathcal{K}_{\Gamma_i}^* & -\dfrac{\partial}{\partial \nu_i}\mathcal{S}_{\Gamma_e} \\ \dfrac{\partial}{\partial \nu_e}\mathcal{S}_{\Gamma_i} & z_e^\delta I + \mathcal{K}_{\Gamma_e}^* \end{bmatrix} \begin{bmatrix} \varphi_i \\ \varphi_e \end{bmatrix} = \begin{bmatrix} \dfrac{\partial F}{\partial \nu_i} \\ -\dfrac{\partial F}{\partial \nu_e} \end{bmatrix} \tag{2.4}$$

on $\mathcal{H}_0 = L_0^2(\Gamma_i) \times L_0^2(\Gamma_e)$, where we set

$$z_i^\delta = \frac{\epsilon_c + \epsilon_s + i\delta}{2(\epsilon_c - \epsilon_s - i\delta)}, \quad z_e^\delta = \frac{1 + \epsilon_s + i\delta}{2(1 - \epsilon_s - i\delta)}. \tag{2.5}$$

Let $\mathcal{H} = L^2(\Gamma_i) \times L^2(\Gamma_e)$ and let the Neumann-Poincaré-type operator $\mathbb{K}^* : \mathcal{H} \to \mathcal{H}$ be defined by

$$\mathbb{K}^* := \begin{bmatrix} -\mathcal{K}_{\Gamma_i}^* & -\dfrac{\partial}{\partial \nu_i}\mathcal{S}_{\Gamma_e} \\ \dfrac{\partial}{\partial \nu_e}\mathcal{S}_{\Gamma_i} & \mathcal{K}_{\Gamma_e}^* \end{bmatrix}, \tag{2.6}$$

and let

$$\Phi := \begin{bmatrix} \varphi_i \\ \varphi_e \end{bmatrix}, \quad g := \begin{bmatrix} \dfrac{\partial F}{\partial \nu_i} \\ -\dfrac{\partial F}{\partial \nu_e} \end{bmatrix}. \tag{2.7}$$

Then, (2.4) can be rewritten in the form

$$(\mathbb{I}^\delta + \mathbb{K}^*)\Phi = g, \tag{2.8}$$

where $\mathbb{I}^\delta$ is given by

$$\mathbb{I}^\delta = \begin{bmatrix} z_i^\delta I & 0 \\ 0 & z_e^\delta I \end{bmatrix}. \tag{2.9}$$

3. Eigenvalues of the NP operator

It is proved in [**2**] that for arbitrary-shaped domains Ω and D the spectrum of the NP operator $\mathbb{K}^*$ lies in $[-1/2, 1/2]$, and if Ω and D are concentric disks, the eigenvalues of $\mathbb{K}^*$ on $\mathcal{H}_0$ are $\pm\rho^n/2$, $n = 1, 2, \ldots$. In this section we compute the eigenvalues $\mathbb{K}^*$ on $\mathcal{H}$ when Ω and D are concentric disks or balls.

3.1. Two dimensions. Let $\Gamma = \{|x| = r_0\}$ in two dimensions. It is known that for each integer n

$$\mathcal{S}_\Gamma[e^{in\theta}](x) = \begin{cases} -\dfrac{r_0}{2|n|}\left(\dfrac{r}{r_0}\right)^{|n|} e^{in\theta} & \text{if } |x| = r < r_0, \\ -\dfrac{r_0}{2|n|}\left(\dfrac{r_0}{r}\right)^{|n|} e^{in\theta} & \text{if } |x| = r > r_0. \end{cases} \tag{3.1}$$

Moreover,

$$\mathcal{K}_\Gamma^*[e^{in\theta}] = 0 \quad \forall n \neq 0, \tag{3.2}$$

and

$$\mathcal{K}_\Gamma[1] = \frac{1}{2}. \tag{3.3}$$

In other words, $\mathcal{K}_\Gamma$ is a rank 1 operator whose only non-zero eigenvalue is $1/2$.

Using (3.2), it is proved that eigenvalues of $\mathbb{K}^*$ on $\mathcal{H}_0$ are $\pm\rho^n/2$ (see [**2**]). We now show that $\pm 1/2$ are also eigenvalues of $\mathbb{K}^*$ on $\mathcal{H}$. These eigenvalues are of interest in relation to estimation of stress concentration [**4**]. Using (3.3) we have

$$\mathcal{S}_\Gamma[1](x) = \begin{cases} \log r_0 & \text{if } |x| = r < r_0, \\ \log|x| & \text{if } |x| = r > r_0, \end{cases} \tag{3.4}$$

and hence

$$\frac{\partial}{\partial r}\mathcal{S}_\Gamma[1](x) = \begin{cases} 0 & \text{if } |x| = r < r_0, \\ \dfrac{1}{r} & \text{if } |x| = r > r_0. \end{cases} \tag{3.5}$$

It then follows that

$$\mathbb{K}^* \begin{bmatrix} a \\ b \end{bmatrix} = \begin{bmatrix} -\frac{1}{2} & 0 \\ \frac{1}{r_e} & \frac{1}{2} \end{bmatrix} \begin{bmatrix} a \\ b \end{bmatrix}, \tag{3.6}$$

where a and b are constants. So $\pm 1/2$ are eigenvalues of $\mathbb{K}^*$.

We summarize our findings in the following proposition.

PROPOSITION 3.1. *The eigenvalues of $\mathbb{K}^*$ defined on concentric circles in two dimensions are*

$$-\frac{1}{2},\ \frac{1}{2},\ -\frac{1}{2}\rho^n,\ \frac{1}{2}\rho^n,\ n = 1, 2, \ldots, \tag{3.7}$$

and corresponding eigenfunctions are

$$\begin{bmatrix} 1 \\ -\frac{1}{r_e} \end{bmatrix},\ \begin{bmatrix} 0 \\ 1 \end{bmatrix},\ \begin{bmatrix} e^{\pm in\theta} \\ \rho e^{\pm in\theta} \end{bmatrix},\ \begin{bmatrix} e^{\pm in\theta} \\ -\rho e^{\pm in\theta} \end{bmatrix},\ n = 1, 2, \ldots. \tag{3.8}$$

3.2. Three dimensions. Let $Y_n^m(\hat{x})$ ($m = -n, -n+1, \ldots, 0, 1, \ldots, n$) be the orthonormal spherical harmonics of degree n. Here $\hat{x} = \frac{x}{|x|}$. Then $|x|^n Y_n^m(\hat{x})$ is harmonic in $\mathbb{R}^3$.

LEMMA 3.2. *Let $\Gamma = \{|x| = r_0\}$ in three dimensions. We have for $n = 0, 1, \ldots$*

$$\mathcal{K}_\Gamma^*[Y_n^m](x) = \frac{1}{2(2n+1)} Y_n^m(\hat{x}), \quad |x| = r_0, \quad m = -n, \ldots, n. \tag{3.9}$$

PROOF. It is proved in [**18**, Lemma 2.3] that

$$\mathcal{K}_\Gamma^*[\varphi](x) = -\frac{1}{2r_0}\mathcal{S}_\Gamma[\varphi](x), \quad |x| = r_0 \tag{3.10}$$

for any function $\varphi \in L^2(\Gamma)$. So it follows from (2.1) that

$$\frac{\partial}{\partial r}\mathcal{S}_\Gamma[\varphi]\big|_-(x) + \frac{1}{2r_0}\mathcal{S}_\Gamma[\varphi](x) = -\frac{1}{2}\varphi(x), \quad |x| = r_0. \tag{3.11}$$

Let $\varphi(x) = Y_n^m(\hat{x})$. Then since $\mathcal{S}_\Gamma[Y_n^m](x)$ and $|x|^n Y_n^m(\hat{x})$ are harmonic functions in $\{|x| < r_0\}$, we have

$$\mathcal{S}_\Gamma[Y_n^m](x) = -\frac{1}{2n+1}\frac{r^n}{r_0^{n-1}} Y_n^m(\hat{x}), \quad \text{for } |x| = r \le r_0, \tag{3.12}$$

and (3.9) follows from (3.10). □

Lemma 3.2 says that the eigenvalues of $\mathcal{K}_\Gamma^*$ on $L^2(\Gamma)$ when Γ is a sphere are $\frac{1}{2(2n+1)}$, $n = 0, 1, \ldots$, and their multiplicities are $2n+1$.

By (3.12), we have

$$\frac{\partial}{\partial \nu_i} \mathcal{S}_{\Gamma_e}[Y_n^m](x) = -\frac{n}{2n+1}\left(\frac{r_i}{r_e}\right)^{n-1} Y_n^m(\hat{x}), \quad |x| = r_i. \tag{3.13}$$

Similarly, we have

$$\mathcal{S}_{\Gamma_i}[Y_n^m](x) = -\frac{1}{2n+1}\frac{r_i^{n+2}}{r^{n+1}} Y_n^m(\hat{x}), \quad |x| = r \geq r_i,$$

and hence

$$\frac{\partial}{\partial \nu_e} \mathcal{S}_{\Gamma_i}[Y_n^m](x) = \frac{n+1}{2n+1}\left(\frac{r_i}{r_e}\right)^{n+2} Y_n^m(\hat{x}), \quad |x| = r_e. \tag{3.14}$$

We now have for constants a and b

$$\mathbb{K}^* \begin{bmatrix} aY_n^m \\ bY_n^m \end{bmatrix} = \begin{bmatrix} \left(-\frac{a}{2(2n+1)} + b\frac{n}{2n+1}\rho^{n-1}\right) Y_n^m \\ \left(a\frac{n+1}{2n+1}\rho^{n+2} + \frac{b}{2(2n+1)}\right) Y_n^m \end{bmatrix} = \begin{bmatrix} -\frac{1}{2(2n+1)} & \frac{n}{2n+1}\rho^{n-1} \\ \frac{n+1}{2n+1}\rho^{n+2} & \frac{1}{2(2n+1)} \end{bmatrix} \begin{bmatrix} aY_n^m \\ bY_n^m \end{bmatrix}. \tag{3.15}$$

Thus we have the following result.

PROPOSITION 3.3. *The eigenvalues of $\mathbb{K}^*$ defined on two concentric spheres are*

$$\pm\frac{1}{2(2n+1)}\sqrt{1+4n(n+1)\rho^{2n+1}}, \quad n = 0, 1, \ldots, \tag{3.16}$$

and corresponding eigenfunctions are

$$\begin{bmatrix} (\sqrt{1+4n(n+1)\rho^{2n+1}}-1)Y_n^m \\ 2(n+1)\rho^{n+2}Y_n^m \end{bmatrix}, \quad \begin{bmatrix} (-\sqrt{1+4n(n+1)\rho^{2n+1}}-1)Y_n^m \\ 2(n+1)\rho^{n+2}Y_n^m \end{bmatrix}, \tag{3.17}$$

for $m = -n, \ldots, n$, respectively.

It is quite interesting to observe that if we let $\frac{1}{2} = \lambda_0 \geq \lambda_1 \geq \ldots$ be the eigenvalues of $\mathcal{K}_\Gamma$ for a circle or a sphere enumerated according to their multiplicities, then the eigenvalues μ_n of $\mathbb{K}^*$ (which is defined on concentric circles or spheres) satisfy

$$\mu_n = \pm\lambda_n + O(\rho^n). \tag{3.18}$$

4. Anomalous localized resonance in two dimensions

In this section we consider CALR when the domains Ω and D are concentric disks. We first observe that z_i^δ and z_e^δ converges to non-zero numbers as δ tends to 0 if $\epsilon_c \neq -\epsilon_s \neq 1$. So, in this case CALR does not occur regardless of the location of the source. Furthermore, if $\epsilon_c = -\epsilon_s = 1$, a thorough study was done in [**2**]. It is proved in [**2**] that if the source f is supported inside the critical radius $r_* = \sqrt{r_e^3/r_i}$, then the weak CALR occurs, namely,

$$\limsup_{\delta \to 0} E_\delta = \infty. \tag{4.1}$$

Moreover, if F is the Newtonian potential of f and the Fourier coefficients g_e^n of $-\frac{\partial F}{\partial \nu_e}$ satisfies the following gap property:

[GP] There exists a sequence $\{n_k\}$ with $|n_1| < |n_2| < \cdots$ such that

$$\lim_{k\to\infty} \rho^{|n_{k+1}|-|n_k|} \frac{|g_e^{n_k}|^2}{|n_k|\rho^{|n_k|}} = \infty,$$

then CALR occurs, namely,

$$\lim_{\delta\to 0} E_\delta = \infty, \tag{4.2}$$

and $V_\delta/\sqrt{E_\delta}$ goes to zero outside the radius $\sqrt{r_e^3/r_i}$.

The remaining two cases are when $\epsilon_c \neq -\epsilon_s = 1$ and $\epsilon_c = -\epsilon_s \neq 1$. In these cases, we have the following theorem.

THEOREM 4.1. (i) *If $\epsilon_c = -\epsilon_s \neq 1$, then CALR does not occur, i.e.,*

$$E_\delta \le C \tag{4.3}$$

for some $C > 0$.

(ii) *If $\epsilon_c \neq -\epsilon_s = 1$, then weak CALR occurs and the critical radius is $r_* = r_e^2 r_i^{-1}$, i.e., if the source function is supported inside r_* (and its Newtonian potential does not extend harmonically to $\mathbb{R}^2$), then*

$$\limsup_{\delta\to 0} E_\delta = \infty, \tag{4.4}$$

and there exists a constant C such that

$$|V_\delta(x)| < C \tag{4.5}$$

for all x with $|x| > r_e^3/r_i^2$.

(iii) *In addition to the assumptions of (ii), the Fourier coefficients g_e^n of $-\frac{\partial F}{\partial \nu_e}$ satisfies the following gap property:*

[GP2] There exists a sequence $\{n_k\}$ with $|n_1| < |n_2| < \cdots$ such that

$$\lim_{k\to\infty} \rho^{2(|n_{k+1}|-|n_k|)} \frac{|g_e^{n_k}|^2}{|n_k|\rho^{|n_k|}} = \infty,$$

then the CALR occurs, i.e.,

$$\lim_{\delta\to 0} E_\delta = \infty, \tag{4.6}$$

and $V_\delta/\sqrt{E_\delta}$ goes to zero outside the radius r_e^3/r_i^2, as implied by (4.5).

Before proving Theorem 4.1 we make some remarks: Even if $\epsilon_c = -\epsilon_s \neq 1$, there will be CALR for appropriately placed sources inside the core, as can be seen from the fact that the equations are invariant under conformal transformations, and in particular under the inverse transformation $1/z$ where $z = x_1 + ix_2$, which in effect interchanges the roles of the matrix and core. We also mention that [GP] is weaker than [GP2], namely, if [GP] holds, so does [GP2].

The rest of this section is devoted to the proof of Theorem 4.1. As was proved in [**2**], we have

$$\frac{\partial}{\partial \nu_e} \mathcal{S}_{\Gamma_i}[e^{in\theta}](x) = \frac{1}{2}\rho^{|n|+1} e^{in\theta},$$

$$\frac{\partial}{\partial \nu_i} \mathcal{S}_{\Gamma_e}[e^{in\theta}](x) = -\frac{1}{2}\rho^{|n|-1} e^{in\theta}.$$

Using these identities, one can see that if g defined by (2.7) has the Fourier series expansion

$$g = \sum_{n \neq 0} \begin{bmatrix} g_i^n \\ g_e^n \end{bmatrix} e^{in\theta},$$

then the integral equations (2.8) are equivalent to

$$\begin{cases} z_i^\delta \varphi_i^n + \dfrac{\rho^{|n|-1}}{2} \varphi_e^n = g_i^n, \\ z_e^\delta \varphi_e^n + \dfrac{\rho^{|n|+1}}{2} \varphi_i^n = g_e^n \end{cases} \tag{4.7}$$

for every $|n| \geq 1$. It is readily seen that the solution $\Phi = (\varphi_i, \varphi_e)$ to (4.7) is given by

$$\varphi_i = 2 \sum_{n \neq 0} \frac{2z_e^\delta g_i^n - \rho^{|n|-1} g_e^n}{4z_i^\delta z_e^\delta - \rho^{2|n|}} e^{in\theta},$$

$$\varphi_e = 2 \sum_{n \neq 0} \frac{2z_i^\delta g_e^n - \rho^{|n|+1} g_i^n}{4z_i^\delta z_e^\delta - \rho^{2|n|}} e^{in\theta}.$$

If the source is located outside the structure, *i.e.*, f is supported in $|x| > r_e$, then the Newtonian potential of f, F, is harmonic in $|x| \leq r_e$ and

$$F(x) = c - \sum_{n \neq 0} \frac{g_e^n}{|n| r_e^{|n|-1}} r^{|n|} e^{in\theta}, \quad |x| \leq r_e. \tag{4.8}$$

Thus we have

$$g_i^n = -g_e^n \rho^{|n|-1}, \tag{4.9}$$

and hence

$$\begin{cases} \varphi_i = -2 \displaystyle\sum_{n \neq 0} \frac{(2z_e^\delta + 1)\rho^{|n|-1} g_e^n}{4z_i^\delta z_e^\delta - \rho^{2|n|}} e^{in\theta}, \\ \varphi_e = 2 \displaystyle\sum_{n \neq 0} \frac{(2z_i^\delta + \rho^{2|n|}) g_e^n}{4z_i^\delta z_e^\delta - \rho^{2|n|}} e^{in\theta}. \end{cases} \tag{4.10}$$

Therefore, from (3.1) we find that

$$\mathcal{S}_{\Gamma_i}[\varphi_i](x) + \mathcal{S}_{\Gamma_e}[\varphi_e](x) = \sum_{n \neq 0} \frac{2(r_i^{2|n|} z_e^\delta - r_e^{2|n|} z_i^\delta)}{|n| r_e^{|n|-1} (4z_i^\delta z_e^\delta - \rho^{2|n|})} \frac{g_e^n}{r^{|n|}} e^{in\theta}, \quad r_e < r = |x|, \tag{4.11}$$

and

$$\mathcal{S}_{\Gamma_i}[\varphi_i](x) = -\sum_{n \neq 0} \frac{r_i^{2|n|}(2z_e^\delta + 1)}{|n| r_e^{|n|-1} (\rho^{2|n|} - 4z_i^\delta z_e^\delta)} \frac{g_e^n}{r^{|n|}} e^{in\theta}, \quad r_i < r = |x| < r_e, \tag{4.12}$$

$$\mathcal{S}_{\Gamma_e}[\varphi_e](x) = \sum_{n \neq 0} \frac{(2z_i^\delta + \rho^{2|n|})}{|n| r_e^{|n|-1} (\rho^{2|n|} - 4z_i^\delta z_e^\delta)} g_e^n r^{|n|} e^{in\theta}, \quad r_i < r = |x| < r_e. \tag{4.13}$$

We obtain the following lemma.

LEMMA 4.2. *There exists δ_0 such that*

$$E_\delta \approx \begin{cases} \displaystyle\sum_{n\neq 0} \frac{\delta |g_e^n|^2}{|n|(\delta^2+\rho^{4|n|})}, & \text{if } \epsilon_c \neq \epsilon_s = 1, \\ \displaystyle\sum_{n\neq 0} \frac{\delta \rho^{2|n|} |g_e^n|^2}{|n|(\delta^2+\rho^{4|n|})}, & \text{if } \epsilon_c = \epsilon_s \neq 1, \end{cases} \tag{4.14}$$

uniformly in $\delta \leq \delta_0$.

PROOF. Using (4.8), (4.12), and (4.13), one can see that

$$V_\delta(x) = c + r_e \sum_{n\neq 0} \left[\frac{r_i^{2|n|}}{r^{|n|}} - 2z_i^\delta r^{|n|} \right] \frac{(2z_e^\delta + 1) g_e^n e^{in\theta}}{|n| r_e^{|n|} (4 z_i^\delta z_e^\delta - \rho^{2|n|})}.$$

We check that

$$\left| \nabla \left(\left(\frac{r_i^{2|n|}}{r^{|n|}} - 2z_i^\delta r^{|n|} \right) e^{in\theta} \right) \right|^2 = \frac{2|n|^2}{r^2} \left| \frac{r_i^{2|n|}}{r^{|n|}} - 2z_i^\delta r^{|n|} \right|^2.$$

Then straightforward computations yield that

$$\int_{B_e \setminus B_i} \delta |\nabla V_\delta|^2 \approx \sum_{n\neq 0} \delta \left| \frac{2z_e^\delta + 1}{4 z_i^\delta z_e^\delta - \rho^{2|n|}} \right|^2 (4|z_i^\delta|^2 + \rho^{2|n|}) \frac{|g_e^n|^2}{|n|}.$$

If δ is sufficiently small, then one can also easily show that

$$|4 z_i^\delta z_e^\delta - \rho^{2|n|}| \approx \delta + \rho^{2|n|}.$$

Therefore we get (4.14) and the proof is complete. □

First, if $\epsilon_c = -\epsilon_s \neq 1$, then

$$E_\delta \approx \sum_{n\neq 0} \frac{\delta \rho^{2|n|} |g_e^n|^2}{|n|(\delta^2+\rho^{4|n|})} \leq \sum_{n\neq 0} \frac{|g_e^n|^2}{2|n|} \leq \frac{1}{2} \left\| \frac{\partial F}{\partial \nu_e} \right\|_{L^2(\Gamma_e)} \leq C \|f\|_{L^2(\mathbb{R}^2)}.$$

Suppose that $\epsilon_c \neq -\epsilon_s = 1$, and let

$$N_\delta = \frac{\log \delta}{2 \log \rho}. \tag{4.15}$$

If $|n| \leq N_\delta$, then $\delta \leq \rho^{2|n|}$, and hence

$$\sum_{n\neq 0} \frac{\delta |g_e^n|^2}{|n|(\delta^2+\rho^{4|n|})} \geq \sum_{0\neq |n| \leq N_\delta} \frac{\delta |g_e^n|^2}{|n|(\delta^2+\rho^{4|n|})} \geq \sum_{0\neq |n| \leq N_\delta} \frac{\delta |g_e^n|^2}{|n| \rho^{4|n|}}. \tag{4.16}$$

If the following holds

$$\limsup_{n\to\infty} \frac{|g_e^n|^2}{|n| \rho^{2|n|}} = \infty, \tag{4.17}$$

then one can show as in [**2**] that there is a sequence $\{|n_k|\}$ such that

$$\lim_{k\to\infty} E_{\rho^{|n_k|}} = \infty. \tag{4.18}$$

Suppose that the source function f is supported inside the critical radius $r_* = r_e^2 r_i^{-1}$ (and outside r_e). Then its Newtonian potential F cannot be extended harmonically in $|x| < r_*$ in general. So, if F is given by

$$F = c - \sum_{n\neq 0} a_n r^{|n|} e^{in\theta}, \quad r < r_e + \epsilon \tag{4.19}$$

for some $\epsilon > 0$, then the radius of convergence of the series is less than r_*. Thus we have

$$\limsup_{|n|\to\infty} |a_n|^2 r_*^{2|n|} = \infty. \tag{4.20}$$

Since $g_e^n = |n| a_n r_e^{|n|-1}$, (4.17) holds.

By (4.11), we know

$$|V_\delta| \leq |F| + C \sum_{n\neq 0} \frac{r_e^{|n|}}{\delta + \rho^{2|n|}} \frac{|g_e^n|}{r^{|n|}} \leq |F| + C \sum_{n\neq 0} \frac{r_e^{3|n|}}{r_i^{2|n|}} \frac{|g_e^n|}{r^{|n|}} \leq C' \tag{4.21}$$

if $r > r_e^3/r_i^2$. Thus (ii) is proved.

We now prove (iii). We first emphasize that [GP2] implies (4.17), but the converse may not be true. On the other hand [GP2] holds if

$$\lim_{n\to\infty} \frac{|g_e^n|^2}{|n|\rho^{2|n|}} = \infty. \tag{4.22}$$

So we may regard the condition [GP2] something between (4.17) and (4.22).

Suppose that [GP2] holds. If we take $\delta = \rho^{2\alpha}$ and let $k(\alpha)$ be the number such that

$$|n_{k(\alpha)}| \leq \alpha < |n_{k(\alpha)+1}|,$$

then

$$\sum_{0\neq|n|\leq N_\delta} \frac{\delta |g_e^n|^2}{|n|\rho^{4|n|}} = \rho^{2\alpha} \sum_{0\neq|n|\leq\alpha} \frac{|g_e^n|^2}{|n|\rho^{4|n|}} \geq \rho^{2(|n_{k(\alpha)+1}|-|n_{k(\alpha)}|)} \frac{|g_e^{n_{k(\alpha)}}|^2}{|n_{k(\alpha)}|\rho^{2|n_{k(\alpha)}|}} \to \infty, \tag{4.23}$$

as $\alpha \to \infty$. Combined with Lemma 4.2 and (4.16), it gives us (iii).

5. Non-occurrence of CALR in three dimensions

In this section we show that CALR does not occur in a radially symmetric three dimensional coated sphere structure when the core, matrix and shell are isotropic. We have the following theorem.

THEOREM 5.1. *Suppose that Γ_e and Γ_i are concentric spheres. For any ϵ_c and ϵ_s, there is a constant C independent of δ such that if V_δ is the solution to (1.3), then*

$$\int_{\Omega\setminus D} \delta |\nabla V_\delta|^2 \leq C \|f\|^2_{L^2(\mathbb{R}^3)}. \tag{5.1}$$

PROOF. Suppose that $\frac{\partial}{\partial \nu_e} F$ has the Fourier series expansion

$$\frac{\partial}{\partial \nu_e} F = -\sum_{n=0}^{\infty} \sum_{m=-n}^{n} g_{mn}^e Y_m^n. \tag{5.2}$$

Then one can show as in (4.9) that

$$\frac{\partial}{\partial \nu_i} F = -\sum_{n=0}^{\infty} \sum_{m=-n}^{n} g_{mn}^e \rho^{n-1} Y_m^n. \tag{5.3}$$

By solving the integral equation (2.4) using (3.15), we obtain

$$\varphi_i = -\sum_{n=0}^{\infty}\sum_{m=-n}^{n} \Delta_n^{-1}\rho^{n-1}\left(z_e^\delta + \frac{1}{2}\right) g_{mn}^e Y_n^m, \tag{5.4}$$

$$\varphi_e = \sum_{n=0}^{\infty}\sum_{m=-n}^{n} \Delta_n^{-1}\left(z_i^\delta - \frac{1}{2(2n+1)} + \frac{n+1}{2(2n+1)}\rho^{2n+1}\right) g_{mn}^e Y_n^m, \tag{5.5}$$

where

$$\Delta_n := \left(z_i^\delta - \frac{1}{2(2n+1)}\right)\left(z_e^\delta + \frac{1}{2(2n+1)}\right) - \frac{n(n+1)}{(2n+1)^2}\rho^{2n+1}.$$

Suppose for simplicity that $\epsilon_c = -\epsilon_s = 1$, so that the expressions for z_i^δ and z_e^δ given by (2.5) are simplified to

$$z_i^\delta = z_e^\delta = \frac{i\delta}{2(2-i\delta)}.$$

Then one can see that if δ is sufficiently small, then

$$|\Delta_n| \approx \delta^2 + n^{-2}.$$

So we have

$$\begin{aligned}
\delta\|\varphi_i\|_{L^2(\Gamma_i)}^2 &\le C\sum_{n=0}^{\infty}\sum_{m=-n}^{n} \frac{\delta\rho^{2n}}{(\delta^2+n^{-2})^2}|g_{mn}^e|^2 \\
&\le C\sum_{n=0}^{\infty}\sum_{m=-n}^{n} n^3\rho^{2n}|g_{mn}^e|^2 \\
&\le C\sum_{n=0}^{\infty}|g_{mn}^e|^2 \le C\|f\|_{L^2(\mathbb{R}^3)}^2,
\end{aligned}$$

and

$$\delta\|\varphi_e\|_{L^2(\Gamma_e)}^2 \le C\sum_{n=0}^{\infty}\sum_{m=-n}^{n} \frac{\delta}{\delta^2+n^{-2}}|g_{mn}^e|^2 \le C\|f\|_{L^2(\mathbb{R}^3)}^2.$$

Therefore we have

$$\begin{aligned}
\int_{\Omega\setminus D}\delta|\nabla V_\delta|^2 &= \int_{\Omega\setminus D}\delta|\nabla F|^2 + \int_{\Omega\setminus D}\delta|\nabla(\mathcal{S}_{\Gamma_i}[\varphi_i] + \mathcal{S}_{\Gamma_e}[\varphi_e])|^2 \\
&\le \int_{\Omega\setminus D}\delta|\nabla F|^2 + \int_{\Omega\setminus D}\delta|\nabla(\mathcal{S}_{\Gamma_i}[\varphi_i] + \mathcal{S}_{\Gamma_e}[\varphi_e])|^2 \\
&\le \int_{\Omega\setminus D}\delta|\nabla F|^2 + \delta(\|\varphi_i\|_{L^2(\Gamma_i)}^2 + \|\varphi_e\|_{L^2(\Gamma_e)}^2) \le C\|f\|_{L^2(\mathbb{R}^3)}^2.
\end{aligned}$$

If $\epsilon_s \neq -1$ or/and $\epsilon_c \neq 1$, then the same argument can be applied to infer (5.1). This completes the proof. □

References

[1] A. Alú and N. Engheta, Achieving transparency with plasmonic and metamaterial coatings, Phys. Rev. E 72 (2005), 0166623.

[2] Habib Ammari, Giulio Ciraolo, Hyeonbae Kang, Hyundae Lee, and Graeme W. Milton, *Spectral theory of a Neumann-Poincaré-type operator and analysis of cloaking due to anomalous localized resonance*, Arch. Ration. Mech. Anal. **208** (2013), no. 2, 667–692, DOI 10.1007/s00205-012-0605-5. MR3035988

[3] H. Ammari, G. Ciraolo, H. Kang, H. Lee and G.W. Milton, Anomalous localized resonance using a folded geometry in three dimensions, Proc. R. Soc. A 469 (2013), 20130048.

[4] Habib Ammari, Giulio Ciraolo, Hyeonbae Kang, Hyundae Lee, and Kihyun Yun, *Spectral analysis of the Neumann-Poincaré operator and characterization of the stress concentration in anti-plane elasticity*, Arch. Ration. Mech. Anal. **208** (2013), no. 1, 275–304, DOI 10.1007/s00205-012-0590-8. MR3021549

[5] Habib Ammari and Hyeonbae Kang, *Polarization and moment tensors*, Applied Mathematical Sciences, vol. 162, Springer, New York, 2007. With applications to inverse problems and effective medium theory. MR2327884 (2009f:35339)

[6] Habib Ammari, Hyeonbae Kang, Hyundae Lee, and Mikyoung Lim, *Enhancement of near cloaking using generalized polarization tensors vanishing structures. Part I: The conductivity problem*, Comm. Math. Phys. **317** (2013), no. 1, 253–266, DOI 10.1007/s00220-012-1615-8. MR3010374

[7] Guy Bouchitté and Ben Schweizer, *Cloaking of small objects by anomalous localized resonance*, Quart. J. Mech. Appl. Math. **63** (2010), no. 4, 437–463, DOI 10.1093/qjmam/hbq008. MR2738456 (2011k:35036)

[8] O.P. Bruno and S. Lintner, Superlens-cloaking of small dielectric bodies in the quasi-static regime, J. Appl. Phys. 102 (2007), 124502.

[9] Kurt Bryan and Tanya Leise, *Impedance imaging, inverse problems, and Harry Potter's cloak*, SIAM Rev. **52** (2010), no. 2, 359–377, DOI 10.1137/090757873. MR2646807 (2012h:35373)

[10] W. Cai, U.K. Chettiar, A.V. Kildishev, V.M. Shalaev and G.W. Milton, Non-magnetic cloak with minimized scattering, Appl. Phys. Lett. 91 (2007), 111105.

[11] J.-W. Dong, H.H. Zheng, Y. Lai, H.-Z. Wang, and C. T. Chan, Metamaterial slab as a lens, a cloak, or an intermediate, Phys. Rev. B 83 (2011), 115124.

[12] Gerald B. Folland, *Introduction to partial differential equations*, Princeton University Press, Princeton, N.J., 1976. Preliminary informal notes of university courses and seminars in mathematics; Mathematical Notes. MR0599578 (58 #29031)

[13] Allan Greenleaf, Yaroslav Kurylev, Matti Lassas, and Gunther Uhlmann, *Cloaking devices, electromagnetic wormholes, and transformation optics*, SIAM Rev. **51** (2009), no. 1, 3–33, DOI 10.1137/080716827. MR2481110 (2010b:35484)

[14] Allan Greenleaf, Matti Lassas, and Gunther Uhlmann, *On nonuniqueness for Calderón's inverse problem*, Math. Res. Lett. **10** (2003), no. 5-6, 685–693, DOI 10.4310/MRL.2003.v10.n5.a11. MR2024725 (2005f:35316)

[15] F. Guevara Vasquez and G. W. Milton and D. Onofrei, Broadband exterior cloaking, Opt. Express 17 (2009), 14800–14805.

[16] Victor Isakov, *Inverse source problems*, Mathematical Surveys and Monographs, vol. 34, American Mathematical Society, Providence, RI, 1990. MR1071181 (92g:35230)

[17] Hyeonbae Kang and Jin Keun Seo, *The layer potential technique for the inverse conductivity problem*, Inverse Problems **12** (1996), no. 3, 267–278, DOI 10.1088/0266-5611/12/3/007. MR1391539 (97d:35242)

[18] Hyeonbae Kang and Jin Keun Seo, *Inverse conductivity problem with one measurement: uniqueness of balls in* $\mathbf{R}^3$, SIAM J. Appl. Math. **59** (1999), no. 5, 1533–1539 (electronic), DOI 10.1137/S0036139997324595. MR1699026 (2000j:35272)

[19] O.D. Kellogg, *Foundations of Potential Theory*, Dover, New York, 1953.

[20] R.V. Kohn, J. Lu, B. Schweizer, and M.I. Weinstein, A variational perspective on cloaking by anomalous localized resonance, preprint, arXiv:1210.4823.

[21] Robert V. Kohn, Daniel Onofrei, Michael S. Vogelius, and Michael I. Weinstein, *Cloaking via change of variables for the Helmholtz equation*, Comm. Pure Appl. Math. **63** (2010), no. 8, 973–1016, DOI 10.1002/cpa.20326. MR2642383 (2011j:78004)

[22] R. V. Kohn, H. Shen, M. S. Vogelius, and M. I. Weinstein, *Cloaking via change of variables in electric impedance tomography*, Inverse Problems **24** (2008), no. 1, 015016, 21, DOI 10.1088/0266-5611/24/1/015016. MR2384775 (2008m:78014)

[23] Y. Lai, H. Chen, Z.-Q. Zhang, and C. T. Chan, Complementary media invisibility cloak that cloaks objects at a distance outside the cloaking shell, Phys. Rev. Lett. 102 (2009), 093901.

[24] Ulf Leonhardt, *Optical conformal mapping*, Science **312** (2006), no. 5781, 1777–1780, DOI 10.1126/science.1126493. MR2237569

[25] U. Leonhardt and T. Tyc, Broadband invisibility by non-euclidean cloaking, Science 323 (2009), 110–111.

[26] Hongyu Liu, *Virtual reshaping and invisibility in obstacle scattering*, Inverse Problems **25** (2009), no. 4, 045006, 16, DOI 10.1088/0266-5611/25/4/045006. MR2482157 (2010d:35044)

[27] R.C. McPhedran, N.-A.P. Nicorovici, L.C. Botten, and G.W. Milton, Cloaking by plasmonic resonance among systems of particles: cooperation or combat? C.R. Phys. 10 (2009), 391–399.

[28] D.A.B. Miller, On perfect cloaking, Opt. Express 14 (2006), 12457–12466.

[29] Graeme W. Milton and Nicolae-Alexandru P. Nicorovici, *On the cloaking effects associated with anomalous localized resonance*, Proc. R. Soc. Lond. Ser. A Math. Phys. Eng. Sci. **462** (2006), no. 2074, 3027–3059, DOI 10.1098/rspa.2006.1715. MR2263683 (2008e:78018)

[30] G.W. Milton, N.-A.P. Nicorovici, and R.C. McPhedran, Opaque perfect lenses, Physica B 394 (2007), 171–175.

[31] G.W. Milton, N.-A.P. Nicorovici, R.C. McPhedran, K. Cherednichenko, and Z. Jacob, Solutions in folded geometries, and associated cloaking due to anomalous resonance, New. J. Phys. 10 (2008), 115021.

[32] Graeme W. Milton, Nicolae-Alexandru P. Nicorovici, Ross C. McPhedran, and Viktor A. Podolskiy, *A proof of superlensing in the quasistatic regime, and limitations of superlenses in this regime due to anomalous localized resonance*, Proc. R. Soc. Lond. Ser. A Math. Phys. Eng. Sci. **461** (2005), no. 2064, 3999–4034, DOI 10.1098/rspa.2005.1570. MR2186014 (2006h:78001)

[33] N.-A. P. Nicorovici, R. C. McPhedran, and L. C. Botten, *Relative local density of states and cloaking in finite clusters of coated cylinders*, Waves Random Complex Media **21** (2011), no. 2, 248–277, DOI 10.1080/17455030.2010.547885. MR2795374 (2012c:78008)

[34] N.-A.P. Nicorovici, R.C. McPhedran, S. Enoch, and G. Tayeb, Finite wavelength cloaking by plasmonic resonance, New. J. Phys. 10 (2008), 115020.

[35] N.-A.P. Nicorovici, R.C. McPhedran, and G.W. Milton, Optical and dielectric properties of partially resonant composites, Phys. Rev. B 49 (1994), 8479–8482.

[36] N.-A.P. Nicorovici, G.W. Milton, R.C. McPhedran, and L.C. Botten, Quasistatic cloaking of two-dimensional polarizable discrete systems by anomalous resonance, Optics Express 15 (2007), 6314–6323.

[37] J.B. Pendry, Negative refraction makes a perfect lens, Phys. Rev. Lett. 85 (2000), 3966–3969.

[38] J. B. Pendry, D. Schurig, and D. R. Smith, *Controlling electromagnetic fields*, Science **312** (2006), no. 5781, 1780–1782, DOI 10.1126/science.1125907. MR2237570

[39] M. Xiao, X. Huang, J.-W. Dong, and C. T. Chan, On the time evolution of the cloaking effect of a metamaterial slab, Opt. Lett. 37 (2012), 4594–4596.

Department of Mathematics and Applications, Ecole Normale Supérieure, 45 Rue d'Ulm, 75005 Paris, France.
E-mail address: `habib.ammari@ens.fr`

Dipartimento di Matematica e Informatica, Università di Palermo Via Archirafi 34, 90123, Palermo, Italy
E-mail address: `g.ciraolo@math.unipa.it`

Department of Mathematics, Inha University, Incheon 402-751, Korea
E-mail address: `hbkang@inha.ac.kr`

Department of Mathematics, Inha University, Incheon 402-751, Korea
E-mail address: `hdlee@inha.ac.kr`

Department of Mathematics, University of Utah, Salt Lake City, Utah 84112
E-mail address: `milton@math.utah.edu`

Contemporary Mathematics
Volume **615**, 2014
http://dx.doi.org/10.1090/conm/615/12289

Hybrid inverse problems and redundant systems of partial differential equations

Guillaume Bal

Dedicated to Gunther Uhlmann's 60th birthday

ABSTRACT. Hybrid inverse problems are mathematical descriptions of coupled-physics (also called multi-waves) imaging modalities that aim to combine high resolution with high contrast. The solution of a high-resolution inverse problem, a first step that is not considered in this paper, provides internal information combining unknown parameters and solutions of differential equations. In several settings, the internal information and the differential equations may be described as a redundant system of nonlinear partial differential equations.

We propose a framework to analyze the uniqueness and stability properties of such systems. We consider the case when the linearization of the redundant system is elliptic and with boundary conditions satisfying the Lopatinskii conditions. General theories of elliptic systems then allow us to construct a parametrix for such systems and derive optimal stability estimates.

The injectivity of the nonlinear problem or its linearization is not guaranteed by the ellipticity condition. We revisit unique continuation principles, such as the Holmgren theorem and the uniqueness theorem of Calderón, in the context of redundant elliptic systems of equations.

The theory is applied to the case of power density measurements, which are internal functionals of the form $\gamma|\nabla u|^2$ where γ is an unknown parameter and u is the solution to the elliptic equation $\nabla\cdot\gamma\nabla u = 0$ on a bounded domain with appropriate boundary conditions.

1. Introduction

A recent class of (mostly medical) imaging modalities, called hybrid, coupled-physics, or multi-wave modalities offers the possibility to reconstruct high-contrast parameters of interest with high resolution. High contrast is important to discriminate between, say, healthy and non-healthy tissues. Resolution is important to detect anomalies at an early stage.

Such hybrid modalities typically involve two steps. In the first step, not considered in this paper, a high resolution modality takes as an input measurements performed at the boundary of a domain of interest and provides as an output internal functionals of the parameters of interest and of specific solutions of underlying partial differential equations describing the probing (medical imaging) mechanism.

2010 *Mathematics Subject Classification.* Primary 35R30, 35G50, 35A02.

This paper is concerned with the second step, involving the quantitative reconstruction of the parameters from knowledge of said internal functionals. For recent books and reviews on hybrid inverse problems, we refer the reader to, e.g., [**4,6,7,32,46,52**].

Most practically used hybrid inverse problems involve internal functionals that are polynomials in the parameters of interest and the specific solutions mentioned above. Combined with the equations describing the latter solutions, we observe that all available information represents a coupled, often redundant, system of nonlinear partial differential equations.

In some instances, local algebraic manipulations allow us to solve such a system explicitly. In the framework of functionals of solutions to second-order equations (and not of their derivatives), we refer the reader to, e.g., [**14, 15, 17–19**]. Such theories find applications in the quantitative step of the imaging modalities Photo-acoustic tomography, Thermo-acoustic tomography, Transient Elastography, and Magnetic Resonance Elastography; see also [**24,25,37,38,45,50**]. In the framework of functionals of the gradients of solutions, which find applications in Ultrasound Modulated tomography and in Current Density Imaging, we refer the reader to, e.g., [**5,8,9,11–13,16,22,27,33,39–41**].

In many cases, explicit algebraic inversions may not be known or may not be applicable because not enough information is available. This paper proposes a framework to address several such problems when the linearization of the coupled system is *elliptic*. Hybrid inverse problems need not be elliptic; see the example of the 0-Laplacian in [**8,22**] (also recalled below in section 2.6.1) or the Photo-acoustic problem as treated in, e.g., [**14, 18**]. However, when the number of internal functionals increases, the resulting hybrid system becomes more redundant and hence more likely to be elliptic. We consider such a setting in section 2. We recall that elliptic systems augmented with boundary conditions that satisfy the Lopatinskii conditions admit left-parametrices. This follows from the theory of Agmon-Douglis-Nirenberg [**2, 3**] and the extensions to redundant systems by Solonnikov [**48**]. The existence of parametrices allows us to solve the linear problem up to possibly a finite dimensional space. Along with the construction of a parametrix, elliptic regularity theory provides optimal stability results for the linearization of the hybrid inverse problem.

The analysis of elliptic hybrid inverse problems was first addressed in [**34**] by means of systems of pseudo-differential operators that were shown to be elliptic in the sense of Douglis and Nirenberg. The differential systems considered in this paper simplify the analysis of boundary conditions and hence of injectivity for the linearized and nonlinear hybrid inverse problems as we now describe.

The possible existence of a finite dimensional kernel for the linearized hybrid inverse problem prevents us from determining whether the available internal functionals uniquely determine the coefficients of interest. Moreover, the dimension of the finite dimensional kernel is not stable with respect to small perturbations, which prevents us from analyzing the uniqueness and stability properties of the nonlinear hybrid problem. A powerful methodology to obtain uniqueness results in the framework of elliptic systems of equations is the notion of *unique continuation*. In section 3, we revisit two classical notions of unique continuation. One is based on the Holmgren theorem, which we generalize to the setting of redundant systems considered in this paper. The second one is based on the use of Carleman estimates as they are formulated in Calderón's uniqueness theorem. See [**20, 21, 29–31, 36, 44, 53**] for references on these unique continuation results. Several extensions of these results, following the presentation in [**44**], are given in the setting of redundant systems in section 3 with proofs postponed to the appendix.

Once a reasonable uniqueness result has been obtained for the linearization of the nonlinear hybrid inverse problem, several statements about uniqueness and iterative reconstruction procedures can be formulated for the nonlinear hybrid inverse problem. A constructive fixed point iteration method and a non-constructive local uniqueness result for the nonlinear problem are presented in section 4.

As an application of the conditions of ellipticity including boundary conditions and the conditions for unique continuation, we consider the case of power density internal functionals $H_j(x) = \gamma(x)|\nabla u_j|^2(x)$, where u_j is the solution to $\nabla\cdot\gamma\nabla u_j = 0$ on an open domain $X \subset \mathbb{R}^n$ with boundary conditions $u_j = f_j$ on ∂X for $1 \le j \le J$. For such a problem, we characterize the conditions under which the redundant problem is elliptic (for $J = 2$ in dimension $n = 2$ and $J = 3$ in higher dimension) and analyze cases in which a unique continuation principle (UCP) applies.

2. Inverse Problems with local internal functionals.

2.1. Systems of nonlinear partial differential equations. Let γ be a set of constitutive parameters in (linear or nonlinear, scalar or systems of) partial differential equations of the form

$$\mathcal{L}(\gamma, u_j) = 0 \quad \text{in } X, \qquad \tilde{\mathcal{B}} u_j = f_j \quad \text{on } \partial X, \tag{1}$$

where $\mathcal{L}$ is a polynomial in the derivatives of the solution u_j and those of γ on the open domain $X \subset \mathbb{R}^n$, with u_j augmented with boundary conditions on ∂X for $1 \le j \le J$.

Let us now assume knowledge of the functionals

$$\mathcal{M}(\gamma, u_j) = H_j \quad \text{in } X, \qquad 1 \le j \le J. \tag{2}$$

where $\mathcal{M}$ is a polynomial in the derivatives of the solution u_j and those of γ.

Several hybrid inverse problems may be recast in this general framework. More generally, we could have knowledge of functionals of the form $\mathcal{M}(\gamma, u_i, u_j) = H_{ij}$, or functionals $\mathcal{M}$ depending on more than two solutions u_j. We restrict ourselves to (2) to simplify notation.

The above problem may thus be recast as a system of nonlinear partial differential equations for $(\gamma, \{u_j\})$:

$$\begin{array}{rcl} \mathcal{L}(\gamma, u_j) &=& 0 \quad \text{in } X, \qquad \tilde{\mathcal{B}} u_j = f_j \quad \text{on } \partial X, \qquad 1 \le j \le J \\ \mathcal{M}(\gamma, u_j) &=& H_j \quad \text{in } X, \qquad 1 \le j \le J. \end{array} \tag{3}$$

The first relevant question for the inverse problem is whether the above system admits a *unique solution.* Since the solutions u_j are uniquely determined by knowledge of γ, we are primarily interested in finding a unique solution to the parameters γ. The strategy followed in [**34**] consists of writing a system of pseudo-differential equations for γ. Considering the higher-dimensional coupled system of equations for $(\gamma, \{u_j\})$ allows us to simplify the analysis of the uniqueness question for (3).

The second question pertains to the stability properties of the reconstruction. Provided that the solution to the inverse problem is unique, we wish to understand how perturbations in the information $\{H_j\}$ propagates to the reconstruction of γ.

The uniqueness and stability properties of the system depend on the number of acquired internal functionals J and on the way the medium was probed via the boundary conditions $\{f_j\}$. Understanding how the uniqueness and stability properties are affected by changes in J and the boundary conditions $\{f_j\}$ is the third question we wish to (very partially) answer.

2.2. Linearization. Some problems of the form (3) can directly be solved as non-linear systems. For instance, when $\mathcal{L}$ is a linear second-order equation in u_j and $\mathcal{M}(\gamma, u_j) = u_j$ the solution itself, the full non-linear problem is analyzed in [**17, 19**].

For many problems in which direct reconstruction procedures may not readily be available, it is fruitful to analyze the linearization of (3). Neglecting boundary conditions at first, this yields

$$\begin{array}{rclll} \partial_\gamma \mathcal{L}(\gamma, u_j)\delta\gamma + \partial_u \mathcal{L}(\gamma, u_j)\delta u_j &=& 0 & \text{in } X, & 1 \le j \le J \\ \partial_\gamma \mathcal{M}(\gamma, u_j)\delta\gamma + \partial_u \mathcal{M}(\gamma, u_j)\delta u_j &=& \delta H_j & \text{in } X, & 1 \le j \le J. \end{array} \tag{4}$$

Note that the above differential operators may all be of different orders. With $v = (\delta\gamma, \{\delta u_j\})$, we may recast the above system as

$$\mathcal{A}v = \mathcal{S}, \tag{5}$$

for an implicitly defined source $\mathcal{S}$. Let us assume that each u_j is a scalar solution and that each H_j is also a scalar information. Then $\mathcal{A}$ is a system of differential operators of size $2J \times (J + M)$, where M is the number of scalar functions describing γ. Note that for $J < M$, the above system is under-determined. We consider here the case $J \ge m$. In several practical problems, $J = M$ gives rise to a determined system $\mathcal{A}$ that is either not invertible, or invertible with non-optimal stability properties. It is therefore also fruitful to consider the setting with $J > M$.

2.3. Ellipticity. In applications, $\mathcal{L}$ is often a linear, elliptic, operator in the variables $\{u_j\}$. Adding the constraints (2), however, may render the coupled system (3) or its linear version (4) non-elliptic. One fruitful strategy to solve (5) on the whole domain X (with appropriate boundary conditions) is therefore to "ellipticize" $\mathcal{A}$, i.e., to find a number of constraints J sufficiently large so that $\mathcal{A}$ is elliptic, provided that such a J exists.

What we mean by elliptic is defined as follows. For each $x \in X$, $\mathcal{A}(x, D)_{ij}$ is a polynomial in $D = (\partial_{x_1}, \ldots, \partial_{x_n})$ for $1 \le i \le 2J$ and $1 \le j \le J + M$. We define the *principal part* $\mathcal{A}_0$ of $\mathcal{A}$ in the sense of Douglis and Nirenberg [**26**]. For each row $1 \le i \le 2J$ of the system, we associate an integer s_i and for each column $1 \le j \le J+M$ of the system an integer t_j. We normalize these integers by assuming that $\max(s_i) = 0$.

We assume that $\mathcal{A}_{ij}(x, D)$ is a polynomial in D of degree not greater than s_i+t_j. Then $\mathcal{A}_{0,ij}(x, D)$ is the part of the polynomial in $\mathcal{A}_{ij}(x, D)$ of degree exactly equal to $s_i + t_j$.

When all differential operators in (4) are of the same order t, then we may choose $s_i = 0$ and $t_j = t$, in which case $\mathcal{A}_0(x, D)$ is composed of entries that are homogeneous polynomials of degree t in D. Many practical problems arise in forms in which the differential operators in (4) have different orders.

We say that $\mathcal{A}$ is *elliptic* when the matrix $\mathcal{A}_0(x, \xi) = \{\mathcal{A}_{0,ij}(x, \xi)\}$, the symbol of the operator $\mathcal{A}_0$, is *full-rank* (i.e., of rank $J + M$ here) for all $\xi \in \mathbb{S}^{n-1}$ the unit sphere and all $x \in \bar{X}$.

Being full-rank is "more likely" when J is large, i.e., when $\mathcal{A}$ is over-determined. It is then useful to acquire redundant information H_j until (5) above is elliptic because elliptic systems enjoy more favorable (and in fact optimal) stability estimates than non-elliptic systems.

2.4. Lopatinskii boundary conditions. Let us assume that we have been able to prove that $\mathcal{A}$ was a redundant elliptic system of equations. Then the system can be solved, up to possibly a finite dimensional subspace, when it is augmented by boundary conditions that satisfy the Lopatinskii criterion. This is defined as follows; see [**48**].

We consider the system

$$\mathcal{A}v = \mathcal{S} \quad \text{in } X, \qquad \mathcal{B}v = \phi \quad \text{on } \partial X, \tag{6}$$

where $\mathcal{B}(x, D)$ is a $Q \times (J + M)$ matrix with entries $\mathcal{B}_{ij}(x, D)$ that are polynomial in D for each $x \in \partial X$. We denote by b_{ij} the order of $\mathcal{B}_{ij}$ and by $\sigma_i = \max_j(b_{ij} - t_j)$. Then $\mathcal{B}_0(x, D)$ is the principal part of $\mathcal{B}$ and consists of entries $\mathcal{B}_{0,ij}(x, D)$ defined as the polynomials of $\mathcal{B}_{ij}(x, D)$ of degree exactly equal to $\sigma_i + t_j$.

The Lopatinskii conditions are defined as follows. For each $x \in \partial X$, we denote by $\nu(x)$ the outward unit normal to X at $x \in \partial X$. We then think of z as the parameterization of the half line $x - z\nu(x)$ for $z \geq 0$. Let $\zeta \in \mathbb{S}^{n-1}$ with $\zeta \cdot \nu(x) = 0$ and consider the system of ordinary differential equations

$$\begin{aligned} \mathcal{A}_0(x, i\zeta + \nu(x)\frac{d}{dz})u(z) &= 0 \quad \text{in } z > 0 \\ \mathcal{B}_0(x, i\zeta + \nu(x)\frac{d}{dz})u(z) &= 0 \quad \text{at } z = 0. \end{aligned} \tag{7}$$

We assume that for each $x \in \partial X$, the only solution to the above system such that $u(z) \to 0$ as $z \to \infty$ is $u \equiv 0$. This is the Lopatinskii condition for $(\mathcal{A}, \mathcal{B})$. We then also say that $\mathcal{B}$ covers $\mathcal{A}$.

The above conditions need to be verified for the specific problems being considered. In some situations, the boundary conditions provided by (1) (or their linearization) generate a cover of $\mathcal{A}$. In other situations, they need to be augmented with additional boundary conditions for δu_j as well as for $\delta\gamma$ as we shall see.

When $\mathcal{A}$ is elliptic and $\mathcal{B}$ covers $\mathcal{A}$, we say that $(\mathcal{A}, \mathcal{B})$ is an elliptic system.

2.5. Parametrices and stability estimates. Following work in [**2, 3, 26**] on determined systems, the case of overdetermined elliptic systems was treated in [**48**]. The salient feature of these works is that the operator $A = (\mathcal{A}, \mathcal{B})$ admits a left-parametrix (a left-regularizer) in the following sense. Let $(\mathcal{S}, \phi)$ in (6) be in the space

$$\mathcal{R}(p, l) = W_p^{l-s_1}(X) \times \ldots \times W_p^{l-s_{2J}}(X) \times W_p^{l-\sigma_1-\frac{1}{p}}(\partial X) \times \ldots \times W_p^{l-\sigma_Q-\frac{1}{p}}(\partial X),$$

for some $l \geq 0$ and $p > 1$ and let us assume that $(\mathcal{A}, \mathcal{B})$ is a bounded operator from

$$v \in \mathcal{U}(p, l) = W_p^{l+t_1}(X) \times \ldots \times W_p^{l+t_{J+M}}(X)$$

to $(\mathcal{A}, \mathcal{B})v = (\mathcal{S}, \phi) \in \mathcal{R}(p, l)$. Such is the case when the coefficients of $\mathcal{A}$ and $\mathcal{B}$ are sufficiently regular. More precisely, with l sufficiently large so that $p(l - s_i) > n$ for all $1 \leq i \leq 2J$ (to simplify; see [**48**] for slight generalizations), we assume that $\mathcal{A}_{ij}$ is a sum of homogeneous operators of degree $s_i + t_j - \kappa$ for $0 \leq \kappa \leq s_i + t_j$ and that the coefficients of these operators are of class $W_p^{l-s_i}(X)$. Moreover, assuming

l large enough so that $p(l-\sigma_q) > n$ as well for $1 \le q \le Q$, we assume that $\mathcal{B}_{qj}$ is a sum of homogeneous operators of degree $\sigma_q + t_j - \kappa$ for $0 \le \kappa \le \sigma_q + t_j$ and that the coefficients of these operators are of class $W^{l-\sigma_q-\frac{1}{p}}(\partial X)$. Here $W_p^s(X)$ is the standard Sobolev space of functions with s derivatives that are p-integrable in X with standard extensions for s not an integer [**1**].

The main result in [**48**] is the existence of a bounded operator R from $\mathcal{R}(p,l)$ to $\mathcal{U}(p,l)$ such that

$$RA = I - T, \tag{8}$$

where I is the indentity operator and T is compact in $\mathcal{U}(p,l)$. When 1 is not in the spectrum of T so that $I-T$ is invertible, then A is invertible with bounded inverse $(I-T)^{-1}R$. However, 1 could very well be in the spectrum of T, in which case dimKer$A$ is finite but positive.

Moreover, we have the following stability estimate

$$\sum_{j=1}^{J+M} \|v_j\|_{W_p^{l+t_j}(X)} \le C\Big(\sum_{i=1}^{2J} \|\mathcal{S}_i\|_{W_p^{l-s_i}(X)} + \sum_{i=1}^{Q} \|\phi_i\|_{W_p^{l-\sigma_i-\frac{1}{p}}(\partial X)}\Big) + C_2 \sum_{t_j>0} \|v_j\|_{L^p(X)}, \tag{9}$$

for some constants $C > 0$ and $C_2 > 0$.

The presence of $C_2 > 0$ indicates the possibility that A may not be invertible. The presence of finite dimensional kernels is a serious difficulty in the analysis of the nonlinear problem (3) because such a dimension is not stable with respect to perturbations. What we can ensure is that for A_1 sufficiently small, then dimKer$(A + A_1) \le$ dimKer$A$; see, e.g., [**31**].

Whether we can choose $C_2 = 0$ above, i.e., whether A is invertible, depends on lower-order terms that are not captured by the principal part $(\mathcal{A}_0, \mathcal{B}_0)$. Their analysis can prove quite complicated in practical settings and we do not follow that route here. Instead, our aim is to modify A so that a unique continuation principle may be applied. In section 3, we augment the properly modified system $(\mathcal{A}, \mathcal{B})v = (\mathcal{S}, \phi)$ with additional boundary conditions, which in some cases allow us to obtain injectivity results.

Note that the parametrix R is clearly not unique. It is theoretically constructive, as can be seen by following the proof in [**3, 48**]. However, its practical, for instance numerical, implementation is not straightforward. The modified, higher-order, systems proposed later in the section offer a more direct numerical inversion procedure.

2.6. Example of power-density measurements. To illustrate the theoretical result of this paper, we consider the example of the reconstruction of a scalar coefficient from knowledge of the so-called power density measurements. Consider the scalar elliptic equation

$$\mathcal{L}(\gamma, u_j) := \nabla\cdot\gamma\nabla u_j = 0 \quad \text{in} \quad X, \qquad u_j = f_j \quad \text{on } \partial X, \qquad 1 \le j \le J. \tag{10}$$

Here, X is an open domain in $\mathbb{R}^n$ for $n \ge 2$ with smooth boundary ∂X. The objective is to reconstruct the scalar coefficient γ, uniformly bounded above and below by positive constants, from knowledge of the power densities

$$H_j(x) = \mathcal{M}(\gamma, u_j) := \gamma(x)|\nabla u_j|^2, \qquad x \in X, \qquad 1 \le j \le J, \tag{11}$$

where u_j is the solution to (10).

This problem and some variations have received significant theoretical and numerical analyses in recent years; see, e.g., [**5, 8, 9, 16, 22, 27, 33, 34, 41**]; generalizations for anisotropic coefficients γ can be found in [**11, 12, 39, 40**]. Explicit reconstruction procedures exist when the number of internal functionals J is sufficiently large; see [**9, 11, 22, 39, 41**]. The case $J = 1$, which does not correspond to an elliptic system, was analyzed in [**8**]. The main features of this analysis are recalled in section 2.6.1. For intermediate values of J, the above hybrid inverse problem may not have an explicit reconstruction but may still be modeled by a redundant elliptic system. Such a problem was also analyzed in [**33, 34**]. The conditions of ellipticity of the system $(\mathcal{A}, \mathcal{B})$ are described in detail in section 2.6.2. A modified system is presented in 2.6.5, whereas optimal stability estimates of the form (9) are presented in section 2.6.6 for the power density measurement problem.

2.6.1. *The* 0*−Laplacian when* $J = 1$. When $J = 1$, the 2×2 system of nonlinear partial differential equations is formally determined with two unknown coefficients (γ, u_1). The elimination of γ from such a system is in fact straightforward and we obtain the equation for $u := u_1$ (with $H := H_1$) given by

$$\nabla \cdot \frac{H(x)}{|\nabla u|^2} \nabla u = 0 \quad \text{in} \quad X, \qquad u = f \quad \text{on } \partial X. \tag{12}$$

The above equation may be transformed as

$$(I - 2\widehat{\nabla u} \otimes \widehat{\nabla u}) : \nabla^2 u + \nabla \ln H \cdot \nabla u = 0 \text{ in } X, \qquad u = f \text{ and } \frac{\partial u}{\partial \nu} = j \text{ on } \partial X. \tag{13}$$

Here $\widehat{\nabla u} = \frac{\nabla u}{|\nabla u|}$ and we introduced Cauchy data on ∂X anticipating the fact that (13) is a quasilinear strictly hyperbolic equation, at least provided that $\widehat{\nabla u}$ is defined. Indeed, we observe that the operator $(I - 2\widehat{\nabla u} \otimes \widehat{\nabla u}) : \nabla^2$ is hyperbolic with respect to the (unknown) direction $\widehat{\nabla u}$.

The above problem is analyzed in [**8**]. The two salient features of that analysis are: (i) unique reconstructions of u, and hence γ, are guaranteed only on part of the domain X; see [**8**]; and (ii) the stability estimates are *sub-elliptic*: first-order derivatives of u are controlled by the gradient of H rather than second-order derivatives as would be the case if $(I - 2\widehat{\nabla u} \otimes \widehat{\nabla u}) : \nabla^2$ was replaced by an elliptic operator. For γ, this translates into an inequality of the following form. Let H and $\tilde{H}$ be two measurements corresponding to the pairs (u, γ) and $(\tilde{u}, \tilde{\gamma})$, respectively. Assume that the Cauchy data of u and $\tilde{u}$ agree on ∂X. Then we find that on an appropriate (see [**8**]) subdomain $\mathcal{O} \subset X$, we have the following stability estimate:

$$\|\gamma - \tilde{\gamma}\|_{L^2(\mathcal{O})} \leq C \|\nabla H - \nabla \tilde{H}\|_{L^2(\mathcal{O})}. \tag{14}$$

As we shall see below, this estimate is sub-optimal (with a loss of one derivative) when compared to elliptic estimates of the form (9). It is however optimal for (principally normal) operators of principal type [**29**].

2.6.2. *Linearization and ellipticity.* The linearization of the above problem with $J = 1$ is a hyperbolic equation. We now wish to show that redundancy in the data ($J \geq 2$) allows us to render the system elliptic under some conditions. We consider two ways to obtain elliptic systems of equations.

We first linearize the coupled system (10)-(11) about solutions (γ, u_j) and obtain

$$\begin{array}{rcll} \nabla\cdot\delta\gamma\nabla u_j + \nabla\cdot\gamma\nabla\delta u_j &=& 0 & \text{in } X \\ \delta\gamma|\nabla u_j|^2 + 2\gamma\nabla u_j\cdot\nabla\delta u_j &=& \delta H_j & \text{in } X \\ \delta u_j &=& 0 & \text{on } \partial X. \end{array} \tag{15}$$

We define

$$F_j = \nabla u_j \tag{16}$$

and *assume* that $|F_j| \geq c_0 > 0$ is bounded from below by a positive constant uniformly. Such an assumption is valid for and appropriate open set of boundary conditions f_j; see, e.g., [**7, 9, 17, 18**] for details of constructions based on complex geometric optics solutions or unique continuation principles, which we do not reproduce here.

Let $\mathcal{A}_J$ be the operator applied to $\delta v = (\delta\gamma, \{\delta u_j\})$ in the above system. Its principal part $\mathcal{P}_J$ has for (principal) symbol $\mathfrak{p}_j$ a $2J\times(J+1)$ matrix given by

$$\mathfrak{p}_J(x,\xi) = \begin{pmatrix} |F_1|^2 & 2\gamma F_1\cdot i\xi & \dots & 0 \\ F_1\cdot i\xi & -\gamma|\xi|^2 & \dots & 0 \\ \vdots & \vdots & \ddots & \vdots \\ |F_J|^2 & 0 & \dots & 2\gamma F_J\cdot i\xi \\ F_J\cdot i\xi & 0 & \dots & -\gamma|\xi|^2 \end{pmatrix}. \tag{17}$$

When $J=1$, the determinant of $\mathfrak{p}_J$ is given by $\gamma|F_1|^2(2(\hat{F}_1\cdot\xi)^2 - |\xi|^2)$, which is hyperbolic with respect to F_1 as seen in section 2.6.1 above. The above system is in Douglis-Nirenberg form for $s_{2k}=1$, $s_{2k+1}=0$, $t_1=0$, $t_j=1$ for $j\geq 2$.

When $J\geq 1$, we observe that the sub-determinants with the largest number of powers of $|\xi|^2$ are of the form

$$\gamma^J|\xi|^{2(J-1)}|F_j|^2 q_j(x,\xi), \qquad q_j(x,\xi) := 2(\hat{F}_j(x)\cdot\xi)^2 - |\xi|^2.$$

We thus obtain that $\mathfrak{p}_J(x,\xi)$ is injective if the quadratic forms $q_j(x,\xi)=0$ for all $1\leq j\leq J$ imply that $\xi=0$.

DEFINITION 2.1. *Define the quadratic forms and operators*

$$q_j(x,\xi) = 2\big(\hat{F}_j\cdot\xi\big)^2 - |\xi|^2, \qquad P_j(x,D) = \Delta - 2\hat{F}_j\otimes\hat{F}_j : \nabla\otimes\nabla. \tag{18}$$

Here $\hat{F}_j(x)$ are vector fields of unit vectors defined on $\bar{X}$. We say that the family $\{q_j\}$ or $\{P_j\}$ is elliptic at x if

$$q_j(x,\xi) = 0 \text{ for all } 1\leq j\leq J \quad \text{implies} \quad \xi = 0. \tag{19}$$

We say that such families are elliptic in X if they are elliptic in at all points $x\in X$.

We can then prove the

LEMMA 2.2. *We assume that $F_j := \nabla u_j$ is such that $|F_j|$ is bounded from below by a positive constant on $\bar{X}$ for all $1\leq j\leq J$.*

The operator $\mathcal{A}_J$ defined in (15) *with principal symbol given in* (17) *is elliptic in $\bar{X}$ if and only if the above family of quadratic forms $\{q_j\}$ is elliptic in $\bar{X}$.*

Proof. We have already seen the sufficiency of the condition. Let us prove its necessity and assume that $\mathfrak{p}_J(x,\xi)$ is maximal rank for $x\in X$ and $\xi\neq 0$. This means that one determinant of $(J+1)\times(J+1)$ sub-matrices of $\mathfrak{p}_J$ is non-vanishing. Each column of $\mathfrak{p}_J$ beyond the first one has two non-vanishing entries.

For all $1 \le j \le J$ except for one entry j_0, then either $-\gamma(x)|\xi|^2$ or $2\gamma F_j \cdot i\xi$ appears as a multiplicative factor in the determinant of the sub-matrix. Since $\gamma(x)|\xi|^2$ never vanishes, we may discard the determinant involving $2\gamma F_j \cdot i\xi$. We thus obtain that if one determinant of a sub-matrix does not vanish, then that determinant may be chosen as $\gamma^J|\xi|^{2(J-1)}|F_{j_0}|^2 q_{j_0}(x,\xi)$. Since by assumption $\gamma^J|\xi|^{2(J-1)}|F_{j_0}|^2$ is bounded away from 0, we observe that the injectivity of $\mathfrak{p}_J$ implies that (at least) one of the quadratic forms $q_j(x,\xi)$ does not vanish. Thus, $\mathcal{A}_J$ being elliptic implies that $\{q_j\}$ is elliptic in the sense of definition 2.1. □

The ellipticity of $\mathcal{A}_J$ is thus a consequence of the fact that the null cones of quadratic forms intersect only at 0. We have the following properties:

PROPOSITION 2.3. *(i) Let $x \in \mathbb{R}^2$ and assume that $\hat{F}_1(x)$ and $\hat{F}_2(x)$ are neither parallel nor orthogonal. Then the corresponding (q_1, q_2) in (18) form an elliptic family at x.*

(ii) In dimension $n \ge 2$, let $\hat{F}_1(x)$ and $\hat{F}_2(x)$ be two different directions and define $\hat{F}_3(x) = \alpha\hat{F}_1(x) + \beta\hat{F}_2(x)$ with $\alpha\beta \neq 0$ such that $|\hat{F}_3(x)| = 1$. Then the corresponding (q_1, q_2, q_3) in (18) form an elliptic family at x.

(iii) In dimension $n \ge 3$, a family (q_1, q_2) is never elliptic at a given point x independent of the choice of $\hat{F}_1, \hat{F}_2$.

Proof. (i) In dimension $n = 2$, it is clear that the null cones (where q_j vanishes, two lines of vectors ξ in $\mathbb{R}^2$) coincide if and only if $\hat{F}_1$ and $\hat{F}_2$ are either parallel or orthogonal.

(iii) The result is obvious if $\hat{F}_1 = \pm\hat{F}_2$. Assume otherwise and let F_3 be a unit vector orthogonal to $\hat{F}_1$ and $\hat{F}_2$. Assume $\hat{F}_1 \cdot \hat{F}_2 \ge 0$ for otherwise change the sign of $\hat{F}_2$ (which does not modify the quadratic forms q_j). ξ belongs to the intersection of the null cones $\{q_j(x,\xi) = 0\}$ for $j = 1, 2$ if $2(\hat{F}_1 \cdot \xi)^2 = 2(\hat{F}_2 \cdot \xi)^2 = |\xi|^2$. Define $\xi = \hat{F}_1 + \hat{F}_2 + \lambda F_3$. The first constraint is satisfied when $2(1 + \hat{F}_1 \cdot \hat{F}_2)^2 = |\hat{F}^1 + \hat{F}^2|^2 + \lambda^2$, i.e., when $\lambda = \pm\big(2\hat{F}_1 \cdot \hat{F}_2 + 2(\hat{F}_1 \cdot \hat{F}_2)^2\big)^{\frac{1}{2}}$. But then it is clear that this $\xi \neq 0$ also belongs to the second cone so that (q_1, q_2) is not elliptic.

(ii) Let us assume that $\hat{F}_3 = \alpha\hat{F}_1 + \beta\hat{F}_2$ with $\alpha\beta \neq 0$. Then ξ belongs to the three null cones if

$$\Big(\frac{\alpha\hat{F}_1 + \beta\hat{F}_2}{|\alpha\hat{F}_1 + \beta\hat{F}_2|} \cdot \xi\Big)^2 = \frac{1}{2}|\xi|^2 = (\hat{F}_1 \cdot \xi)^2 = (\hat{F}_2 \cdot \xi)^2.$$

Expanding the first constraint, we get

$$\alpha^2(\hat{F}_1 \cdot \xi)^2 + \beta^2(\hat{F}_2 \cdot \xi)^2 + 2\alpha\beta\hat{F}_1 \cdot \xi\hat{F}_2 \cdot \xi = \frac{1}{2}|\xi|^2(\alpha^2 + \beta^2 + 2\alpha\beta\hat{F}_1 \cdot \hat{F}_2).$$

The last two constraints imply that $2\hat{F}_1 \cdot \xi\hat{F}_2 \cdot \xi = \varepsilon|\xi|^2$ with $\varepsilon = \pm 1$. Combined with the latter, they also imply that

$$\alpha\beta\varepsilon = \alpha\beta\hat{F}_1 \cdot \hat{F}_2.$$

Since $\alpha\beta \neq 0$, this can only occur if $\hat{F}_1 = \pm\hat{F}_2$, which is a contradiction. □

From the practical point of view, this result says that if $F_1 = \nabla u_1$ and $F_2 = \nabla u_2$ are not parallel, then the internal functionals H_j for u_1, u_2, and $u_1 + u_2$ (with $F_3 = \nabla(u_1 + u_2) = \nabla u_1 + \nabla u_2$) generate three quadratic forms q_j, $1 \le j \le 3$ that form an elliptic family. Such internal functionals are obtained by choosing three boundary conditions of the form f_1, f_2, and $f_3 = f_1 + f_2$. This result holds for all $n \ge 2$.

2.6.3. *Sufficient conditions for ellipticity.* We have thus obtained the following result. In dimension $n = 2$, $J \geq 2$ is necessary for $\mathcal{A}_J$ to be elliptic. Moreover, $J = 2$ is sufficient when $n = 2$ if ∇u_1 and ∇u_2 are nowhere parallel or orthogonal. In dimension $n \geq 3$, $J \geq 3$ is necessary for $\mathcal{A}_J$ to be elliptic. Moreover, $J = 3$ is sufficient for $\mathcal{A}_J$ to be elliptic in all dimensions $n \geq 2$ by choosing as boundary conditions, e.g., $(f_1, f_2, f_1 + f_2)$ provided that ∇u_1 and ∇u_2 are nowhere parallel.

2.6.4. *Boundary conditions and Lopatinskii condition.* In order to obtain an optimal theory of stability estimates, the system needs to be augmented with boundary conditions that satisfy the Lopatinskii condition. Dirichlet conditions on δu_j and no condition on $\delta\gamma$ satisfy such conditions. Indeed, we need to show that $v(z) = (\delta\gamma(z), \ldots, \delta u_J(z)) \equiv 0$ is the only solution to

$$\begin{aligned} &\delta u_j(0) = 0, \qquad (iF_j \cdot \zeta + F_j \cdot N\partial_z)\delta\gamma + \gamma(i\zeta + \partial_z)^2 \delta u_j = 0, \\ &|F_j|^2 \delta\gamma + 2\gamma(iF_j \cdot \zeta + F_j \cdot N\partial_z)\delta u_j = 0, \ z > 0 \end{aligned} \tag{20}$$

with $v(z)$ vanishing as $z \to \infty$ for $N = \nu(x)$ at $x \in \partial X$ and z coordinate along $-N$. Eliminating $\delta\gamma$ as earlier, we deduce that

$$\Big(|F_j|^2(i\zeta + \partial_z)^2 - 2(iF_j \cdot \zeta + F_j \cdot N\partial_z)^2\Big)\delta u_j = 0$$

The leading term of the above second order equation with constant coefficients is $|F_j|^2 q_j(x, N)\partial_z^2$. If $q_j(x, N) \neq 0$ for some $j = j_0$, which is the condition for joint ellipticity described in definition 2.1, then the same proof showing that Dirichlet conditions cover the Laplace operator show that $\delta u_{j_0} = 0$. We then deduce that $\delta\gamma = 0$ from the second line in (20) and by ellipticity on the first line in (20) that all $\delta u_j = 0$ and hence $v \equiv 0$.

Let us define the spaces $\mathcal{X} = W^l_p(X) \times W^{l+1}_p(X; \mathbb{R}^J)$ and $\mathcal{Y}^i = W^l_p(X; \mathbb{R}^{2J})$ with l large enough so that the latter spaces are all algebras; i.e., $pl > n$. We also define $\mathcal{Y}^\partial = W_p^{l+1-\frac{1}{p}}(X; \mathbb{R}^J)$ for the traces of δu_j on ∂X, which vanish by construction. Then we observe that $(\mathcal{A}, \mathcal{B})$ in (15) maps $\mathcal{U}(p, l) = \mathcal{X}$ to $\mathcal{R}(p, l) = \mathcal{Y}^i \times \mathcal{Y}^\partial$. Moreover, the coefficients γ and u_j appearing in the definition of $\mathcal{A}$ belong to $W^l_p(X)$ and $W^{l+1}_p(X; \mathbb{R}^J)$, respectively, by assumption for γ and by elliptic regularity for u_j solution of the elliptic problem (10).

2.6.5. *Elimination and ellipticity.* The above system involves $J + 1$ unknowns. Strategies to lower the dimension of the system include: (i) eliminating γ as we did in obtaining (12); or (ii) eliminating all u_j. The second strategy follows from the observation

$$\delta u_j = L_\gamma^{-1}(\nabla \cdot \delta\gamma \nabla u_j), \qquad L_\gamma = -\nabla \cdot \gamma \nabla,$$

with L_γ^{-1} defined by solving L_γ with vanishing Dirichlet conditions.

The result is a redundant system of the form $P_j \delta\gamma = \delta H_j$ where P_j is a pseudo-differential operator with principal symbol given by $|F_j|^2 q_j(x, \xi)$. The redundant system for $\delta\gamma$ is therefore elliptic under the same conditions as those for (17) above. The main difficulty is that P_j is no longer local (no longer a (system of) partial differential equation). It is not clear how one may approach the question of uniqueness for such a system. The unique continuation principles presented in section 3 do not apply directly. See [**34**] for an analysis of such a method.

The first strategy based on the elimination of $\delta\gamma$ preserves the differential structure of the original system since $\delta\gamma$ appears undifferentiated in the second equation in (15). We find that

$$\nabla\cdot\gamma\big(-\nabla\delta u_j+2\hat F_j\hat F_j\cdot\nabla\delta u_j\big)=\nabla\cdot\frac{\delta H_j}{|F_j|^2}F_j. \tag{21}$$

The symbol for such an equation is then given by $\gamma q_j(x,\xi)$ so that the above equation is not elliptic, as we already know. However, the elimination of $\delta\gamma$ also provides the constraint

$$2\gamma\frac{1}{|F_j|^2}F_j\cdot\nabla\delta u_j-\frac{\delta H_j}{|F_j|^2}=2\gamma\frac{1}{|F_k|^2}F_k\cdot\nabla\delta u_k-\frac{\delta H_k}{|F_k|^2}\quad 1\le j<k\le J. \tag{22}$$

It turns out that the combination of (21) with (22) makes the redundant system for the $\{u_j\}$ elliptic provided $q_j(x,\xi)=0$ for all j implies that $\xi=0$ as above. To see this, assume that

$$q_j(x,\xi)\delta u_j=0,\qquad |F_k|^2F_j\cdot\xi\delta u_j=|F_j|^2F_k\cdot\xi\delta u_k.$$

Let $\xi\neq0$. Then, not all $q_j(x,\xi)$ vanish. Assume that $q_1(x,\xi)\neq0$. Then $\delta u_1=0$ so that $F_j\cdot\xi\delta u_j=0$ for all $j\ge2$. However, $F_j\cdot\xi$ and $q_j(x,\xi)$ cannot vanish at the same time so that $\delta u_j=0$. This shows the injectivity of the symbol of the redundant system, which is easily found to be in Douglis-Nirenberg form.

Moreover, unlike the redundant system of strategy (ii), the above system (21)-(22) is differential. It may therefore be augmented with boundary conditions that satisfy the Lopatisnkii conditions. We leave the details to the reader to verify, as we did in (20), that such conditions are satisfied for Dirichlet conditions $\delta u_j=0$ on ∂X under the same conditions guaranteeing ellipticity in X. The above system for the $(\{\delta u_j\})$ is therefore elliptic when the family of quadratic forms $\{q_j\}$ is elliptic.

2.6.6. *Stability estimates.* Both systems of differential equations for $(\delta\gamma,\{\delta u_j\})$ and for $(\{\delta u_j\})$ after elimination of $\delta\gamma$ are elliptic with Dirichlet conditions for δu_j under the conditions stated, e.g., in Proposition 2.3. We may therefore apply the general theory of elliptic redundant systems described above and obtain a parametrix for both systems. Moreover, in both cases, we obtain the following stability estimates

$$\begin{aligned}\|\delta\gamma-\delta\tilde\gamma\|_{W^l_p(X)}+\sum_{j=1}^J\|\delta u_j-\delta\tilde u_j\|_{W^{l+1}_p(X)}&\le C\sum_{j=1}^J\|\delta H_j-\delta\tilde H_j\|_{W^l_p(X)}\\&\quad+C_2\sum_{j=1}^J\|\delta u_j-\delta\tilde u_j\|_{L_p(X)}.\end{aligned} \tag{23}$$

In other words, $\delta\gamma$ and $\nabla\delta u_j$ are reconstructed from δH_j with *no loss* of derivative, unlike the construction in (14). However, we are not guaranteed that any of the linear systems is indeed invertible, and hence the presence of the last term on the right-hand-size in (23). Proving the injectivity of the above systems is much more delicate than proving their ellipticity. The following section proposes some generic strategies to do so.

2.6.7. *Generalization to similar models.* The results presented above generalize to the setting where the internal functionals are of the form

$$H_j(x)=\gamma^\alpha|\nabla u_j|^2, \tag{24}$$

where $\alpha\ge0$. The case $\alpha=1$ was treated above. The case $\alpha=2$ corresponds to the setting of CDII and MREIT [**42, 47**]. As shown, e.g., in [**34, 41**], the coupled problem is elliptic with $J = 1$ for $\alpha > 2$, is degenerate elliptic for $\alpha = 2$, and is hyperbolic for $\alpha < 2$. The cases $\alpha \geq 2$ become elliptic for J chosen sufficiently large. We do not consider these extensions further here.

2.7. Sufficient conditions for ellipticity. To summarize the above derivation, we have modeled the nonlinear hybrid inverse problem as a coupled system of nonlinear partial differential equations (3). Its linearization is then given by (4). These systems have $J + M$ unknowns for $2J \geq J + M$ equations. When J is large compared to M, we expect the principal part $\mathcal{A}_0$ of $\mathcal{A}$ in the Douglis-Nirenberg sense to be elliptic since a redundant matrix is "more likely" to be full-rank than a less elongated matrix.

The matrix $\mathcal{A}_0$ is full rank if we can prove that, for well chosen boundary conditions f_j, the internal functionals H_j are sufficiently independent. Such a result is problem-dependent. In the setting of power-density measurements, we have obtained that the boundary conditions f_j were well-chosen if the quadratic form $q_j(x, \xi)$ were jointly elliptic at each point $x \in \bar{X}$. A sufficient condition to do so is to choose two boundary conditions f_1 and f_2 such that $F_1 = \nabla u_1$ and $F_2 = \nabla u_2$ are nowhere co-linear (for the quadratic forms corresponding to the three boundary conditions f_1, f_2, and $f_1 + f_2$ are then jointly elliptic).

Strategies to ensure that, e.g., ∇u_1 and ∇u_2 are nowhere co-linear have been presented in, e.g., [**8**, **9**, **17**, **18**], see also the review [**7**]. Such strategies are based on the use of Complex Geometrical Optics when the latter are available, or on the use on local construction and unique continuation properties of the operators in (1) as described in [**17**]. In both settings, it is proved that qualitative properties of solutions to elliptic equations, such as for instance the independence of gradients of solutions, hold for an open set of boundary conditions f_j. More explicit constructions of such boundary conditions are also proposed in [**10**].

3. Injectivity results for the linearized problem

In this section, we consider two methods to obtain injectivity of an elliptic operator $\mathcal{A}$ augmented with appropriate boundary conditions $\mathcal{B}$. Both are based on replacing the redundant system $\mathcal{A}$ by its normal, determined, form $\mathcal{A}^t\mathcal{A}$, and augmenting it with appropriate Dirichlet boundary conditions that ensure its injectivity under certain assumptions. The first method invokes a Holmgren unique continuation principle while the second method is based on unique continuation principles that are consequences of Carleman estimates.

Here, the operator $\mathcal{A}^t$ is defined such that $(\mathcal{A}^t)_{ki} = (\mathcal{A}_{ik})^t =: \mathcal{A}^t_{ik}$, where $(\mathcal{A}_{ik})^t$ is the formal adjoint to $\mathcal{A}_{ik}$ for the usual inner product $(\cdot, \cdot)$ on $L^2(X)$.

One difficulty with operators $\mathcal{A}$ that are elliptic in the Douglis-Nirenberg (DN) sense is that the normal operator $\mathcal{A}^t\mathcal{A}$ need not be elliptic, even in the DN sense. Consider for instance the 2×2 system in one independent variable defined by $\mathcal{A}_{11} = a$, $\mathcal{A}_{12} = \mathcal{A}_{21} = \partial_x$ and $\mathcal{A}_{22} = \partial_x^2$, which is DN elliptic with $s_1 = t_1 = 0$ and $s_2 = t_2 = 1$ when $a \neq 1$. Defining $\mathcal{C} = \mathcal{A}^T\mathcal{A}$, the principal term in $\mathcal{C}$ is $\mathcal{C}_{11} = \partial_x^2$, $\mathcal{C}_{12} = \mathcal{C}_{21} = \partial_x^3$, $\mathcal{C}_{22} = \partial_x^4$, which is independent of a and not elliptic.

One way to ensure that $\mathcal{A}^t\mathcal{A}$ is elliptic when $\mathcal{A}$ is elliptic is to assume that $\tau = t_j$ independent of j and $s_i = 0$ independent of i. We then verify that the leading term in $(\mathcal{A}^t\mathcal{A})_{kl}$ is a differential operator of degree equal to 2τ and that $\mathcal{A}^t\mathcal{A}$ is a strongly elliptic system of size $(J+M)\times(J+M)$; see, e.g., [**51**, Proposition 4.1.16]. The principal part of $\mathrm{Det}(\mathcal{A}^t\mathcal{A})(x, \xi)$ is equal to $\mathrm{Det}(\mathcal{A}^t_0\mathcal{A}_0)(x, \xi)$ and is a polynomial of degree $2(J+M)\tau$ that is uniformly bounded from below for all $x \in \bar{X}$ and $\xi \in \mathbb{S}^{n-1}$ by assumption of ellipticity.

The results in [**23**] show that an elliptic system in DN form can always be transformed into an elliptic overdetermined system of first-order equations. The procedure increases the order of the system and differentiates any row involving a term with $s_i + t_j = 0$. This forces us to impose boundary conditions on the parameters $\delta\gamma$ and the solutions δu_j that may not be necessary in the definition of $(\mathcal{A}, \mathcal{B})$ above.

For the rest of the section, we *assume* that $\mathcal{A}$ has been recast into a form where $\tau = t_j$ is independent of j and $s_i = 0$ is independent of i; for instance with $\tau = 1$ or $\tau = 2$ as described in [**23**]. We still keep the notation $2J$ and $J+M$ for the size of the system $\mathcal{A}$ so that with this notation, $\mathcal{A}^t\mathcal{A}$ is a system of size $(J+M)\times(J+M)$.

Let us augment $\mathcal{A}^t\mathcal{A}$ with the Dirichlet boundary conditions

$$(25)\qquad \Big(\frac{\partial}{\partial\nu}\Big)^q v_j = \phi_{qj} \quad \text{on } \partial X, \qquad 0 \le q \le \tau - 1, \quad 1 \le j \le J+M.$$

We recast the above constraints as $\mathcal{D}v = \phi$ on ∂X. It is proved in [**3**, p.43-44] that such boundary conditions cover $\mathcal{A}^t\mathcal{A}$, i.e., that the Lopatinskii conditions are satisfied. We thus consider the problem

$$(26)\qquad \mathcal{A}^t\mathcal{A}v = \mathcal{A}^t\mathcal{S} \quad \text{in } X, \qquad \mathcal{D}v = \phi \quad \text{on } \partial X.$$

Since $N := (\mathcal{A}^t\mathcal{A}, \mathcal{D})$ is elliptic, the above system admits a left parametrix G such that $GN = I - T$ with T compact in $\mathcal{U}(p,l)$. Moreover, we have the stability estimate

$$(27)\qquad \begin{aligned} \sum_{j=1}^{J+M} \|v_j\|_{W_p^{l+\tau}(X)} \le C\Big(\sum_{j=1}^{J+M} \|(\mathcal{A}^t\mathcal{S})_j\|_{W_p^{l-\tau}(X)} + \sum_{j,q} \|\phi_{qj}\|_{W_p^{l-\tau+q-\frac{1}{p}}(\partial X)}\Big) \\ + C_2 \sum_{j=1}^{J+M} \|v_j\|_{L^p(X)}, \end{aligned}$$

We verify that

$$\|\sum_{i=1}^{2J} \mathcal{A}^t_{ji} S_i\|_{W_p^{l-\tau}(X)} \le C \sum_{i=1}^{2J} \|S_i\|_{W_p^l(X)},$$

so that (27) may also be seen as an analog of (9).

Our objective is to find sufficient conditions under which the above system is injective, and hence invertible, so that $C_2 = 0$ in the above estimates. We start with the following simple lemma:

LEMMA 3.1. *Let us assume that v is a solution of*

$$(28)\qquad \mathcal{A}^t\mathcal{A}v = 0 \quad \textit{in } X, \qquad \mathcal{D}v = 0 \quad \textit{on } \partial X.$$

Then v is a solution of

$$(29)\qquad \mathcal{A}v = 0 \quad \textit{in } X, \qquad \mathcal{D}v = 0 \quad \textit{on } \partial X.$$

Note that the Dirichlet conditions associated to $\mathcal{A}^t\mathcal{A}$ are now redundant for (29). For example, consider $\mathcal{A} = \Delta + c(x)$ a scalar operator. For some choices of $c(x)$ (for instance $c(x) = \lambda$ an eigenvalue of $-\Delta$ on X with Dirichlet conditions), $\mathcal{A}$ is not invertible. However, (29) corresponds to $(\Delta + c(x))v = 0$ with both $v = 0$ and

$\partial_\nu v = 0$ on ∂X. It is then known that the unique solution to the above constraints is $v = 0$ and hence $(\mathcal{A}^t\mathcal{A}, \mathcal{D})$ is injective.

Proof. We observe that the differential operator $(\mathcal{A}^t)_{ki} = (\mathcal{A}_{ik})^t =: \mathcal{A}^t_{ik}$ is of the same order τ as $\mathcal{A}_{ik}$. Since $\partial^q_\nu v_k = 0$ for $0 \le q \le \tau - 1$, we obtain by integrations by parts that

$$0 = \sum_{ijk}(\mathcal{A}^t_{ik}\mathcal{A}_{ij}v_j, v_k) = \sum_i(\sum_j \mathcal{A}_{ij}v_j, \sum_k \mathcal{A}_{ik}v_k) = \sum_i \|(\mathcal{A}v)_i\|^2.$$

This implies the result. □

3.1. Holmgren unique continuation and generic results. The above result shows that the injectivity of (26) is a consequence of the injectivity of (29) with redundant boundary conditions. The operator $\mathcal{A}$ is therefore augmented with more boundary conditions than is necessary to render it elliptic. This is a possible price to pay to guarantee that the solution to the normal system (26) is uniquely defined.

In general, however, even with redundant boundary conditions, it is not entirely straightforward to ensure that $v = 0$ is the only solution to (29). In this section, we consider the case where $\mathcal{A}$ is well-approximated by an operator with analytic coefficients. We have the following result adapted from [**30**].

PROPOSITION 3.2. *Let $\mathcal{A}_A(x, D)$ be a $2J \times (J + M)$ system of differential equations with analytic coefficients in X such that $\mathcal{A}^t_A\mathcal{A}_A(x, D)$ is an elliptic operator for all $x \in X$. Assume that $\mathcal{A}_A v = 0$ in X. Then v is analytic in X.*

Proof. Since $\mathcal{A}_A v = 0$, we also have that

$${}^{co}(\mathcal{A}^t_A\mathcal{A}_A)(\mathcal{A}^t_A\mathcal{A}_A)v = \mathrm{Diag}\big(\det(\mathcal{A}^t_A\mathcal{A}_A), \ldots, \det(\mathcal{A}^t_A\mathcal{A}_A)\big)v = 0,$$

where ${}^{co}(\mathcal{A}^t_A\mathcal{A}_A)$ is the matrix of co-factors of $\mathcal{A}^t_A\mathcal{A}_A$. Since $\mathcal{A}^t_A\mathcal{A}_A$ is elliptic (a consequence of the fact that the leading term in $\mathcal{A}_A(x, \xi)$ is of full rank $J + M$ for all $(x, \xi) \in \bar{X} \times \mathbb{S}^{n-1}$), then $P = \det(\mathcal{A}^t_A\mathcal{A}_A)$ is an elliptic operator such that $\mathrm{Char}P = \emptyset$, where the characteristic set of P is $\{(x, \xi) \in \bar{X} \times \mathbb{S}^{n-1}, P_m(x, \xi) = 0\}$ and P_m is the principal part of P. We deduce from [**30**, Theorem 6.1] that $WF_A(v_j) \subset \mathrm{Char}P \cup WF_A(Pv_j) = \emptyset$ so that each v_j is analytic in X. □

From this, we deduce the following unique continuation result:

THEOREM 3.3 (Holmgren). *Let $\mathcal{A}_A$ be as in Proposition 3.2 and let us assume that $\mathcal{A}_A v = 0$ in X. Then we have:*

(i) Assume that $v = 0$ in an open set $\Omega \subset\subset X$. Then $v = 0$ in X.

(ii) Assume that $\mathcal{D}v = 0$ on an open set Σ of ∂X. Then $v = 0$ in X.

Proof. Let us first prove (i). We know from Proposition 3.2 that the functions v_j are analytic. Since they vanish on an open set, they have to vanish everywhere.

Now (ii) is a standard consequence of (i), which we write in detail for systems. Let $x_0 \in \Sigma$ and V a sufficiently small open ball around x_0 where the coefficients of $\mathcal{A}_A$ are analytic and where $\mathcal{D}v = 0$ on $\Sigma \cap V \subset \Sigma$. Since $\mathcal{A}_A$ is injective, we deduce that $(\partial_n)^{\tau_j} v_j = 0$ on $\Sigma \cap V$ as well, as for any higher-order derivative in fact. Let us extend v by 0 on V. Then we verify that $\mathcal{A}_A v = 0$ in V (all that needed verification was that the equation was satisfied point-wise on $\Sigma \cap V$). But now (i) implies that $v = 0$ on $V \cup X$. □

The above result shows that $A_A = \mathcal{A}^t_A\mathcal{A}$ is injective as well. Since $R_A A_A = I - T_A$ for some left-parametrix R_A and compact operator T_A, we deduce that 1 is not in the spectrum of T_A so that $A_A^{-1} = (I - T_A)^{-1} R_A$. As a consequence, any operator A sufficiently close to A_A is also invertible with a bounded inverse. We summarize this result as:

COROLLARY 3.4. *Let $\mathcal{A} = \mathcal{A}_A + \mathcal{A}_1$ with $\mathcal{A}_1$ sufficiently small in operator norm from $\prod_i W_p^l(X)$ to $\prod_j W_p^{l+\tau}(X)$. Then $\mathcal{A}$ augmented with $\mathcal{D}v = 0$ is injective and $\mathcal{A}^t\mathcal{A}$ augmented with the same boundary conditions is invertible with bounded inverse. In other words,* (27) *holds with $C_2 = 0$.*

The invertibility of $\mathcal{A}^t\mathcal{A}$ is therefore *generic*, i.e., holds for an open and dense set of coefficients in the definition of the elliptic operator $\mathcal{A}^t\mathcal{A}$; see [**49**]. Note, however, that the size of norm of $\mathcal{A}_1$ for which $\mathcal{A}$ is invertible depends on $\mathcal{A}_A$.

Another similar result states that the invertibility of $\mathcal{A}^t\mathcal{A}$ is guaranteed when X is a sufficiently small domain.

THEOREM 3.5. *Let us assume that $\mathcal{A}(0, D)$ is an elliptic operator. Then for X sufficiently small, we have that $\mathcal{A}$ is injective and that $\mathcal{A}^t\mathcal{A}$ is invertible when augmented with Dirichlet conditions $\mathcal{D}v = 0$. As a consequence,* (27) *holds with $C_2 = 0$.*

As in the previous corollary, the size of the domain X depends on the bound for $(\mathcal{A}(0, D)^t\mathcal{A}(0, D))^{-1}$ with Dirichlet conditions on ∂X, which is independent of X as we shall see. It is therefore possible to estimate the size of the domain X for which the above theory applies; see also [**13**] for a similar result with a different method of proof.

Proof. We assume that all coefficients of $\mathcal{A}$ are sufficiently smooth and that $0 \in X$. Let $P = \mathcal{A}(0, D)$ be the operator of order $m = \tau$ with coefficients frozen at $x_0 = 0$. Then P^tP with Dirichlet conditions on ∂X is invertible as an application of Theorem 3.3. For constant coefficients, a more precise theory applies. Let $u \in H^m(X)$ (we restrict ourselves to Hilbert spaces to simplify notation) be such that $u = \partial_\nu^j u = 0$ for $1 \leq j \leq m - 1$, i.e., $\mathcal{D}u = 0$. We assume ∂X sufficiently smooth that we can extend u by 0 outside of X and obtain a function in $H^m(\mathbb{R}^n)$ with the same norm. Let Y be an open set such that $\bar{X} \subset Y \subset \mathbb{R}^n$.

The elliptic theory for constant coefficient operators [**30**] provides the existence of a fundamental solution E such that

$$E * P^*Pu = u.$$

This implies that

$$\|u\|_{H^m(X)} = \|u\|_{H^m(Y)} = \|E * P^*Pu\|_{H^m(Y)} \leq \|E * P^*\|_{\mathcal{L}(L^2(X), H^m(Y))} \|Pu\|_{L^2(X)}.$$

This proves the existence of a constant C independent of X such that

$$\|Pu\|_{L^2(X)} \geq C\|u\|_{H^m(X)}.$$

This holds for any operator with constant coefficients.

Now let Q be an operator of order at most m and of small norm ε from $L^2(X)$ to $H^m(X)$. Then

$$\|(P + Q)u\|_{L^2(X)} \geq (C - \varepsilon)\|u\|_{H^m(X)} \geq C_1\|u\|_{H^m(X)},$$

which implies that $P + Q$ is injective for ε sufficiently small.

This is applied as follows. Let $P = \mathcal{A}(x, D)$ be an elliptic operator with not-necessarily constant coefficients and define $P_0 = \mathcal{A}(0, D)$ the operator with coefficients frozen at $x_0 = 0$. Then $P - P_0$ is an operator bounded from $H^m(X_\varepsilon)$ to

$L^2(X_\varepsilon)$ with bound $C_1\varepsilon$ for C_1 independent of ε and $X_\varepsilon \subset B(x_0,\varepsilon)$. From the above, we deduce that

$$\|P_0 u\|_{L^2(X_\varepsilon)} \geq C\|u\|_{H^m(X)}$$

for all function $u \in H^m(X)$ such that $\mathcal{D}u = 0$. For $C_1\varepsilon < C$, we find that

$$\|Pu\|_{L^2(X_\varepsilon)} \geq C_2\|u\|_{H^m(X)}$$

for some $C_2 > 0$. This proves that $P = \mathcal{A}$ is injective on sufficiently small domains. □

3.2. Unique continuation principle. When the domain X is large or when the operator $\mathcal{A}$ is not sufficiently close to an operator $\mathcal{A}_A$ with analytic coefficients, then the unique continuation results must rely on an other principle than analyticity.

Unique continuation in the absence of analyticity in the coefficients is not always guaranteed, even for scalar elliptic operators, although the construction of counter-examples is not straightforward. For general references on unique continuation principles (UCP) for mostly scalar, not necessarily elliptic, operators, as well as counter-examples, see [**31**, **43**, **44**, **53**] and their references.

In this section, we revisit Calderón's [**20**, **21**] result of uniqueness from Cauchy boundary data closely following the presentation in [**44**] and extend it to redundant systems. Our objective is to adapt [**44**, Theorem 5] to specific settings of redundant systems. We first present local uniqueness results, whose proofs are postponed to the appendix, and then use classical arguments to extend them to global uniqueness results.

3.2.1. *Local uniqueness result.* Let x be a point in $\mathbb{R}^{n+1}$ and N be a unit vector equal to $(0,\dots,0,1)$ in an appropriate system of coordinates. We want to address the local uniqueness of the Cauchy problem. Assume that L_q for $1 \leq q \leq Q$ are differential operators of order m and that

$$L_q u = 0 \text{ in } V \cap \{x_{n+1} > 0\}, \qquad \partial^j_{n+1} u = 0 \text{ on } V \cap \{x_{n+1} = 0\}, \tag{30}$$

for $1 \leq q \leq Q$ and $1 \leq j \leq m-1$, where V is a neighborhood of 0. We assume that N is non characteristic (at x) for all L_q, i.e., $p_q(x,N) \neq 0$ with p_q the principal symbol of L_q. This and (30) imply that all derivatives of u vanish on $\{x_{n+1} = 0\}$ and that u can be extended by 0 on $V \cap \{x_{n+1} < 0\}$. The uniqueness problem may therefore be recast as: if u satisfies

$$L_q u = 0 \text{ in } V \quad \text{for } 1 \leq q \leq Q \qquad \text{and} \qquad u = 0 \text{ in } \{x_{n+1} < 0\}, \tag{31}$$

then $u \equiv 0$ in a full neighborhood of 0.

When $Q = 1$, it is known that $L = L_1$ needs to satisfy several restrictive assumptions in order for the result to hold; see [**44**]. The main advantage of the redundancy in the above system is that such assumptions need to be valid only locally (in the Fourier variable ξ) for each operator, and globally collectively.

Changing notation, we define $t = x_{n+1}$ and still call $x = (x_1,\dots,x_n)$. Then $p_q = p_q(x,t,\xi,\tau)$ is the principal symbol of L_q for this choice of coordinates. Sufficient conditions for the uniqueness to the above Cauchy problem involve the properties of the roots τ of the above polynomials $\tau \mapsto p_q(\cdot,\tau)$ as a function of ξ. In the setting of redundant measurements, these conditions may hold for different values of q for different values of ξ. This justifies the following definition of our assumptions.

We assume the existence of a finite covering $\{\Omega_\nu\}$ of the unit sphere $\mathbb{S}^{n-1}$ (corresponding to $|\xi|=1$) such that the following holds. For each ν, there exists $q=q(\nu)$ and $\varepsilon>0$ such that for each (x,t) close to 0 and each $\xi\in\Omega_\nu$, we have:

(*i*) $p_q(x,t,\xi,\tau)$ has at most simple real roots τ and at most double complex roots

(*ii*) distinct roots τ_1 and τ_2 satisfy $|\tau_1-\tau_2|\geq\varepsilon>0$

(*iii*) non-real roots τ satisfy $|\Im\tau|\geq\varepsilon$.

Then we have the following result:

THEOREM 3.6. *[Calderón's result for redundant systems.] Assume that N is non characteristic for the operators L_q at the origin and that for a finite covering $\{\Omega_\nu\}$ of the unit sphere, (i)-(ii)-(iii) above are satisfied. Then* (31) *implies that $u=0$ in a full neighborhood of* 0.

The proof, which closely follows that of Theorem 5 in [**44**], is presented in the appendix.

In the above theorem, u is scalar. The above proof extends to vector-valued functions u when the system for u is diagonally dominant. For a determined system given by a matrix L_{ij} of operators of order $m\geq1$, this means that the operators L_{ii} are of order m and satisfy the hypotheses of the above theorem and the operators L_{ij} for $i\neq j$ are of order at most $m-1$. This generalizes to the setting of redundant systems $L_{ij}u_j=S_i$ where for each i, only one operator L_{ij} is of order m and for each j, the operators L_{kj} of order m collectively satisfy the hypotheses of the above theorem; see Theorem A.2 in the appendix, which we do not reproduce here.

The above theorem may not apply directly in applications. However, it serves as a component to obtain more general results. We consider one such result that finds applications in the framework of power density measurements.

We consider a setting where the leading term in the system may not be diagonal but rather upper-triangular. Unique continuation properties may still be valid provided that (a sufficiently large number of) the diagonal operators are elliptic. However, the corresponding complex roots may no longer be double. Instead, we need the stronger condition for some $\varepsilon>0$:

(*iv*) $p_q(x,t,\xi,\tau)$ has at most simple roots τ that satisfy $|\Im\tau|\geq\varepsilon$.

Then we have the following result.

THEOREM 3.7. *Consider the redundant system of equations*

$$(32)\quad \begin{pmatrix} L_1 & L_0\\ L_3 & L_2\end{pmatrix}\begin{pmatrix}u_1\\ u_2\end{pmatrix}=0 \text{ in } V\cap\{x_{n+1}>0\},\qquad \partial^j_{n+1}u_k=0 \text{ on } V\cap\{x_{n+1}=0\},$$

for $1\leq j\leq m-1$ and $1\leq k\leq 2$, with the following assumptions. The operators L_1 and L_0 are (vector-valued) $Q_1\times1$ operators of order m, where L_1 satisfies the hypotheses of Theorem 3.6 *with $Q=Q_1$. The operators L_2 and L_3 are $Q_2\times 1$ operators of order m and at most $m-1$, respectively. Moreover, L_2 satisfies the ellipticity hypothesis (i)-(ii)-(iv) with $Q=Q_2$. Then $(u_1,u_2)=0$ in a full neighborhood of* 0.

The same result extends to systems of the form

$$(33)\quad (L_{ij})_{1\leq i,j\leq R}\,u=0 \text{ in } V\cap\{x_{n+1}>0\},\qquad \partial^j_{n+1}u_k=0 \text{ on } V\cap\{x_{n+1}=0\},$$

for $1 \leq k \leq R$, *where* L_{ij} *is (a vector-valued operator) of order* $m-1$ *for* $i > j$, L_{ii} *is (a vector-valued operator) of order* m *that satisfies the hypotheses of Theorem* 3.6 *with an appropriate value of* Q *when all* L_{ik} *are of order* $m-1$ *for* $k \neq i$ *and the ellipticity hypothesis (i)-(ii)-(iv) with an appropriate value of* Q *when at least one operator* L_{ik} *for* $k < i$ *is of order* m.

The proof of this theorem is also given in the appendix. Note that *(i)-(ii)-(iv)*, as opposed to *(i)-(ii)-(iii)*, may be seen as an ellipticity condition since the symbol $p_{q(\nu)}$ does not vanish on Ω_ν. It is known that Carleman estimates are sharper in such a setting; see, e.g., [**35**, **44**].

REMARK 3.8. Let $\mathcal{Q}$ be an invertible $Q_1 \times Q_1$ matrix and define the linear operators $\tilde{L}_1 = \mathcal{Q}L_1$ and $\tilde{L}_0 = \mathcal{Q}L_0$. Then the conclusions of Theorem 3.7 are the same as those obtained when L_1 and L_0 are replaced by $\tilde{L}_1$ and $\tilde{L}_0$. In other words, the results of Theorem 3.6 hold if L_q is replaced by an equivalent linear combination. The condition that N be non characteristic for L_q may then be replaced by the condition that it be non characteristic for the linear combinations $\tilde{L}_q = \mathcal{Q}_{q,q'}L_{q'}$.

REMARK 3.9. If L_0 is an operator of order $m-1$ in the latter theorem so that the above system is diagonally dominant, then it is sufficient to assume that L_1 and L_2 satisfy the less constraining hypotheses of Theorem 3.6; see Theorem A.2 and the discussion below Theorem 3.6.

3.2.2. *Global uniqueness result.* The above local results extend to global unique continuation results such as [**30**, Sections 8.5-8.6]:

THEOREM 3.10. *Let* $\mathcal{A}$ *be a (redundant) system of operators of order* m, *let* $R_{\mathcal{A}} \subset \{(x,N) \in \bar{X} \times \mathbb{S}^{n-1}\}$ *be the set of points where the hypotheses of either Theorem* 3.6 *or Theorem* 3.7, *(and possibly their extensions in Remark* 3.9*) are satisfied and let us define* $\Sigma_{\mathcal{A}} = \bar{X} \times \mathbb{S}^{n-1} \backslash R_{\mathcal{A}}$. *Let*

$$\mathcal{A}u = 0 \quad \text{in } X, \qquad \partial_\nu^j u = 0 \quad \text{on } \partial X,\ 0 \leq j \leq m-1.$$

Let $\bar{N}(\mathrm{supp}(u))$ *be the subset of* $(\bar{x},\xi) \in X \times \mathbb{S}^{n-1}$ *composed of the closure of the normal set of the support of* $\{u_j\}$; *see* [**30**]. *Heuristically, that set may be seen as the points* (x,ξ) *where* x *belongs to the boundary of* supp(u) *and* ξ *is a normal (inward or outward) unit vector to that boundary.*

Then $\bar{N}(\mathrm{supp}(u)) \subset \Sigma_{\mathcal{A}}$. *When the latter set is empty, then* $u \equiv 0$ *in* X.

The proof follows from the local results noting that the boundary to the support of u cannot occur at a point and for a direction where a local UCP principle holds; see [**30**, Sections 8.5-8.6] for the definitions and derivations.

REMARK 3.11. Let $\mathcal{A}$ be a $p \times q$ system with $p \geq q$ and let us assume that $\mathcal{A}u = 0$ in the above theorem is replaced by

$$|(\mathcal{A}u)_k|(x) \leq C \sum_{|\alpha| \leq m-1} \sum_{l=1}^{q} |D^\alpha u_l|(x), \tag{34}$$

for a constant C independent of $x \in \bar{X}$ and $1 \leq k \leq p$. Let, as above, $\Sigma_{\mathcal{A}}$ be the set of points where the hypotheses of Theorem 3.6 do not hold (with obvious modifications for Theorems 3.7). Then the conclusions of the above theorem, namely that $\bar{N}(\mathrm{supp}(u)) \subset \Sigma_{\mathcal{A}}$, still hold; see, e.g., [**31**] and the proofs in the appendix.

3.3. Application to power-density measurements. We now apply the unique continuation results of sections 3.1 and 3.2 to the setting of power density measurements described in section 2.6.

3.3.1. *Second-order linear systems of equations.* Both unique continuation results require that the system $\mathcal{A}$ be elliptic in the regular sense, i.e., with $\tau = t_j$ independent of j and $s_i = 0$ for all i. The systems (15) and (21)-(22) are not in this form. We propose two modifications of the latter systems for which uniqueness results can be proven.

3.3.2. *System after elimination.* Let us first consider the system (21)-(22). The latter constraints are first-order. However, they easily become second-order by differentiation. We thus consider the system

(35)
$$\begin{aligned}\nabla\cdot\gamma\big(-\nabla\delta u_j+2\hat{F}_j\hat{F}_j\cdot\nabla\delta u_j\big) &= \nabla\cdot\frac{\delta H_j}{|F_j|^2}F_j.\\ \nabla\big(2\gamma\frac{1}{|F_j|^2}F_j\cdot\nabla\delta u_j-2\gamma\frac{1}{|F_k|^2}F_k\cdot\nabla\delta u_k\big) &= \nabla\big(\frac{\delta H_j}{|F_j|^2}-\frac{\delta H_k}{|F_k|^2}\big)\quad 1\le j<k\le J,\end{aligned}$$

in X with boundary conditions $\delta u_j = \partial_\nu \delta u_j = 0$ on ∂X. The above system is clearly elliptic under the same hypotheses as (21)-(22). Moreover, it satisfies the above ellipticity condition with $\tau = 2$.

3.3.3. *System in triangular form.* Let us come back to (15). We apply the operator $K_j = 2\gamma|F_j|^{-2}F_j\cdot\nabla$ to the first line and $L_\gamma|F_j|^{-2}$ to the second line with $L_\gamma := \nabla\cdot\gamma\nabla$ to obtain after subtraction

$$\begin{aligned}\nabla\cdot\delta\gamma F_j+\nabla\cdot\gamma\nabla\delta u_j &= 0\\ L_\gamma\delta\gamma-K_j\nabla\cdot\delta\gamma F_j+[L_\gamma K_j,K_jL_\gamma]\delta u_j &= L_\gamma\frac{\delta H_j}{|F_j|^2}.\end{aligned}\tag{36}$$

Since L_γ is second order and K_j is first-order, the commutator $[L_\gamma K_j, K_j L_\gamma]$ is a second-order differential operator. We thus again obtain a system that is elliptic when (15) is (we leave this proof to the reader; see also (39) below) and that is in the appropriate form with $\tau = 2$.

3.3.4. *Holmgren unique continuation result.* The results of section 3.1 apply to the power density measurement systems (35) and (36). Let $\mathcal{A}$ denote the linear operator in one of the aforementioned systems. Then $\mathcal{A}^t\mathcal{A}$ augmented with Dirichlet conditions is injective provided that γ and u_j are sufficiently close to analytic coefficients or provided that the problem is posed on a sufficiently small domain, and provided of course that the quadratic forms $\{q_j\}$ are collectively elliptic on $\bar{X}$ in the sense of Definition 2.1.

The main difference between (35) and (36) pertains to the boundary conditions. In the case of (36), vanishing Dirichlet and Neumann boundary conditions are imposed on $\delta\gamma$ and δu_j. This means that the value of $(\gamma, \{u_j\})$ and of its (normal) derivatives is known on ∂X. In the case of (35), only the values of $(\{u_j\})$ and of its (normal) derivatives need to be known.

In fact, the boundary conditions for $\delta\gamma$ at $x \in \partial X$ may be deduced from those for δu_1, say, provided that $\nu(x)$ is non characteristic for P_1. Indeed, we observe that

$$\delta H_1 = \delta\gamma|\nabla u_1|^2 + 2\gamma\nabla u_j\cdot\nabla\delta u_1.$$

This provides boundary conditions for $\delta\gamma$ when δu_1 and $\partial_\nu\delta u_1$ are known on ∂X. The derivation of $\partial_\nu\delta\gamma$ is obtained as follows. Using the equation

$$\nabla\cdot\delta\gamma\nabla u_j + \nabla\cdot\gamma\nabla\delta u_j = 0,$$

and replacing $\delta\gamma$ above using the expression of δH_1, we obtain after straightforward eliminations that

$$(1-2(\widehat{\nabla u_1}\cdot\nu(x))^2)\partial_\nu^2 u_1 = f_1,$$

where f_1 is a known expression involving $\partial_\nu u_1$ and tangential derivatives of u_1 of degree up to 2 at x, which are known since u_1 is known on ∂X. Since ν is non characteristic for P_1, this provides an expression for $\partial_\nu^2 u_1$, and hence for $\partial_\nu\delta\gamma$ using the above expression for δH_1.

So knowledge of Cauchy data for $\delta\gamma$ is in fact a consequence of knowledge of Cauchy data for the solutions δu_j. Thus, the major requirement in solving $\mathcal{A}^t\mathcal{A}v = \mathcal{A}^t\mathcal{S}$ is that we impose that $\partial_\nu \delta u_j = g_j$ for a known function g_j, which implies that $\partial_\nu u_j$ is known on ∂X.

3.3.5. *Calderón unique continuation result.* The symbol of the operator (15) is elliptic in the Douglis-Nirenberg sense. Recall that the results of unique continuation principle (UCP) of the preceding sections apply to operators that are elliptic in a classical sense. We thus consider the modification of the system (36). We were not able to obtain a satisfactory global UCP using Carleman-type estimates for the system (35).

UCP for such the modified system (36) is based on proving a collective UCP for quadratic forms, which we now address in detail.

LEMMA 3.12. *Let $P_j(x,D)$ be second-order operators with principal symbols given by $q_j(x,\xi) = 2(\hat F_j\cdot\xi)^2 - |\xi|^2$, where $\hat F_j(x)\in\mathbb{S}^{n-1}$ for $1\le j\le J$. Let $(x,N)\in\bar X\times\mathbb{S}^{n-1}$. For $\xi'\in\mathbb{R}^n$ such that $\xi'\cdot N=0$, we define*

$$q_{j;N}(x,\xi') = 2(\hat F_j\cdot\xi')^2 - \big(1-2(\hat F_j\cdot N)^2\big)|\xi'|^2.$$

Let us assume that the quadratic forms are independent in the following UCP sense:

(37)
$$q_{j;N}(x,\xi')=0 \quad \textit{for all } 1\le j\le J \textit{ s.t. } q_j(x,N)\ne 0 \qquad \textit{implies that} \qquad \xi'=0.$$

Then $\{P_j\}$ collectively satisfies a UCP for (x,N) in the sense that we can find a partition $\{\Omega_\nu\}$ such that (i)-(ii)-(iii) holds.

Proof. The proof in dimension $n=2$ is trivial: P_j satisfies UCP except for four directions on the null cone, i.e., for ξ such that $q_j(x,\xi)=0$. Since (37) implies that N is not in the intersection of the null cones, then UCP holds for one of the operators P_j and we can choose $\Omega_1=\mathbb{S}^1$ in the partition.

Let $(x,N)\in\bar X\times\mathbb{S}^{n-1}$ and $n\ge 3$. Let $\xi'\in\mathbb{S}^{n-1}$ such that $\xi'\cdot N=0$. Then

$$\begin{aligned} q_j(x,\xi'+\tau N) &= 2(\hat F_j\cdot(\xi'+\tau N))^2 - |\xi'+\tau N|^2 \\ &= \big(2(\hat F_j\cdot N)^2-1\big)\tau^2 + 4\hat F_j\cdot N\hat F_j\cdot\xi'\tau + 2(\hat F_j\cdot\xi')^2 - 1. \end{aligned}$$

This quadratic polynomial (because $q_j(x,N)=2(\hat F_j\cdot N)^2-1\ne 0$) with real-valued coefficients has a double root when the determinant

$$\Delta_j' = -1 + 2\big((\hat F_j\cdot N)^2 + (\hat F_j\cdot\xi')^2\big) = q_{j;N}(x,\xi') = 0. \tag{38}$$

For ξ' away from such points, then the UCP condition holds for P_j since the roots cannot be simple and complex roots (which come in conjugate pairs) cannot be close to each-other. By assumption, $\Delta_j'=0$ for all j is not possible and by continuity, $\max_j|\Delta_j'|$ is bounded away from 0 so that different roots τ are separated by some $\varepsilon>0$ (as well as the imaginary part of such roots since they come in conjugate pairs). This generates a finite partition $\{\Omega_\nu\}$. On each patch, one P_j satisfies the UCP properties *(i)-(ii)-(iii)*. □

In dimension $n = 2$, ellipticity of (P_1, P_2) is equivalent to UCP of (P_1, P_2) at each (x, N). In dimension $n \geq 3$, collective ellipticity and collective UCP are different notions. As soon as there is one j such that $2(\hat{F}_j \cdot N)^2 \geq 1$, the UCP is satisfied at (x, ξ) while ellipticity does not necessarily holds. However, we have the result:

LEMMA 3.13. *In dimension $n = 2$, collective UCP and collective ellipticity of operators of the form P_j above is equivalent. In dimension $n = 3$, collective ellipticity of $\{P_j\}$ such that $q_j(N) \neq 0$ implies collective UCP at (x, N).*

Proof. The case $n = 2$ is handled as in the proof of Lemma 3.12. Consider now the case $n = 3$. Non-UCP at (x, N) implies from (38) and the fact that N is non characteristic for P_j that

$$(\hat{F}_j \cdot N)^2 + (\hat{F}_j \cdot \xi')^2 = \frac{1}{2}.$$

If we decompose $\hat{F}_j = \hat{F}_j \cdot NN + \hat{F}_j \cdot \xi'\xi' + \hat{F}_j''$ with $\hat{F}_j''$ orthogonal to $\mathrm{span}(N, \xi')$, then we find that

$$|\hat{F}_j''|^2 = \frac{1}{2}.$$

In dimension $n = 3$, we find that $\hat{F}_j'' = \pm|\hat{F}_j| N \times \xi'$. This implies that $N \times \xi'$ belongs to the null cone of q_j, i.e., $q_j(x, N \times \xi') = 0$. This holds for every j, which implies that $N \times \xi' = 0$, a contradiction. □

DEFINITION 3.14. *We say that the family of operators $P_j(x, D)$ or of quadratic forms $q_j(x, \xi)$ collectively satisfy a global UCP on $\bar{X}$ if for every $(x, N) \in \bar{X} \times \mathbb{S}^{n-1}$, we can find a family $\tilde{q}_j(x, \xi) = \mathcal{Q}_{jk} q_k(x, \xi)$ with $\mathcal{Q}$ an invertible matrix such that $\{\tilde{q}_j(x, \xi)\}$ collectively satisfies a UCP at (x, N) in the sense of Lemma* 3.12.

The above definition involves a linear change of quadratic forms for each (x, N) that allows us to obtain a family of modified forms $\tilde{q}_j$ with similar properties to those of q_j and such that $\tilde{q}_j(N) \neq 0$; see Remark 3.8. This will prove useful in the analysis of reconstruction from power density functionals. We will also need the following Lemma in dimension $n = 3$.

LEMMA 3.15. *Let F_1 and F_2 be three-dimensional vector fields on $\bar{X}$ such that for each $x \in \bar{X}$, $\mathrm{rank}(F_1, F_2) = 2$. Let $F_3 = F_1 + F_2$. We denote by $\{P_j\}$ and $\{q_j\}$ the corresponding operators and quadratic forms defined in* (18) *for $1 \leq j \leq 3$. Then $\{q_j\}$ satisfies a global UCP on $\bar{X}$.*

Proof. Let us fix $(x, N) \in \bar{X} \times \mathbb{S}^2$ and choose a basis of $\mathbb{R}^3$ such that $\hat{F}_1(x) = e_1$ and $\mathrm{span}(F_1, F_2) = (e_1, e_2)$. The three quadratic forms q_j allow us to obtain by linear combination any q_c corresponding to $F_c = ce_1 + se_2$ with $s = \sqrt{1 - c^2}$. It is then not difficult to find three values of c such that the corresponding $\tilde{q}_j$ for $1 \leq j \leq 3$ form an elliptic family and such that $\tilde{q}_j(N) \neq 0$. From Lemma 3.13, we obtain that $\{\tilde{q}_j\}$ satisfies a UCP at (x, N). The hypotheses of global UCP in Definition 3.14 are met. □

3.3.6. *Elliptic system in triangular form.* We now present a modification of the linear system for $(\delta\gamma, \{\delta u_j\})$ for which a global UCP result as described in the above definition can be obtained.

We recast (36) as

$$\begin{pmatrix} P_j & \tilde{P}_j \\ 0 & \Delta \end{pmatrix} \begin{pmatrix} \delta\gamma \\ \delta u_j \end{pmatrix} = l.o.t. \begin{pmatrix} \delta\gamma \\ \delta u_j \end{pmatrix} + S_j, \tag{39}$$

where $l.o.t.$ means a system of differential operators of order at most 1. Let P and $\tilde{P}$ denote the $J \times 1$ columns of the second-order, homogeneous, operators P_j and $\tilde{P}_j$, respectively. The symbol of P_j is proportional to $q_j(x, \xi)$. The operator Δ satisfies hypothesis *(iv)* of Theorem 3.7 while the operator $\tilde{P}$ is of order 2. If the operators P collectively satisfy the hypotheses of Theorem 3.6, then the above system (36) satisfies a UCP, as does the fourth-order operator $\mathcal{A}^t\mathcal{A}$ as constructed in the preceding section:

THEOREM 3.16. *Let $\mathcal{A}v = \mathcal{S}$ in X be the system for $v = (\delta\gamma, \{\delta u_j\})$ described in* (39) *and augmented with boundary conditions $v = \partial_\nu v = 0$ on ∂X. Assume that the coefficients in $\mathcal{A}$ are sufficiently smooth (see proof of Theorem* A.1*). Let the operators P_j above satisfy a UCP on $\bar{X}$ as described in Definition* 3.14 *and Lemma* 3.12*. Then any solution to that system, or to the system $\mathcal{A}^t\mathcal{A}v = \mathcal{A}^t\mathcal{S}$ with the same boundary conditions, is unique.*

Proof. The proof is a direct application of Theorem 3.7 since Δ satisfies hypothesis *(iv)*, $\tilde{P}$ is of order 2, and P collectively satisfies a UCP at each (x, N) in $\bar{X} \times \mathbb{S}^{n-1}$. □

In dimensions $n = 2$ and $n = 3$, we have the following sufficient conditions on the internal functionals to guaranty injectivity of $\mathcal{A}^t\mathcal{A}$:

COROLLARY 3.17. *Assume that (u_1, u_2) are solutions such that $F_1 = \nabla u_1$ and $F_2 = \nabla u_2$ are such that* $\text{rank}(F_1, F_2) = 2$ *for each $x \in \bar{X}$.*

In dimension $n = 2$, assume moreover that $F_1 \cdot F_2 \neq 0$ for each $x \in \bar{X}$. Then $\{P_1, P_2\}$ satisfies a global UCP property on $\bar{X}$ and the results of Theorem 3.16 *hold.*

In dimension $n = 3$, define $F_3 = F_1 + F_2$. Then $\{P_1, P_2, P_3\}$ satisfies a global UCP property on $\bar{X}$ and the results of Theorem 3.16 *hold.*

The proof is immediate using Lemma 3.15, remark 3.8 and the preceding Theorem.

We have thus exhibited a system of equations $\mathcal{A}^t\mathcal{A}v = \mathcal{A}^t\mathcal{S}$ with boundary conditions v and $\partial_\nu v$ prescribed on ∂X, which admits a unique solution that verifies the stability estimate (23) (generalized as in (27) for non-homogeneous boundary conditions) with $C_2 = 0$.

As in the preceding section devoted to the Holmgren uniqueness result, this result comes at the cost of having to impose boundary conditions of the form v and $\partial_\nu v$ to both $v = \delta u_j$ and $v = \delta\gamma$. As we saw in the preceding section, the boundary conditions for $\delta\gamma$ at $x \in \partial X$ may be deduced from those for δu_j.

Comparing ellipticity and uniqueness criteria for $\mathcal{A}$ in (39), we observe that ellipticity is obtained by imposing conditions $\delta u_j = g_j$ on ∂X and ensuring that $\{q_j\}$ is an elliptic family of quadratic forms. UCP is obtained by imposing the additional boundary conditions $\partial_\nu \delta u_j = \psi_j$ and by ensuring that $\{q_j\}$ satisfy a global UCP on $\bar{X}$. In dimension $n = 2$, global ellipticity and global UCP for $\{q_j\}$

are equivalent. In dimension $n = 3$, both are satisfied for the family of three internal functionals described in Corollary 3.17.

4. Local uniqueness for nonlinear inverse problem

Once the injectivity of a linear system can be established, standard theories may be applied to obtain local uniqueness results for the nonlinear inverse problems. Let us recast the nonlinear system of PDE (3) as

$$\begin{aligned} \tilde{\mathcal{F}}(v) &= \tilde{\mathcal{H}} \quad \text{in } X \\ \tilde{\mathcal{B}}v &= \tilde{g} \quad \text{on } \partial X. \end{aligned} \tag{40}$$

Here $v = (\gamma, \{u_j\}_{1\le j\le J})$ and $\tilde{\mathcal{B}}$ is an operator that maps v to the trace of $\{u_j\}$ on ∂X. We assume here, as is the case for hybrid inverse problems, that the boundary condition operator $\tilde{\mathcal{B}}$ is linear.

The linearization of the above system involves the operator $\tilde{\mathcal{F}}'(v_0)$ for a fixed v_0. The analysis of the preceding sections did not allow us to show that the differential operator $\tilde{\mathcal{F}}'(v_0)$ augmented with the boundary conditions $\tilde{\mathcal{B}}$ was invertible. Rather, we obtained that for an appropriate linear differential operator $\mathcal{T}(v_0)$, then $\mathcal{A} := \mathcal{T}(v_0)\tilde{\mathcal{F}}'(v_0)$ augmented with the augmented boundary conditions $\mathcal{B}$ was invertible. More precisely, we obtained that

$$\begin{aligned} \mathcal{A}w := \mathcal{T}(v_0)\tilde{\mathcal{F}}'(v_0)w &= \mathcal{S} \quad \text{in } X \\ \mathcal{B}w &= h \quad \text{on } \partial X, \end{aligned} \tag{41}$$

admitted a unique solution with a stability estimate given by

$$\|w\|_{\mathcal{X}} \le C\|(\mathcal{S}, h)\|_{\mathcal{Y}}. \tag{42}$$

The above linearized operator is the linearization of the nonlinear operator

$$\mathcal{F}(v_0 + w) := \mathcal{T}(v_0)\tilde{\mathcal{F}}(v_0 + w). \tag{43}$$

We thus modify the original inverse problem and replace it with an inverse problem that is necessarily satisfied by the exact solution $v = v_0 + w$ we wish to reconstruct and is given by

$$\begin{aligned} \mathcal{F}(v_0 + w) &= \mathcal{H} := \mathcal{T}(v_0)\tilde{\mathcal{H}} \quad \text{in } X \\ \mathcal{B}(v_0 + w) &= g \quad \text{on } \partial X. \end{aligned} \tag{44}$$

Borrowing the notation of section 2.5, we observe that the nonlinear operator $(\mathcal{F}(\cdot), \mathcal{B}\cdot)$ maps $\mathcal{X} = \mathcal{U}(p, l)$ to $\mathcal{Y} = \mathcal{R}(p, l)$ for appropriately defined spaces $\mathcal{U}$ and $\mathcal{R}$ for $(\mathcal{A}, \mathcal{B}) = (\mathcal{F}'(v_0), \mathcal{B})$.

Let us pause on the definition of the boundary condition g for $v_0 + w$. We cannot expect $\mathcal{B}v_0 = g$ with g known so that $\mathcal{B}w = 0$ on ∂X. The reason is that v_0 is typically constructed by guessing γ_0 and solving the linear elliptic problems for $\{u_{j,0}\}$ with imposed Dirichlet conditions. It is for such a construction of v_0 that we were able to show that $A = (\mathcal{A}, \mathcal{B})$ above was injective for $\mathcal{A} = \mathcal{F}'(v_0)$. The boundary condition $g_0 := \mathcal{B}v_0$ thus partially depends on solving the above problem and is not in general given by the measurements g (unless v_0 is the solution v). We thus recast the (modified) nonlinear hybrid inverse problem as

$$\begin{aligned} \mathcal{F}(v_0 + w) &= \mathcal{H} := \mathcal{T}(v_0)\tilde{\mathcal{H}} \quad \text{in } X \\ \mathcal{B}w &= g - g_0 \quad \text{on } \partial X. \end{aligned} \tag{45}$$

The objective of the following section is to provide an iterative algorithm to reconstruct w provided that v_0 is sufficiently close to the solution we wish to obtain in the sense that $g - g_0$ and $\mathcal{H} - \mathcal{H}_0$ are sufficiently small with $\mathcal{H}_0 := \mathcal{F}(v_0)$. In section 4.2, we obtain an injectivity result stating that if $\mathcal{F}(v_0 + w) = \mathcal{F}(v_0 + \tilde{w})$ and $\mathcal{B}w = \mathcal{B}\tilde{w}$, then $w = \tilde{w}$.

REMARK 4.1. *In the power density setting, we recast* (10) *and* (11) *as*

$$\mathcal{F}_{2j-1}(v_0 + w) = 0, \qquad \tilde{\mathcal{F}}_{2j}(v_0 + w) = H_j, \qquad 1 \le j \le J,$$

respectively. Let then L_γ, F_j, and M_j constructed from $v_0 = (\gamma, \{u_j\})$ as in the derivation of (36). *Finally, let us define*

$$\mathcal{F}_{2j}(v_0+w) = L_\gamma \tilde{\mathcal{F}}_{2j-1}(v_0+w) - M_j \mathcal{F}_{2j-1}(v_0+w) = |F_j|^{-2} H_j =: K_j, \quad 1 \le j \le J.$$

The system $\mathcal{F}_{2j-1}(v_0 + w) = 0$ and $\mathcal{F}_{2j}(v_0 + w) = K_j$ for $1 \le j \le J$ is recast as

$$\mathcal{F}(v_0 + w) = \mathcal{H},$$

in the notation of (45) *and implicitly defines the linear operator $\mathcal{T}(v_0)$. We denote by $\mathcal{A} = \mathcal{F}'(v_0)$ the linearization of $\mathcal{F}$ at v_0, which agrees with the differential operator defined in* (36).

Whereas the operator $\tilde{\mathcal{B}}$ maps v_0 to the traces of $\{u_j\}$ on ∂X, the extended operator $\mathcal{B}$ maps v_0 to the traces of v_0 and $\nu \cdot \nabla v_0$ on ∂X.

4.1. Iterative fixed point and reconstruction procedure. Let us define

$$\mathcal{F}(v_0 + w) = \mathcal{F}(v_0) + \mathcal{F}'(v_0)w + \mathcal{G}(w; v_0), \tag{46}$$

where $\mathcal{G}(w; v_0)$ is quadratic in the first variable in the sense that

$$\|(\mathcal{G}(w; v_0), 0)\|_{\mathcal{Y}} \le C\|w\|_{\mathcal{X}}^2. \tag{47}$$

The latter estimate comes from the fact that $\mathcal{F}(v)$ is polynomial in v and the partial derivatives D and that $\mathcal{X}$ is an algebra. We may thus recast the nonlinear system of equations for w as

$$\begin{aligned} \mathcal{A}w &= \mathcal{H} - \mathcal{H}_0 - \mathcal{G}(w; v_0) \quad \text{in } X \\ \mathcal{B}w &= g - g_0 \quad \text{on } \partial X. \end{aligned} \tag{48}$$

Since the linear operator $A = (\mathcal{A}, \mathcal{B})$ is invertible, we may recast the above equation as

$$w = \mathcal{I}(w) := A^{-1}(\mathcal{H} - \mathcal{H}_0, g - g_0) - A^{-1}(\mathcal{G}(w; v_0), 0). \tag{49}$$

From the polynomial structure of $\mathcal{F}$and the boundedness of A^{-1} from $\mathcal{Y}$ to $\mathcal{X}$, we deduce in addition to (47) that

$$\begin{aligned} \|A^{-1}(\mathcal{G}(w; v_0), 0)\|_{\mathcal{X}} &\le C_1\|w\|_{\mathcal{X}}^2 \\ \|\mathcal{I}(w) - \mathcal{I}(\tilde{w})\|_{\mathcal{X}} &\le C_2(\|w\|_{\mathcal{X}} + \|\tilde{w}\|_{\mathcal{X}})\|w - \tilde{w}\|_{\mathcal{X}}. \end{aligned} \tag{50}$$

As a consequence, for $\|w\|_{\mathcal{X}} \le R$ and $\|A^{-1}(\mathcal{H} - \mathcal{H}_0, g - g_0)\|_{\mathcal{X}} \le \eta$ sufficiently small so that $\eta + C_1R^2 \le R$ and $2C_2R < 1$, we find that $\mathcal{I}$ is a contraction from the ball $B = \{w, \|w\|_{\mathcal{X}} \le R\}$ onto itself.

This shows that for $(\mathcal{H} - \mathcal{H}_0, g - g_0)$ sufficiently small, then the solution to (48) is unique and is obtained by the converging algorithm $w^{k+1} = \mathcal{I}(w^k)$ initialized with $w^0 = 0$.

4.2. Injectivity results. The fixed point algorithm of the preceding section provides us with an explicit local reconstruction procedure for the nonlinear hybrid inverse problem. A similar methodology allows us to obtain local uniqueness results that are more general but not constructive. Using the same notation as in the preceding section, let us assume that

$$\mathcal{F}(v_0 + w) = \mathcal{F}(v_0 + \tilde{w}) = \mathcal{H} \quad \text{in } X, \qquad \mathcal{B}w = \mathcal{B}\tilde{w} \text{ on } \partial X, \tag{51}$$

for a fixed $v_0 \in \mathcal{X}$. In other words, the measurements associated with $v_0 + w$ and $v_0 + \tilde{w}$ are identical. Injectivity of the nonlinear problem $\mathcal{F}$ locally in the vicinity of v_0 then means proving that $w = \tilde{w}$.

Since $\mathcal{F}$ is a polynomial in w, then there is a polynomial function in w and $\tilde{w}$ such that

$$\mathcal{F}(v_0 + \tilde{w}) - \mathcal{F}(v_0 + w) = \mathfrak{J}(w, \tilde{w}) \cdot (\tilde{w} - w). \tag{52}$$

When $\tilde{w} - w$ is sufficiently small, then $\mathfrak{J}(w, \tilde{w})$ satisfies the same properties as $\mathcal{F}'(v_0)$, the Fréchet derivative of $\mathcal{F}$. For instance, the ellipticity and unique continuation properties of $\mathcal{F}'(v_0)$ still hold for $\mathfrak{J}(w, \tilde{w})$ for w and $\tilde{w}$ small in $\mathcal{X}$. As a consequence, we obtain that (51) implies that $w = \tilde{w}$ and hence that $\mathcal{F}$ is injective. More generally, $\mathfrak{J}(w, \tilde{w})$ may still admit a left-inverse for w and $\tilde{w}$ not necessarily close to each-other, in which case we also deduce that $w = \tilde{w}$ for w and $\tilde{w}$ not necessarily small.

Let us assume more generally that we have two measurements

$$\mathcal{F}(v_0 + w) = \mathcal{H}, \quad \mathcal{B}w = g - g_0; \qquad \mathcal{F}(v_0 + \tilde{w}) = \tilde{\mathcal{H}}, \quad \mathcal{B}\tilde{w} = \tilde{g} - g_0.$$

Assume that $\mathcal{F}'(v_0)$ is elliptic and injective. Then, $\mathfrak{J}(w, \tilde{w})$ is elliptic and hence admits a left-inverse (since it is injective), at least for w and $\tilde{w}$ sufficiently close to 0. As a consequence, we obtain as above the stability estimate

$$\|w - \tilde{w}\|_{\mathcal{X}} \leq C\|(\mathcal{H} - \tilde{\mathcal{H}}, g - \tilde{g})\|_{\mathcal{Y}}. \tag{53}$$

Note that the nonlinear problem (51) is invertible generically. Indeed, for v_0 analytic, $\mathcal{A}^t\mathcal{A}$ augmented with Dirichlet boundary conditions $\mathcal{D}$ is invertible. As a consequence, the above result shows that (51) may be inverted for $(v_0 + w)$ in an open set including v_0. Since analytic coefficients v_0 in X_0 restricted to X are dense in the set of sufficiently smooth coefficients on X, we obtain that the inverse problem may be inverted *generically*; see [**49**]. When v_0 is not analytic, then we need to find a unique continuation principle based on Carleman estimates to obtain an estimate of the form (53) for the fully nonlinear hybrid inverse problem.

Acknowledgment

The author would like to thank the organizers of the Irvine conference for their invitation to a meeting that was a befitting tribute to the exceptional contributions to analysis and inverse problems of Gunther Uhlmann. The author is glad to acknowledge many insightful discussions with Shari Moskow on the content of section 4, which borrows some ideas of our joint work in [**13**]. The author is also indebted to Thomas Widlak for noticing and proposing solutions to inconsistencies in the presentation of the Lopatinskii conditions in section 2.6.2. This paper was partially funded by NSF grant DMS-1108608 and AFOSR Grant NSSEFF-FA9550-10-1-0194.

Appendix A. Unique continuation for redundant systems

A.1. Redundant system for a scalar function. Let x be a point in $\mathbb{R}^{n+1}$ and N be a unit vector equal to $(0,\dots,0,1)$ in an appropriate system of coordinates. We want to address the uniqueness of the Cauchy problem. Assume that L_q for $1\leq q\leq Q$ are operators of order m and that

$$L_q u = 0 \text{ in } V\cap\{x_{n+1}>0\},\qquad \partial^j_{n+1}u=0 \text{ on } V\cap\{x_{n+1}=0\}, \tag{54}$$

for $1\leq q\leq Q$ and $1\leq j\leq m-1$, where V is a neighborhood of 0. Here, u is a *scalar* function solution of a *redundant* system of equations. We assume that N is non characteristic for all L_q. This implies, as in section 3.2.1, that all derivatives of u vanish on $\{x_{n+1}=0\}$ and that u can be extended by 0 on $V\cap\{x_{n+1}<0\}$. The uniqueness problem may therefore be recast as: if u satisfies

$$L_q u = 0 \text{ in } V\quad \text{for } 1\leq q\leq Q\qquad\text{and}\qquad u=0 \text{ in } \{x_{n+1}<0\}, \tag{55}$$

then $u\equiv 0$ in a full neighborhood of 0.

When $Q=1$, it is known that $L=L_1$ needs to satisfy some restrictive assumptions in order for the result to hold; see [**29, 36, 44, 53**]. The main advantage of the redundancy in the above system is that such assumptions need to be valid only locally (in the Fourier domain) for each operator, and globally collectively in a sense defined below. Our derivations closely follow the presentation in [**44**], to which we refer to additional details.

Changing notation, we define $t=x_{n+1}$ and still call $x=(x_1,\dots,x_n)$. We then define $p_q=p_q(x,t,\xi,\tau)$ as the principal symbol of L_q for this choice of coordinates. Sufficient conditions for the uniqueness to the above Cauchy problem involve the properties of the roots τ of the above polynomials $\tau\mapsto p_q(\cdot,\tau)$ as a function of ξ. In the setting of redundant measurements, these conditions may hold for different values of q depending on the value of ξ. This justifies the following definition.

We assume the existence of a finite covering $\{\Omega_\nu\}$ of the unit sphere $\mathbb{S}^{n-1}$ (corresponding to $|\xi|=1$) such that the following holds. For each ν, there exists $q=q(\nu)$ and $\varepsilon>0$ such that for each (x,t) close to 0 and each $\xi\in\Omega_\nu$, we have

- *(i)* $\tau\to p_q(x,t,\xi,\tau)$ has at most simple real roots and at most double complex roots
- *(ii)* distinct roots τ_1 and τ_2 satisfy $|\tau_1-\tau_2|\geq\varepsilon>0$
- *(iii)* non-real roots τ satisfy $|\Im\tau|\geq\varepsilon$.

Then we have the following result:

THEOREM A.1. *[Calderón's result for redundant systems.] Assume that N is non characteristic for the operators L_q at the origin and that for a finite covering $\{\Omega_\nu\}$ of the unit sphere, (i)-(ii)-(iii) above are satisfied. Then* (55) *implies that $u=0$ in a full neighborhood of* 0.

Proof. The proof very closely follows that of Theorem 5 in [**44**]. Note that the proof there requires all coefficients to be of class C^∞. It has been shown in [**28**] (see also [**53**, Notes of Chapter 2]) that such results also hold under less restrictive assumptions on smoothness. We do not track such results and assume that the coefficients are sufficiently smooth so that the technology developed in [**44**] applies.

The differences between the proof of our theorem and that in [**44**] are mostly notational. To present these differences in a somewhat self-contained manner, we need to introduce most of the notation in [**44**], to which we refer the reader for details that are not reproduced here.

Let $u(x,t)$ be of class C^m with support contained in $|x| \le r$ and $0 \le t \le T$. Define $w = e^{k(T-t)^2}$. Then there is a constant C independent of u such that for r, T, and k^{-1} sufficiently small, we have the Carleman-type inequality

$$\sum_{|\alpha|<m} |||D^\alpha u|||^2 \le C(k^{-1}+T^2)\sum_{q=1}^{Q} |||L_q u|||^2, \tag{56}$$

where we have defined

$$|||u|||^2 = \int_0^T w(t)\|u\|^2 dt, \qquad w(t) = e^{k(T-t)^2}, \tag{57}$$

with $\|\cdot\|$ the L^2 norm in the variable x. It is shown in [**44**, p.31] that the theorem follows from the above inequality, which we now prove.

Since N is non characteristic for p_q, the leading term of p_q in τ^m does not vanish and we may divide p_q by such a term to consider leading symbols (still called p_q) of the form

$$p_q(x,t,\xi,\tau) = \tau^m + \sum_{j=1}^{m} Q_{qj}(x,t,\xi)\tau^{m-j},$$

with Q_{qj} homogeneous polynomials in ξ of degree j.

Let $\Lambda = (1-\Delta_x)^{\frac12}$ be the operator with symbol $(1+|\xi|^2)^{\frac12}$. For u with (small) compact support, we define

$$u_j = \Lambda^{m-j} D_t^{j-1} u, \quad 1 \le j \le m.$$

Let then $U = (u_1,\dots,u_m)^t$. We may recast $L_q u = f_q$ as

$$\begin{aligned} &D_t u_j - \Lambda u_{j+1} = 0, \quad j<m \\ &D_t u_m + \sum_{j=1}^{m} Q_{qj}(x,t,D_x)\Lambda^{1-j} u_{m-j+1} + R_q U = f_q, \end{aligned}$$

for R_q a pseudo-differential of order 0 in x while $Q_{qj}(x,t,D_x)\Lambda^{1-j}$ is a pseudo-differential of order 1 in x. The leading symbol of the above system for each q is therefore a $m\times m$ matrix $\tau I + h_q$ with h_q homogeneous of degree 1 in ξ. We denote by $H_q(t,x,D_x)$ the pseudo-differential operator with symbol $h_q(t,x,\xi)$. Note that $\mathrm{Det}(\tau I + h_q) = p_q$ so that the roots of the system $\tau I + h_q$ are those of p_q.

From the derivation in [**44**, p.33 & p.35], (56) follows (for k^{-1} and T^2 sufficiently small) from

$$|||U|||^2 \le C(k^{-1}+T^2)\sum_{q=1}^{Q} |||(D_t+H_q)U|||^2 + C\int_0^T w(t)\|U\|_{-1}^2 dt. \tag{58}$$

It thus remains to analyze the terms $\|(D_t+H_q)U\|$. The main idea of Calderón in [**20**, **21**] is to diagonalize h_q and analyze the corresponding diagonalized first-order system. For ξ unit vectors close to some ξ_0, we find a non-singular smooth $m\times m$ matrix $r_q(x,t,\xi)$ such that for all (x,t) close to 0,

$$r_q h_q r_q^{-1} = J_q,$$

with J_q in Jordan form. From the assumption (i), we obtain that for $q = q(\nu)$, then J_q is either a 1×1 matrix $\lambda(x,t,\xi)$ or a 2×2 matrix with $\lambda(x,t,\xi)$ on the diagonal,

0 on the (2,1) entry, and 1 on the (1,2) entry. Once r_q is defined for $|\xi| = 1$, we extend it smoothly to all ξ as a homogeneous function of degree 0. This may be done locally on cones in $|\xi| \geq 1$ but not necessarily globally. Following [**44**, p.34], the global obstructions may be overcome by replacing our finite covering $\{\Omega_\nu\}$ by a finer (sub)covering, which we still call $\{\Omega_\nu\}$ such that for each $\xi \in \Omega_\nu$, there is a non-singular smooth matrix $r_{q\nu}$ such that

$$r_{q\nu} h_q r_{q\nu}^{-1} \text{ is in Jordan canonical form.}$$

We then define $\phi_\nu(\xi)$ a C^∞ partition of unity subordinate to Ω_ν such that $\sum_\nu \phi_\nu^2(\xi) = 1$ for $|\xi| = 1$ and such that each ϕ_ν is extended to $|\xi| \geq 1$ smoothly as a homogeneous function of degree 0 and such that $\sum_\nu \phi_\nu^2(\xi) = 1$ for $|\xi| > 1$. Finally, we define Φ_ν as the corresponding pseudo-differential operator of order 0 and $U_\nu := \Phi_\nu U$.

Since $[\Phi_\nu, H_q]$ is an operator of order 0, we verify as in [**44**, p.35] that (58) is a consequence of

$$|||U_\nu|||^2 \leq C(k^{-1} + T^2)|||(D_t + H_{q(\nu)})U_\nu|||^2 + C\int_0^T w(t)\|U\|_{-1}^2 dt, \tag{59}$$

which is to hold for each ν. Note that the summation over q in (58) is a clear consequence of us obtaining the above inequality for $q = q(\nu)$. For the rest of the derivation, $q = q(\nu)$.

Now, $H_q U_\nu$ is unchanged if $r_{q\nu}$ and $j_{q\nu}$ are changed outside of the support of ϕ_ν. We thus construct symbols h_ν that agree with h_q on Ω_ν but that are globally reducible to a Jordan canonical form. Each h_ν is then extended to all ξ by a smooth function still called h_ν that is homogeneous of degree 1 for $|\xi| \geq 1$. We thus have globally defined symbols such that $r_\nu h_\nu r_\nu^{-1} = j_\nu$. We denote by H_ν, J_ν, R_ν, and S_ν the pseudo-differential operators of order 1, 1, 0, 0, respectively, corresponding to the symbols h_ν, j_ν, r_ν, and r_ν^{-1}.

We finally define $V_\nu = R_\nu U_\nu$. The rest of the proof is then as in [**44**]. Following [**44**, p.35], we deduce that (59) is a consequence of

$$|||V_\nu|||^2 \leq C(k^{-1} + T^2)|||(D_t + J_\nu)V_\nu|||^2. \tag{60}$$

Now J_ν obtained from $p_{q(\nu)}$ for $\xi \in \Omega_\nu$ has diagonal entries of the form $A(t) + iB(t)$ with A and B pseudo-differential operators of order 1 and B elliptic or vanishing when J_ν is a 1×1 matrix. Thus (60) is a consequence of [**44**, Lemma 2] and this concludes the proof of the theorem. □

A.2. Redundant system for vector-valued functions. The proof of the above theorem provides a local control encapsulated in (56). Above, u is a scalar function. Such results may be extended to systems when u is vector-valued provided that the leading term of the system is diagonal. Indeed, (56) may then be used component by component to provide a uniqueness result for the system. We now consider such an extension.

Consider operators L_{qr} for $1 \leq q \leq Q$ and $1 \leq r \leq R$ and the system

$$L_{qr} u_r = 0 \ \text{ in } V \cap \{x_{n+1} > 0\}, \qquad \partial_{n+1}^j u_r = 0 \ \text{ on } V \cap \{x_{n+1} = 0\}, \tag{61}$$

with summation over repeated indexes and holding for all $1 \leq q \leq Q$ in the first equation and for all $1 \leq j \leq m-1$ and all $1 \leq r \leq R$ in the second equation. Here V is still a neighborhood of 0. We still call $t = x_{n+1}$ and $x = (x_1, \ldots, x_n)$ in an appropriate system with N as before. We denote by $p_{qr} = p_{qr}(x, t, \xi, \tau)$ the principal symbol of L_{qr} for this choice of coordinates. As above, this uniqueness problem may be recast as: Assume that

$$L_{qr}u_r = 0 \text{ in } V \quad \text{for } 1 \leq q \leq Q \qquad \text{and} \qquad u_r = 0 \text{ in } \{x_{n+1} < 0\}, \tag{62}$$

for $1 \leq r \leq R$. Then $u_r \equiv 0$ in a full neighborhood of 0.

We sort the equations for $1 \leq q \leq Q$ such that the following holds. We assume that $\{1, \ldots, Q\}$ is decomposed into R connected components I_r of the form $l(r-1)+1 \leq q \leq l(r)$ for a strictly increasing function $l(r)$ from $\{1, \ldots, R\}$ to $\mathbb{N}^*$ with $l(0) := -1$ and $l(R) = Q$. Then we assume that p_{qr} is of order m for $q \in I_r$ and of order at most $m-1$ otherwise. When $Q = R$, then p_{qr} forms a determined systems with diagonal entries of order m and extra-diagonal entries of order at most $m-1$. More generally, the matrix p_{qr} is composed of operators of order m that are block-diagonal with blocks of size $I_r \times 1$ and of other operators of order at most $m-1$.

Then, as we did before, we assume the existence of a finite covering $\{\Omega_\nu\}$ of the unit sphere $\mathbb{S}^{n-1}$ such that the following holds. For each ν and r, there exists a $q = q(\nu, r) \in I_r$ such that for each (x,t) close to 0 and each $\xi \in \Omega_\nu$, we have

(*i*) $p_{qr}(x,t,\xi,\tau)$ has at most simple real roots and at most double complex roots
(*ii*) distinct roots τ_1 and τ_2 satisfy $|\tau_1 - \tau_2| \geq \varepsilon > 0$
(*iii*) non-real roots τ satisfy $|\Im\tau| \geq \varepsilon$.

Then we have the following result:

THEOREM A.2. *[Calderóns result for Redundant systems.] Assume that N is non characteristic for the operators L_{qr} for $q \in I_r$ at the origin and that for a finite covering $\{\Omega_\nu\}$ of the unit sphere, (i)-(ii)-(iii) above are satisfied. Then* (55) *implies that $u = 0$ in a full neighborhood of* 0.

Proof. The proof is similar to that of the preceding theorem. It is based on the following Carleman estimate

$$\sum_{r=1}^{R} \sum_{|\alpha|<m} |||D^\alpha u_r|||^2 \leq C(k^{-1} + T^2) \sum_{q=1}^{Q} \sum_{r=1}^{R} |||L_{qr}u_r|||^2, \tag{63}$$

whose proof is the same as that of (56). We leave the details to the reader. □

A.3. Redundant systems in upper-triangular form. In some applications, the leading term in the system may not be diagonal (or block-diagonal) but rather upper-triangular. In such a setting, unique continuation properties may still be valid provided that (a sufficiently large number of) the diagonal operators are elliptic. However, the corresponding complex roots may not longer be double. We have the following result, which as above, is stated for possibly redundant systems:

THEOREM A.3. *Assume that N is non characteristic for the operators L_q at the origin and that there exists a finite covering $\{\Omega_\nu\}$ of the unit sphere such that for each ν, there exists $q = q(\nu)$ such that*

$p_q(x,t,\xi,\tau)$ *has at most simple complex roots τ satisfying* $|\Im\tau| \geq \varepsilon$

for some $\varepsilon > 0$. Then there exists a constant C independent of u such that for k^{-1} and T sufficiently small, we have

$$\sum_{|\alpha|\le m} |||D^\alpha u|||^2 \le C(k^{-1}+T^2)\sum_{q=1}^{Q} |||L_q u|||^2. \tag{64}$$

Proof. The main difference with respect to the preceding theorem is that we have a full control of the mth-order derivatives of u. We follow the proof of Theorem A.1 until the very last step. The additional control of $D^\alpha u$ for $|\alpha| = m$ comes from the control of ΛU_ν and $D_t U_\nu$, or equivalently, that of ΛV_ν and $D_t V_\nu$. Now, $h_{q(\nu)}$ is diagonalized with simple eigenvalues $\lambda(x,t,\xi) = -a(x,t,\xi) - ib(x,t,\xi)$. By assumption $b(x,t,\xi)$ is uniformly bounded away from 0. We may then apply [**44**, (7.12)] to obtain that

$$|||\Lambda V_\nu|||^2 + |||D_t V_\nu|||^2 \le C(1+kT^2)|||(D_t + J_\nu)V_\nu|||^2.$$

The proof of (64) follows. □

As an application of the preceding result, let us consider the following corollary:

THEOREM A.4. *Consider the redundant system of equations*

$$\begin{pmatrix} L_1 & L_0 \\ L_3 & L_2 \end{pmatrix}\begin{pmatrix} u_1 \\ u_2 \end{pmatrix} = 0 \text{ in } V\cap\{x_{n+1}>0\}, \qquad \partial^j_{n+1} u_k = 0 \text{ on } V\cap\{x_{n+1}=0\}, \tag{65}$$

for $1 \le j \le m-1$ and $1 \le k \le 2$, with the following assumptions. The operators L_1 and L_0 are (vector-valued) $Q_1 \times 1$ operators of order m, where L_1 satisfies the hypotheses of Theorem A.1 *with $Q = Q_1$. The operators L_2 and L_3 are $Q_2 \times 1$ operators of order m and at most $m-1$, respectively. Moreover, L_2 satisfies the hypotheses of Theorem* A.3 *with $Q = Q_2$. Then $(u_1, u_2) = 0$ in a full neighborhood of* 0.

The same result extends systems of the form

$$(L_{ij})_{1\le i,j\le R}\, u = 0 \text{ in } V\cap\{x_{n+1}>0\}, \qquad \partial^j_{n+1} u_k = 0 \text{ on } V\cap\{x_{n+1}=0\}, \tag{66}$$

for $1 \le k \le R$, where L_{ij} is (a vector-valued operator) of order $m-1$ for $i > j$, L_{ii} is (a vector-valued operator) of order m that satisfies the hypotheses of Theorem A.1 *with an appropriate value of Q when all L_{ik} are of order $m-1$ for $k \ne i$ and the hypotheses of Theorem* A.3 *with an appropriate value of Q when at least one operator L_{ik} for $k < i$ is of order m.*

Proof. First of all, since N is non characteristic for L_1 and L_2, we observe that all derivatives of u_k vanish on $\{x_{n+1}\}$ so that both functions can be extended by 0 for $\{x_{n+1} < 0\}$. The above system of PDEs is thus satisfied on V with $u_k = 0$ for $x_{n+1} < 0$ and we wish to show that $u_k = 0$ for $x_{n+1} \ge 0$.

The proof is based on a modification of the method showing Theorem A.1 from (56). Indeed, we claim that for (u_1, u_2) with support contained in $|x| \le r$ and $0 \le t \le T$, then the theorem follows from

$$\sum_{|\alpha|<m}\sum_{k=1}^{2} |||D^\alpha u_k|||^2 \le C(k^{-1}+T^2)|||Lu|||^2, \tag{67}$$

where we denote by Lu the left-hand side in (65). The proof is a straightforward modification of [**44**, p.31], which we reproduce here. Let ζ be a smooth function defined for $t \ge 0$ equal to 1 for $t \le 2T/3$ and 0 for $t \ge T$. Let $u_k = \zeta v_k$ for

$v = (v_1, v_2)$ solution of $Lv = 0$. Then from (67), using only the estimate with $|\alpha| = 0$, we obtain that

$$\int_0^{\frac{2T}{3}} \|v\|^2 w dt \le C(k^{-1} + T^2) \int_{\frac{2T}{3}}^T \|L(\zeta v)\|^2 w dt \le C'(k^{-1} + T^2) \int_{\frac{2T}{3}}^T w dt,$$

with C' independent of k. This shows that

$$e^{kT^2/4} \int_0^{\frac{T}{2}} \|v\|^2 dt \le C'(k^{-1} + T^2) T e^{kT^2/9},$$

and as $k \to \infty$ that $\int_0^{\frac{T}{2}} \|v\|^2 dt \to 0$.

It remains to show (67). Using Theorem A.1, we obtain that the left-hand-side is bounded by

$$C(k^{-1} + T^2)\big(|||L_1 u_1|||^2 + |||L_2 u_2|||^2\big) \le C(k^{-1} + T^2)\big(|||Lu|||^2 + |||L_0 u_2|||^2 + |||L_3 u_1|||^2\big).$$

Now from Theorem A.3, we obtain that

$$|||L_0 u_2|||^2 \le C(k^{-1} + T^2)|||L_2 u_2|||^2.$$

Moreover, we obtain that

$$|||L_3 u_1|||^2 \le C \sum_{|\alpha| < m} |||D^\alpha u_1|||^2.$$

Therefore, for $(k^{-1} + T^2)$ sufficiently small, we deduce (67). The generalization to (66) is mostly notational and is left to the reader. □

References

[1] Robert A. Adams, *Sobolev spaces*, Academic Press [A subsidiary of Harcourt Brace Jovanovich, Publishers], New York-London, 1975. Pure and Applied Mathematics, Vol. 65. MR0450957 (56 #9247)

[2] S. Agmon, A. Douglis, and L. Nirenberg, *Estimates near the boundary for solutions of elliptic partial differential equations satisfying general boundary conditions. I*, Comm. Pure Appl. Math. **12** (1959), 623–727. MR0125307 (23 #A2610)

[3] ———, *Estimates near the boundary for solutions of elliptic partial differential equations satisfying general boundary conditions. II*, Comm. Pure Appl. Math., 17 (1964), pp. 35–92.

[4] Habib Ammari, *An introduction to mathematics of emerging biomedical imaging*, Mathématiques & Applications (Berlin) [Mathematics & Applications], vol. 62, Springer, Berlin, 2008. MR2440857 (2010j:44002)

[5] H. Ammari, E. Bonnetier, Y. Capdeboscq, M. Tanter, and M. Fink, *Electrical impedance tomography by elastic deformation*, SIAM J. Appl. Math. **68** (2008), no. 6, 1557–1573, DOI 10.1137/070686408. MR2424952 (2009h:35439)

[6] S. R. Arridge and O. Scherzer, *Imaging from coupled physics*, Inverse Problems, 28 (2012), p. 080201.

[7] G. Bal, *Hybrid inverse problems and internal functionals*, Inside Out II, MSRI Publications, G. Uhlmann Editor, Cambridge University Press, Cambridge, UK, 2012.

[8] Guillaume Bal, *Cauchy problem for ultrasound-modulated EIT*, Anal. PDE **6** (2013), no. 4, 751–775, DOI 10.2140/apde.2013.6.751. MR3092728

[9] Guillaume Bal, Eric Bonnetier, François Monard, and Faouzi Triki, *Inverse diffusion from knowledge of power densities*, Inverse Probl. Imaging **7** (2013), no. 2, 353–375, DOI 10.3934/ipi.2013.7.353. MR3063538

[10] G. Bal and M. Courdurier, *Boundary control of elliptic solutions to enforce local constraints*, J. Differential Equations **255** (2013), no. 6, 1357–1381, DOI 10.1016/j.jde.2013.04.039. MR3065291

[11] G. Bal, C. Guo, and F. Monard, *Inverse anisotropic conductivity from internal current densities*, Submitted.

[12] ———, *Linearized internal functionals for anisotropic conductivities*, Submitted.

[13] Guillaume Bal, *Cauchy problem for ultrasound-modulated EIT*, Anal. PDE **6** (2013), no. 4, 751–775, DOI 10.2140/apde.2013.6.751. MR3092728

[14] Guillaume Bal and Kui Ren, *Multi-source quantitative photoacoustic tomography in a diffusive regime*, Inverse Problems **27** (2011), no. 7, 075003, 20, DOI 10.1088/0266-5611/27/7/075003. MR2806367 (2012h:65255)

[15] Guillaume Bal, Kui Ren, Gunther Uhlmann, and Ting Zhou, *Quantitative thermo-acoustics and related problems*, Inverse Problems **27** (2011), no. 5, 055007, 15, DOI 10.1088/0266-5611/27/5/055007. MR2793826

[16] G. Bal and J. C. Schotland, *Inverse Scattering and Acousto-Optics Imaging*, Phys. Rev. Letters, 104 (2010), p. 043902.

[17] G. Bal and G. Uhlmann, *Reconstruction of coefficients in scalar second-order elliptic equations from knowledge of their solutions*, Commun. Pure Appl. Math., 66 (2013), 1629–1652. MR3084700

[18] Guillaume Bal and Gunther Uhlmann, *Inverse diffusion theory of photoacoustics*, Inverse Problems **26** (2010), no. 8, 085010, 20, DOI 10.1088/0266-5611/26/8/085010. MR2658827

[19] Guillaume Bal and Gunther Uhlmann, *Reconstructions for some coupled-physics inverse problems*, Appl. Math. Lett. **25** (2012), no. 7, 1030–1033, DOI 10.1016/j.aml.2012.03.005. MR2915119

[20] A. Calderón, *Uniqueness in the Cauchy problem for Partial Differential Equations*, Amer. J. Math., 80(1) (1958), pp. 16–36.

[21] ———, *Existence and uniqueness theorems for systems of partial differential equations*, Proc. Sympos. Fluid Dynamics and Appl. Math. (University of Maryland, 1961), Gordon and Breach, New York, (1962), pp. 147–195.

[22] Y. Capdeboscq, J. Fehrenbach, F. de Gournay, and O. Kavian, *Imaging by modification: numerical reconstruction of local conductivities from corresponding power density measurements*, SIAM J. Imaging Sci. **2** (2009), no. 4, 1003–1030, DOI 10.1137/080723521. MR2559157 (2011c:35611)

[23] Chris Cosner, *On the definition of ellipticity for systems of partial differential equations*, J. Math. Anal. Appl. **158** (1991), no. 1, 80–93, DOI 10.1016/0022-247X(91)90268-5. MR1113401 (92h:35060)

[24] B. T. Cox, S. R. Arridge, and P. C. Beard, *Photoacoustic tomography with a limited-aperture planar sensor and a reverberant cavity*, Inverse Problems **23** (2007), no. 6, S95–S112, DOI 10.1088/0266-5611/23/6/S08. MR2441001 (2009f:92045)

[25] B. T. Cox, J. G. Laufer, and P. C. Beard, *The challenges for quantitative photoacoustic imaging*, Proc. of SPIE, 7177 (2009), p. 717713.

[26] Avron Douglis and Louis Nirenberg, *Interior estimates for elliptic systems of partial differential equations*, Comm. Pure Appl. Math. **8** (1955), 503–538. MR0075417 (17,743b)

[27] Bastian Gebauer and Otmar Scherzer, *Impedance-acoustic tomography*, SIAM J. Appl. Math. **69** (2008), no. 2, 565–576, DOI 10.1137/080715123. MR2465856 (2009j:35381)

[28] Lars Hörmander, *On the uniqueness of the Cauchy problem. II*, Math. Scand. **7** (1959), 177–190. MR0121569 (22 #12306)

[29] ———, *Linear Partial Differential Operators*, Springer-Verlag, Berlin, 1963.

[30] ———, *The Analysis of Linear Partial Differential Operators I: Distribution Theory and Fourier Analysis*, Springer Verlag, 1983.

[31] ———, *The Analysis of Linear Partial Differential Operators III: Pseudo-Differential Operators*, Springer Verlag, 1994.

[32] Peter Kuchment and Leonid Kunyansky, *Mathematics of thermoacoustic tomography*, European J. Appl. Math. **19** (2008), no. 2, 191–224, DOI 10.1017/S0956792508007353. MR2400720 (2009c:92026)

[33] Peter Kuchment and Leonid Kunyansky, *2D and 3D reconstructions in acousto-electric tomography*, Inverse Problems **27** (2011), no. 5, 055013, 21, DOI 10.1088/0266-5611/27/5/055013. MR2793832 (2012b:65166)

[34] Peter Kuchment and Dustin Steinhauer, *Stabilizing inverse problems by internal data*, Inverse Problems **28** (2012), no. 8, 084007, 20, DOI 10.1088/0266-5611/28/8/084007. MR2956563

[35] J. Le Rousseau, *On Carleman estimates with two large parameters*, J. Phys. Conf. Ser., 290 (2011), p. 012010.

[36] Hans Lewy, *An example of a smooth linear partial differential equation without solution*, Ann. of Math. (2) **66** (1957), 155–158. MR0088629 (19,551d)

[37] Joyce R. McLaughlin and Jeong-Rock Yoon, *Unique identifiability of elastic parameters from time-dependent interior displacement measurement*, Inverse Problems **20** (2004), no. 1, 25–45, DOI 10.1088/0266-5611/20/1/002. MR2044604 (2005b:74070)

[38] Joyce R. McLaughlin, Ning Zhang, and Armando Manduca, *Calculating tissue shear modules and pressure by 2D log-elastographic methods*, Inverse Problems **26** (2010), no. 8, 085007, 25, DOI 10.1088/0266-5611/26/8/085007. MR2658824

[39] F. Monard and G. Bal, *Inverse anisotropic conductivity from power density measurements in dimensions $n \geq 3$*, Comm. Partial Differential Equations, 38(7) (2013), 1183–1207.

[40] François Monard and Guillaume Bal, *Inverse anisotropic diffusion from power density measurements in two dimensions*, Inverse Problems **28** (2012), no. 8, 084001, 20, DOI 10.1088/0266-5611/28/8/084001. MR2956557

[41] ———, *Inverse diffusion problem with redundant internal information*, Inverse Problems and Imaging, 6(2) (2012), pp. 289–313.

[42] Adrian Nachman, Alexandru Tamasan, and Alexander Timonov, *Current density impedance imaging*, Tomography and inverse transport theory, Contemp. Math., vol. 559, Amer. Math. Soc., Providence, RI, 2011, pp. 135–149, DOI 10.1090/conm/559/11076. MR2885199

[43] Louis Nirenberg, *Uniqueness in Cauchy problems for differential equations with constant leading coefficients*, Comm. Pure Appl. Math. **10** (1957), 89–105. MR0086232 (19,147a)

[44] ———, *Lectures on Linear Partial Differential Equations*, Amer. Math. Soc., Providencem R.I., 1973.

[45] Sarah K. Patch and Otmar Scherzer, *Guest editors' introduction: Photo- and thermo-acoustic imaging*, Inverse Problems **23** (2007), no. 6, S1–S10, DOI 10.1088/0266-5611/23/6/S01. MR2440994

[46] O. Scherzer, *Handbook of Mathematical Methods in Imaging*, Springer Verlag, New York, 2011.

[47] Jin Keun Seo and Eung Je Woo, *Magnetic resonance electrical impedance tomography (MREIT)*, SIAM Rev. **53** (2011), no. 1, 40–68, DOI 10.1137/080742932. MR2785879 (2012d:35412)

[48] V. A. Solonnikov, *Overdetermined elliptic boundary-value problems*, J. Sov. Math., 1 (1973), pp. 477–512.

[49] Plamen Stefanov and Gunther Uhlmann, *Linearizing non-linear inverse problems and an application to inverse backscattering*, J. Funct. Anal. **256** (2009), no. 9, 2842–2866, DOI 10.1016/j.jfa.2008.10.017. MR2502425 (2010f:47028)

[50] ———, *Multi-wave methods by ultrasounds*, Inside out, Cambridge University Press (G. Uhlmann, Ed.), (2012).

[51] Nikolai N. Tarkhanov, *The Cauchy problem for solutions of elliptic equations*, Mathematical Topics, vol. 7, Akademie Verlag, Berlin, 1995. MR1334094 (96d:35024)

[52] L. V. Wang and H. Wu, *Biomedical Optics: Principles and Imaging*, Wiley, 2007.

[53] Claude Zuily, *Uniqueness and nonuniqueness in the Cauchy problem*, Progress in Mathematics, vol. 33, Birkhäuser Boston Inc., Boston, MA, 1983. MR701544 (85d:35009)

Department of Applied Physics and Applied Mathematics, Columbia University, New York New York 10027

E-mail address: gb2030@columbia.edu

Contemporary Mathematics
Volume **615**, 2014
http://dx.doi.org/10.1090/conm/615/12264

A Direct Imaging Method for Inverse Scattering Using the Generalized Foldy–Lax Formulation

Gang Bao, Kai Huang, Peijun Li, and Hongkai Zhao

ABSTRACT. Consider the scattering of a time-harmonic plane wave incident on a two-scale heterogeneous medium, which consists of scatterers that are much smaller than the wavelength and extended scatterers that are comparable to the wavelength. In addition to the standard inverse obstacle scattering problem, i.e., to image the shape of extended scatterers as well as the location of the point scatterers where strong multiple scattering are present, there are a few other interesting issues such as imaging a target in a cluttered environment without resolving the clutters, and increasing the effective aperture by utilizing the multiple scattering between extended scatterers and point scatterers whose location are known. Some preliminary computational study will be given in this paper. To simulate the wave propagation in the heterogeneous medium with both point and extended scatterers, a generalized Foldy–Lax formulation and a physically based block Gauss–Seidel iterative method are used to solve the two-scale multiple scattering problem. Based on the singular value decomposition of the response matrix constructed from the far-field pattern, imaging functions are designed to visualize the location of the point scatterers and the shape of the extended obstacle scatterers. The method leads to a direct imaging algorithm which is simple and efficient since no direct solver or iteration is needed. The imaging functions are robust with respect to the measurement noise. Numerical experiments are presented for uniformly and randomly distributed point scatterers and multiple extended obstacle scatterers in both two- and three-dimensional cases.

1. Introduction

Scattering problems are about how an inhomogeneous medium scatters an incident wave field. The direct scattering problem is to determine the scattered field from the incident field and the differential equation governing the wave motion; the inverse scattering problem is to determine the nature of the inhomogeneity, such as

---

1991 *Mathematics Subject Classification.* Primary 78A46, 78M15; Secondary 65N21.

Key words and phrases. Foldy–Lax formulation, point scatterers, boundary integral equation, inverse obstacle scattering problem, Helmholtz equation, direct imaging method.

The research was supported in part by the NSF grants DMS-0908325, DMS-0968360, DMS-1211292, the ONR grant N00014-12-1-0319, a Key Project of the Major Research Plan of NSFC (No. 91130004), and a special research grant from Zhejiang University.

The research was supported in part by NSF grants DMS-0914595, DMS-1042958, and DMS-1151308.

The research was supported in part by NSF grant DMS-1115698 and ONR grant N00014-11-1-0602.

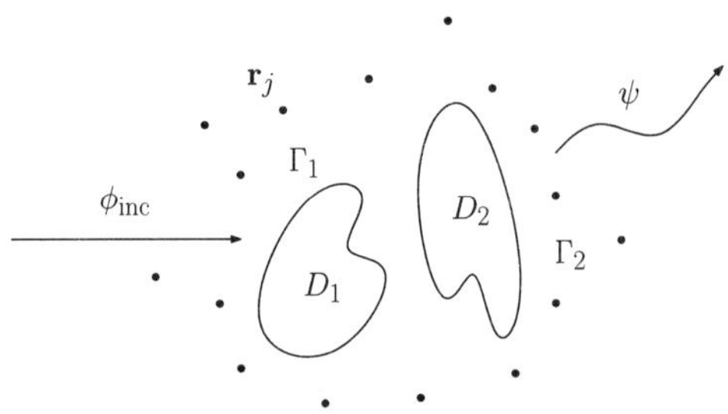

FIGURE 1. Schematic of problem geometry. The scattered wave ψ is generated from the incidence of a plane wave ϕ_{inc} on the heterogeneous medium consisting a group of point scatterers centered at $\mathbf{r}_j$ and an extended obstacle scatterer with possible multiple disjoint components represented by the domain D_k with boundary Γ_k.

location, geometry, and material property, from a knowledge of the scattered field. These problems have played a fundamental role in diverse scientific areas such as radar and sonar, geophysical exploration, medical imaging, near-field and nano-optics. This work is devoted to an inverse scattering problem where the considered scatterers are of two-scale nature: it consists of isotropic point scatterers and an extended obstacle scatterer with possible multiple disjoint components. "Isotropic point" refers that the scale of the scatterer is much smaller than the wavelength of the incident field so that the scatterer can be represented by a source point within it when scattering an incident wave; "extended" means that the scale of the obstacle scatterer is comparable with the wavelength of the incident field. More precisely, we consider the inverse scattering of a time-harmonic plane wave incident on the heterogeneous medium, and intend to reconstruct an image of the shape for the extended obstacle, represented by the boundary Γ of the domain D, and the location of the point scatterers at $\mathbf{r}_j$ from the far-field pattern of the scattered wave, as seen in Figure 1. The major motivation is to image a target in a cluttered environment. Here the target is modeled by an impenetrable obstacle and the cluttered environment is modeled by a set of isolated point scatterers surrounding the obstacle. In this work, we think that there are two contributions to the field of inverse scattering theory: (1) based on the generalized Foldy–Lax formulation, the proposed imaging functions can simultaneously show the shape of the extended obstacle scatterers and the location of the point scatterers; (2) computational study of the interaction and separation between point scatterers and extended scatterers in the imaging process.

The inverse scattering problem is challenging due to the inherently high nonlinearity and, more seriously from the point of view of numerical computations, ill-posedness, i.e., small variations in the measured data can lead to large errors in the reconstruction. A number of methods have been proposed for solving the inverse scattering problem and they can be roughly categorized into two groups: nonlinear optimization based iterative methods and imaging based direct methods. The former methods are usually more quantitative and attempt to recover the properties or parameters as well as the geometry of the obstacle. On the other hand, they are more computationally intensive since iterations that requires solving a sequence of direct and adjoint scattering problems are needed, see e.g., [**18**, **19**, **30**, **35**–**37**]. While the latter approaches are more qualitative and less computationally intensive, which give a characterization and/or visualization of the geometry by designing imaging functions which can highlight target boundary, see e.g., the MUSIC algorithm [**3**, **14**, **17**, **21**, **29**, **42**, **44**], the linear sampling method [**6**, **7**, **10**, **11**], the factorization method [**31**, **32**], the probe or enclosure method [**8**, **15**, **27**, **28**], and the method of singular sources [**40**, **41**]. In [**4**, **5**], a hybrid approach of combining the direct imaging method and the recursive linearization algorithm was used to solve an inverse medium scattering problem; a direct imaging method was adopted to generate an initial guess while the recursive linearization algorithm was applied to updates the approximations. We refer to the monograph [**33**] for a survey of all the non-iterative methods and the books [**12**, **13**] for comprehensive accounts of the direct and inverse scattering problems.

In this paper, we present a direct imaging method to simultaneously locate the point scatterers and reconstruct the shape of the extended scatterers. Specifically, we consider the inverse scattering problem of time-harmonic plane waves incident on the heterogeneous medium from a range of incident directions in two- or three-dimensional space. The far-field patterns of the scattered waves are recorded as data at various observation directions from plane incident waves sending at different directions of incidence. These measurements form the so-called response matrix which can be regarded as a discrete version of the scattering operator. Based on the singular value decomposition (SVD) of the response matrix, we design direct imaging functions to show the location of the point scatterers and the shape of the extended scatterers. The proposed direct imaging method is simple and efficient since no direct solver or iteration is needed. It can handle both full and limited aperture data. The imaging functions are robust with respect to the measurement noise. The method is closely related to the direct imaging methods in [**20**–**22**] and more recently in [**23**], where the medium was considered to be either point scatterers or extended obstacle, i.e., the medium is of the same scale nature. Here we consider the situation where both extended scatterers and a set of point scatterers with strong multiple scattering among them are present. To simulate the wave propagation in the heterogeneous medium with presence of both types of scatterers, it is required to accurately solve the forward scattering problem where the multiple scattering occurs among all the scatterers. We mention that the classic Foldy–Lax formulation is efficient to solve the multiple scattering of scalar waves by a distribution of small isotropic scatterers [**16**, **38**], and boundary integral equation methods have been considered as appropriate approaches for solving the scattering of acoustic or electromagnetic waves by bounded obstacles [**36**, **37**]. The generalized Foldy–Lax formulations have been developed in [**24**–**26**] to solve the multiple scattering problem when both the point scatterers and the extended obstacles are present. We briefly introduce the generalized Foldy–Lax formulation in the context of the inverse obstacle scattering and present a physically motivated block Gauss–Seidel iterative method. The detailed discussion may be found in [**25**] on the uniqueness and existence of the solution for the generalized Foldy–Lax formulation, as well as the convergence analysis of the block Gauss–Seidel iterative method. We refer to the proceedings [**2**] for an overview of mathematical and statistical methods for scattering problems related imaging.

The outline of the paper is as follows. In Section 2, the classical Foldy–Lax formulation is briefly reviewed for the scattered field from the interaction among a set of well-separated point scatterers; based on combined single- and double-layer potential representations, a uniquely solvable boundary integral equation is introduced for the obstacle scattering problem, where the scatterer is modeled as a wavelength comparable or an extended sound-soft obstacle; a generalized Foldy–Lax formulation is described for the multiple scattering problem involving both small-scale point scatterers and extended scatterers; and a block Gauss–Seidel iterative method is given to solve the coupled scattering system. Section 3 is devoted to the derivation of the far-field patterns and the construction of the imaging functions for three types of scatterers: point scatterers, extended obstacle scatterers, and mixed scatterers with both point and extended scatterers. Numerical results are shown in Section 4 for uniformly and randomly distributed point scatterers and multiple extended obstacle scatterers in both two- and three-dimensional cases. The paper is concluded with comments and directions for future work in Section 5.

2. Generalized Foldy–Lax formulation

In this section, we briefly introduce the original Foldy–Lax formulation for the point scatterers; the boundary integral equation for the extended obstacle scattering problem; the generalized Foldy–Lax formulation for the multiple scattering problem involving both extended obstacle scatterers and a set of isotropic point scatterers; and an efficient physically motivated block Gauss–Seidel algorithm for solving the coupled scattering system.

2.1. Foldy–Lax formulation for point scatterers. We briefly introduce the original Foldy–Lax formulation to model the scattering of isotropic point scatterers, which are named as delta-potentials in Rudnick and Ueberschär [**43**], and Colin de Verdière[**9**]. We refer to the book by Martin [**39**] for detailed discussions on the Foldy–Lax formulation.

Consider a collection of m separated isotropic point scatterers, which are represented by point sources located at $\mathbf{r}_1, \dots, \mathbf{r}_m$. Let ϕ_{inc} be the plane incident wave, given explicitly

$$\phi_{\text{inc}}(\mathbf{r}) = e^{\mathrm{i}\kappa\mathbf{r}\cdot\mathbf{d}} \quad \text{in } \mathbb{R}^\mu, \tag{2.1}$$

where i is the imaginary unit, κ is the wavenumber, $\mathbf{d} \in \mathbb{S}^{\mu-1}$ is the propagation direction defined on the unit sphere, and the dimension $\mu = 2, 3$. Clearly, the incident field satisfies the Helmholtz equation in the free space

$$\Delta\phi_{\text{inc}} + \kappa^2\phi_{\text{inc}} = 0 \quad \text{in } \mathbb{R}^\mu. \tag{2.2}$$

The total field is represented as the sum of the incident field and the scattered field

$$\phi(\mathbf{r}) = \phi_{\text{inc}}(\mathbf{r}) + \sum_{j=1}^{m} \sigma_j\phi_j G(\mathbf{r}, \mathbf{r}_j), \tag{2.3}$$

where σ_j is referred to as the scattering coefficient for the j-th point scatterer, ϕ_j is the external field acting on the j-th point scatterer, and G is the free-space Green's function given as

$$G(\mathbf{r},\mathbf{r}') = \begin{cases} \dfrac{\mathrm{i}}{4} H_0^{(1)}(\kappa|\mathbf{r}-\mathbf{r}'|), & \text{for } \mu = 2, \\[2ex] \dfrac{1}{4\pi} \dfrac{e^{\mathrm{i}\kappa|\mathbf{r}-\mathbf{r}'|}}{|\mathbf{r}-\mathbf{r}'|}, & \text{for } \mu = 3. \end{cases}$$

Here $H_0^{(1)}$ is the Hankel function of the first kind with order zero.

Evaluating (2.3) at $\mathbf{r}_i$ and excluding the self-interaction yield a linear system of algebraic equation for $\phi_j, j = 1, \ldots, m$:

$$\phi_i - \sum_{\substack{j=1 \\ j\neq i}}^{m} \sigma_j \phi_j G(\mathbf{r}_i, \mathbf{r}_j) = \phi_{\mathrm{inc}}(\mathbf{r}_i), \tag{2.4}$$

which is known as the Foldy–Lax formulation. Once (2.4) is solved, the scattered field can be computed as

$$\psi(\mathbf{r}) = \sum_{j=1}^{m} \sigma_j \phi_j G(\mathbf{r}, \mathbf{r}_j). \tag{2.5}$$

2.2. Boundary integral formulation for extended obstacle scatterers. This section is concerned with a brief introduction to the method of the boundary integral equation for solving the extended obstacle scattering problem. We refer to the monograph by Colton and Kress [**12**] for a comprehensive account of the boundary integral equation methods for solving the obstacle scattering problem.

We assume that the extended obstacle scatterer is represented by the domain D with boundary Γ, which is the open complement of an unbounded domain of class C^2, i.e., scattering from more than one component is included in our analysis.

Consider the Helmholtz equation

$$\Delta\phi + \kappa^2\phi = 0 \quad \text{in } \mathbb{R}^\mu \setminus \overline{D}, \tag{2.6}$$

along with the sound-soft boundary condition

$$\phi = 0 \quad \text{on } \Gamma, \tag{2.7}$$

where ϕ is the total field and κ is the wavenumber.

The extended obstacle scatterer is illuminated by the plane incident wave given in (2.1). The total field ϕ consists of the incident field ϕ_{inc} and the scattered field ψ:

$$\phi = \phi_{\mathrm{inc}} + \psi. \tag{2.8}$$

It follows from (2.2) and (2.6)–(2.8) that the scattered field satisfies

$$\Delta\psi + \kappa^2\psi = 0 \quad \text{in } \mathbb{R}^\mu \setminus \overline{D}, \tag{2.9}$$

together with the boundary condition

$$\psi = -\phi_{\mathrm{inc}} \quad \text{on } \Gamma. \tag{2.10}$$

In addition, the scattered field is required to satisfy the Sommerfeld radiation condition

$$\lim_{\rho\to\infty} \rho^{\frac{\mu-1}{2}} \left(\frac{\partial\psi}{\partial\rho} - \mathrm{i}\kappa\psi \right) = 0, \quad \rho = |\mathbf{r}|. \tag{2.11}$$

Based on the Green's representation theorem, it can be shown that the scattered field has the integral representation in terms of the normal derivative of the total field:

$$\psi(\mathbf{r}) = -\int_{\Gamma} G(\mathbf{r}, \mathbf{r}')\partial_{\mathbf{n}'}\phi(\mathbf{r}')ds(\mathbf{r}'), \quad \mathbf{r} \in \mathbb{R}^{\mu} \setminus \overline{D}, \tag{2.12}$$

where $\mathbf{n}$ is the unit outward normal on Γ and is assumed to be directed into the exterior of D.

To compute the scattered field, it is required to determine $\partial_{\mathbf{n}}\phi$ on Γ. We adopt the following uniquely solvable boundary integral equation:

$$\frac{1}{2}\partial_{\mathbf{n}}\phi(\mathbf{r}) + \int_{\Gamma} \left[\partial_{\mathbf{n}}G(\mathbf{r}, \mathbf{r}') - \mathrm{i}\eta G(\mathbf{r}, \mathbf{r}')\right] \partial_{\mathbf{n}'}\phi(\mathbf{r}')ds(\mathbf{r}') = (\partial_{\mathbf{n}} - \mathrm{i}\eta)\phi_{\mathrm{inc}}(\mathbf{r}), \tag{2.13}$$

where η is the coupling parameter. We refer to Kress [**34**] for an investigation on the proper choice of the coupling parameter η with respect to the condition number of the coefficient matrix for the integral equation.

REMARK 2.1. If the extended obstacle scatterer is consisted of n disjoint components, i.e., $\Gamma = \Gamma_1 \cup \cdots \cup \Gamma_n$, $\Gamma_k \cap \Gamma_{k'} = \emptyset$, $k \neq k'$, then the boundary integral equation (2.13) can be written as

$$\frac{1}{2}\varphi(\mathbf{r}) + \sum_{k=1}^{n} \int_{\Gamma_k} \left[\partial_{\mathbf{n}}G(\mathbf{r}, \mathbf{r}') - \mathrm{i}\eta G(\mathbf{r}, \mathbf{r}')\right] \varphi_k(\mathbf{r}')ds(\mathbf{r}') = (\partial_{\mathbf{n}} - \mathrm{i}\eta)\phi_{\mathrm{inc}}(\mathbf{r}), \tag{2.14}$$

where $\varphi(\mathbf{r}) = \partial_{\mathbf{n}}\phi(\mathbf{r})$ and $\varphi_k(\mathbf{r}) = \varphi(\mathbf{r})|_{\mathbf{r}\in\Gamma_k}$.

2.3. Generalized Foldy–Lax formulation. This section introduces the generalized Foldy–Lax formulation for the multiple scattering among the extended obstacle scatterer and the group of point scatterers. The existence and uniqueness of the solution for the formulation may be found in [**25**].

Viewing the external field acting on the point scatterers as point sources for the extended obstacle scatterer, we consider the equation for the total field

$$\Delta\phi + \kappa^2\phi = -\sum_{j=1}^{m} \sigma_j\phi_j\delta(\mathbf{r} - \mathbf{r}_j) \quad \text{in } \mathbb{R}^{\mu} \setminus \overline{D}, \tag{2.15}$$

along with the sound-soft boundary condition

$$\phi = 0 \quad \text{on } \Gamma, \tag{2.16}$$

where ϕ_j is again the external field acting on the j-th point scatterer and δ is the Dirac delta function. Subtracting the incident field from the total field, we may obtain the equation for the scattered field

$$\Delta\psi + \kappa^2\psi = -\sum_{j=1}^{m} \sigma_j\phi_j\delta(\mathbf{r} - \mathbf{r}_j) \quad \text{in } \mathbb{R}^{\mu} \setminus \overline{D}. \tag{2.17}$$

The scattered field is also required to satisfy the Sommerfeld radiation condition

$$\lim_{\rho\to\infty} \rho^{\frac{\mu-1}{2}} \left(\frac{\partial\psi}{\partial\rho} - \mathrm{i}\kappa\psi\right) = 0, \quad \rho = |\mathbf{r}|. \tag{2.18}$$

Similarly, it can be verified that the scattered field has the representation given in terms of the normal derivative of the total field and the external fields acting on the point scatterers:

$$\psi(\mathbf{r}) = \sum_{j=1}^{m} \sigma_j \phi_j G(\mathbf{r}, \mathbf{r}_j) - \int_\Gamma G(\mathbf{r}, \mathbf{r}')\partial_{\mathbf{n}'}\phi(\mathbf{r}')ds(\mathbf{r}'), \quad \mathbf{r} \in \mathbb{R}^\mu \setminus \overline{D}. \tag{2.19}$$

Adding the incident field on both sides of (2.19) yields

$$\phi(\mathbf{r}) = \phi_{\text{inc}}(\mathbf{r}) + \sum_{j=1}^{m} \sigma_j \phi_j G(\mathbf{r}, \mathbf{r}_j) - \int_\Gamma G(\mathbf{r}, \mathbf{r}')\partial_{\mathbf{n}'}\phi(\mathbf{r}')ds(\mathbf{r}'), \quad \mathbf{r} \in \mathbb{R}^\mu \setminus \overline{D}. \tag{2.20}$$

To compute the scattered field, it is required to compute $\partial_{\mathbf{n}}\phi$ and $\phi_j, j = 1, \ldots, m$.

Evaluating (2.20) on both sides at $\mathbf{r}_i$ and excluding the self-interaction for the point scatterers give

$$\phi_i = \phi_{\text{inc}}(\mathbf{r}_i) + \sum_{\substack{j=1 \\ j\neq i}}^{m} \sigma_j \phi_j G(\mathbf{r}_i, \mathbf{r}_j) - \int_\Gamma G(\mathbf{r}_i, \mathbf{r}')\partial_{\mathbf{n}'}\phi(\mathbf{r}')ds(\mathbf{r}'). \tag{2.21}$$

Using the jump relation for the single-layer and double-layer potentials and taking the normal derivative, we may obtain a boundary integral equation on Γ:

$$\begin{aligned} \frac{1}{2}\partial_{\mathbf{n}}\phi(\mathbf{r}) = {} & (\partial_{\mathbf{n}} - \mathrm{i}\eta)\,\phi_{\text{inc}}(\mathbf{r}) + \sum_{j=1}^{m} \sigma_j \phi_j \partial_{\mathbf{n}} G(\mathbf{r}, \mathbf{r}_j) \\ & - \int_\Gamma \left[\partial_{\mathbf{n}} G(\mathbf{r}, \mathbf{r}') - \mathrm{i}\eta G(\mathbf{r}, \mathbf{r}')\right] \partial_{\mathbf{n}'}\phi(\mathbf{r}')ds(\mathbf{r}'). \end{aligned} \tag{2.22}$$

The coupled scattering system (2.21) and (2.22) form the self-consistent generalized Foldy–Lax formulation, which take full account of the multiple scattering among the extended obstacle and the point scatterers.

REMARK 2.2. If the extended obstacle scatterer is consisted of n multiple components, then the generalized Foldy–Lax formulation can be written as the following coupled system

$$\phi_i = \phi_{\text{inc}}(\mathbf{r}_i) + \sum_{\substack{j=1 \\ j\neq i}}^{m} \sigma_j \phi_j G(\mathbf{r}_i, \mathbf{r}_j) - \sum_{k=1}^{n} \int_{\Gamma_k} G(\mathbf{r}_i, \mathbf{r}')\varphi_k(\mathbf{r}')ds(\mathbf{r}'), \tag{2.23}$$

$$\begin{aligned} \frac{1}{2}\varphi(\mathbf{r}) = {} & (\partial_{\mathbf{n}} - \mathrm{i}\eta)\,\phi_{\text{inc}}(\mathbf{r}) + \sum_{j=1}^{m} \sigma_j \phi_j \partial_{\mathbf{n}} G(\mathbf{r}, \mathbf{r}_j) \\ & - \sum_{k=1}^{n} \int_{\Gamma_k} \left[\partial_{\mathbf{n}} G(\mathbf{r}, \mathbf{r}') - \mathrm{i}\eta G(\mathbf{r}, \mathbf{r}')\right] \varphi_k(\mathbf{r}')ds(\mathbf{r}'), \end{aligned} \tag{2.24}$$

where $\varphi(\mathbf{r}) = \partial_{\mathbf{n}}\phi(\mathbf{r})$ and $\varphi_k(\mathbf{r}) = \varphi(\mathbf{r})|_{\mathbf{r}\in\Gamma_k}$.

2.4. Block Gauss-Seidel iteration. We briefly describe an efficient physically motivated block Gauss–Seidel iterative method to solve the generalized Foldy–Lax formulation (2.23)–(2.24). The convergence analysis may be found in [**25**].

A block Gauss–Seidel iterative algorithm proceeds as follows: Let $\left(\phi_1^{(0)}, \ldots, \phi_m^{(0)}\right)^\top = (0, \ldots, 0)^\top$ and $\varphi_k^{(0)}(\mathbf{r}) = 0$ for $k = 1, \ldots, n$, define $\left(\phi_1^{(\nu)}, \ldots, \phi_m^{(\nu)}\right)^\top$ and

$\varphi_k^{(\nu)}$ for $\nu \geq 1$ by the solutions of the following system of equations

$$\phi_i^{(\nu)} - \sum_{\substack{j=1 \\ j\neq i}}^{m} \sigma_j \phi_j^{(\nu)} G(\mathbf{r}_i, \mathbf{r}_j) = \phi_{\text{inc}}(\mathbf{r}_i) - \sum_{k=1}^{n} \int_{\Gamma_k} G(\mathbf{r}_i, \mathbf{r}')\varphi_k^{(\nu-1)}(\mathbf{r}')ds(\mathbf{r}') \tag{2.25}$$

and

$$\begin{aligned} \frac{1}{2}\varphi_k^{(\nu)}(\mathbf{r}) + \int_{\Gamma_k} \left[\partial_{\mathbf{n}} G(\mathbf{r}, \mathbf{r}') - \mathrm{i}\eta G(\mathbf{r}, \mathbf{r}')\right] \varphi_k^{(\nu)}(\mathbf{r}')ds(\mathbf{r}') = \left(\partial_{\mathbf{n}} - \mathrm{i}\eta\right) \phi_{\text{inc}}(\mathbf{r}) \\ + \sum_{j=1}^{m} \sigma_j \phi_j^{(\nu)} \partial_{\mathbf{n}} G(\mathbf{r}, \mathbf{r}_j) - \sum_{j=1}^{k-1} \int_{\Gamma_j} \left[\partial_{\mathbf{n}} G(\mathbf{r}, \mathbf{r}') - \mathrm{i}\eta G(\mathbf{r}, \mathbf{r}')\right] \varphi_j^{(\nu)}(\mathbf{r}')ds(\mathbf{r}') \\ - \sum_{j=k+1}^{n} \int_{\Gamma_j} \left[\partial_{\mathbf{n}} G(\mathbf{r}, \mathbf{r}') - \mathrm{i}\eta G(\mathbf{r}, \mathbf{r}')\right] \varphi_j^{(\nu-1)}(\mathbf{r}')ds(\mathbf{r}'), \end{aligned} \tag{2.26}$$

for $k = 1, \ldots, n$.

As can be seen from the block Gauss–Seidel iterative algorithm, it is only required to solve a linear system of algebraic equations and a boundary integral equation for a single obstacle at each step of iteration.

3. Response matrix

In this section, we introduce the response matrix and design imaging functions to characterize the location of the point scatterers and the geometry of the extended obstacle scatterer. The imaging function as proposed in **[23]** is made of those singular vectors corresponding to dominant singular values of the response matrix, which is constructed from the far-field pattern of the scattered waves.

3.1. Far-field pattern. The far-field pattern of scattered wave plays a fundamental role in the inverse scattering theory due to the fact that it induces the important geometrical and physical information, e.g., location, shape, and the impedance of the boundary, on the scattering object. More specifically, given an incident field with incident direction $\mathbf{d}$, if ψ is the scattered field, then ψ has the asymptotic behavior

$$\psi(\mathbf{r}, \mathbf{d}) = \frac{e^{\mathrm{i}\kappa|\mathbf{r}|}}{|\mathbf{r}|^{\frac{\mu-1}{2}}} \left[\psi_\infty(\hat{\mathbf{r}}, \mathbf{d}) + O(|\mathbf{r}|)^{-1}\right] \quad \text{as } |\mathbf{r}| \to \infty \tag{3.1}$$

uniformly in all directions $\hat{\mathbf{r}} = \mathbf{r}/|\mathbf{r}|$, where the function ψ_∞ is called as the far-field pattern of the scattered field ψ, and $\hat{\mathbf{r}} \in \mathbb{S}^{\mu-1}$ is known as the observation direction.

Recall the asymptotic behavior for the Hankel function for large arguments

$$H_0^{(1)}(z) = \sqrt{\frac{2}{\pi z}}\, e^{\mathrm{i}\left(z - \frac{\pi}{4}\right)} \left[1 + O(z^{-1})\right]$$

and the following identity

$$|\mathbf{r} - \mathbf{r}'| = \sqrt{|\mathbf{r}|^2 - 2\mathbf{r}\cdot\mathbf{r}' + |\mathbf{r}'|^2} = |\mathbf{r}| - \hat{\mathbf{r}}\cdot\mathbf{r}' + O(|\mathbf{r}|^{-1}) \quad \text{as } |\mathbf{r}| \to \infty.$$

Following (3.1) and (2.5), we obtain the far-field pattern of the scattered field for the scattering problem of a set of m point scatterers:

$$\psi_{\infty,\,\mathrm{FL}}(\hat{\mathbf{r}}, \mathbf{d}) = \gamma \sum_{j=1}^{m} \sigma_j \phi_j(\mathbf{d}) e^{-\mathrm{i}\kappa\hat{\mathbf{r}}\cdot\mathbf{r}_j}, \tag{3.2}$$

where the constant γ is defined as

$$\gamma = \begin{cases} \dfrac{e^{i\frac{\pi}{4}}}{\sqrt{8\pi\kappa}}, & \text{for } \mu = 2, \\ \dfrac{1}{4\pi}, & \text{for } \mu = 3. \end{cases}$$

It follows from the integral representation of the scattered field (2.12) and the asymptotic expansion of Green's function that the far-field pattern for the extended obstacle scatterer is given by

$$\psi_{\infty,\mathrm{E}}(\hat{\mathbf{r}}, \mathbf{d}) = -\gamma \int_\Gamma \partial_{\mathbf{n}'}\phi(\mathbf{r}'; \mathbf{d}) e^{-i\kappa\hat{\mathbf{r}}\cdot\mathbf{r}'} ds(\mathbf{r}'). \tag{3.3}$$

Based on the integral equation representation (2.19) for the multiple scattering problem and the asymptotic behavior for Green's function, we may also obtain the far-field pattern of the scattered field for the generalized Foldy–Lax formulation

$$\psi_{\infty,\mathrm{GFL}}(\hat{\mathbf{r}}, \mathbf{d}) = \gamma \left[\sum_{j=1}^{m} \sigma_j \phi_j(\mathbf{d}) e^{-i\kappa\hat{\mathbf{r}}\cdot\mathbf{r}_j} - \int_\Gamma \partial_{\mathbf{n}'}\phi(\mathbf{r}'; \mathbf{d}) e^{-i\kappa\hat{\mathbf{r}}\cdot\mathbf{r}'} ds(\mathbf{r}') \right]. \tag{3.4}$$

REMARK 3.1. For the extended obstacle scatterer with n disjoint components, the far-field patterns for the extended obstacle scatterer and the generalized Foldy–Lax formulation can be written as

$$\psi_{\infty,\mathrm{E}}(\hat{\mathbf{r}}, \mathbf{d}) = -\gamma \sum_{k=1}^{n} \int_{\Gamma_k} \partial_{\mathbf{n}'}\phi(\mathbf{r}'; \mathbf{d}) e^{-i\kappa\hat{\mathbf{r}}\cdot\mathbf{r}'} ds(\mathbf{r}').$$

and

$$\psi_{\infty,\mathrm{GFL}}(\hat{\mathbf{r}}, \mathbf{d}) = \gamma \left[\sum_{j=1}^{m} \sigma_j \phi_j(\mathbf{d}) e^{-i\kappa\hat{\mathbf{r}}\cdot\mathbf{r}_j} - \sum_{k=1}^{n} \int_{\Gamma_k} \partial_{\mathbf{n}'}\phi(\mathbf{r}'; \mathbf{d}) e^{-i\kappa\hat{\mathbf{r}}\cdot\mathbf{r}'} ds(\mathbf{r}') \right].$$

These far-field patterns will be recorded as measurement and used to construct the response matrices for the imaging functions.

3.2. Imaging functions. Consider an array of transmitters that can send out plane incident waves and record the far-field pattern of the scattered waves. Assume that we have a set of incident plane waves with incident directions $\mathbf{d}_1, \ldots, \mathbf{d}_N$ and the far-field patterns are recorded at observation direction $\hat{\mathbf{r}}_1, \ldots, \hat{\mathbf{r}}_M$. These measurement of the far-field patterns form an $M \times N$ response matrix P with each entry given by

$$P_{ij} = \psi_\infty(\hat{\mathbf{r}}_i, \mathbf{d}_j), \tag{3.5}$$

where the far-field pattern ψ_∞ represents any one of the far-field patterns for (3.2), (3.3), or (3.4). The response matrix can be regarded as a discrete version of the scattering operator.

Here we give some physical interpretation of the response matrix to motivate the construction of our imaging functions. First, from the observation point of view, the scattered far field pattern can be viewed as the superposition of far field pattern of point scatterers distributed (with the total field ϕ as the weight function) at locations where scattering occurs, i.e., at the point scatterers as well as at the boundary of extended scatterers based on equation (3.4). Although this interpretation sounds similar to Born approximation, which simply replaces the total field in (3.4) by the incidence field, here we try not to assume or use explicit quantitative knowledge of the total field in the direct imaging algorithm. In the discrete setting,

$$\begin{aligned} P_{ij} &= \psi_\infty(\hat{\mathbf{r}}_i, \mathbf{d}_j) \\ &= \gamma \left[\sum_{k=1}^{m} \sigma_k \phi_k(\mathbf{d}_j) e^{-\mathrm{i}\kappa \hat{\mathbf{r}}_i \cdot \mathbf{r}_k} - \int_\Gamma \partial_{\mathbf{n}'} \phi(\mathbf{r}'; \mathbf{d}_j) e^{-\mathrm{i}\kappa \hat{\mathbf{r}}_i \cdot \mathbf{r}'} ds(\mathbf{r}') \right], \end{aligned}$$

which means the column space of the response matrix is a subspace of the space spanned by the illumination vectors with respect to the observation array/directions

$$\mathbf{g}_{\mathrm{r}}(\mathbf{r}) = (e^{-\mathrm{i}\kappa \mathbf{r}\cdot\hat{\mathbf{r}}_1}, \dots, e^{-\mathrm{i}\kappa \mathbf{r}\cdot\hat{\mathbf{r}}_M})^\top, \ \mathbf{r} \in \{\mathbf{r}_1, \dots, \mathbf{r}_m\} \cup \Gamma.$$

If we introduce far field pattern of the Green's function $G_\infty(\hat{\mathbf{r}}, \mathbf{r})$, which is the far field pattern of the scattered wave due to a point source at $\mathbf{r}$ with presence of both point and extended scatterers that can be defined by a generalized Foldy–Lax formulation similar to the one we defined for the total field (2.21) and (2.22). Then we can rewrite far field pattern for the scattered field for an incident wave $e^{\mathrm{i}\kappa \mathbf{r}\cdot\mathbf{d}}$ as

$$\psi_{\infty,\mathrm{GFL}}(\hat{\mathbf{r}}, \mathbf{d}) = \sum_{k=1}^{m} G_\infty(\hat{\mathbf{r}}, \mathbf{r}_k) e^{\mathrm{i}\kappa \mathbf{r}_k\cdot\mathbf{d}} - \int_\Gamma \partial_{\mathbf{n}'} G_\infty(\hat{\mathbf{r}}, \mathbf{r}') e^{\mathrm{i}\kappa \mathbf{r}'\cdot\mathbf{d}} ds(\mathbf{r}'). \tag{3.6}$$

From the incident point of view, we can view the far field pattern as superposition of scattered wave excited at point scatterers and extended scatterer boundary by incident wave $e^{\mathrm{i}\kappa \mathbf{r}\cdot\mathbf{d}}$, although we do not know the weight function G_∞. In the discrete setting,

$$\begin{aligned} P_{ij} &= \psi_\infty(\hat{\mathbf{r}}_i, \mathbf{d}_j) \\ &= \sum_{k=1}^{m} G_\infty(\hat{\mathbf{r}}_i, \mathbf{r}_k) e^{\mathrm{i}\kappa \mathbf{r}_k\cdot\mathbf{d}_j} - \int_\Gamma \partial_{\mathbf{n}'} G_\infty(\hat{\mathbf{r}}_i, \mathbf{r}') e^{\mathrm{i}\kappa \mathbf{r}'\cdot\mathbf{d}_j} ds(\mathbf{r}'), \end{aligned}$$

which means the row space of the response matrix is a subspace of the space spanned by the illumination vectors with respect to the incident directions

$$\mathbf{g}_{\mathrm{t}}(\mathbf{r}) = (e^{\mathrm{i}\kappa \mathbf{r}\cdot\mathbf{d}_1}, \dots, e^{\mathrm{i}\kappa \mathbf{r}\cdot\mathbf{d}_N})^\top, \ \mathbf{r} \in \{\mathbf{r}_1, \dots, \mathbf{r}_m\} \cup \Gamma.$$

In general, the knowledge of the incident wave and/or observation point gives a good clue on the form of illumination vector one should use to approximate the row and/or column space of the response matrix. How well the column or the row space can approximate the space of $\mathbf{g}_{\mathrm{r}}(\mathbf{r})$ or $\mathbf{g}_{\mathrm{t}}(\mathbf{r})$ respectively for $\mathbf{r} \in \{\mathbf{r}_1, \dots, \mathbf{r}_m\} \cup \Gamma$ depends on the wavenumber κ , the sampling of incident directions $\mathbf{d}_1, \dots, \mathbf{d}_N$ and the far-field observation directions $\hat{\mathbf{r}}_1, \dots, \hat{\mathbf{r}}_M$, and the geometry. For discrete point scatterers whose number is fewer than $\min(M, N)$, the two spaces are the same. In general, for a point $\mathbf{r} \in \{\mathbf{r}_1, \dots, \mathbf{r}_m\} \cup \Gamma$ that is well illuminated by the incident waves and well observed by the receiver array, $\mathbf{g}_{\mathrm{r}}(\mathbf{r})$ and $\mathbf{g}_{\mathrm{t}}(\mathbf{r})$ have a strong signature in the column space and row space respectively. Highly concave part of the obstacle boundary often can not be well illuminated or observed.

Now we turn the imaging problem into finding the location where scattering happens as the following. The motivation is: at each point $\mathbf{r}$ in the imaging domain, how well $\mathbf{g}_{\mathrm{r}}(\mathbf{r})$ and $\mathbf{g}_{\mathrm{t}}(\mathbf{r})$ are approximated by the significant component of column and row space of the response matrix respectively. First, compute the singular value decomposition (SVD) of the response matrix (3.5).

$$P = \sum_{k=1}^{\min(M,N)} \lambda_k \mathbf{u}_k \overline{\mathbf{v}}_k^\top,$$

where $\mathbf{u}_k$ and $\mathbf{v}_k$ are the unit left and right singular vectors, and λ_k are the singular values of P. This matrix factorization turns imaging passive targets into imaging active sources with respect to the receiver and transmitter array since $\mathbf{u}_k$ and $\mathbf{v}_k$ form the orthogonal basis for the column and row spaces respectively with λ_k being the signal strength. The following quantity measures projections in each pair of left and right singular vectors $\mathbf{u}_k$ and $\mathbf{v}_k$:

$$I_k(\mathbf{r}) = \left[\hat{\mathbf{g}}_\mathrm{r}^\top(\mathbf{r}) \cdot \mathbf{u}_k\right] \left[\overline{\hat{\mathbf{g}}}_\mathrm{t}^\top(\mathbf{r}) \cdot \overline{\mathbf{v}}_k\right], \tag{3.7}$$

where the $\hat{\mathbf{g}}_\mathrm{t}$ and $\hat{\mathbf{g}}_\mathrm{r}$ are the normalized illumination vectors with respect to the transmitter and the receiver, given explicitly as

$$\hat{\mathbf{g}}_\mathrm{r}(\mathbf{r}) = \frac{1}{\sqrt{M}}(e^{\mathrm{i}\kappa\mathbf{r}\cdot\hat{\mathbf{r}}_1}, \ldots, e^{\mathrm{i}\kappa\mathbf{r}\cdot\hat{\mathbf{r}}_M})^\top \quad \text{and} \quad \hat{\mathbf{g}}_\mathrm{t}(\mathbf{r}) = \frac{1}{\sqrt{N}}(e^{\mathrm{i}\kappa\mathbf{r}\cdot\mathbf{d}_1}, \ldots, e^{\mathrm{i}\kappa\mathbf{r}\cdot\mathbf{d}_N})^\top.$$

This quantity removes the phase ambiguity of the SVD of the response matrix too. The two dot product terms in (3.7) is phase matching with respect to the receiver array and the transmitter array respectively. The product of these two terms is just a summation of the two phase matching. If further superposed in frequency, it becomes travel time matching with respect to the two arrays, which is the Kirchhoff migration used in geophysics context. Furthermore, it can be easily shown that

$$\left|\sum_{k=1}^{\min(M,N)} I_k(\mathbf{r})\right| \leq 1.$$

Define

$$\lambda_{\max} = \max_{1\leq k\leq \min(M,N)} \{\lambda_k\} \quad \text{and} \quad \eta_k = \lambda_k/\lambda_{\max}.$$

The first imaging function is defined as

$$\mathcal{I}_1(\mathbf{r}) = \sum_{k=1}^{\min(M,N)} \eta_k I_k(\mathbf{r}), \tag{3.8}$$

where the singular values play the role of the weights and thus the regularization in the imaging function. As we mention, if further superposed in frequency, it becomes the Kirchhoff migration in geophysics context which matches the travel time between each pair of transmitter and receiver.

The second imaging function is defined as

$$\mathcal{I}_2(\mathbf{r}) = \sum_{k=1}^{K} I_k(\mathbf{r}), \tag{3.9}$$

where the truncation number $K \leq \min(M,N)$ is a threshold of the number of significant singular vectors. For m point scatterers, then $K = m$. In general, for extended obstacles, K may depend on the size of the obstacle and physical resolution of the array, which depends on the wavenumber κ, aperture and array setup. The truncation number K can also be used as a regularization parameter that is determined by the noise level [**20**, **46**]. We refer to Ammari et al. [**1**, **3**] for stability and resolution analysis for these imaging functionals.

We define the following transformed imaging functions, which seem to produce better visualization results.

$$\widetilde{\mathcal{I}}_i(\mathbf{r}) = [1 - \mathcal{I}_i(\mathbf{r})]^{-1}, \quad i = 1, 2. \tag{3.10}$$

In Section 4, all these imaging functions are demonstrated and compared for imaging the point and extended obstacle scatterers. They essentially produce the similar quality of imaging results; while the imaging function $\widetilde{\mathcal{I}}_i$ displays a little better visualization effect.

Since we transform the inverse scattering problem into an active source problem with respect to both transmitter and receiver arrays using the SVD of the response matrix, we avoid the need to estimate the total field quantitatively which becomes a nonlinear and nonlocal problem in terms of the location and shape of the scatterers. However, consequent issues related to this formulation are:

(1) Poor illumination due to partial illumination or occlusion, may cause the total field weak at certain locations. Similarly, scattering at certain locations may not be measured due to partial observation aperture. Imaging resolution at these locations will be compromised as we will see from some of the tests below.
(2) Multiple scattering information, which is implicitly embedded in the total field, is not used if one uses homogeneous Green's function as the illumination vector.

4. Numerical experiments

In this section, we present numerical examples for the inverse scattering problem, which contains both point scatterers and extended obstacle scatterers. As discussed in Section 3.2, the response matrix is based on the far-field pattern of the scattered field with row index corresponding to the observation direction $\hat{\mathbf{r}}$ and column index corresponding to the incident direction $\mathbf{d}$. To test the stability and robustness of the method, the response matrix is perturbed by multiplicative noise as

$$P_{ij}^{\text{noise}} = a\,\text{Re}P_{ij} + \mathrm{i}\,b\,\text{Im}P_{ij},$$

where a and b are uniformly distributed random numbers in $[1 - c, 1 + c]$ with $c = 20\%$. In the computation of the imaging functions, we use those singular vectors corresponding to singular values greater than some threshold which gives robustness to noise. As we mentioned before this threshold is the only parameter here which plays the role of physical regularization that depends on the physical setup such as used frequency, desired resolution, size of the obstacle and/or number of point scatterers, as well as signal noise ratio. On the other hand, the threshold is not very sensitive and can take a range of values that render very similar results. In all of the following examples, the wavenumber κ is taken to be 4π, i.e., the wavelength is taken to be $\lambda = 2\pi/\kappa = 0.5$.

Example 1. This is a 2D example. The incident and observation directions can be represented by the incident angle α_{inc} and the observation angle α_{obs}, respectively, i.e., $\mathbf{d} = (\cos\alpha_{\text{inc}}, \sin\alpha_{\text{inc}}) \in \mathbb{S}^1$ and $\hat{\mathbf{r}} = (\cos\alpha_{\text{obs}}, \sin\alpha_{\text{obs}}) \in \mathbb{S}^1$. First, we consider the far-field pattern with full aperture data, i.e., incident angle $\alpha_{\text{inc}} \in [0, 2\pi]$ and observation angle $\alpha_{\text{obs}} \in [0, 2\pi]$. The size of the response matrix is 120×120, i.e., the intervals for the incident angle and observation angle are equally subdivided into 120 subintervals. In this example, we consider an extended

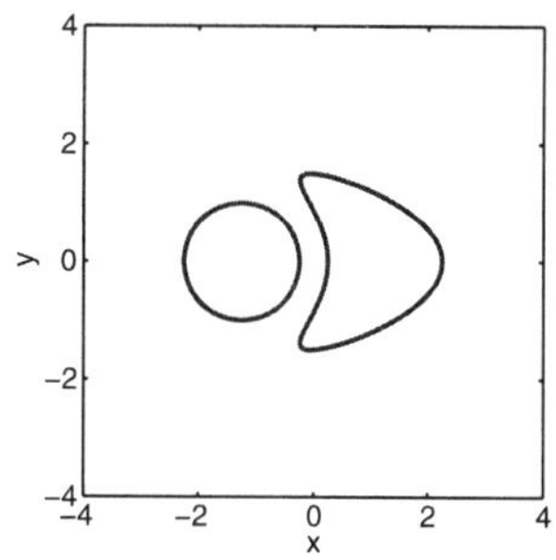

FIGURE 2. An extended obstacle scatterer with two disjoint components: one is a circle-shaped scatterer and another is kite-shaped scatterer in Example 1.

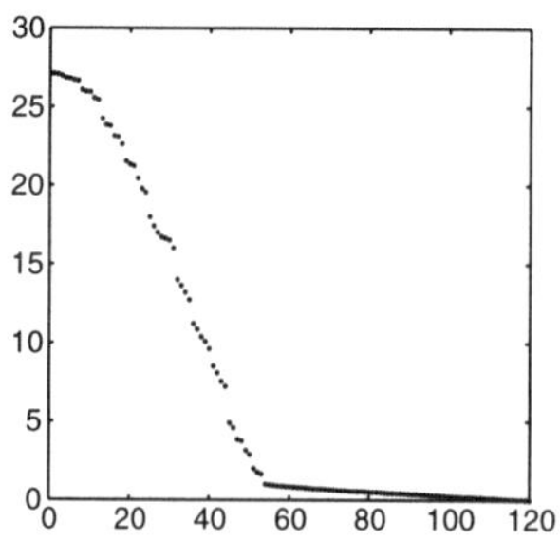

FIGURE 3. The singular value plot of the response matrix for the extended obstacle scatterers in Example 1.

obstacle scatterer with two disjoint components: one is a circle-shaped extended scatterer with the parametric representation given by

$$\mathbf{r}_1(t) = (\cos t - 1.25,\ \sin t), \quad 0 \le t \le 2\pi;$$

and another is a kite-shaped extended scatterer with the parametric representation given by

$$\mathbf{r}_2(t) = (\cos t + 0.65\cos 2t + 0.6,\ 1.5\sin t), \quad 0 \le t \le 2\pi,$$

as seen in Figure 2. The separation distance between these two components is only about one wavelength, which makes it hard to distinguish these two components for the imaging, especially for the concave part of the kite-shaped scatterer and the adjacent part of the circle-shaped scatterer.

Figure 3 plots the singular value pattern of the response matrix from the far-field pattern $\psi_{\infty,\mathrm{E}}$ by using the boundary integral equation for the extended obstacle scatterers. Figure 4 shows, from top to bottom and from left to right, the imaging of the two isolated extended obstacle scatterers by using imaging functions $\mathcal{I}_1, \mathcal{I}_2, \widetilde{\mathcal{I}}_1$, and $\widetilde{\mathcal{I}}_2$, respectively. It can be seen from the imaging results that all the four imaging functions essentially produce similar results, and the two modified imaging functions $\widetilde{\mathcal{I}}_i$ displays a little better result from the visualization point of view. From now on, we will use the imaging function $\widetilde{\mathcal{I}}_2$ to present the imaging results instead of showing the results from all the imaging functions.

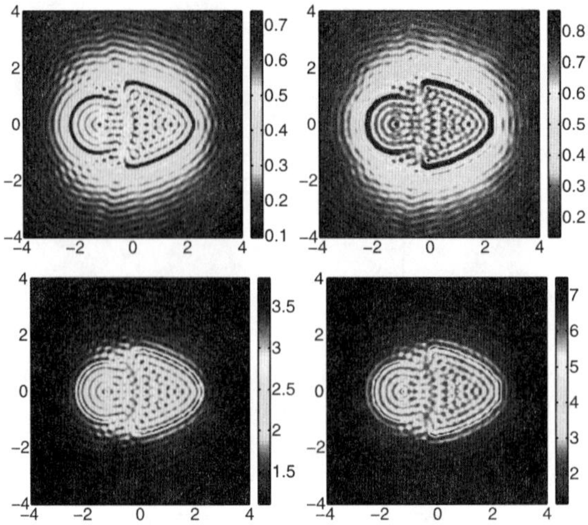

FIGURE 4. Imaging of the extended obstacle scatterers by using the boundary integral equation in Example 1. (top left) imaging result by $\mathcal{I}_1$; (top right) imaging result by $\mathcal{I}_2$; (bottom left) imaging result by $\widetilde{\mathcal{I}}_1$; (bottom right) imaging result by $\widetilde{\mathcal{I}}_2$.

Next we show the imaging results from the far-field pattern with limited aperture data and demonstrate that the proposed imaging method can also handle the missing data case. We consider so-called limited back-scattered aperture data with incident angle $\alpha_{\text{inc}} \in [-\pi/2, \pi/2]$ and observation angle $\alpha_{\text{obs}} \in [-\pi/2, \pi/2]$, and limited forward-scattered aperture data with incident angle $\alpha_{\text{inc}} \in [-\pi/2, \pi/2]$ and observation angle $\alpha_{\text{obs}} \in [\pi/2, 3\pi/2]$. The size of the response matrix is still the same as the full aperture case 120×120, i.e., the intervals for the incident angle and observation angle are also equally subdivided into 120 subintervals. Figure 5 shows the singular value distribution of the response matrices for the far-field pattern by using back-scattered limited aperture data and forward-scattered limited aperture data. Based on the singular value distributions, we can make the following two observations: (i) the extended scatterers is not a superposition of point scatterers, i.e., each singular vector does not corresponds to a point on the boundary; (ii) the response matrix obtained from full aperture data contains more leading singular values and hence more information than that from limited aperture data. Figure 6 shows the imaging of the two obstacle scatterers by using back-scattered limited aperture data and forward-scattered limited aperture data. Evidently, comparing 4 with 6, the full aperture data produces a much better result than those using limited aperture data. The back-scattered limited aperture data yields a blurred profile for the two extended obstacle scatterers. Using Born approximation, one can see that back-scattered data contains mainly low frequency information of the shape. The forward-scattered limited aperture gives a clear image of the convex arc part of the circle-shaped obstacle, which is illuminated best by the array of transducers.

Example 2. This example has the same two extended obstacles as Example. Besides, the extended obstacles are surrounded by a set of 100 randomly distributed point scatterers in an annulus with radii bounded below by 3 and above by 4, as

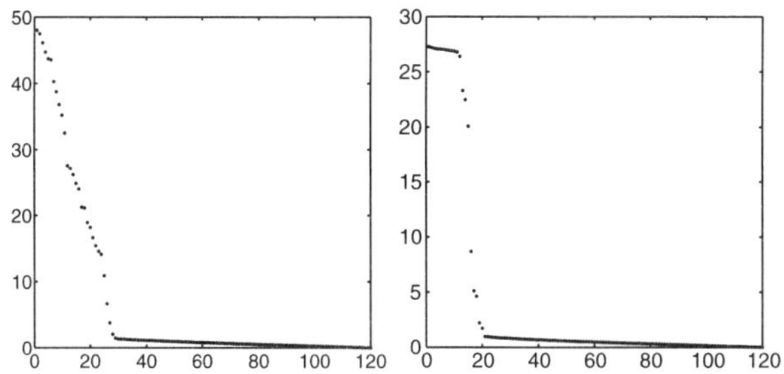

FIGURE 5. The singular value plot of the response matrices for the extended obstacle scatterers in Example 1. (left) limited back-scattered aperture data; (right) limited forward-scattered aperture data.

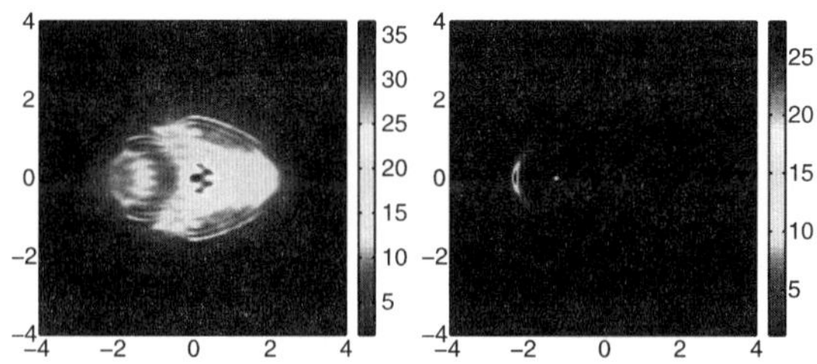

FIGURE 6. Imaging of the extended obstacle scatterers by using the boundary integral equation in Example 1. (left) limited back-scattered aperture data; (right) limited forward-scattered aperture data.

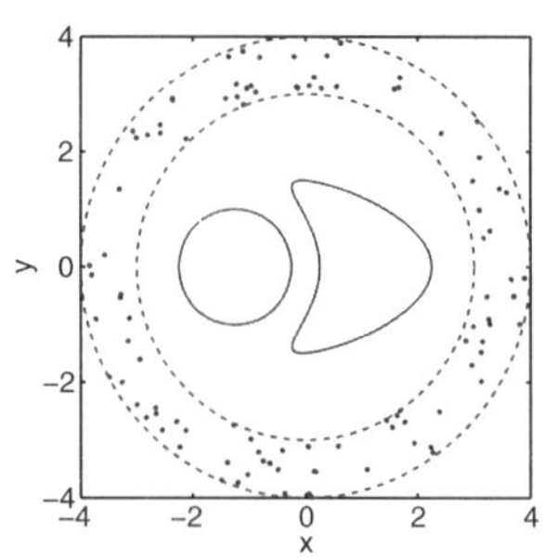

FIGURE 7. A set of one hundred of randomly distributed point scatterers in the annulus with radii bounded blow by 3 and above by 4, and an extended obstacle scatterer with two disjoint components: one is a circle-shaped scatterer and another is kite-shaped scatterer in Example 2.

seen in Figure 7. The scattering coefficient σ_j is taken as a random number from the interval $(0, 1)$ for $j = 1, \ldots, 100$. We consider the full aperture data and the size of the response matrix is 120×120.

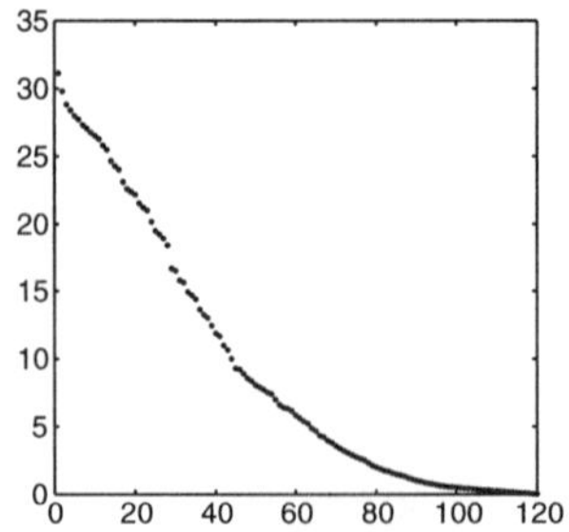

FIGURE 8. The singular value plot of the response matrix for the point and extended obstacle scatterers in Example 2.

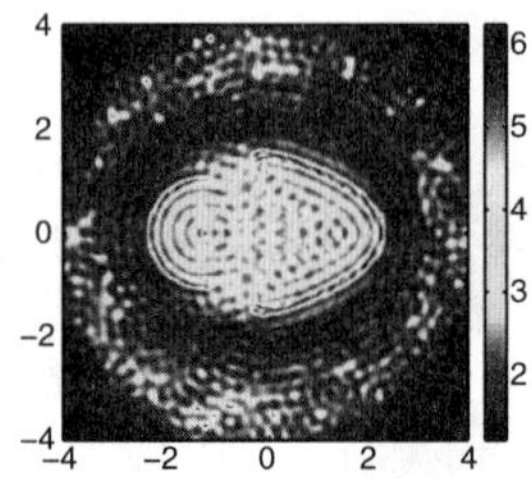

FIGURE 9. Imaging of the point and extended obstacle scatterers by using the generalized Foldy–Lax formulation by $\widetilde{\mathcal{I}}_2$ in Example 2.

Figure 8 plots the singular value pattern of the response matrix from the far-field pattern $\psi_{\infty,\mathrm{GFL}}$ by using the generalized Foldy–Lax formulation for the mixed point and extended scatterers. Figure 9 shows the imaging of the 100 randomly distributed point scatterers and two scatterers by using full aperture data and imaging function $\widetilde{\mathcal{I}}_2$. The imaging highlights not only the profile of the two extended scatterers but also the location of the point scatterers. Since the scattering coefficients are randomly chosen from the interval $(0, 1)$, they may not be all strong scatterers, which explains some point scatterers are clearly located but some are not so clearly spotted.

To investigate how each far-field pattern contributes to the overall imaging, Figure 10 shows the imaging results based on the projection method of the far-field patterns. The method of the projection is given in Appendix, which describes how to project the column space of the response matrix P_{GFL} out of the column spaces spanned by the columns of the response matrices of P_{FL} and P_{E}. Figure 10 shows, from left to right, the imaging results by using the response matrix P_1, P_2, and P_3, respectively. The left figure in Figure 10 looks like the one in Figure 4 since the 100 point scatterers are taken from the overall imaging in Figure 9. The middle figure in Figure 9 shows the imaging for the 100 randomly distributed point scatterers by taking out the contribution of the extended scatterers from the response matrix. The right figure in Figure 10 shows the imaging result arising from the multiple interaction between the point scatterers and the extended obstacle scatterers.

Example 3. This example is a three-dimensional case. The incident and observation directions can be represented by the incident angles α_{inc} and β_{inc} and

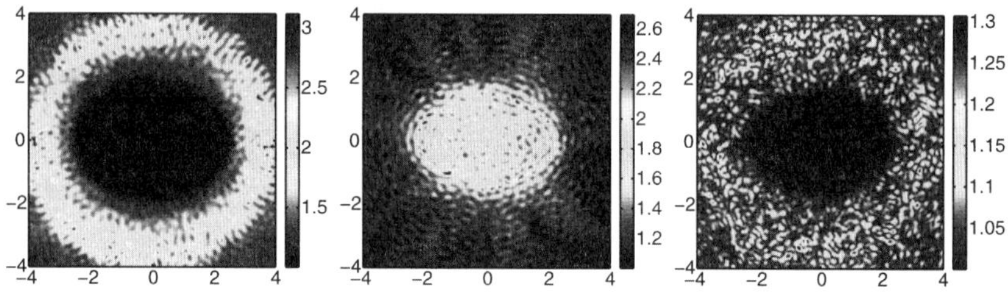

FIGURE 10. Imaging of the point and extended obstacle scatterers by using different far-field patterns in Example 2. (left) imaging from the response matrix P_1; (middle) imaging from the response matrix P_2; (right) imaging from the response matrix P_3.

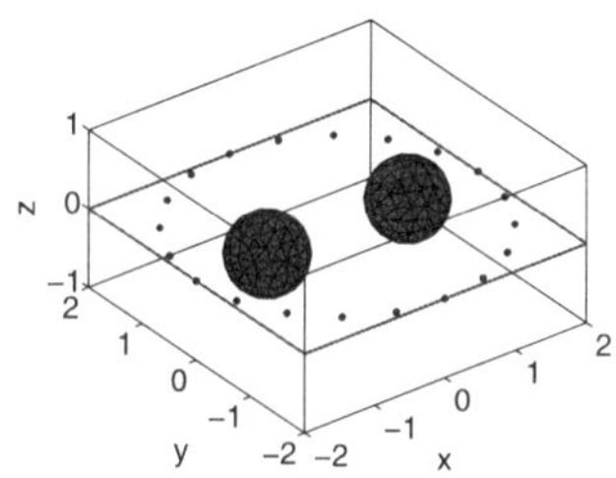

FIGURE 11. A set of twenty equally spaced point scatterers on a circle with radius 2 on the xy-plane and two spherical obstacle scatterers with centers at $(-1.0, 0.0, 0.0)$ and $(1.0, 0.0, 0.0)$ and the same radius of 0.5 in Example 3.

the observation angles α_{obs} and β_{obs}, i.e.,

$$\mathbf{d} = (\cos\alpha_{\text{inc}}\sin\beta_{\text{inc}},\ \sin\alpha_{\text{inc}}\sin\beta_{\text{inc}},\ \cos\beta_{\text{inc}}) \in \mathbb{S}^2,$$

$$\hat{\mathbf{r}} = (\cos\alpha_{\text{obs}}\sin\beta_{\text{obs}},\ \sin\alpha_{\text{obs}}\sin\beta_{\text{obs}},\ \cos\beta_{\text{obs}}) \in \mathbb{S}^2.$$

In this example, we only show the results for full aperture data, where $\alpha_{\text{inc}} \in [0, 2\pi], \beta_{\text{inc}} \in [0, \pi]$ and $\alpha_{\text{obs}} \in [0, 2\pi], \beta_{\text{obs}} \in [0, \pi]$, since the results for limited aperture data are analogous to the comparison between the full and the limited aperture data for the two-dimensional case. The size of the response matrix is 800×800, i.e., the intervals for the angles α_{inc} and α_{obs} are equally subdivided into 40, and the angles β_{inc} and β_{obs} are equally subdivided into 20.

In this example, we consider a group of twenty equally distributed point scatterers on a circle with radius 2 on the xy-plane, as seen in Figure 11. The scattering coefficients σ_j is chosen to be all equal, i.e., $\sigma_j = 1.0, j = 1, \ldots, 20$. The centers of two spherical obstacle scatterers with the same radius of 0.5 are located at $(-1.0, 0.0, 0.0)$ and $(1.0, 0.0, 0.0)$.

Figure 12 plots the singular value distribution of the response matrix of the scattered field for the point and two spherical obstacle scatterers. Figures 13 shows the imaging result for the point and two spherical obstacle scatterers on the planes defined by $z = 0$, $y = 0$, and $x = 1$, respectively. As we can see, the imaging results accurately locate the point scatterers and plot the profiles of the two spherical extended scatterers.

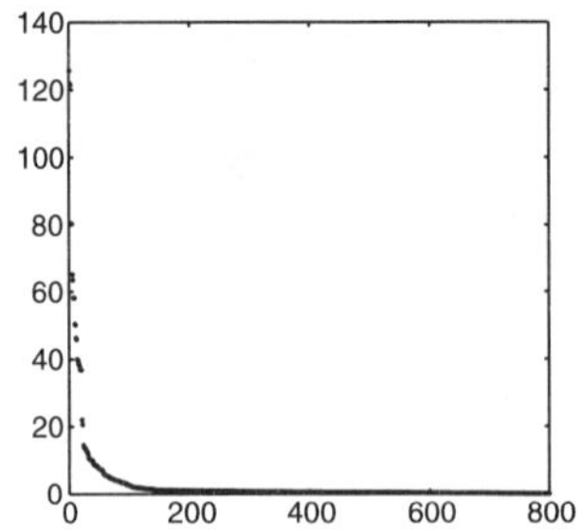

FIGURE 12. The singular value plots of the response matrix for the scattered field in Example 3.

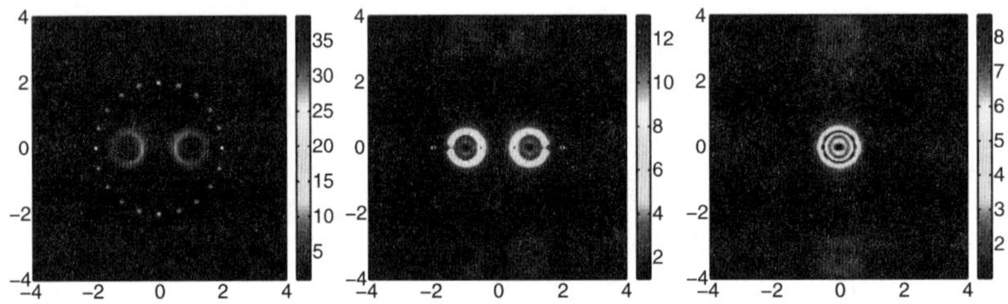

FIGURE 13. Imaging of the point and two spherical obstacle scatterers by using the Foldy–Lax formulation in Example 3. (left) slice $z = 0$; (middle) slice $y = 0$; (right) slice $x = 1$.

5. Conclusion

We use a direct imaging method to study the inverse scattering problem which contains multiple scattering between point scatterers and extended scatterers. The method is simple and efficient since no direct solver or iteration is needed, and can handle full and limited synthetic aperture data. The motivation of the problem formulation is to simulate the wave propagation and image a target in a cluttered environment, where small particles are modeled as point scatterers. A generalized Foldy–Lax formulation is used to resolve the multiple scattering between the small scale point scatterers and the extended obstacle scatterers. The direct imaging method uses the singular value decomposition of the response matrix from the far-field pattern with physically based regularization and/or thresholding to visualize the location of strong scattering. The imaging functions are robust and stable with respect to the measurement noise. In future work, we will investigate the extended obstacle scatterers with different boundary conditions, such as sound hard and mixed type of sound soft and sound hard, and the three-dimensional electromagnetic wave propagation which is governed by Maxwell's equations.

Appendix A. Orthogonal projection of response matrix

Let $\mathbf{v}_1, \ldots, \mathbf{v}_k$ be a family of orthonormal vectors in $\mathbb{C}^n$, and let

$$V = \text{span}(\mathbf{v}_1, \ldots, \mathbf{v}_k).$$

Obviously, the vectors $\mathbf{v}_1, \ldots, \mathbf{v}_k$ form an orthonormal basis for the k-dimensional subspace V of $\mathbb{C}^n$. Given a vector $\mathbf{u} \in \mathbb{C}^n$, we next describe how to find the projection of the vector $\mathbf{u}$ onto the subspace V, i.e., to find a vector $\mathbf{w} \in V$ such that $\mathbf{u} - \mathbf{w} \perp V$ for all vectors in V.

Construct a $n \times k$ matrix A by using the vectors $\mathbf{v}_1, \ldots, \mathbf{v}_k$ as the columns

$$A = [\mathbf{v}_1 \cdots \mathbf{v}_k].$$

To find the project $\mathbf{w}$ onto the subspace V is equivalent to the following equation

$$A^{\mathrm{H}}(\mathbf{u} - \mathbf{w}) = 0,$$

which is equivalent to

$$A^{\mathrm{H}}\mathbf{u} = A^{\mathrm{H}}\mathbf{w}. \tag{A.1}$$

Here H denotes the complex conjugate of the transpose. Meanwhile, since the projection $\mathbf{w}$ is on the subspace V, the vector $\mathbf{w}$ can be written as the linear combination of the vectors $\mathbf{v}_1, \ldots, \mathbf{v}_k$:

$$\mathbf{w} = x_1\mathbf{v}_1 + x_2\mathbf{v}_2 + \cdots + x_k\mathbf{v}_k = A\mathbf{x},$$

where $\mathbf{x} = (x_1 \ldots x_k)^\top$ is a k-dimensional column vector. Hence (A.1) can be written as

$$A^{\mathrm{H}}\mathbf{u} = A^{\mathrm{H}}A\mathbf{x}.$$

Noticing $A^{\mathrm{H}}A = I_{k\times k}$ since $\mathbf{v}_1, \ldots, \mathbf{v}_k$ are orthonormal. The above equation gives

$$\mathbf{x} = A^{\mathrm{H}}\mathbf{u},$$

which yields the projection

$$\mathbf{w} = AA^{\mathrm{H}}\mathbf{u}.$$

Let $Q = AA^{\mathrm{H}}$. The matrix Q is called the projection matrix for the subspace V.

Define the response matrix

$$P_{ij} = \psi_\infty(\hat{\mathbf{r}}_i, \mathbf{d}_j),$$

where ψ_∞ is the far-field pattern, $\mathbf{r}_i$ is the observation direction vector, and $\mathbf{d}_j$ is the incident direction vector. For convenience, we introduce three response matrices $P_{\mathrm{FL}}, P_{\mathrm{E}}$, and P_{GFL}, which stand for the response matrix for the point scatterers, extended obstacle, and mixed scatterers with both point scatterers and extended obstacle, respectively.

Next we describe how to project the columns of the response matrix P_{GFL} out of the response matrix P_{FL} and/or the response matrix P_{E}. We can do the SVD for P_{FL} and P_{E}, and pick the set of leading left singular vectors $\mathbf{u}_1, \ldots, \mathbf{u}_m$ and $\mathbf{v}_1, \ldots, \mathbf{v}_n$ from the SVD of P_{FL} and P_{E}, respectively. Using these orthonormal vectors, we can construct

$$A_{\mathrm{FL}} = (\mathbf{u}_1 \ldots \mathbf{u}_m)$$

and

$$A_{\mathrm{E}} = (\mathbf{v}_1 \ldots \mathbf{v}_n).$$

Then we can construct the project matrices

$$Q_{\mathrm{FL}} = A_{\mathrm{FL}}A_{\mathrm{FL}}^{\mathrm{H}}$$

and

$$Q_{\mathrm{E}} = A_{\mathrm{E}}A_{\mathrm{E}}^{\mathrm{H}}.$$

Finally we can study the projection of the response matrix P_{GFL}:

$$P_1 = P_{\mathrm{GFL}} - Q_{\mathrm{E}}P_{\mathrm{GFL}} = (I - Q_{\mathrm{E}})P_{\mathrm{GFL}}, \tag{A.2}$$

$$P_2 = P_{\mathrm{GFL}} - Q_{\mathrm{FL}}P_{\mathrm{GFL}} = (I - Q_{\mathrm{FL}})P_{\mathrm{GFL}}, \tag{A.3}$$

$$P_3 = P_{\mathrm{GFL}} - Q_{\mathrm{E}}P_{\mathrm{GFL}} - Q_{\mathrm{FL}}P_{\mathrm{GFL}} = (I - Q_{\mathrm{E}} - Q_{\mathrm{FL}})P_{\mathrm{GFL}}. \tag{A.4}$$

The direct imaging function can be construct from the SVD of P_1, P_2, and P_3.

References

[1] Habib Ammari, Josselin Garnier, Hyeonbae Kang, Mikyoung Lim, and Knut Sølna, *Multistatic imaging of extended targets*, SIAM J. Imaging Sci. **5** (2012), no. 2, 564–600, DOI 10.1137/10080631X. MR2971173

[2] *Mathematical and statistical methods for imaging*, Contemporary Mathematics, vol. 548, American Mathematical Society, Providence, RI, 2011. Papers from the NIMS Thematic Workshop held at Inha University, Incheon, August 10–13, 2010; Edited by Habib Ammari, Josselin Garnier, Hyeonbae Kang and Knut Sølna. MR2868483 (2012h:65006)

[3] Habib Ammari, Hyeonbae Kang, Eunjoo Kim, Mikyoung Lim, and Kaouthar Louati, *A direct algorithm for ultrasound imaging of internal corrosion*, SIAM J. Numer. Anal. **49** (2011), no. 3, 1177–1193, DOI 10.1137/100784710. MR2812563 (2012m:65386)

[4] Gang Bao, Songming Hou, and Peijun Li, *Inverse scattering by a continuation method with initial guesses from a direct imaging algorithm*, J. Comput. Phys. **227** (2007), no. 1, 755–762, DOI 10.1016/j.jcp.2007.08.020. MR2361544 (2008h:65038)

[5] G. Bao, S. Hou, and P. Li, *Recent studies on inverse medium scattering problems*, Modeling and computations in electromagnetics, Lect. Notes Comput. Sci. Eng. vol. 59, Springer, Berlin, 2008, pp. 165–186, DOI 10.1007/978-3-540-73778-0_6. MR2766829 (2012f:78011)

[6] Fioralba Cakoni and David Colton, *Qualitative methods in inverse scattering theory*, Interaction of Mechanics and Mathematics, Springer-Verlag, Berlin, 2006. An introduction. MR2256477 (2008c:35334)

[7] Margaret Cheney, *The linear sampling method and the MUSIC algorithm*, Inverse Problems **17** (2001), no. 4, 591–595, DOI 10.1088/0266-5611/17/4/301. Special issue to celebrate Pierre Sabatier's 65th birthday (Montpellier, 2000). MR1861470

[8] J. Cheng, J. J. Liu, and G. Nakamura, *The numerical realization of the probe method for the inverse scattering problems from the near-field data*, Inverse Problems **21** (2005), no. 3, 839–855, DOI 10.1088/0266-5611/21/3/004. MR2144808 (2006b:35335)

[9] Yves Colin de Verdière, *Pseudo-laplaciens. I* (French, with English summary), Ann. Inst. Fourier (Grenoble) **32** (1982), no. 3, xiii, 275–286. MR688031 (84k:58221)

[10] David Colton, Joe Coyle, and Peter Monk, *Recent developments in inverse acoustic scattering theory*, SIAM Rev. **42** (2000), no. 3, 369–414, DOI 10.1137/S0036144500367337. MR1786932 (2001f:76066)

[11] David Colton and Andreas Kirsch, *A simple method for solving inverse scattering problems in the resonance region*, Inverse Problems **12** (1996), no. 4, 383–393, DOI 10.1088/0266-5611/12/4/003. MR1402098 (97d:35032)

[12] David L. Colton and Rainer Kress, *Integral equation methods in scattering theory*, Pure and Applied Mathematics (New York), John Wiley & Sons Inc. New York, 1983. A Wiley-Interscience Publication. MR700400 (85d:35001)

[13] David Colton and Rainer Kress, *Inverse acoustic and electromagnetic scattering theory*, 2nd ed. Applied Mathematical Sciences, vol. 93, Springer-Verlag, Berlin, 1998. MR1635980 (99c:35181)

[14] A. Devaney, *Super-resolution processing of multi-static data using time-reversal and MUSIC*, J. Acoust. Soc. Am., submitted.

[15] Klaus Erhard and Roland Potthast, *A numerical study of the probe method*, SIAM J. Sci. Comput. **28** (2006), no. 5, 1597–1612 (electronic), DOI 10.1137/040607149. MR2272180 (2007h:35349)

[16] Leslie L. Foldy, *The multiple scattering of waves. I. General theory of isotropic scattering by randomly distributed scatterers*, Phys. Rev. (2) **67** (1945), 107–119. MR0011845 (6,224e)

[17] F. Gruber, E. Marengo, and A. Devaney, *Time-reversal imaging with multiple signal calssification considering multiple scattering between the targets*, J. Acoust. Soc. Am. **115** (2004), 3042–3047.

[18] Frank Hettlich, *Fréchet derivatives in inverse obstacle scattering*, Inverse Problems **11** (1995), no. 2, 371–382. MR1324650 (95k:35217)

[19] Thorsten Hohage, *Convergence rates of a regularized Newton method in sound-hard inverse scattering*, SIAM J. Numer. Anal. **36** (1999), no. 1, 125–142 (electronic), DOI 10.1137/S0036142997327750. MR1654583 (99j:65107)

[20] Songming Hou, Knut Solna, and Hongkai Zhao, *Imaging of location and geometry for extended targets using the response matrix*, J. Comput. Phys. **199** (2004), no. 1, 317–338, DOI 10.1016/j.jcp.2004.02.010. MR2081006 (2005k:78016)

[21] Songming Hou, Knut Solna, and Hongkai Zhao, *A direct imaging algorithm for extended targets*, Inverse Problems **22** (2006), no. 4, 1151–1178, DOI 10.1088/0266-5611/22/4/003. MR2249458 (2007b:78009)

[22] Songming Hou, Knut Solna, and Hongkai Zhao, *A direct imaging method using far-field data*, Inverse Problems **23** (2007), no. 4, 1533–1546, DOI 10.1088/0266-5611/23/4/010. MR2348080 (2009b:65142)

[23] S. Hou, K. Huang, K. Solna, and H. Zhao, *A phase and space coherent direct imaging method*, J. Acoust. Soc. Am. **125** (2009), 227–238.

[24] Kai Huang and Peijun Li, *A two-scale multiple scattering problem*, Multiscale Model. Simul. **8** (2010), no. 4, 1511–1534, DOI 10.1137/090771090. MR2718270 (2011k:35160)

[25] Kai Huang, Peijun Li, and Hongkai Zhao, *An efficient algorithm for the generalized Foldy-Lax formulation*, J. Comput. Phys. **234** (2013), 376–398, DOI 10.1016/j.jcp.2012.09.027. MR2999783

[26] Kai Huang, Knut Solna, and Hongkai Zhao, *Generalized Foldy-Lax formulation*, J. Comput. Phys. **229** (2010), no. 12, 4544–4553, DOI 10.1016/j.jcp.2010.02.021. MR2643662 (2011b:78011)

[27] Masaru Ikehata, *Reconstruction of an obstacle from the scattering amplitude at a fixed frequency*, Inverse Problems **14** (1998), no. 4, 949–954, DOI 10.1088/0266-5611/14/4/012. MR1642552 (99e:35242)

[28] M. Ikehata, *Reconstruction of obstacle from boundary measurements*, Wave Motion **30** (1999), no. 3, 205–223, DOI 10.1016/S0165-2125(99)00006-2. MR1708203 (2000i:35214)

[29] E. Kerbrat, C. Prada, and M. Fink, *Imaging in the presence of grain noise using the decomposition of the time reversal operator*, J. Acoust. Soc. Am. **113** (2003), 1230–1240.

[30] A. Kirsch, *The domain derivative and two applications in inverse scattering theory*, Inverse Problems **9** (1993), no. 1, 81–96. MR1203018 (93m:35182)

[31] Andreas Kirsch, *Characterization of the shape of a scattering obstacle using the spectral data of the far field operator*, Inverse Problems **14** (1998), no. 6, 1489–1512, DOI 10.1088/0266-5611/14/6/009. MR1662460 (99k:35193)

[32] Andreas Kirsch, *The MUSIC algorithm and the factorization method in inverse scattering theory for inhomogeneous media*, Inverse Problems **18** (2002), no. 4, 1025–1040, DOI 10.1088/0266-5611/18/4/306. MR1929280 (2003f:35286)

[33] Andreas Kirsch and Natalia Grinberg, *The factorization method for inverse problems*, Oxford Lecture Series in Mathematics and its Applications, vol. 36, Oxford University Press, Oxford, 2008. MR2378253 (2009k:35322)

[34] R. Kress, *Minimizing the condition number of boundary integral operators in acoustic and electromagnetic scattering*, Quart. J. Mech. Appl. Math. **38** (1985), no. 2, 323–341, DOI 10.1093/qjmam/38.2.323. MR788479 (86g:76021)

[35] Rainer Kress, *Newton's method for inverse obstacle scattering meets the method of least squares*, Inverse Problems **19** (2003), no. 6, S91–S104, DOI 10.1088/0266-5611/19/6/056. Special section on imaging. MR2036523 (2004m:65075)

[36] Rainer Kress and William Rundell, *A quasi-Newton method in inverse obstacle scattering*, Inverse Problems **10** (1994), no. 5, 1145–1157. MR1296365 (95h:65098)

[37] Rainer Kress and William Rundell, *Inverse obstacle scattering using reduced data*, SIAM J. Appl. Math. **59** (1999), no. 2, 442–454 (electronic), DOI 10.1137/S0036139997316598. MR1654431 (99h:35029)

[38] Melvin Lax, *Multiple scattering of waves*, Rev. Modern Physics **23** (1951), 287–310. MR0046273 (13,708a)

[39] P. A. Martin, *Multiple scattering*, Encyclopedia of Mathematics and its Applications, vol. 107, Cambridge University Press, Cambridge, 2006. Interaction of time-harmonic waves with N obstacles. MR2259988 (2007k:35348)

[40] Roland Potthast, *Stability estimates and reconstructions in inverse acoustic scattering using singular sources*, J. Comput. Appl. Math. **114** (2000), no. 2, 247–274, DOI 10.1016/S0377-0427(99)00201-0. MR1737077 (2000k:35297)

[41] Roland Potthast, *A new non-iterative singular sources method for the reconstruction of piecewise constant media*, Numer. Math. **98** (2004), no. 4, 703–730, DOI 10.1007/s00211-004-0524-y. MR2099318 (2006b:35346)

[42] C. Prada and J.-L. Thomas, *Experimental subwavelength localization of scatterers by decomposition of the time reversal operator interpreted as a covariance matrix*, J. Acoust. Soc. Am. **114** (2003), 235–243.

[43] Zeév Rudnick and Henrik Ueberschär, *Statistics of wave functions for a point scatterer on the torus*, Comm. Math. Phys. **316** (2012), no. 3, 763–782, DOI 10.1007/s00220-012-1556-2. MR2993932

[44] R. Schmidt, *Multiple emitter location and signal parameter estimation*, IEEE Trans. Antennas Propag. **34** (1986), 276–280.

[45] J. Sun, S. Carney, and J. Schotland, *Strong tip effects in near-field optical tomography*, J. App. Phys. **102** (2007), 103103.

[46] Hongkai Zhao, *Analysis of the response matrix for an extended target*, SIAM J. Appl. Math. **64** (2004), no. 3, 725–745, DOI 10.1137/S0036139902415282. MR2068119 (2005a:74050)

Department of Mathematics, Zhejiang University, Hangzhou 310027, China and Department of Mathematics, Michigan State University, East Lansing, Michigan 48824

E-mail address: drbaogang@gmail.com

Department of Mathematics and Statistics, Florida International University, Miami, Florida 33199

E-mail address: khuang@fiu.edu

Department of Mathematics, Purdue University, West Lafayette, Indiana 47907

E-mail address: lipeijun@math.purdue.edu

Department of Mathematics, University of California at Irvine, Irvine, California 92697

E-mail address: zhao@math.uci.edu

Contemporary Mathematics
Volume **615**, 2014
http://dx.doi.org/10.1090/conm/615/12246

The Inverse Scattering Problem for a Penetrable Cavity with Internal Measurements

Fioralba Cakoni, David Colton, and Shixu Meng

Dedicated to Gunther Uhlmann on the occasion of his 60th birthday

ABSTRACT. We consider the inverse scattering problem for a cavity that is bounded by a penetrable inhomogeneous medium of compact support and seek to determine the shape of the cavity from internal measurements on a curve or surface inside the cavity. We prove uniqueness and establish a linear sampling method for determining the shape of the cavity. A central role in our analysis is played by an unusual non-selfadjoint eigenvalue problem which we call the exterior transmission eigenvalue problem.

1. Introduction

The use of sampling methods and transmission eigenvalues has played an important role in inverse scattering theory for the past fifteen years and for a survey of recent results in this area we refer the reader to [**3**] and [**6**]. These methods are concerned with the inverse scattering problem for an inhomogeneous medium and seek to determine the support and bounds on the constitutive parameters of the scattering object by solving a linear integral equation of the first kind called the far field equation. A central role in this approach is an investigation of a class of non-selfadjoint eigenvalue problems called interior transmission eigenvalue problems. On the other hand, in the case of scattering by an impenetrable obstacle with Dirichlet, Neumann or impedance boundary conditions, there has been a recent interest in the inverse scattering problem with measured data inside a cavity [**12**], [**13**], [**21**]-[**23**]. In this class of problems the object is to determine the shape of the cavity from the use of sources and measurements along a curve or surface inside the cavity. A possible motivation for studying such a problem is to determine the shape of an underground reservoir by lowering receivers and transmitters into the reservoir through a bore hole drilled from the surface of the earth. In this paper we will combine the above two directions of research and consider the inverse scattering problem for a cavity that is bounded by a penetrable inhomogeneous medium of compact support and seek to determine the shape of the cavity from internal

1991 *Mathematics Subject Classification.* Primary 35R30, 35P25; Secondary 35Q60, 76Q05.
The research of the first author was supported in part by the Air Force Office of Scientific Research Grant FA9550-13-1-0199 and NSF Grant DMS-1106972.
The research of the second author was supported in part by the Air Force Office of Scientific Research Grant FA9550-13-1-0199 and NSF Grant DMS-1106972.
The research of the third author was supported in part by the NSF Grant DMS-1106972.

measurements. Of particular interest in this investigation is the central role played by an unusual non-selfajoint eigenvalue problem called the *exterior transmission eigenvalue problem.*

The plan of our paper is as follows. In the next section we will formulate both the direct and inverse scattering problems for a cavity bounded by a penetrable inhomogeneous medium of compact support. In Section 3 we will formulate the exterior transmission eigenvalue problem and establish the Fredholm property of the associated exterior transmission problem. These results will prove to be central to our analysis of both the uniqueness of the solution to our inverse scattering problem (discussed in Section 4) as well as our use of the linear sampling method in Section 5 to recover the shape of the cavity.

2. The direct and inverse scattering problems

Let $D \subset \mathbb{R}^d$, $d = 2,3$, be a simply connected bounded region of $\mathbb{R}^d$ with Lipschitz boundary ∂D and denote by ν the outward unit normal to ∂D. We assume the medium inside D is homogeneous with refractive index scaled to one and denote by k the corresponding wave number. The medium outside D is assumed to be inhomogeneous and possibly anisotropic such that outside a large ball B_R it is homogeneous with the same wave number as the medium in D. More specifically, the physical properties of the medium in $\mathbb{R}^d \backslash \overline{D}$ are described by the $d \times d$ symmetric matrix valued function A with $L^\infty(\mathbb{R}^d \setminus \overline{D})$ entries and the bounded function $n \in L^\infty(\mathbb{R}^d \backslash \overline{D})$ such that $\overline{\xi} \cdot \Re(A)\xi \geq \alpha \|\xi\|^2$, $\overline{\xi} \cdot \Im(A)\xi \leq 0$, for all $\xi \in \mathbb{C}$ and $n \geq n_0 > 0$ in $\mathbb{R}^d \setminus \overline{D}$. Furthermore we assume that $A \equiv I$ and $n \equiv 1$ in $\mathbb{R}^d \setminus B_R$ where B_R is a large ball containing D.

In acoustic scattering ($d = 3$) or electromagnetic scattering ($d = 2$, for an H-polarized infinite cylinder) D represents the support of a cavity filled e.g. with air which is assumed to be the reference media with wave number k. Let $\Phi(x, y)$ be a point source located at a point $y \in D$ inside the cavity given by

$$\Phi(x,y) = \begin{cases} \frac{i}{4} H_0^{(1)}(k|x-y|) & \text{in } \mathbb{R}^2 \\ \\ \dfrac{1}{4\pi} \dfrac{e^{ik|x-y|}}{|x-y|} & \text{in } \mathbb{R}^3. \end{cases} \tag{2.1}$$

and consider the scattering of this point source by the inhomogeneous media. The total field $u = u^s + \Phi(\cdot, y)$ inside the cavity satisfies

$$\Delta_x u + k^2 u = \delta(x-y) \qquad x \in D$$

whereas the total field w outside the cavity satisfies

$$\nabla \cdot A(x) \nabla w + k^2 n(x) w = 0 \qquad x \in \mathbb{R}^d \setminus \overline{D} \tag{2.2}$$

and across the interface ∂D both the total field and its normal derivative are continuos, i.e.

$$u = w \qquad \text{and} \qquad \frac{\partial u}{\partial \nu} = \frac{\partial w}{\partial \nu_A} \qquad \text{on } \partial D \tag{2.3}$$

together with the Sommerfeld radiation condition

$$\lim_{r \to \infty} r^{\frac{d-1}{2}} \left(\frac{\partial w}{\partial r} - ikw \right) = 0$$

uniformly with respect to $\hat{x} = x/r$, $r = |x|$, where $\frac{\partial w}{\partial \nu_A} := A\nabla w \cdot \nu$. We recall that $\operatorname{supp}(A - I) \subset B_R \setminus D$ and $\operatorname{supp}(n-1) \subset B_R \setminus D$. Written in terms of the scattered field u^s in D and the total field w in $\mathbb{R}^d \setminus \overline{D}$ the above scattering problem is a particular case of the following boundary value problem in $\mathbb{R}^d$: Find $u^s \in H^1(D)$ and $u \in H^1_{loc}(\mathbb{R}^d \setminus \overline{D})$ such that

$$\nabla \cdot A\nabla w + k^2 nw = 0 \quad \text{in} \quad \mathbb{R}^d \setminus \overline{D} \tag{2.4}$$

$$\Delta u^s + k^2 u^s = 0 \quad \text{in} \quad D \tag{2.5}$$

$$w - u^s = f \quad \text{on} \quad \partial D \tag{2.6}$$

$$\frac{\partial w}{\partial \nu_A} - \frac{\partial u^s}{\partial \nu} = h \quad \text{on} \quad \partial D \tag{2.7}$$

$$\lim_{r\to\infty} r^{\frac{d-1}{2}} \left(\frac{\partial w}{\partial r} - ikw\right) = 0 \tag{2.8}$$

where $f := \Phi(\cdot, y)|_{\partial D}$ and $h := \frac{\partial \Phi(\cdot,y)}{\partial \nu}|_{\partial D}$. In general $f \in H^{\frac{1}{2}}(\partial D)$ and $h \in H^{-\frac{1}{2}}(\partial D)$. Using a variational approach it is shown in [**3**] that the forward scattering problem (2.4)-(2.8) has a unique solution which depends continuously on the data f, h.

Now assume that C is a smooth $(d-1)$-manifold entirely included in D which is referred to as the measurement manifold. We place the point source at every $y \in C$ and measure the corresponding scattered field $u^s(x) := u^s(x, y)$ for $x \in C$. The *inverse problem* we consider in this paper is for fixed (but not necessarily known) A and n satisfying the above assumptions, determine the boundary of the cavity ∂D from a knowledge of $u^s(x, y)$ for all $x, y \in C$. (Note that if C is chosen to be an analytic manifold by the analyticity of the solution u^s to the Helmholtz equation in D it suffices to know $u^s(x, y)$ for x, y on a open arc $C_0 \subset C$.) Throughout this paper we make the following assumption:

ASSUMPTION 2.1. The measurement manifold C is such that k^2 is not a Dirichlet eigenvalue for $-\Delta$ in the bounded region circumscribed by C.

Note that since the interrogating wave number k is known, it is easy to choose C to satisfy Assumption 2.1. The main goal of this paper is to prove a uniqueness result and develop a solution method for solving the above inverse problem.

3. The exterior transmission eigenvalue problem

In this section we will formulate and study the so-called exterior transmission problem which will play a fundamental role in our uniqueness proof and the justification of the linear sampling method. To make an analogy with the exterior scattering problem, the exterior transmission problem here plays the same role as the interior transmission problem does for the exterior scattering problem for an inhomogeneous media [**4**], [**6**], [**11**]. As a physical motivation of the exterior transmission problem we ask the question if it is possible to send an outgoing incident field u^i from inside the cavity D that does not produce any scattered field in D and all the energy is transmitted to the exterior of D. Since this outgoing incident field satisfies the Helmholtz equation outside D, the scattering problem (2.4)-(2.8) implies that $v := u^i|_{\mathbb{R}^d\setminus\overline{D}}$ and the total field w satisfies the homogenous problem

$$\nabla\cdot A\nabla w + k^2 nw = 0 \quad \text{in} \quad \mathbb{R}^d\setminus\overline{D} \tag{3.1}$$

$$\Delta v + k^2 v = 0 \quad \text{in} \quad \mathbb{R}^d\setminus\overline{D} \tag{3.2}$$

$$w = v \quad \text{on} \quad \partial D \tag{3.3}$$

$$\frac{\partial w}{\partial \nu_A} = \frac{\partial v}{\partial \nu} \quad \text{on} \quad \partial D \tag{3.4}$$

$$\lim_{r\to\infty} r^{\frac{d-1}{2}}(\frac{\partial w}{\partial r} - ikw) = 0 \quad \text{and} \quad \lim_{r\to\infty} r^{\frac{d-1}{2}}(\frac{\partial v}{\partial r} - ikv) = 0 \tag{3.5}$$

DEFINITION 3.1. *Values of $k \in C$ with $\Re(k) > 0$ for which the homogeneous problem (3.1)-(3.5) has a nontrivial solution are called exterior transmission eigenvalues. (For complex k the radiation condition (3.5) is replaced by the condition that v and w behave asymptotically like $e^{ikr}r^{\frac{1}{2}-\frac{d}{2}}$.)*

In next section we will see that exterior transmission eigenvalues are related to the injectivity of the near field (data) operator. For later use we need the non-homogeneous version of (3.1)-(3.5) which we formulate in the following.

The *exterior transmission problem* is given $f \in H^{\frac{1}{2}}(\partial D)$, $h \in H^{\frac{1}{2}}(\partial D)$, $\ell_1 \in L^2(B_R\setminus\overline{D})$ and $\ell_2 \in L^2(B_R\setminus\overline{D})$, find $w \in H^1_{loc}(\mathbb{R}^d\setminus\overline{D})$, $v \in H^1_{loc}(\mathbb{R}^d\setminus\overline{D})$ such that

$$\nabla\cdot A\nabla w + k^2 nw = \ell_1 \quad \text{in} \quad \mathbb{R}^d\setminus\overline{D} \tag{3.6}$$

$$\Delta v + k^2 v = \ell_2 \quad \text{in} \quad \mathbb{R}^d\setminus\overline{D} \tag{3.7}$$

$$w - v = f \quad \text{on} \quad \partial D \tag{3.8}$$

$$\frac{\partial w}{\partial \nu_A} - \frac{\partial v}{\partial \nu} = h \quad \text{on} \quad \partial D \tag{3.9}$$

$$\lim_{r\to\infty} r^{\frac{d-1}{2}}(\frac{\partial w}{\partial r} - ikw) = 0 \quad \text{and} \quad \lim_{r\to\infty} r^{\frac{d-1}{2}}(\frac{\partial v}{\partial r} - ikv) = 0, \tag{3.10}$$

where ℓ_1 and ℓ_2 vanish in $\mathbb{R}^d\setminus\overline{B_R}$ and R is the radius of the ball B_R outside of which $A = I$ and $n = 1$. We use a variational approach to study this problem. To this end we take $v_l \in H^1_{loc}(\mathbb{R}^d\setminus\overline{D})$ to be the unique solution of the exterior Dirichlet problem

$$\Delta v_l + k^2 v_l = 0 \quad in \quad \mathbb{R}^d\setminus\overline{D}$$

$$v_l = f \quad on \quad \partial D$$

$$\lim_{r\to\infty} r^{\frac{d-1}{2}}(\frac{\partial v_l}{\partial r} - ikv_l) = 0$$

and set $v_0 = v + v_l$. Then (w, v_0) satisfies (3.6)-(3.10) with $(f, h) = (0, \tilde{h} := h - \frac{\partial v_l}{\partial \nu})$. Therefore it suffices to study (3.6)-(3.10) with $f = 0$. We can now rewrite (3.6)-(3.10) as an equivalent problem in the bounded domain $B_R\setminus\overline{D}$, namely find $w \in$

$H^1(B_R\backslash\overline{D})$, $v \in H^1(B_R\backslash\overline{D})$ such that

$$\nabla\cdot A\nabla w + k^2 nw = \ell_1 \quad \text{in} \quad B_R\backslash\overline{D} \tag{3.11}$$

$$\Delta v + k^2 v = \ell_2 \quad \text{in} \quad B_R\backslash\overline{D} \tag{3.12}$$

$$w - v = 0 \quad \text{on} \quad \partial D \tag{3.13}$$

$$\frac{\partial w}{\partial \nu_A} - \frac{\partial v}{\partial \nu} = h \quad \text{on} \quad \partial D \tag{3.14}$$

$$\frac{\partial w}{\partial \nu} = T_k w \quad \text{on} \quad \partial B_R \tag{3.15}$$

$$\frac{\partial v}{\partial \nu} = T_k v \quad \text{on} \quad \partial B_R \tag{3.16}$$

where $T_k : H^{\frac{1}{2}}(\partial B_R) \to H^{-\frac{1}{2}}(\partial B_R)$ is the exterior Dirichlet to Neuman map defined by

$$T_k : g \to \frac{\partial v}{\partial \nu}|_{\partial B_R}, \qquad g \in H^{\frac{1}{2}}(\partial B_R)$$

where u is the radiating solution to the Helmholtz equation $\Delta u + k^2 u = 0$ outside B_R with boundary data $u = g$ on ∂B_R, and ν is the outward unit normal to ∂B_R [**3**]. Next we define

$$H = \{(w,v) \in H^1(B_R\backslash\overline{D}) \times H^1(B_R\backslash\overline{D}), w - v = 0 \quad on \quad \partial D\}.$$

Taking a test function $(w', v') \in H$, multiplying both sides of (3.11) by $\overline{w'}$ and (3.12) by $\overline{v'}$, and integrating by parts we obtain

$$\int\limits_{\partial B_R} T_k w\overline{w'}ds - \int\limits_{\partial D} \frac{\partial w}{\partial \nu_A}\overline{w'}ds - \int\limits_{B_R\backslash\overline{D}} A\nabla w\cdot\nabla\overline{w'}dx + \int\limits_{B_R\backslash\overline{D}} nk^2 w\overline{w'}dx = \int\limits_{B_R\backslash\overline{D}} \ell_1\overline{w'}dx,$$

and

$$\int\limits_{\partial B_R} T_k v\overline{v'}ds - \int\limits_{\partial D} \frac{\partial v}{\partial \nu}\overline{v'}ds - \int\limits_{B_R\backslash\overline{D}} \nabla v\cdot\nabla\overline{v'}dx + \int\limits_{B_R\backslash\overline{D}} k^2 v\overline{v'}dx = \int\limits_{B_R\backslash\overline{D}} \ell_2\overline{v'}dx,$$

respectively. Now taking the difference and and using the fact that $w' = v'$ on ∂D together with (3.14) we have that

$$\begin{aligned}&\int\limits_{B_R\backslash\overline{D}} A\nabla w\cdot\nabla\overline{w'}dx - \int\limits_{B_R\backslash\overline{D}} \nabla v\cdot\nabla\overline{v'}dx + \int\limits_{B_R\backslash\overline{D}} (-nk^2 w\overline{w'} + k^2 v\overline{v'})dx \\ &- \int\limits_{\partial B_R} T_k w\overline{w'}ds + \int\limits_{\partial B_R} T_k v\overline{v'}ds = -\int_{\partial D} h\overline{w'} - \int\limits_{B_R\backslash\overline{D}} \ell_1\overline{w'}dx + \int\limits_{B_R\backslash\overline{D}} \ell_1\overline{v'}dx.\end{aligned} \tag{3.17}$$

We define the sesquilinear form $a_k(\cdot,\cdot) : H \times H \to \mathbb{C}$ by

$$\begin{aligned}a_k((w,v),(w',v')) = &\int\limits_{B_R\backslash\overline{D}} A\nabla w\cdot\nabla\overline{w'}dx - \int\limits_{B_R\backslash\overline{D}} \nabla v\cdot\nabla\overline{v'}dx \\ &+ \int\limits_{B_R\backslash\overline{D}} (-nk^2 w\overline{w'} + k^2 v\overline{v'})dx - \int\limits_{\partial B_R} T_k w\overline{w'} + \int_{\partial B_R} T_k v\overline{v'}dx\end{aligned}$$

and the conjugate linear functional $F(\cdot) : H \to \mathbb{C}$ by

$$F(w',v') := -\int_{\partial D} h\overline{w'} - \int_{B_R\backslash\overline{D}} \ell_1\overline{w'}dx + \int_{B_R\backslash\overline{D}} \ell_1\overline{v'}dx.$$

Conversely, assume that $(w,v) \in H$ satisfies $a_k((w,v),(w',v')) = F(w',v')$ for all $(w',v') \in H$. Taking $v' = 0$, $w' \in C_0^\infty(B_R\backslash\overline{D})$, we have (3.11) and in a similar way we have (3.12). Taking $(w',v') \in H$ such that $w' = v' = 0$ on ∂B_R, we recover (3.12). Finally, a choice of $(w',0) \in H$ implies (3.15) and in a similar way we obtain (3.16). Hence we have proven the following theorem.

THEOREM 3.2. *The exterior transmission problem (3.1)-(3.5) is equivalent to the following problem: Find $(w,v) \in H$ such that for all $(w',v') \in H$*

$$a_k((w,v),(w',v')) = F(w',v'). \tag{3.18}$$

Note that by means of the Riesz representation theorem we can define the operator $\mathcal{A}_k : H \to H$ by

$$(\mathcal{A}_k(w,v),(w',v'))_H = a_k((w,v),(w',v')) \qquad \text{for all } ((w,v),(w',v')) \in H \times H.$$

We would like to show that $\mathcal{A}_{i\kappa} : H \to H$ for $\kappa > 0$ is invertible. To prove this we use the T-coercivity approach introduced in [**2**] and [**7**], following the ideas in [**1**]. The idea behind the T-coercivity method is to consider an equivalent formulation of (3.18) where a_k is replaced by $a_k^{\mathcal{T}}$ defined by

$$a_k^{\mathcal{T}}((w,v),(w',v')) := a_k((w,v),\mathcal{T}(w',v')), \quad \forall((w,v),(w',v')) \in H \times H, \tag{3.19}$$

with $\mathcal{T}$ being an *ad hoc* isomorphism of H. Indeed, $(w,v) \in H$ satisfies

$$a_k((w,v),(w',v')) = 0 \qquad \text{for all} \qquad (w',v') \in H$$

if, and only if, it satisfies $a_k^{\mathcal{T}}((w,v),(w',v')) = 0$ for all $(w',v') \in H$. Assume that $\mathcal{T}$ and k are chosen so that $a_k^{\mathcal{T}}$ is coercive. Then using the Lax-Milgram theorem and the fact that $\mathcal{T}$ is an isomorphism of H, one deduces that $\mathcal{A}_k$ is an isomorphism on H.

In the following, in addition to the assumptions on A and n stated at the beginning of Section 2, we assume that there exists a neighborhood Ω of ∂D where both $\Im(A) = 0$ and $\Im(n) = 0$ in $B_R\backslash\overline{D} \cap \Omega$. Setting $\mathcal{N} := B_R\backslash\overline{D} \cap \Omega$, we denote by

$$\begin{aligned} A_* &:= \inf_{x\in\mathcal{N}} \inf_{|\xi|=1} \overline{\xi} \cdot A(x)\xi > 0, & A^* &:= \sup_{x\in\mathcal{N}} \sup_{|\xi|=1} \overline{\xi} \cdot A(x)\xi < \infty, \\ n_* &:= \inf_{x\in\mathcal{N}} n(x) > 0, & n^* &:= \sup_{x\in\mathcal{N}} n(x) < \infty. \end{aligned} \tag{3.20}$$

for $\xi \in \mathbb{C}^d$. Then we can prove the following result.

LEMMA 3.1. *Assume that either $A^* < 1$ and $n^* < 1$ or $A_* > 1$ and $n_* > 1$. Then there exists $\kappa > 0$ such that $\mathcal{A}_{i\kappa}$ is invertible.*

PROOF. We first consider the case when $A^* < 1$ and $n^* < 1$. Take $\chi \in C^\infty(\overline{B_R\backslash\overline{D}})$ to be a cut off function equal to 1 in a neighbourhood of ∂D with support in $\mathcal{N} := (B_R\backslash\overline{D}) \cap \Omega$ and let $\mathcal{T}(w,v) = (w - 2\chi v, -v)$. We then have that

$$\begin{aligned} a_{i\kappa}^{\mathcal{T}}((w,v),(w,v)) &= (A\nabla w, \nabla w)_{B_R\backslash\overline{D}} + (\nabla v, \nabla v)_{B_R\backslash\overline{D}} - 2(A\nabla w, \nabla(\chi v))_{B_R\backslash\overline{D}} \\ &\quad + \kappa^2((nw,w)_{B_R\backslash\overline{D}} + (v,v)_{B_R\backslash\overline{D}} - 2(nw,\chi v)_{B_R\backslash\overline{D}}) \\ &\quad - (T_{i\kappa}w,w)_{\partial B_R} - (T_{i\kappa}v,v)_{\partial B_R} + 2(T_{i\kappa}w,\chi v)_{\partial B_R} \end{aligned}$$

where $(\cdot,\cdot)_X$ denotes the L^2-inner product in the generic space X. By Young's inequality we have

$$\begin{aligned}2|(A\nabla w,\nabla\chi v)_{B_R\setminus\overline{D}}| &\leq 2|(\chi A\nabla w,\nabla v)_{\mathcal{N}}|+2|(A\nabla w,\nabla(\chi)v)_{\mathcal{N}}|\\ &\leq \alpha(A\nabla w,\nabla w)_{\mathcal{N}}+\alpha^{-1}(A\nabla v,\nabla v)_{\mathcal{N}}\\ &+ \beta(A\nabla w,\nabla w)_{\mathcal{N}}+\beta^{-1}(A\nabla(\chi)v,\nabla(\chi)v)_{\mathcal{N}},\end{aligned}$$

and

$$2|(nw,\chi v)_{B_R\setminus\overline{D}}|\leq 2|(nw,v)_{\mathcal{N}}|\leq \eta(nw,w)_{\mathcal{N}}+\eta^{-1}(nv,v)_{\mathcal{N}}$$

for some $\alpha>0,\beta>0,\eta>0$. Recall that A and n are real in $\mathcal{N}$. Furthermore, due to the exponential decay of w and v at ∞ we have that

$$-(T_{i\kappa}w,w)_{\partial B_R}=\int\limits_{\mathbb{R}^d\setminus\overline{B_R}}\left(|\nabla w|^2+\kappa^2|w|^2\right)\,dx$$

with a similar expression for $-(T_{i\kappa}v,v)_{\partial B_R}$. Note also that

$$(T_{i\kappa}w,\chi v)_{\partial B_R}=0.$$

Using all the above estimates we finally obtain that

$$\begin{aligned}|a^{\mathcal{T}}_{i\kappa}((w,v),(w,v))| &\geq \Re\left(a^{\mathcal{T}}_{i\kappa}((w,v),(w,v))\right)\\ &\geq \Re(A\nabla w,\nabla w)_{\{B_R\setminus\overline{D}\}\setminus\overline{\Omega}}+(\nabla v,\nabla v)_{\{B_R\setminus\overline{D}\}\setminus\overline{\Omega}}\\ &+ \kappa^2\left(\Re(nw,w)_{\{B_R\setminus\overline{D}\}\setminus\overline{\Omega}}+(v,v)_{\{B_R\setminus\overline{D}\}\setminus\overline{\Omega}}\right)\\ &+ (1-\alpha-\beta)(A\nabla w,\nabla w)_{\mathcal{N}}+((I-\alpha^{-1}A)\nabla v,\nabla v)_{\mathcal{N}}\\ &+ \kappa^2(1-\eta)(nw,w)_{\mathcal{N}}+(\kappa^2(1-\eta^{-1}n)-sup|\nabla\chi|^2A_+)v,v)_{\mathcal{N}}.\end{aligned}$$

Taking α,β,η,κ such that $A^*<\alpha$, $n^*<\eta$, $\beta<1-\alpha$, and κ large enough yields that $a^{S}_{i\kappa}$ is coercive.

The case when $A^*>1$ and $n^*>1$ can be proven the same way using $\mathcal{T}(w,v)=(w,-v+2\chi w)$. □

REMARK 3.3. *In Lemma* 3.1 *the assumption that A and n are real in a neighborhood $\mathcal{N}$ of ∂D can be relaxed. In particular, the proof of Lemma* 3.1 *goes through if we only assume that $-\Im(A)<\Re(A)$ snd $\Im(n)<\Re(n)$ in $\mathcal{N}$.*

THEOREM 3.4. *Assume that A and n satisfies the assumptions of Lemma* 3.1. *Then if k is not an exterior transmission eigenvalue the exterior transmission problem* (3.6)*-*(3.10) *has a unique solution which depends continuously on the data f, h, ℓ_1 and ℓ_2.*

PROOF. From Lemma 3.1, we can choose κ such that $\mathcal{A}_{i\kappa}$ is invertible. Since the embedding from H to $L^2(B_R\setminus\overline{D})\times L^2(B_R\setminus\overline{D})$ is compact and $T_k-T_{i\kappa}$ is a compact operator from $H^{\frac{1}{2}}(\partial B_R)$ to $H^{-\frac{1}{2}}(\partial B_R)$ [**3**], we can conclude that $\mathcal{A}_k-\mathcal{A}_{i\kappa}$ is a compact, and hence the result follows from the Fredholm alternative. □

We can now prove the following discreteness result for exterior transmission eigenvalues.

THEOREM 3.5. *Assume that A and n satisfies the assumptions of Lemma* 3.1. *Then the set of exterior transmission eigenvalues is discrete.*

PROOF. Since T_k depends analytically on $k \in \mathbb{C}$, $\Re(k) > 0$, we have the mapping $\mathcal{A}_k - \mathcal{A}_{i\kappa} : k \to \mathcal{L}(H)$ is analytic. We can choose κ such that $\mathcal{A}_{i\kappa}$ is invertible. The theorem follows from the analytic Fredholm theory [**8**]. □

4. Uniqueness of the inverse problem

In this section we prove that the boundary of the cavity is uniquely determined from a knowledge of the scattered field $u^s(x, y)$ for all $x, y \in C$ where C is the measurement manifold introduced in Section 2. It is not necessary to know the physical properties of the inhomogeneous exterior medium as long as they satisfy appropriate a priori assumptions. The proof of uniqueness for the inverse penetrable cavity is more complicated than for the case of scattering by an impenetrable cavity considered in [**22**]. The idea of the uniqueness proof for the inverse medium scattering problem originates from [**14**], [**15**]. Here we make use of the exterior transmission problem inspired by the idea in [**11**]. Since we are using some regularity results, in this section we assume more regularity of the boundary ∂D and material properties A and n than in previous sections.

Let C be the smooth closed $d-1$ manifold of measurement satisfying Assumption 2.1 and let us define the admissible set of cavities

$$\mathbb{S} := \{D \subset \mathbb{R}^d : \partial D \text{ is of class } C^1,\ D \text{ contains } C \text{ in its interior.}\}$$

Furthermore, we assume that the media outside the cavity has the material properties (A, n) which belong to

$$\mathcal{A} := \left\{ \begin{array}{l} A, n \in C^1(\Omega_{\partial D} \setminus \overline{D}) \cap L^\infty(\mathbb{R}^d \setminus \overline{D}),\ \Omega_{\partial D} \text{ is a neighborhood of } \partial D \\ \text{and } A, n \text{ satisfy the assumptions in Section 2 and in Theorem 3.4.} \end{array} \right\}$$

We begin with a simple lemma.

LEMMA 4.1. *Assume that $A, n \in \mathcal{A}$. Let $\{v_n, w_n\} \in H^1(\mathbb{R}^d \setminus \overline{D}) \times H^1(\mathbb{R}^d \setminus \overline{D})$, $n \in \mathbb{N}$, be a sequence of solutions to the exterior transmission problem* (3.6)*-*(3.10) *with boundary data $f_n \in H^{\frac{1}{2}}(\partial D)$, $h_n \in H^{-\frac{1}{2}}(\partial D)$. If the sequences $\{f_n\}$ and $\{h_n\}$ converge in $H^{\frac{1}{2}}(\partial D)$ and $H^{-\frac{1}{2}}(\partial D)$ respectively, and if the sequences $\{v_n\}$ and $\{w_n\}$ are bounded in $H^1(B_R \setminus \overline{D})$, then there exists a subsequence $\{v_{n_k}\}$ which converges in $H^1(B_R \setminus \overline{D})$.*

PROOF. Let $\{v_n, w_n\}$ be as in the statement of the lemma. Due to the compact imbedding of $H^1(B_R \setminus \overline{D})$ into $L^2(B_R \setminus \overline{D})$ we can select L^2-convergent subsequences $\{v_{n_k}\}$ and $\{w_{n_k}\}$. Hence, $\{v_{n_k}\}$ and $\{w_{n_k}\}$ satisfy

$$\nabla \cdot A \nabla w_{n_k} - \kappa^2 n w_{n_k} = -(\kappa^2 + k^2) w_{n_k} \quad \text{in } B_R \setminus \overline{D} \tag{4.1}$$

$$\Delta v_{n_k} - \kappa^2 v_{n_k} = -(\kappa^2 + k^2) v_{n_k} \quad \text{in } B_R \setminus \overline{D} \tag{4.2}$$

$$w_{n_k} - v_{n_k} = f_{n_k} \quad \text{on } \partial D \tag{4.3}$$

$$\frac{\partial w_{n_k}}{\partial \nu_A} - \frac{\partial v_{n_k}}{\partial \nu} = h_{n_k} \quad \text{on } \partial D \tag{4.4}$$

$$\frac{\partial w_{n_k}}{\partial \nu} - T_{i\kappa} w_{n_k} = (T_k - T_{i\kappa}) w_{n_k} \quad \text{on } \partial B_R \tag{4.5}$$

$$\frac{\partial v_{n_k}}{\partial \nu} - T_{i\kappa} v_{n_k} = (T_k - T_{i\kappa}) v_{n_k} \quad \text{on } \partial B_R \tag{4.6}$$

for $\kappa > 0$ chosen as in Lemma 3.1. Note that the left hand side of (4.1)-(4.6) in the variational setting is equivalent to the bounded invertible map $\mathcal{A}_{i\kappa}$. Thus v_{n_k} and

w_{n_k} are bounded by the right hand side with respect to the appropriate norm. Now, due to compactly embedding of H^1 into L^2, there is a subsequence of the right hand sides of (4.1) and (4.2) that converge in L^2. Since $T_k - T_{i\kappa}$ is a compact operator there is a subsequence of the right hand side of (4.5) and (4.6) that converge in $H^{-\frac{1}{2}}(\partial B_R)$. Hence the result follows from the boundeness of $\mathcal{A}_{ik}$. □

Note that Lemma 4.1 allows us to prove the uniqueness result without assuming that k is not an exterior transmission eigenvalue.

THEOREM 4.1. *Assume that $D_1, D_2 \in \mathbb{S}$ are two penetrable cavities having material properties $A_1, n_1 \in \mathcal{A}$ and $A_2, n_2 \in \mathcal{A}$ in the exterior of D_1 and D_2, respectively, such that the corresponding scattered fields coincide on C for all point sources located in C and any fixed wave number k. Then $D_1 = D_2$.*

PROOF. We denote by G the connected component of $D_1 \cap D_2$ which contains the region bounded by C. Let $u_j^s(\cdot, z)$ be the solution of (2.4)-(2.8) corresponding to D_j, A_j, n_j, $j = 1, 2$. We have that $u_1^s(x, z) = u_2^s(x, z)$ for $x, z \in C$. Following the argument in [**21**], the latter implies that $u_1^s(x, z) = u_2^s(x, z)$ for $x, z \in \overline{G}$. Next, assume that $\bar{D}_1$ is not included in $\bar{D}_2$. We can find a point $z \in \partial D_1$ and $\epsilon > 0$ with the following properties, where $\Omega_\delta(z)$ denotes the ball of radius δ centered at z:

(1) $\Omega_{8\epsilon}(z) \cap \bar{D}_2 = \emptyset$.
(2) The intersection $\bar{D}_1 \cap \Omega_{8\epsilon}(z)$ is contained in the connected component of $\bar{D}_1$ to which z belongs.
(3) There are points from this connected component of $\bar{D}_1$ to which z belongs which are not contained in $\bar{D}_1 \cap \bar{\Omega}_{8\epsilon}(z)$.
(4) The points $z_n := z + \dfrac{\epsilon}{n}\nu(z)$ lie in G for all $n \in \mathbb{N}$, where $\nu(z)$ is the innerward unit normal to ∂D_1 at z.

Due to the singular behavior of $\Phi(\cdot,\, z_n)$ at the point z_n, it is easy to show that

$$\|\Phi(\cdot,\, z_n)\|_{H^1(B_R \setminus \overline{D_1})} \to \infty \quad \text{as} \quad n \to \infty$$

where B_R is a large ball of radius R containing D_1 and D_2. We now define

$$v_n(x) := \frac{1}{\|\Phi(\cdot,\, z_n)\|_{H^1(B_R \setminus \overline{D_1})}} \Phi(x,\, z_n), \qquad x \in \mathbb{R}^d \setminus \overline{G}$$

and let $w_{1,n}, u_{1,n}^s$ and $w_{2,n}, u_{2,n}^s$ be the solutions of the scattering problem (2.4)-(2.8) with boundary data $f := v_n$ and $h := \partial v_n / \partial \nu$ corresponding to D_1 and D_2, respectively. Note that for each n, v_n is a radiating solution of the Helmholtz equation outside D_1 and D_2. Our aim is to prove that if $\bar{D}_1 \not\subset \bar{D}_2$ then the equality $u_1(\cdot, z) = u_2(\cdot, z)$ for $z \in G$ allows the selection of a subsequence $\{v_{n_k}\}$ from $\{v_n\}$ that converges to zero with respect to $H^1(B_R \setminus \overline{D_1})$. This certainly contradicts the definition of $\{v^n\}$ as a sequence of functions with $H^1(B_R \setminus \overline{D_1})$-norm equal to one. Note that as mentioned above we have $u_{1,n}^s = u_{2,n}^s$ in G.

We begin by noting that, since the functions $\Phi(\cdot,\, z_n)$ together with their derivatives are uniformly bounded in every compact subset of $\mathbb{R}^2 \setminus \Omega_{2\epsilon}(z)$ and $\|\Phi(\cdot,\, z_n)\|_{H^1(B_R \setminus \overline{D_1})} \to \infty$ as $n \to \infty$, then $\|v_n\|_{H^1(B_R \setminus \overline{D_2})} \to 0$ as $n \to \infty$. Hence, $\|u_{2,n}^s\|_{H^1(D_2)} \to 0$ as $n \to \infty$ from the well-posedness of the forward scattering problem. Since $u_{1,n}^s = u_{2,n}^s$ in G then $\|u_{1,n}\|_{H^1(G)} \to 0$ as $n \to \infty$ as well. Now, with the help of a cutoff function $\chi \in C_0^\infty(\Omega_{8\epsilon}(z))$ satisfying $\chi(x) = 1$ in $\Omega_{7\epsilon}(z)$, we see that $\|u_{1,n}\|_{H^1(G)} \to 0$ implies that

$$(4.7) \qquad (\chi u_{1,n}) \to 0, \quad \frac{\partial(\chi u_{1,n})}{\partial\nu} \to 0, \qquad \text{as } n \to \infty$$

with respect to the $H^{\frac{1}{2}}(\partial D_1)$-norm and $H^{-\frac{1}{2}}(\partial D_1)$-norm, respectively. Indeed, for the first convergence we simply apply the trace theorem while for the convergence of $\partial(\chi u_{1,n})/\partial\nu$, we first deduce the convergence of $\Delta(\chi u_{1,n})$ in $L^2(D_1)$, which follows from $\Delta(\chi u_{1,n}) = \chi\Delta u_{1,n} + 2\nabla\chi\cdot\nabla u_{1,n} + u_{1,n}\Delta\chi$, and then apply Theorem 5.5 in [**3**]. Note here that we need conditions 2 and 4 on z to ensure $\Omega_{8\epsilon}(z)\cap D_1^e = \Omega_{8\epsilon}(z)\cap G$.

We next note that in the exterior of $\Omega_{2\epsilon}(z)$ the $H^2(\Omega_R\setminus\Omega_{2\epsilon}(z))$-norms of v_n remain uniformly bounded. Then using the interior elliptic regularity and localization techniques as in Theorem 8.8 in [**10**] we can conclude that $u_{1,n}^s$ is uniformly bounded with respect to the $H^2((\Omega_{\partial D}\cap D_1)\setminus\Omega_{4\epsilon}(z))$-norm, where $\Omega_{\partial D}$ is an open neighborhood of ∂D. Therefore, using the compact imbedding of H^2 into H^1, we can select a $H^1(\Omega_{\partial D}\cap D_1)$ convergent subsequence $\{(1-\chi)u_{1,n_k}^s\}$ from $\{(1-\chi)u_{1,n}^s\}$. Hence, $\{(1-\chi)u_{1,n_k}^s\}$ is a convergent sequence in $H^{\frac{1}{2}}(\partial D_1)$ and similarly to the above reasoning we also have that $\{\partial((1-\chi)u_{1,n_k}^s)/\partial\nu\}$ converges in $H^{-\frac{1}{2}}(\partial D_1)$. This, together with (4.7), implies that the sequences

$$\{u_{1,n_k}^s\} \qquad \text{and} \qquad \left\{\frac{\partial u_{1,n_k}^s}{\partial\nu}\right\}$$

converge in $H^{\frac{1}{2}}(\partial D_1)$ and $H^{-\frac{1}{2}}(\partial D_1)$, respectively.

Finally, the functions v_{n_k} and w_{1,n_k} are solutions to the exterior transmission problem (3.6)-(3.10) for the domain $\mathbb{R}^d\setminus\overline{D_1}$ with boundary data $f = u_{1,n_k}^s$ and $h = \partial u_{1,n_k}^s/\partial\nu$. Since, the $H^1(B_r\setminus\overline{D_1})$-norms of v_{n_k} and w_{1,n_k} remain uniformly bounded, from Lemma 4.1 we can select a subsequence of $\{v_{n_k}\}$, denoted again by $\{v_{n_k}\}$, which converges in $H^1(B_r\setminus\overline{D_1})$ to some v. As H^1-limit of weak solutions to the Helmholtz equation, v is a distributional solution to the Helmholtz equation. We also have that $v|_{B_R\setminus(D_1\cup\Omega_{2\epsilon}(z))} = 0$ because the functions v_{n_k} converge uniformly to zero in the exterior of $\Omega_{2\epsilon}(z)$. Hence, v must be zero in all of $B_R\setminus\overline{D_1}$ (here we make use of condition 3). This contradicts the fact that $\|v_{n_k}\|_{H^1(B_R\setminus\overline{D_1})} = 1$. Hence the assumption $\bar{D}_1 \not\subset \bar{D}_2$ is false.
Since we can derive the analogous contradiction for the assumption $\bar{D}_2 \not\subset \bar{D}_1$, we have proved that $D_1 = D_2$. □

REMARK 4.2. *The assumptions of Theorem* 3.4 *required for* A *and* n *can be replaced by any other assumptions that guaranty the well-posedness of the exterior transmission problem. Also the assumption that* ∂D *is smooth can be relaxed as long as it guaranties* $H^{1+\epsilon}$*-regularity near the boundary of the solution of the corresponding transmission problem (e.g. piecewise smooth* [**9**]*).*

5. The solution of the inverse problem

Now we turn our attention to reconstructing the boundary of the cavity D from a knowledge of the scattered field $u^s(x,y)$ for $x\in C$ corresponding to all point sources for $y\in C$. We will develop the linear sampling method which allows us to reconstruct D without any a priori knowledge about the physical properties of the media outside D, i.e. of A and n. The basic assumptions are Assumption

2.1, the assumptions of Theorem 3.4 and that k is not an exterior transmission eigenvalue.

Our data set defines the data operator N: $L^2(C) \to L^2(C)$ by

$$(Ng)(x) = \int_C u^s(x,y)g(y)ds(y) \quad g \in L^2(C),\ x \in C \tag{5.1}$$

which is obviously compact since it is an integral operator with analytic kernel. If we define the single layer potential v_g by

$$v_g(x) := \int_C \Phi(x,y)g(y)ds(y), \quad x \in \mathbb{R}^2 \setminus C, \tag{5.2}$$

then by linearity Ng is the scattered field evaluated on C due to v_g as incident field.

THEOREM 5.1. *$N : L^2(C) \to L^2(C)$ is injective with dense range if and only if there does not exist a non-zero $g \in L^2(C)$ such that $w \in H^1_{loc}(\mathbb{R}^d \backslash \overline{D})$ and $v := v_g$ solve the homogeneous exterior transmission problem, i.e. (3.6)-(3.10) with $f = 0$ and $h = 0$.*

PROOF. In a similar way as in Theorem 2.1 in [**21**], we can prove that the scattered field u^s satisfies the reciprocity condition $u^s(x,y) = u^s(y,x)$ for $x, y \in C$. Indeed

$$u^s(x,y) = \int\limits_{\partial D} \left\{ \frac{\partial u^s(\cdot,y)}{\partial \nu}\Phi(x,\cdot) - u^s(\cdot,y)\frac{\partial \Phi(x,\cdot)}{\partial \nu} \right\} ds, \qquad x \in C \tag{5.3}$$

and

$$u^s(y,x) = \int\limits_{\partial D} \left\{ \frac{\partial u^s(\cdot,x)}{\partial \nu}\Phi(y,\cdot) - u^s(\cdot,x)\frac{\partial \Phi(y,\cdot)}{\partial \nu} \right\} ds, \qquad y \in C. \tag{5.4}$$

Applying Green's second identity we have that

$$0 = \int\limits_{\partial D} \left\{ \frac{\partial u^s(\cdot,y)}{\partial \nu}u^s(\cdot,x) - u^s(\cdot,y)\frac{\partial u^s(\cdot,x)}{\partial \nu} \right\} ds \tag{5.5}$$

and since $\Phi(\cdot,\cdot)$ satisfies the radiation condition

$$0 = \int\limits_{\partial D} \left\{ \frac{\partial \Phi(x,\cdot)}{\partial \nu}\Phi(y,\cdot) - \Phi(x,\cdot)\frac{\partial \Phi(y,\cdot)}{\partial \nu} \right\} ds \tag{5.6}$$

Since $\Phi(\cdot,\cdot)$ is symmetric, subtracting (5.4) from (5.3) and adding to the result the sum of (5.5) and (5.6) we obtain

$$u^s(y,x) - u^s(x,y) = \int\limits_{\partial D} \left\{ \frac{\partial u(\cdot,y)}{\partial \nu}u(\cdot,x) - u(\cdot,y)\frac{\partial u(\cdot,x)}{\partial \nu} \right\} ds$$

where u is the total field. Now using the transmission conditions (2.3), the fact that A is symmetric, the assumptions that $A - I$ and $n - 1$ are zero in $\mathbb{R}^d \setminus B_R$ and the equation (2.2) we have that

(5.7)

$$
\begin{aligned}
u^s(y,x) - u^s(x,y) &= \int\limits_{\partial D} \left\{ \frac{\partial w(\cdot,y)}{\partial \nu_A} w(\cdot,x) - w(\cdot,y)\frac{\partial w(\cdot,x)}{\partial \nu_A} \right\} ds \\
&= - \int\limits_{B_R\setminus\overline{D}} \{A\nabla w(\cdot,y)\cdot\nabla w(\cdot,x) - A\nabla w(\cdot,x)\cdot\nabla w(\cdot,y)\}\, dv \\
&- \int\limits_{B_R\setminus\overline{D}} \{\nabla\cdot A\nabla w(\cdot,y) w(\cdot,x) - \nabla A\nabla w(\cdot,x) w(\cdot,y)\}\, dv \\
&+ \int\limits_{\partial B_R} \left\{ \frac{\partial w(\cdot,y)}{\partial \nu} w(\cdot,x) - w(\cdot,y)\frac{\partial w(\cdot,x)}{\partial \nu} \right\} ds = 0,
\end{aligned}
$$

since the first volume integral is zero due to the symmetry of A, the second volume integral is zero due the fact that $w(\cdot,x)$ and $w(\cdot,y)$ satisfy the same equation and the last integral is zero due to the fact that $w(\cdot,x)$ and $w(\cdot,y)$ are radiating solutions to the Helmholtz equation outside B_R.

The symmetry property of u^s implies that $N^*h = \overline{N\overline{h}}$, where N^* is the L^2-adjoint of N. Hence N is injective if and only if N^* is injective, Since $\mathrm{Ker}(N^*)^\perp = \overline{\mathrm{Range}(N)}$ to prove the theorem we must only prove that N is injective. To this end, let a non-zero $g \in L^2(C)$ be such that $(Ng)(x) = 0, x \in C$. Let $v_g(x) = \int_C \phi(x,z)g(z)ds(z)$, and consider $(\tilde{w},\tilde{v})$ the unique solution of (3.6)-(3.10) with $f := v_g$ and $g := \frac{\partial v_g}{\partial \nu}$. By superposition $\tilde{v}(x) = (Ng)(x)$, which means that $(Ng)(x) = 0, x \in C$, is equivalent to the fact that $\tilde{v}(x) = 0, x \in C$. Furthermore we have $\Delta\tilde{v} + k^2\tilde{v} = 0$ in the domain bounded by C and since k satisfies Assumption 2.1 we have $\tilde{v} = 0$ inside C. But $\Delta\tilde{v} + k^2\tilde{v} = 0$ in D and hence by analyticity $\tilde{v} = 0$ in D. The latter implies that $\tilde{w}$ and v_g satisfy the homogeneous exterior transmission problem. This proves the theorem. □

The above theorem implies:

COROLLARY 5.1. *If k is not an exterior transmission eigenvalue then the operator $N : L^2(C) \to L^2(C)$ is injective with dense range*

For the rest of the paper we need to assume that k is not an exterior transmission eigenvalue in addition to Assumption 2.1.

We now introduce the data equation

$$(Ng)(x) = \Phi(x,z) \quad \forall \quad x \in C \quad \textit{and} \quad x \neq z \tag{5.8}$$

where z is a sampling point in $\mathbb{R}^d$. This is an ill-posed linear equation whose regularized solution will be the indicator function of the cavity. To this end we investigate the solvability of (5.8).

We first define $\overline{\mathcal{U}}$ to be the closure of the set

$$\mathcal{U} := \left\{ \int_C \phi(\cdot,z)g(y)ds(y),\ g \in L^2(C) \right\} \text{ with respect to } H^1_{loc}(\mathbb{R}^d\setminus\overline{D}).$$

LEMMA 5.1. *Let*

$$\mathcal{U}_0 = \left\{ u \in H^1_{loc}(\mathbb{R}^d\setminus\overline{D}) : \Delta u + k^2 u = 0 \text{ in } \mathbb{R}^d\setminus\overline{D},\ \lim_{r\to\infty} r^{\frac{d-1}{2}}\left(\frac{\partial v}{\partial r} - ikv\right) = 0 \right\}.$$

Then $\overline{\mathcal{U}} = \mathcal{U}_0$

PROOF. By the well-posedness of the problem

$$\Delta u + k^2 u = 0 \quad in \quad \mathbb{R}^d \backslash \overline{D}$$
$$u = g \quad on \quad \partial D$$
$$\lim_{r\to\infty} r^{\frac{d-1}{2}} (\frac{\partial v}{\partial r} - ikv) = 0$$

we have $||u||_{H^1_{loc}(\mathbb{R}^d\backslash\overline{D})} \le c||g||_{H^{\frac{1}{2}}(\partial D)}$, where c is some constant. Then $\mathcal{U}$ is dense in $\mathcal{U}_0$ if we can show that $\{\int_C \phi(\cdot,z)g(z)ds(z)|_{\partial D}, g \in L^2(C)\}$ is dense in $H^{\frac{1}{2}}(\partial D)$. In fact, let $f \in H^{-\frac{1}{2}}(\partial D)$ be such that for any $g \in L^2(C)$

$$\int_{\partial D}\int_C \phi(x,y)g(y)ds(y)f(x)ds(x) = 0,$$

that is

$$\int_C\int_{\partial D} \phi(x,y)f(x)ds(x)g(y)ds(y) = 0.$$

Then

$$\int_{\partial D} \phi(x,y)f(x)ds(x) = 0, \qquad \forall y \in C. \tag{5.9}$$

Then the single layer potential $v_f(x) = \int_{\partial D}\phi(x,y)f(y)ds(y)$ satisfies $v_f|_C = 0$ and the Helmholtz equation in the bounded domain circumscribed by C. Since k satisfies Assumption 2.1, we have that $v_f = 0$ inside C and by analyticity we have that $v_f = 0$ in D. From the jump conditions across ∂D

$$v_f^- = v_f^+ \qquad on \quad \partial D$$
$$f = \frac{\partial v_f^-}{\partial \nu} - \frac{\partial v_f^+}{\partial \nu} \qquad on \quad \partial D$$

where + and − denote approaching the boundary from outside and inside ∂D, respectively, we now have that $v_f^+ = 0$. Since v_f is a radiating solution to the Helmholtz equation, from uniqueness of the exterior Dirichlet problem we have $v_f = 0$ in $\mathbb{R}^d\backslash\overline{D}$ and hence we have $f = \frac{\partial v_f^-}{\partial \nu} - \frac{\partial v_f^+}{\partial \nu} = 0$. Thus the set $\{v_g,\, g \in L^2(C)\}$, v_g defined by (5.2), is dense in $H^{\frac{1}{2}}(\partial D)$. Finally, note that since $\mathcal{U}_0$ is closed in $H^1_{loc}(\mathbb{R}^d\backslash\overline{D})$ and $\{v_g, g \in L^2(C)\}$ is dense in $\mathcal{U}_0$ we have that $\overline{\mathcal{U}} = \mathcal{U}_0$. □

Now we define $\mathcal{U}(\partial D) := \{(u|_{\partial D}, \frac{\partial u}{\partial \nu}|_{\partial D}), u \in \overline{\mathcal{U}}\}$.

LEMMA 5.2. *$\mathcal{U}(\partial D)$ is closed in $H^{\frac{1}{2}}(\partial D) \times H^{-\frac{1}{2}}(\partial D)$ and hence is a Hilbert space.*

PROOF. Let $(f,h) \in H^{\frac{1}{2}}(\partial D) \times H^{-\frac{1}{2}}(\partial D)$. If $(f,h) \in \overline{U}(\partial D)$ then there exists a sequence $\{u_n\}_{n=1}^\infty$ in $\mathcal{U}$ such that

$$\left(u_n|_{\partial D}, \frac{\partial u_n}{\partial \nu}|_{\partial D}\right) \to (f,h) \quad as \quad n \to \infty.$$

Clearly, $(u_n|_{\partial D}, \frac{\partial u_n}{\partial \nu}|_{\partial D})$ is bounded in $H^{\frac{1}{2}}(\partial D) \times H^{-\frac{1}{2}}(\partial D)$ and u_n satisfies

$$\Delta u_n + k^2 u_n = 0 \quad in \quad \mathbb{R}^d\backslash\overline{D}$$
$$u_n = u_n|_{\partial D} \quad on \quad \partial D$$

$$\lim_{r\to\infty} r^{\frac{d-1}{2}}(\frac{\partial u_n}{\partial r} - iku_n) = 0.$$

Thus, from the well-posedness of the exterior Dirichlet problem $||u_n||_{H^1_{loc}(\mathbb{R}^d\setminus\overline{D})}$ is bounded by $||u_n||_{H^{\frac{1}{2}}(\partial D)}$ and therefore $\{u_n\}$ is bounded in $H^1_{loc}(\mathbb{R}^d\setminus\overline{D})$. Hence there exists $u \in \overline{\mathcal{U}}$ such that u_n converges to u weakly. Since the trace operator $H^1_{loc}(\mathbb{R}^d\setminus\overline{D}) \to H^{\frac{1}{2}}(\partial D)$ and $H^1_{loc,\Delta}(\mathbb{R}^d\setminus\overline{D}) \to H^{-\frac{1}{2}}(\partial D)$ are bounded [**3**], we obtain

$$(u_n|_{\partial D}, \frac{\partial u_n}{\partial \nu}|_{\partial D}) \text{ converges to } (u|_{\partial D}, \frac{\partial u}{\partial \nu}|_{\partial D}) \text{ weakly.}$$

Hence $f = u|_{\partial D}, h = \frac{\partial u}{\partial \nu}|_{\partial D}$, which implies that $\mathcal{U}(\partial D)$ is closed in $H^{\frac{1}{2}}(\partial D) \times H^{-\frac{1}{2}}(\partial D)$. □

DEFINITION 5.2. *The operator* $B : \mathcal{U}(\partial D) \to L^2(C)$ *maps* $\left(v|_{\partial D}, \frac{\partial v}{\partial \nu}|_{\partial D}\right)$, $v \in \overline{\mathcal{U}}$, *to* $u^s_v|_C$ *where* (u^s_v, w_v) *is the unique solution of (2.4)-(2.8) with* $f := v|_{\partial D}$ *and* $h := \frac{\partial v}{\partial \nu}|_{\partial D}$.

THEOREM 5.3. *Assume k is not an exterior transmission eigenvalue. Then* $B : \mathcal{U}(\partial D) \to L^2(C)$ *is compact, injective and has dense range in* $L^2(C)$.

PROOF. The solution $u^s_v \in H^1(D)$ depends continuously on $\left(v|_{\partial D}, \frac{\partial v}{\partial \nu}|_{\partial D}\right)$. Since $u^s_v|_C \in H^{\frac{1}{2}}(C)$ and the imbedding $H^{\frac{1}{2}}(C) \to L^2(C)$ is compact, we have B is compact.

Next, if $B(v|_{\partial D}, \frac{\partial v}{\partial \nu}|_{\partial D}) = 0$, we have $u^s_v|_C = 0$. But in addition we have $\Delta u^s_v + k^2 u^s_v = 0$ inside C and since k satisfies Assumption 2.1 we have $u^s_v = 0$ inside C, and by the unique continuation principle $u^s_v = 0$ in D. Then w_v and v satisfy

$$\nabla \cdot A\nabla w_v + k^2 n w_v = 0 \qquad in \quad \mathbb{R}^d\setminus\overline{D}$$

$$\Delta v + k^2 v = 0 \qquad in \quad \mathbb{R}^d\setminus\overline{D}$$

$$w_v = v \qquad on \quad \partial D$$

$$\frac{\partial w_v}{\partial \nu_A} = \frac{\partial v}{\partial \nu} \qquad on \quad \partial D$$

$$\lim_{r\to+\infty} r^{\frac{d-1}{2}}(\frac{\partial w_v}{\partial r} - ikw_v) = \lim_{r\to+\infty} r^{\frac{d-1}{2}}(\frac{\partial v}{\partial r} - ikv) = 0.$$

Since k is not an exterior transmission eigenvalue, we have $v = 0$ in $\mathbb{R}^d\setminus\overline{D}$ and thus $(v|_{\partial D}, \frac{\partial v}{\partial \nu}|_{\partial D}) = 0$. Hence B is injective.

Finally, since Range$(N) \subset$ Range(B), from Corollary 5.1 we can conclude that the range of B is also dense in $L^2(C)$. □

THEOREM 5.4. *Assume that k is not an exterior transmission eigenvalue. Then* $\Phi(\cdot, z)$ *is in the range of B if and only if* $z \in \mathbb{R}^d\setminus\overline{D}$.

PROOF. If $z \in \mathbb{R}^d\setminus\overline{D}$ and k is not an exterior transmission eigenvalue then from Theorem 3.4, we have that the exterior transmission problem

$$\nabla \cdot A\nabla w_z + k^2 n w_z = 0 \qquad in \quad \mathbb{R}^d\setminus\overline{D} \tag{5.10}$$

$$\Delta v_z + k^2 v_z = 0 \qquad in \quad \mathbb{R}^d\setminus\overline{D} \tag{5.11}$$

$$w_z - v_z = \Phi(\cdot, z) \qquad on \quad \partial D \tag{5.12}$$

$$\frac{\partial w_z}{\partial \nu} - \frac{\partial v_z}{\partial \nu} = \frac{\partial \Phi(\cdot, z)}{\partial \nu} \qquad on \quad \partial D \tag{5.13}$$

$$\lim_{r\to+\infty} r^{\frac{d-1}{2}}\left(\frac{\partial v_z}{\partial r}-ikv_z\right)=\lim_{r\to+\infty} r^{\frac{d-1}{2}}\left(\frac{\partial w_z}{\partial r}-ikw_z\right)=0 \tag{5.14}$$

has a unique solution $(w_z,v_z)\in H^1_{loc}(\mathbb{R}^d\backslash\overline{D})\times H^1_{loc}(\mathbb{R}^d\backslash\overline{D})$. Then $(w_z,\Phi(\cdot,z))$ satisfies (2.4)-(2.8) with $(f,h)=(v_z,\frac{\partial v_z}{\partial\nu})|_{\partial D}$. Since $v_z\in\overline{\mathcal{U}}$, we have $B(v_z,\frac{\partial v_z}{\partial\nu})=\Phi(\cdot,z)|_C$, which means that $\Phi(x,z)$ for $x\in C$ is in the range of B.

Now assume that, for $z\in D$, $\Phi(\cdot,z)$ is in the range of B. Then there exists $v\in\overline{\mathcal{U}}$ such that

$$B(v|_{\partial D},\frac{\partial v}{\partial\nu}|_{\partial D})=\Phi(x,z),\quad x\in C.$$

Let w_v,u^s_v be the solution to (2.4)-(2.8) with $(f,h)=(v|_{\partial D},\frac{\partial v}{\partial\nu}|_{\partial D})$. By definition of B, $u^s_v=\Phi(\cdot,z)$ in D but this is not possible since $\Phi(\cdot,z)\notin H^1(D_0)$. □

As the last ingredient to the main theorem of this section we define the bounded linear operator $S:L^2(C)\to\mathcal{U}(\partial D)$ by

$$(Sg)(x)=\left(v_g|_{\partial D},\frac{\partial v_g}{\partial\nu}|_{\partial D}\right),\qquad\text{where } v_g \text{ is defined by (5.2).}$$

Obviously we have that the data operator N can be factorized as

$$Ng=BSg.$$

We can prove the following denseness result for the operator S.

THEOREM 5.5. *The bounded linear operator $S:L^2(C)\to\mathcal{U}(\partial D)$ is injective with dense range.*

PROOF. If g is such that $Sg=0$ then $v_g(x)=\int_C\phi(x,y)g(y)ds(y)$ satisfies

$$\Delta v_g+k^2v_g=0\qquad in\quad \mathbb{R}^d\backslash\overline{D}$$

$$v_g=0\quad on\quad \partial D$$

$$\lim_{r\to+\infty} r^{\frac{d-1}{2}}\left(\frac{\partial v_g}{\partial r}-ikv_g\right)=0.$$

Then $v_g=0$ in $\mathbb{R}^d\backslash\overline{D}$, and since $\Delta v_g+k^2v_g=0$ in $\mathbb{R}^d\backslash C$, by the unique continuation principle $v_g=0$ outside C. In particular the single layer boundary integral operator

$$\int_C\phi(x,y)g(y)ds(y),\qquad g\in L^2(C),\qquad x\in C$$

is invertible as long as k^2 is not Dirichlet eigenvalue for $-\Delta$ inside C [**19**]. Hence $g=0$.

Next, since $\{v_g,\quad g\in L^2(C)\}$ is dense in $\overline{\mathcal{U}}$ by definition, we have that S has dense range in $\mathcal{U}(\partial D)$. □

Now we are ready to prove the main theorem of this section which provides the basis for the linear sampling method.

THEOREM 5.6. *Assume that k is not an exterior transmission eigenvalue eigenvalue and satisfies Assumption 2.1. Let u^s be the scattered field corresponding to the scattering problem (2.4)-(2.8) and N is the associated data operator. Then the following hold:*

(1) *For $z \in \mathbb{R}^d \setminus \bar{D}$ and a given $\epsilon > 0$ there exists a function $g_z^\epsilon \in L^2(C)$ such that*

$$\|Ng_z^\epsilon - \Phi(\cdot, z)\|_{L^2(C)} < \epsilon,$$

and as $\epsilon \to 0$, the potential $v_{g_z^\epsilon}$ given by (5.2) with kernel g_z^ϵ converges to the solution v_z in the $H^1(B_R \setminus \overline{D})$-norm where (w_z, v_z) is the solution of (5.10)-(5.14).

(2) *For $z \in D \setminus C$ and a given $\epsilon > 0$, every $g_z^\epsilon \in L^2(C)$ that satisfies*

$$\|Ng_z^\epsilon - \Phi(\cdot, z)\|_{L^2(C)} < \epsilon,$$

is such that

$$\lim_{\epsilon \to 0} \|v_{g_z^\epsilon}\|_{H^1(B_R \setminus \overline{D})} = \infty.$$

PROOF. (1) Let $z \in \mathbb{R}^d \setminus \overline{D}$. Then from Theorem 5.4, $\Phi(\cdot, z)$ is in the range of B and

$$B(v_z|_{\partial D_0}, \frac{\partial v_z}{\partial \nu}|_{\partial D}) = \Phi(\cdot, z),$$

where (w_z, v_z) is the solution of (5.10)-(5.14). Now, for $\epsilon > 0$, since S has dense range in $\mathcal{U}(\partial D)$ by Theorem 5.5, there exists $g_z^\epsilon \in L^2(C)$ satisfying

$$\left\| Sg_z^\epsilon - \left(v_z|_{\partial D}, \frac{\partial v_z}{\partial \nu}|_{\partial D} \right) \right\|_{H^{\frac{1}{2}}(\partial D) \times H^{-\frac{1}{2}}(\partial D)} < \frac{\epsilon}{\|B\|} \tag{5.15}$$

which yields

$$\left\| BSg_z^\epsilon - B(v_z|_{\partial D}, \frac{\partial v_z}{\partial \nu}|_{\partial D}) \right\|_{L^2(C)} < \epsilon.$$

The latter can be re-written as

$$\|Ng_z^\epsilon - \Phi(\cdot, z)\|_{L^2(C)} < \epsilon.$$

Furthermore,

$$\lim_{\epsilon \to 0} \left\| Sg_z^\epsilon - \left(v_z|_{\partial D}, \frac{\partial v_z}{\partial \nu}|_{\partial D} \right) \right\|_{H^{\frac{1}{2}}(\partial D) \times H^{-\frac{1}{2}}(\partial D)} = 0$$

and hence

$$\lim_{\epsilon \to 0} \|v_{g_z^\epsilon} - v_z\|_{H^1_{loc}(B_R \setminus \overline{D})} = 0.$$

Furthermore for a fixed $\epsilon > 0$, we observe that $u^s := \Phi(\cdot, z)$ and $w := w_z$ satisfy the scattering problem (2.4)-(2.8) with data $f := v_z|_{\partial D}$ and $h := \frac{\partial v_z}{\partial \nu}|_{\partial D}$. From the well-posedness of (2.4)-(2.8) and the fact that $\|\Phi(\cdot, z)\|_{H^1(D)}$ goes to infinity as $z \to \partial D$, we obtain that

$$\lim_{z \to \partial D} \left\| \left(v_z|_{\partial D}, \frac{\partial v_z}{\partial \nu}|_{\partial D} \right) \right\|_{H^{\frac{1}{2}}(\partial D) \times H^{-\frac{1}{2}}(\partial D)} = \infty$$

and hence

$$\lim_{z \to \partial D} \|Sg_z^\epsilon\|_{H^{\frac{1}{2}}(\partial D) \times H^{-\frac{1}{2}}(\partial D)} = \infty.$$

Since $\|Sg_z^\epsilon\|_{H^{\frac{1}{2}}(\partial D) \times H^{-\frac{1}{2}}(\partial D)}$ is bounded bellow by $\|v_{g_z^\epsilon}\|_{H^1(B_R \setminus \overline{D})}$, we can conclude that

$$\lim_{z \to \partial D} \|v_{g_z^\epsilon}\|_{H^1(B_R \setminus \overline{D})} = \infty \quad \text{and} \quad \lim_{z \to \partial D} \|g_z^\epsilon\|_{L^2(C)} = \infty.$$

(2). In order to prove the second statement, for $z \in D \setminus C$ assume to the contrary that there exists a sequence $\{\epsilon_n\} \to 0$ and corresponding functions v_{g_n} with kernels $g_n := g_z^{\epsilon_n}$ satisfying $\|Ng_n - \Phi(\cdot, z)\|_{L^2(C)} < \epsilon_n$ (i.e. $Ng_n \to \Phi(\cdot, z)$

in $L^2(C)$ as $n \to \infty$) such that $\|v_n\|_{H^1(B_R\setminus\overline{D})}$ remains bounded. Then without loss of generality we may assume weak convergence v_n to some $v \in H^1(B_R\setminus\overline{D})$. Let us define $\tau : v \to \left(v|_{\partial D}, \frac{\partial v}{\partial \nu}|_{\partial D}\right)$ which is obviously a bounded operator from $H^1(B_R\setminus\overline{D})$ to $H^{\frac{1}{2}}(\partial D) \times H^{-\frac{1}{2}}(\partial D)$. Since $B\tau$ is also bounded, we can conclude the weak convergence $(B\tau)v_{g_n} \rightharpoonup (B\tau)v$ in $L^2(C)$ as $n \to \infty$. But $(B\tau)v_{g_n} = Ng_n$ converges strongly to $\Phi(\cdot, z)|_C$ as $n \to \infty$, which means $\Phi(\cdot, z) = B(\tau v)$. This contradicts Theorem 5.4. □

This theorem can be used to reconstruct the boundary ∂D, since roughly it says that if g_z^ϵ is the approximate solution of $Ng_z^\epsilon = \Phi(\cdot, z)$ provided by Theorem 5.6 then $\|v_{g_z^\epsilon}\|_{H^1(B_R\setminus\overline{D})}$ is large z in D and small for z outside D, for fixed ϵ. Unfortunately, $\|v_{g_z^\epsilon}\|_{H^1(B_R\setminus\overline{D})}$ can not be used as indicator function for D since it depends on D. Instead in practice we use the indicator function $I(z) := \|g_z^\epsilon\|_{L^2(C)}$. Since the data equation (5.8) is ill-posed, it is necessary to use regularization techniques, e.g. Tikhonov regularization. The question if the Tikhonov regularized solution of (5.8) captures the approximate solution g_z^ϵ provided by Theorem 5.6 remains open.

The linear sampling method for the reconstruction of ∂D can now be described as follows.

- Choose a set of sampling points in a region covering the expected obstacle.
- For each sampling point z, solve the regularized version of the data equation,
$$\alpha g + N^*Ng = N^*\Phi(\cdot, z)$$
with a regularization parameter $\alpha > 0$.
- Calculate the indicator function $I(z)$.
- Plot $I(z)$. Then the cavity D is the region containing points z for which $I(z) > C$ for a cut-off value C chosen by ad-hoc procedure (some procedures for choosing C are available in the literature (see e.g. [**5**] and the references therein).

References

[1] AS. Bonnet-BenDhia, L. Chesnel, and H. Haddar. On the use of t-coercivity to study the interior transmission eigenvalue problem. *C. R. Acad. Sci., Ser. I*, 340, 2011.

[2] A. S. Bonnet-Ben Dhia, P. Ciarlet Jr., and C. M. Zwölf, *Time harmonic wave diffraction problems in materials with sign-shifting coefficients*, J. Comput. Appl. Math. **234** (2010), no. 6, 1912–1919, DOI 10.1016/j.cam.2009.08.041. MR2644187 (2011m:78009)

[3] Fioralba Cakoni and David Colton, *Qualitative methods in inverse scattering theory*, Interaction of Mechanics and Mathematics, Springer-Verlag, Berlin, 2006. An introduction. MR2256477 (2008c:35334)

[4] Fioralba Cakoni, David Colton, and Houssem Haddar, *The linear sampling method for anisotropic media*, J. Comput. Appl. Math. **146** (2002), no. 2, 285–299, DOI 10.1016/S0377-0427(02)00361-8. MR1925961 (2003i:35219)

[5] Fioralba Cakoni, David Colton, and Peter Monk, *The linear sampling method in inverse electromagnetic scattering*, CBMS-NSF Regional Conference Series in Applied Mathematics, vol. 80, Society for Industrial and Applied Mathematics (SIAM), Philadelphia, PA, 2011. MR2731610 (2012e:78019)

[6] F. Cakoni and H. Haddar, Transmission eigenvalues in inverse scattering theory, *Inside Out II*, G. Uhlmann editor, MSRI Publications, Volume 60, 527-578, (2012)

[7] L. Chesnel, Étude de quelques problémes de transmission avec changement de signe. Application aux métamatériaux, *Ph.D. Thesis*. École Doctorale de l'École Polytechnique, France, 2012.

[8] David Colton and Rainer Kress, *Inverse acoustic and electromagnetic scattering theory*, 3rd ed., Applied Mathematical Sciences, vol. 93, Springer, New York, 2013. MR2986407

[9] Monique Dauge, *Elliptic boundary value problems on corner domains*, Lecture Notes in Mathematics, vol. 1341, Springer-Verlag, Berlin, 1988. Smoothness and asymptotics of solutions. MR961439 (91a:35078)

[10] David Gilbarg and Neil S. Trudinger, *Elliptic partial differential equations of second order*, 2nd ed., Grundlehren der Mathematischen Wissenschaften [Fundamental Principles of Mathematical Sciences], vol. 224, Springer-Verlag, Berlin, 1983. MR737190 (86c:35035)

[11] Peter Hähner, *On the uniqueness of the shape of a penetrable, anisotropic obstacle*, J. Comput. Appl. Math. **116** (2000), no. 1, 167–180, DOI 10.1016/S0377-0427(99)00323-4. MR1741793 (2001b:35294)

[12] Masaru Ikehata, *An inverse acoustic scattering problem inside a cavity with dynamical back-scattering data*, Inverse Problems **28** (2012), no. 9, 095016, 24, DOI 10.1088/0266-5611/28/9/095016. MR2972465

[13] Masaru Ikehata and Hiromichi Itou, *On reconstruction of a cavity in a linearized viscoelastic body from infinitely many transient boundary data*, Inverse Problems **28** (2012), no. 12, 125003, 19, DOI 10.1088/0266-5611/28/12/125003. MR2997012

[14] Victor Isakov, *On uniqueness in the inverse transmission scattering problem*, Comm. Partial Differential Equations **15** (1990), no. 11, 1565–1587, DOI 10.1080/03605309908820737. MR1079603 (91i:35203)

[15] Victor Isakov, *Inverse problems for partial differential equations*, Applied Mathematical Sciences, vol. 127, Springer-Verlag, New York, 1998. MR1482521 (99b:35211)

[16] Peter Jakubik and Roland Potthast, *Testing the integrity of some cavity—the Cauchy problem and the range test*, Appl. Numer. Math. **58** (2008), no. 6, 899–914, DOI 10.1016/j.apnum.2007.04.007. MR2420625 (2009g:45024)

[17] Andreas Kirsch, *An introduction to the mathematical theory of inverse problems*, Applied Mathematical Sciences, vol. 120, Springer-Verlag, New York, 1996. MR1479408 (99c:34023)

[18] Andreas Kirsch and Natalia Grinberg, *The factorization method for inverse problems*, Oxford Lecture Series in Mathematics and its Applications, vol. 36, Oxford University Press, Oxford, 2008. MR2378253 (2009k:35322)

[19] William McLean, *Strongly elliptic systems and boundary integral equations*, Cambridge University Press, Cambridge, 2000. MR1742312 (2001a:35051)

[20] Hai-Hua Qin and Fioralba Cakoni, *Nonlinear integral equations for shape reconstruction in the inverse interior scattering problem*, Inverse Problems **27** (2011), no. 3, 035005, 17, DOI 10.1088/0266-5611/27/3/035005. MR2772524 (2012j:78004)

[21] Hai-Hua Qin and David Colton, *The inverse scattering problem for cavities with impedance boundary condition*, Adv. Comput. Math. **36** (2012), no. 2, 157–174, DOI 10.1007/s10444-011-9179-2. MR2886186

[22] Hai-Hua Qin and David Colton, *The inverse scattering problem for cavities*, Appl. Numer. Math. **62** (2012), no. 6, 699–708, DOI 10.1016/j.apnum.2010.10.011. MR2908789

[23] Fang Zeng, Fioralba Cakoni, and Jiguang Sun, *An inverse electromagnetic scattering problem for a cavity*, Inverse Problems **27** (2011), no. 12, 125002, 17, DOI 10.1088/0266-5611/27/12/125002. MR2861032

Department of Mathematical Sciences, University of Delaware, Newark, Delaware 19716

E-mail address: `cakoni@math.udel.edu`

Department of Mathematical Sciences, University of Delaware, Newark, Delaware 19716

E-mail address: `colton@math.udel.edu`

Department of Mathematical Sciences, University of Delaware, Newark, Delaware 19716

E-mail address: `sxmeng@math.udel.edu`

Contemporary Mathematics
Volume **615**, 2014
http://dx.doi.org/10.1090/conm/615/12262

A Neumann series based method for photoacoustic tomography on irregular domains

Eric Chung, Chi Yeung Lam, and Jianliang Qian

ABSTRACT. Recently, a Neumann series based numerical method is developed for photoacoustic tomography in a paper by Qian, Stefanov, Uhlmann, and Zhao [*An efficient neumann series-based algorithm for thermoacoustic and photoacoustic tomography with variable sound speed.* SIAM J. Imag. Sci., 4:850–883, 2011]. It is an efficient and convergent numerical scheme that recovers the initial condition of an acoustic wave equation with non-constant sound speeds by boundary measurements. In practical applications, the domains of interest typically have irregular geometries and contain media with discontinuous sound speeds, and these issues pose challenges for the development of efficient solvers. In this paper, we propose a new algorithm which is based on the use of the staggered discontinuous Galerkin method for solving the underlying wave propagation problem. It gives a convenient way to handle domains with complex geometries and discontinuous sound speeds. Our numerical results show that the method is able to recover the initial condition accurately.

1. Introduction

Mathematical imaging is an important research field in applied mathematics. There have been many significant progresses in both mathematical theories and medical applications; see [**1, 3, 7–10, 12–20, 22, 23, 25, 26, 29, 31, 32**] and references therein. Theoretically, one is interested in uniqueness and stability of the solution for the inverse problem; numerically, one is interested in designing efficient numerical algorithms to recover the solution of the inverse problem. Naturally, the above two aspects have been well studied in the case of the sound speed being constant. In fact, if the sound speed is constant and the observation surface $\partial\Omega$ is of some special geometry, such as planar, spherical or cylindrical surface, there are explicit closed-form inversion formulas; see [**11, 12, 14, 15, 28**] and references therein. In practice the constant sound speed model is inaccurate in many situations [**18, 21, 29, 30**]. For instance in breast imaging, the different components of the breast, such as the glandular tissues, stromal tissues, cancerous tissues and other fatty issues, have different acoustic properties. The variations between their acoustic speeds can be as great as 10% [**18**].

The research of Eric Chung is partially supported by the CUHK Focused Investment Scheme 2012-14. The research of Jianliang Qian is partially supported by NSF..

In this paper, we will focus on photoacoustic tomography which is a very important field in mathematical imaging. Photoacoustic tomography has recently attracted much attention due to its applications in medical imaging. It is based on the non-destructive testing methodology to construct high resolution medical images needed for important diagnostic processes. The physical mechanism involved is the so-called photoacoustic effect, which can be briefly described as follows. Initially, a short pulse of electromagnetic wave is injected into the patient's body. Then the body is heated up which generates some acoustic waves. Different parts of the body have different absorption rates, and this information is contained in the acoustic waves generated by this process. The body structure is then determined by measuring the acoustic waves outside of the patient's body. For more details about this, see for example [**27**, **29**].

Now, we will present the mathematical formulation of photoacoustic tomography. Let $\Omega \subset \mathbb{R}^n$ be an open set having smooth and strictly convex boundary $\partial\Omega$. This domain Ω is understood as the body of interest. As mentioned previously, a pulse of electromagnetic signals will generate some heat and then acoustic waves, the heating process is modeled by the initial condition of the wave propagation problem. More precisely, given a source function $f(x)$ with support in $\overline{\Omega}$ initially, it will generate acoustic signals. The photoacoustic tomography problem is to determine the unknown source function $f(x)$ by boundary measurements of these acoustic signals. The forward problem can be described as follows. Given the initial condition $f(x)$, the acoustic pressure $u(t,x)$ satisfies

$$\frac{\partial^2 u}{\partial t^2} - c^2 \Delta u = 0, \quad \text{in} \quad (0,T) \times \mathbb{R}^n \tag{1}$$

subject to the following initial conditions

$$u(0,x) = f(x), \qquad u_t(0,x) = 0, \qquad \text{on} \quad \mathbb{R}^n. \tag{2}$$

In the above wave equation (1), the function $c(x)$ is the acoustic sound speed. We assume that $c(x)$ is a given, possibly discontinuous, function inside Ω and takes the value one outside Ω. Our measurement can be represented by an operator Λ defined by

$$\Lambda f := u\big|_{[0,T]\times\partial\Omega} \tag{3}$$

which is the value of the acoustic pressure $u(t,x)$ along the boundary of the domain $\partial\Omega$ for all times.

In this paper, we propose a new numerical algorithm that works for irregular domains by following [**24**–**26**]. In [**24**], the method is applied to rectangular domains; in the current work, we extend the idea to unstructured domains so that the methodology is applicable to more practical situations. To achieve our goals, we will apply the staggered discontinuous Galerkin method [**5**, **6**] for the numerical approximation of the wave propagation problem. It gives a systematic way to handle domains with complicated geometries and discontinuous sound speeds. Moreover, there are distinctive advantages of using the staggered discontinuous Galerkin method; namely, the method is an explicit scheme which allows very efficient time stepping. Besides, the method is able to preserve the wave energy and gives smaller dispersion errors compared with non-staggered schemes [**2**, **4**]. In addition, we also need a Poisson solver on irregular domains for our reconstruction algorithm, which will be based on an integral equation approach so that we can handle a wide class of boundary curves. Combining the above methodologies, the resulting method is very efficient and allows us to solve problems arising from realistic imaging applications.

The paper is organized as follows. In Section 2, we will present some background materials, and in Section 3, the reconstruction method together with the implementation details will be presented. Moreover, a brief account of the staggered discontinuous Galerkin method is included. Numerical results are shown in Section 4 to demonstrate the performance of our method.

2. Background

Assume for now that $c > 0$ is smooth. The speed c defines a Riemannian metric $c^{-2}\mathrm{d}x^2$. For any piecewise smooth curve $\gamma : t \in [a, b] \mapsto \gamma(t) \in \mathbb{R}^n$, the length of γ in that metric is given by

$$\text{length}(\gamma) = \int_a^b \frac{|\dot{\gamma}(t)|}{c(\gamma(t))}\mathrm{d}t.$$

The so-defined length is independent of the parameterization of γ. The distance function $\text{dist}(x, y)$ is then defined as the infimum of the lengths of all such curves connecting x and y.

For any $(x, \theta) \in \mathbb{R}^n \times S^{n-1}$ we denote by $\gamma_{x,\theta}(t)$ the unit speed (i.e., $|\dot{\gamma}| = c(\gamma)$) geodesics issued at x in the direction θ.

Similar to the settings in [**25, 26**], the energy of $u(t, x)$ in a domain $U \subset \mathbb{R}^n$ is given by

$$E(u(t)) = \int_U \left(|\nabla_x u|^2 + c^{-2}|u_t|^2\right) \mathrm{d}x,$$

where $u(t) = u(t, \cdot)$. The energy of any Cauchy data (f, g) for equation (1) is given by

$$E(f, g) = \int_U \left(|\nabla_x f|^2 + c^{-2}|g|^2\right) \mathrm{d}x.$$

The energy norm is defined as the square root of the energy. In particular, the energy of $(f, 0)$ in U is given by the square of the Dirichlet norm

$$\|f\|^2_{H_D(U)} := \int_U |\nabla_x f|^2 \,\mathrm{d}x,$$

where the Hilbert space $H_D(U)$ is the completion of $C_0^\infty(U)$ under the above Dirichlet norm. We always assume below that the initial condition $f \in H_D(\Omega)$. We will denote by $\|\cdot\|$ the norm in $H_D(\Omega)$, and in the same way we denote the operator norm in that space.

There are two main geometric quantities that are crucial for the results below. First we set

$$T_0 := \max\{\text{dist}(x, \partial\Omega) :\ x \in \bar{\Omega}\}, \tag{4}$$

where $\text{dist}(x, \partial\Omega)$ is the distance in the given Riemannian metric $c^{-2}\mathrm{d}x^2$. Let $T_1 \leq \infty$ be the supremum of the lengths of all maximal geodesics lying in $\bar{\Omega}$. Clearly, $T_0 < T_1$; however, while the first number is always finite, the second one can be infinite. It can be shown actually that

$$T_0 \leq T_1/2. \tag{5}$$

3. The reconstruction method

In this section, we will present the numerical reconstruction method for the photoacoustic tomography. In [**25**], it is proved that the solution can be represented by a convergent Neumann series. Our method is based on a truncation of this Neumann series, and the addition of each term provides a refinement of the recovered solution. Thus, depending on the error tolerance, typically a few terms are needed to obtain a reasonable solution.

Assume that $f(x)$ is the unknown initial condition and that the boundary data Λf defined on $\partial\Omega$ has been given. Note that Λf is the measurement we obtained. One major step of our reconstruction method is to solve a backward in time wave propagation problem by using the boundary condition Λf. Let $v(t,x)$ be the solution of this problem. To find the solution v, we will need to specify the values of v and v_t at the final time T. For v_t, we will take $u_t = 0$ at the final time T. For v, since we only know the boundary values at the final time T, we will use a function that minimizes the energy $\|\cdot\|_{H_D(\Omega)}$. Thus, we will use the harmonic extension of Λf. To better present our ideas, for a given ϕ defined on $\partial\Omega$, we define $P\phi$ to be the harmonic extension of ϕ.

We then solve the following modified back projection problem. Given a function h defined on $\partial\Omega$, we find $v(0,\cdot)$ such that

$$\frac{\partial^2 v}{\partial t^2} - c^2\Delta v = 0, \quad \text{in} \quad (0,T)\times\Omega \tag{6}$$

subject to the boundary condition

$$v(t,x) = h, \quad \text{on} \quad [0,T]\times\partial\Omega \tag{7}$$

and the final time conditions

$$v(T,x) = Ph, \quad , v_t(T,x) = 0, \qquad \text{on} \quad \Omega. \tag{8}$$

We can then define an operator $Ah = v(0,\cdot)$. Note that, the operator A is not an actual inverse of the operator Λ, but it gives some kind of approximation.

As in [**24**], we have

$$A\Lambda = I - K \tag{9}$$

where K is an error operator. Under suitable conditions, it is proved in [**25**] that

$$\|Kf\|_{H_D(\Omega)} \le \|f\|_{H_D(\Omega)} \tag{10}$$

and that

$$\|K\|_{H_D(\Omega)\to H_D(\Omega)} < 1. \tag{11}$$

Therefore, one can write the following Neumann series [**25**]

$$f = \sum_{m=0}^{\infty} K^m A\Lambda f. \tag{12}$$

This is the key of our reconstruction algorithm. We remark that it is important to choose the final time T in a suitable way. We will use the idea described in [**24**].

Now we summarize the following properties proved in [**25**], which provides some guidances in choosing the final time T.

(i) $T < T_0$.

Λf does not recover f uniquely. Then $\|K\| = 1$, and for any f supported in the inaccessible region, $Kf = f$.

(ii) $T_0 < T < T_1/2$.

This can happen only if there is a strict inequality in (5). Then we have uniqueness but not stability. In this case, $\|K\| = 1$, $\|Kf\| < \|f\|$, and we do not know if the Neumann series (12) converges. If it does, it converges to f.

(iii) $T_1/2 < T < T_1$.

This assumes that Ω is non-trapping for c. The Neumann series (12) converges exponentially but may be not as fast as in the next case. There is stability, and $\|K\| < 1$.

(iv) $T_1 < T$.

This also assumes that Ω is non-trapping for c. The Neumann series (12) converges exponentially. There is stability, $\|K\| < 1$, and K is compact.

Now, we will present some implementation details. In (12), we can evaluate the operator A by solving the modified back-projection problem defined in (6), (7) and (8). Then, for a given function ψ defined on Ω, we can evaluate K by

$$K\psi = \psi - A(\Lambda\psi). \tag{13}$$

This means that, we have to solve the forward in time wave propagation problem with initial condition ψ and then obtain the operator $\Lambda\psi$, which is the boundary values for all times. Then using this boundary function, we solve the modified back-projection problem defined in (6), (7) and (8) to obtain $A(\Lambda\psi)$.

To solve the forward in time wave equation (1) and (2), we write it as a first order form

$$\begin{aligned} \rho\frac{\partial u}{\partial t} - \nabla \cdot p &= 0 \\ \frac{\partial p}{\partial t} - \nabla u &= 0 \end{aligned} \tag{14}$$

where $\rho = c^{-2}$ and $p = \nabla u$. To solve (14) on unstructured grid, we use the staggered discontinuous Galerkin method [**4**–**6**], which gives an explicit and energy conserving forward solver. We remark that explicit solver gives a very fast time-marching process. Besides, the method produces smaller dispersion errors compared with non-staggered methods [**2**, **4**]. Moreover, the staggered discontinuous Galerkin method can be seen as an extension to unstructured grid of the finite difference method used in [**24**]. For completeness, we will give a brief account of the method in the next subsection. Notice that the problem (14) is posed on the whole $\mathbb{R}^n$. Thus, some artificial boundary condition is needed. In this paper, the perfectly matched layer is used as the artificial boundary condition and we use $\widehat{\Omega}$ to represent the computational domain. Finally, the values of the pressure $u(t, x)$ from (14) can then be obtained on the domain boundary $\partial\Omega$.

Another step of our reconstruction method is to generate a final time condition for the problem (6), (7) and (8). To do this, we need to find the harmonic extension for a function ϕ defined on $\partial\Omega$. To perform this step in an efficient way, we will apply the standard integral equation approach, which will be briefly accounted in the next section. Once the final time conditions are known, we can then solve the modified back-projection problem (6), (7) and (8) by the staggered discontinuous Galerkin method [**4**–**6**] together with the given boundary data h.

3.1. The staggered discontinuous Galerkin method. In this section, we briefly summarize the method developed in [**6**]. We start with the triangulation of the domain.

Assume that the domain Ω is triangulated by a family of triangles $\mathcal{T}$ so that $\Omega = \cup\{\tau \mid \tau \in \mathcal{T}\}$. Let $\tau \in \mathcal{T}$. We define h_τ as the diameter of τ and ρ_τ as the supremum of the diameters of the circles inscribed in τ. The mesh size h is defined as $h = \max_{\tau\in\mathcal{T}} h_\tau$. We will assume that the set of triangles $\mathcal{T}$ forms a regular family of triangulation of Ω so that there exist a uniform constant K independent of the mesh size such that

$$h_\tau \leq K\rho_\tau \quad \forall\tau \in \mathcal{T}.$$

Let $\mathcal{E}$ be the set of all edges and let $\mathcal{E}^0 \subset \mathcal{E}$ be the set of all interior edges of the triangles in $\mathcal{T}$. The length of $\sigma \in \mathcal{E}$ will be denoted by h_σ. We also denote by $\mathcal{N}$ the set of all interior nodes of the triangles in $\mathcal{T}$. Here, by interior edge and interior node, we mean any edge and node that does not lie on the boundary $\partial\Omega$. Let $\nu \in \mathcal{N}$. We define

$$\mathcal{S}(\nu) = \cup\{\tau \in \mathcal{T} \mid \nu \in \tau\}. \tag{15}$$

That is, $\mathcal{S}(\nu)$ is the union of all triangles having vertex ν. We will assume that the triangulation of Ω satisfies the following condition.

Assumption on triangulation: There exists a subset $\mathcal{N}_1 \subset \mathcal{N}$ such that

(A1) $\Omega = \cup\{\mathcal{S}(\nu) \mid \nu \in \mathcal{N}_1\}$.

(A2) $\mathcal{S}(\nu_i) \cap \mathcal{S}(\nu_j) \in \mathcal{E}^0$ for all distinct $\nu_i, \nu_j \in \mathcal{N}_1$.

Let $\nu \in \mathcal{N}_1$. We define

$$\mathcal{E}_u(\nu) = \{\sigma \in \mathcal{E} \mid \nu \in \sigma\}. \tag{16}$$

That is, $\mathcal{E}_u(\nu)$ is the set of all edges that have ν as one of their endpoints. We further define

$$\mathcal{E}_u = \cup\{\mathcal{E}_u(\nu) \mid \nu \in \mathcal{N}_1\} \quad \text{and} \quad \mathcal{E}_v = \mathcal{E}\backslash\mathcal{E}_u. \tag{17}$$

Notice that $\mathcal{E}_u$ contains only interior edges since one of the endpoints of edges in $\mathcal{E}_u$ has a vertex from $\mathcal{N}_1$. On the other hand, $\mathcal{E}_v$ has both interior and boundary edges. So, we also define $\mathcal{E}_v^0 = \mathcal{E}_v \cap \mathcal{E}^0$ which contains elements from $\mathcal{E}_v$ that are interior edges. Notice that we have $\mathcal{E}_v\backslash\mathcal{E}_v^0 = \mathcal{E} \cap \partial\Omega$. Furthermore, for $\sigma \in \mathcal{E}_v^0$, we will let $\mathcal{R}(\sigma)$ be the union of the two triangles sharing the same edge σ. For $\sigma \in \mathcal{E}_v\backslash\mathcal{E}_v^0$, we will let $\mathcal{R}(\sigma)$ be the only triangle having the edge σ.

In practice, triangulations that satisfy assumptions (A1)–(A2) are not difficult to construct. In Figure 1, we illustrate how this kind of triangulation is generated. First, the domain Ω is triangulated by a family of triangles, called $\tilde{\mathcal{T}}$. Each triangle in this family is then subdivided into three sub-triangles by connecting a point inside the triangle with its three vertices. Then we define the union of all these sub-triangles to be our triangulation $\mathcal{T}$. Each triangle in $\tilde{\mathcal{T}}$ corresponds to an $\mathcal{S}(\nu)$ for some ν inside the triangle. In Figure 1, we show two of the triangles, enclosed by solid lines, in this family $\tilde{\mathcal{T}}$. This corresponds to 6 triangles in the triangulation $\mathcal{T}$. The dotted lines represent edges in the set $\mathcal{E}_u$ while solid lines represent edges in the set $\mathcal{E}_v$.

Now, we will discuss the FE spaces. Let $k \geq 0$ be a nonnegative integer. Let $\tau \in \mathcal{T}$ and $\kappa \in \mathcal{F}$. We define $P^k(\tau)$ and $P^k(\kappa)$ as the spaces of polynomials of degree less than or equal to k on τ and κ, respectively. The method is based on the following local conforming spaces.

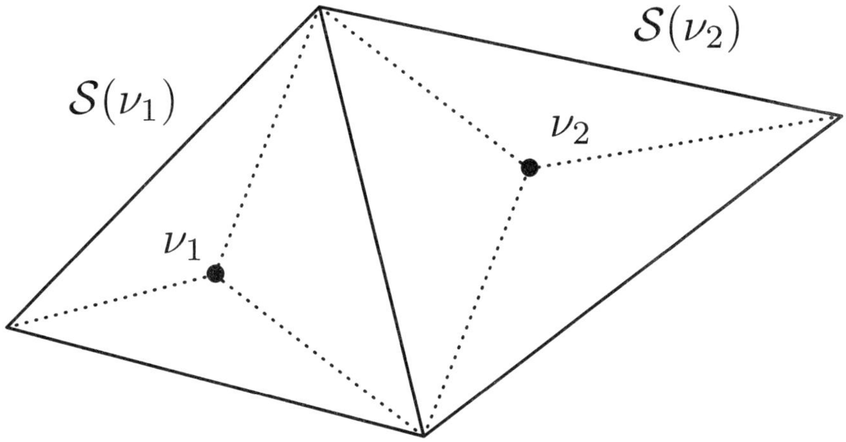

FIGURE 1. Triangulation.

Local $\mathbf{H}^1(\boldsymbol{\Omega})$-conforming FE space:

(18) $$U_h = \{v \mid v|_\tau \in P^k(\tau);\ v \text{ is continuous on } \kappa \in \mathcal{F}_u^0;\ v|_{\partial\Omega} = 0\}.$$

Notice that if $v \in U_h$, then $v|_{\mathcal{R}(\kappa)} \in H^1(\mathcal{R}(\kappa))$ for each face $\kappa \in \mathcal{F}_u$. Furthermore, the condition $v|_{\partial\Omega} = 0$ is equivalent to $v|_\kappa = 0$, $\forall \kappa \in \mathcal{F}_u \backslash \mathcal{F}_u^0$, since $\mathcal{F}_u$ contains all boundary faces. Next, we define the following space.

Local $H(\mathbf{div};\Omega)$-conforming FE space:

(19) $$W_h = \{\mathbf{q} \mid \mathbf{q}|_\tau \in P^k(\tau)^3 \text{ and } \mathbf{q}\cdot n \text{ is continuous on } \kappa \in \mathcal{F}_p\}.$$

Notice that if $\mathbf{q} \in W_h$, then $\mathbf{q}|_{\mathcal{S}(\nu)} \in H(\text{div}; \mathcal{S}(\nu))$ for each $\nu \in \mathcal{N}_1$.

With all the above notations, the staggered discontinuous Galerkin method [**6**] is then stated as: find $u_h \in U_h$ and $\mathbf{p}_h \in W_h$ such that

(20) $$\int_\Omega \rho \frac{\partial u_h}{\partial t} v\, dx + B_h(\mathbf{p}_h, v) = 0,$$

(21) $$\int_\Omega \frac{\partial \mathbf{p}_h}{\partial t} \cdot \mathbf{q}\, dx - B_h^*(u_h, \mathbf{q}) = 0,$$

for all $v \in U_h$ and $\mathbf{q} \in W_h$, where

(22) $$B_h(\mathbf{p}_h, v) = \int_\Omega \mathbf{p}_h \cdot \nabla v\, dx - \sum_{\kappa\in\mathcal{F}_p} \int_\kappa \mathbf{p}_h \cdot n\, [v]\, d\sigma,$$

(23) $$B_h^*(u_h, \mathbf{q}) = -\int_\Omega u_h\, \nabla\cdot\mathbf{q}\, dx + \sum_{\kappa\in\mathcal{F}_u^0} \int_\kappa u_h\, [\mathbf{q}\cdot n]\, d\sigma,$$

where $[v]$ represents the jump of the function v. We remark that the important energy conservation property comes from the fact that $B_h(\mathbf{p}_h, u_h) = B_h^*(u_h, \mathbf{p}_h)$.

3.2. The boundary integral method. In this section, we will give a brief overview of the boundary integral method for finding the harmonic extension of ϕ defined on $\partial\Omega$. Recall that $\Omega \subset \mathbb{R}^n$ for $n \geq 2$. Let

(24) $$K(x) = \begin{cases} \frac{1}{2\pi}\log r & n = 2 \\ \frac{1}{(2-n)\omega_n} r^{2-n} & n \geq 3 \end{cases}$$

where ω_n is the area of the boundary of the unit sphere in $\mathbb{R}^n$ and $r = |x|$. Then it is well known that

$$P\phi = \int_{\partial\Omega} \phi(x) \frac{\partial K(x-\xi)}{\partial \nu_x}\, d\sigma_x \tag{25}$$

where $d\sigma_x$ is the surface measure on $\partial\Omega$ and ν_x is the unit outward normal vector defined on $\partial\Omega$.

4. Numerical examples

In this section, we will present some numerical examples. We will test our numerical algorithm on some domains with irregular shapes. In all cases below, the computational domain $\widehat{\Omega}$ is $[-1.5, 1.5]^2$. Moreover, the perfectly matched layer is imposed in the region $[-1.5, 1.5]^2 \backslash [-1.05, 1.05]^2$. The mesh size for the spatial domain is taken as 0.02 and the time step size is taken according to the CFL condition which allows stability of the wave propagation solver. The final time T is taken as 4 which is large enough to guarantee the convergence of the Neumann series (12).

4.1. Example 1. In our first example, we consider the imaging of the Shepp-Logan phantom contained in the domain Ω which is a circle centered at $(0, 0)$ with radius 1. In the first test case, we take $c_1(x, y) = 1 + 0.2 \sin(2\pi x) + 0.1 \cos(2\pi y)$ as the sound speed inside Ω. In Figure 2, the exact solution is shown, where we see that the Shepp-Logan phantom is located inside a circular domain. Moreover, a coarse triangulation of this circular domain is also shown in Figure 2. Here, we use a coarse triangulation for display purpose, and the actual triangulation for our computation is finer than this.

The numerical reconstruction results are shown in Figure 3. From these figures, we see that the use of the first two terms in the Neumann series (12) is sufficient to give very promising results. In particular, with the use of one term in the Neumann series, we obtain a reconstruction with relative error of 4.46% while the use of two terms in the Neumann series gives a reconstruction with relative error of 2.14%.

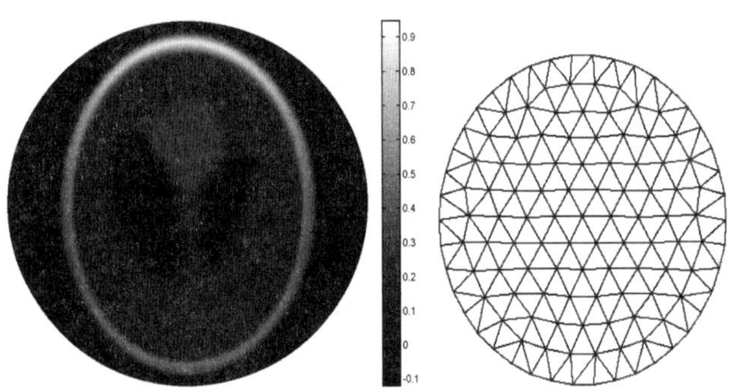

FIGURE 2. Left: exact solution. Right: an example of the mesh used for the domain.

In the second case, we take a piecewise constant sound speed $c_2(x, y)$ with the value 1.2 in the circle centered at $(0, 0)$ and radius 0.6 and the value 1.1 elsewhere in the domain Ω. The numerical reconstruction results are shown in Figure 4 while the exact solution is shown in Figure 2. From these figures, we see that the use of

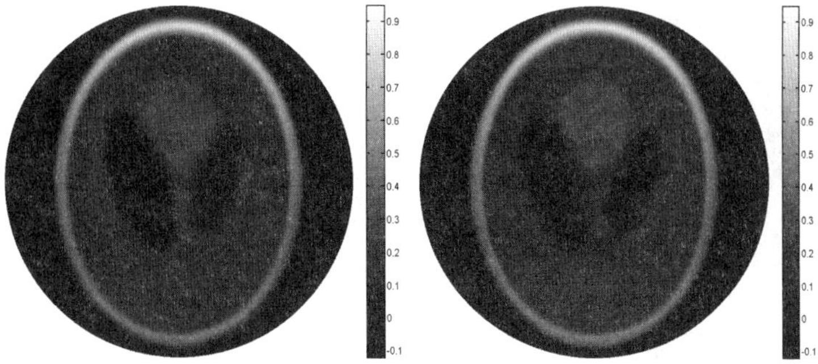

FIGURE 3. Numerical solutions. Left: one term approximation, relative error is 4.46%. Right: two term approximation, relative error is 2.14%.

the first two terms in the Neumann series (12) is sufficient to give very promising results. In particular, with the use of one term in the Neumann series, we obtain a reconstruction with relative error of 2.86% while the use of two terms in the Neumann series gives a reconstruction with relative error of 2.18%.

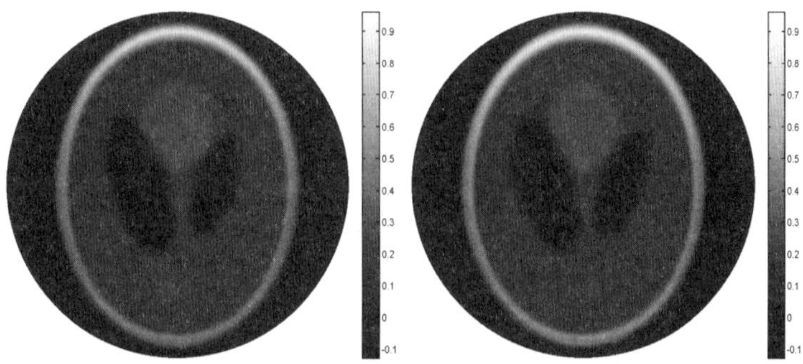

FIGURE 4. Numerical solutions. Left: one term approximation, relative error is 2.86%. Right: two term approximation, relative error is 2.18%.

For our last test case with this domain, we take the sound speed $c_2(x, y)$ defined above and add 2% noise in the data. The numerical reconstruction results are shown in Figure 5 while the exact solution is shown in Figure 2. From these figures, we see that the use of the first two terms in the Neumann series (12) is sufficient to give very promising results. In particular, with the use of one term in the Neumann series, we obtain a reconstruction with relative error of 3.18% while the use of two terms in the Neumann series gives a reconstruction with relative error of 3.10%.

4.2. Example 2. In our second example, we consider a domain with irregular shape, shown in Figure 6. Moreover, a sample triangulation of this domain is also shown in Figure 6. We use $c_1(x, y)$ as the sound speed in this test case.

We will first consider the imaging of a point source in this domain. The exact point source is shown in Figure 6. The numerical reconstruction results are shown in Figure 7. From these figures, we see that the use of the first two terms in the Neumann series (12) is sufficient to give very promising results. In particular,

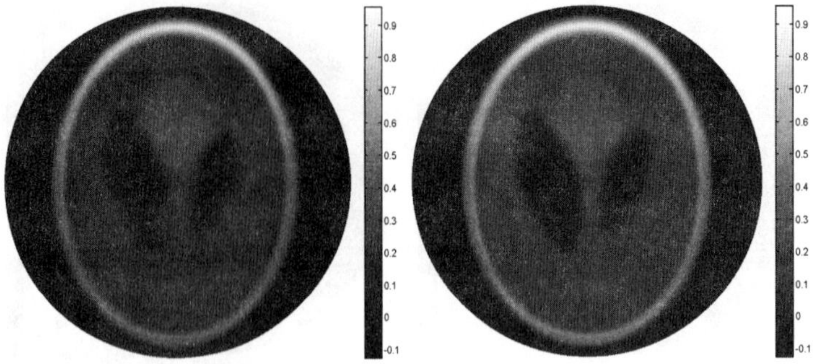

FIGURE 5. Numerical solutions. Left: one term approximation, relative error is 3.18%. Right: two term approximation, relative error is 3.10%.

with the use of one term in the Neumann series, we obtain a reconstruction with relative error of 4.10% while the use of two terms in the Neumann series gives a reconstruction with relative error of 0.82%.

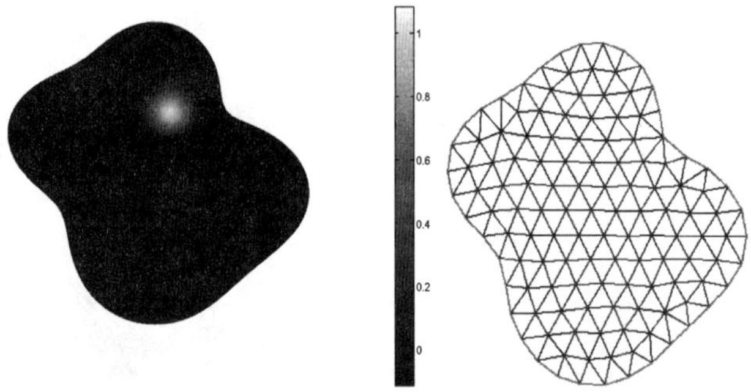

FIGURE 6. Left: exact solution. Right: an example of the mesh used for the domain.

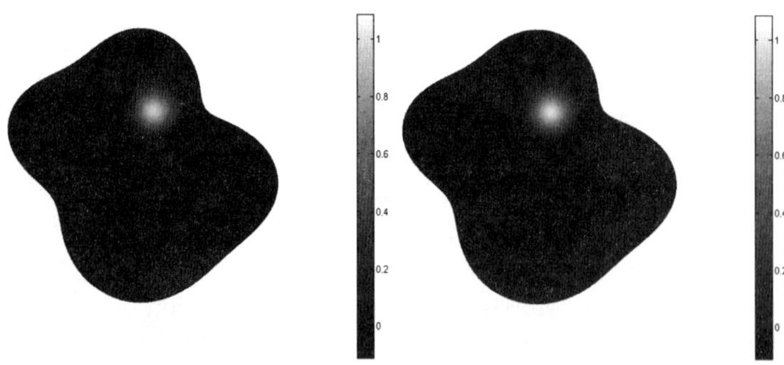

FIGURE 7. Numerical solutions. Left: one term approximation, relative error is 4.10%. Right: two term approximation, relative error is 0.82%.

Next, We will consider the imaging of the Shepp-Logan phantom in this domain. The exact solution is shown in Figure 8. The numerical reconstruction results are shown in Figure 9. From these figures, we see that the use of the first two terms in the Neumann series (12) is sufficient to give very promising results. In particular, with the use of one term in the Neumann series, we obtain a reconstruction with relative error of 9.23% while the use of two terms in the Neumann series gives a reconstruction with relative error of 6.92%.

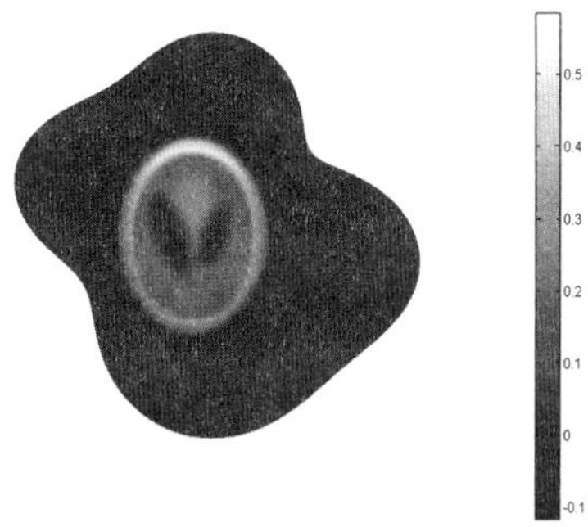

FIGURE 8. The exact solution.

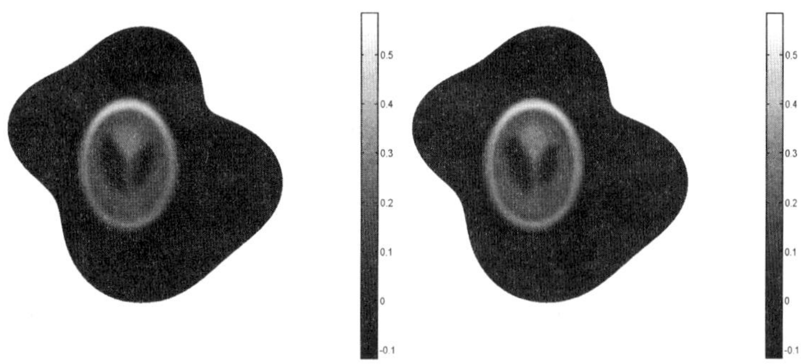

FIGURE 9. Numerical solutions. Left: one term approximation, relative error is 9.23%. Right: two term approximation, relative error is 6.92%.

4.3. Example 3. In our third example, we consider a domain with irregular shape, shown in Figure 10. Moreover, a sample triangulation of this domain is also shown in Figure 10. We use $c_1(x, y)$ as the sound speed in this test case.

We will first consider the imaging of a single Shepp-Logan phantom in this domain. The exact solution is shown in Figure 10. The numerical reconstruction results are shown in Figure 11. From these figures, we see that the use of the first two terms in the Neumann series (12) is sufficient to give very promising results. In particular, with the use of one term in the Neumann series, we obtain a reconstruction with relative error of 4.76% while the use of two terms in the Neumann series gives a reconstruction with relative error of 3.41%.

We next consider the imaging of a single Shepp-Logan phantom together with two circular objects in the domain shown in Figure 10. The exact solution is shown

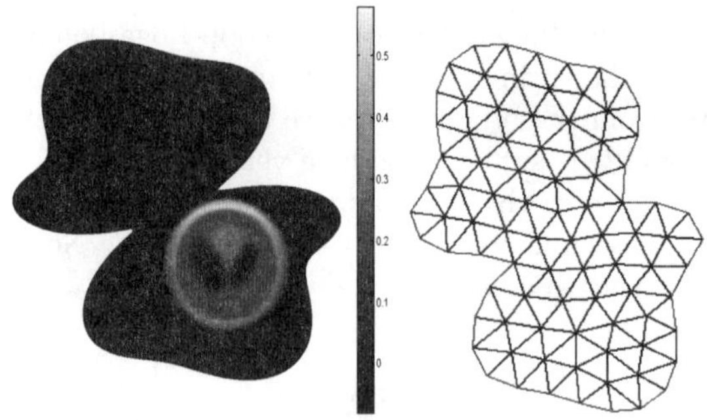

FIGURE 10. Left: exact solution. Right: an example of the mesh used for the domain.

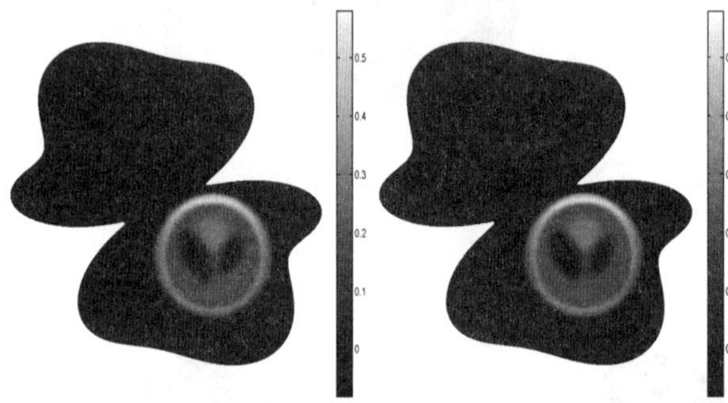

FIGURE 11. Numerical solutions. Left: one term approximation, relative error is 4.76%. Right: two term approximation, relative error is 3.41%.

in Figure 12. The numerical reconstruction results are shown in Figure 13. From these figures, we see that the use of the first two terms in the Neumann series (12) is sufficient to give very promising results. In particular, with the use of one term in the Neumann series, we obtain a reconstruction with relative error of 4.62% while the use of two terms in the Neumann series gives a reconstruction with relative error of 2.61%.

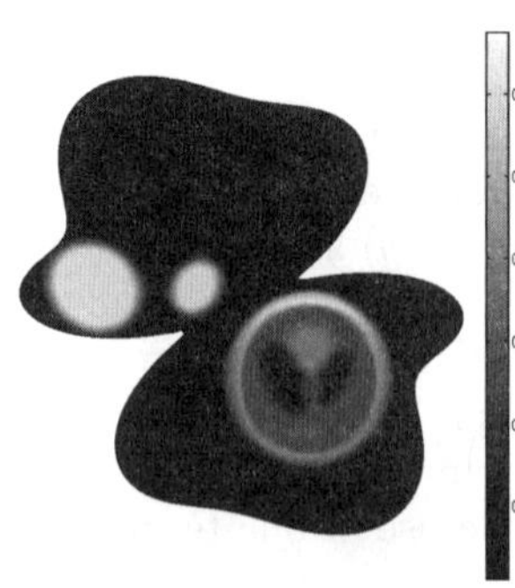

FIGURE 12. The exact solution.

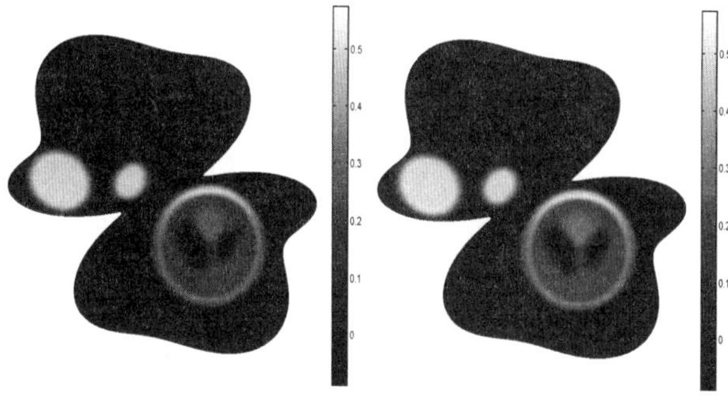

FIGURE 13. Numerical solutions. Left: one term approximation, relative error is 4.62%. Right: two term approximation, relative error is 2.61%.

Next we consider the same example with 2% noise added in the data. The exact solution is shown in Figure 12. The numerical reconstruction results are shown in Figure 13. From these figures, we see that the use of the first two terms in the Neumann series (12) is sufficient to give very promising results. In particular, with the use of one term in the Neumann series, we obtain a reconstruction with relative error of 4.81% while the use of two terms in the Neumann series gives a reconstruction with relative error of 3.39%.

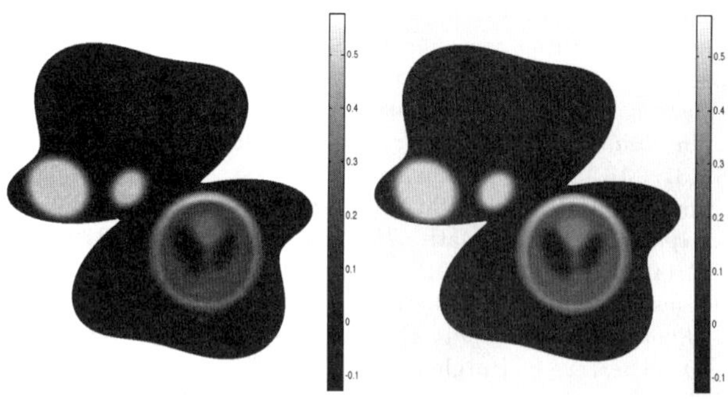

FIGURE 14. Numerical solutions. Left: one term approximation, relative error is 4.81%. Right: two term approximation, relative error is 3.39%.

5. Conclusion

In this paper, we propose an efficient and accurate method for photoacoustic tomography. The method is based on a convergent Neumann series and is applicable to domains with complicated geometries and discontinuous sound speeds. The use of the staggered discontinuous Galerkin method allows a very efficient time-stepping and conservation of wave energy. Our numerical results show that the method has superior performance, and provides a solver for realistic imaging applications.

References

[1] M. Agranovsky, P. Kuchment, and L. Kunyansky. On reconstruction formulas and algorithms for the thermoacoustic tomography. *Photoacoustic Imaging and Spectroscopy, CRC Press*, pages 89–101, 2009.

[2] Hiu Ning Chan, Eric T. Chung, and Gary Cohen, *Stability and dispersion analysis of the staggered discontinuous Galerkin method for wave propagation*, Int. J. Numer. Anal. Model. **10** (2013), no. 1, 233–256. MR3011870

[3] Eric T. Chung, Tony F. Chan, and Xue-Cheng Tai, *Electrical impedance tomography using level set representation and total variational regularization*, J. Comput. Phys. **205** (2005), no. 1, 357–372, DOI 10.1016/j.jcp.2004.11.022. MR2132313 (2005m:78024)

[4] Eric T. Chung, Patrick Ciarlet Jr., and Tang Fei Yu, *Convergence and superconvergence of staggered discontinuous Galerkin methods for the three-dimensional Maxwell's equations on Cartesian grids*, J. Comput. Phys. **235** (2013), 14–31, DOI 10.1016/j.jcp.2012.10.019. MR3017583

[5] Eric T. Chung and Björn Engquist, *Optimal discontinuous Galerkin methods for wave propagation*, SIAM J. Numer. Anal. **44** (2006), no. 5, 2131–2158 (electronic), DOI 10.1137/050641193. MR2263043 (2008e:65297)

[6] Eric T. Chung and Björn Engquist, *Optimal discontinuous Galerkin methods for the acoustic wave equation in higher dimensions*, SIAM J. Numer. Anal. **47** (2009), no. 5, 3820–3848, DOI 10.1137/080729062. MR2576522 (2011f:65195)

[7] Eric Chung, Jianliang Qian, Gunther Uhlmann, and Hongkai Zhao, *A new phase space method for recovering index of refraction from travel times*, Inverse Problems **23** (2007), no. 1, 309–329, DOI 10.1088/0266-5611/23/1/017. MR2302976 (2008j:35183)

[8] E. Chung, J. Qian, G. Uhlmann, and H. Zhao. A phase-space formulation for elastic-wave traveltime tomography. *J. Phys.: Conf. Ser.*, 124:012018, 2008.

[9] Eric Chung, Jianliang Qian, Gunther Uhlmann, and Hongkai Zhao, *An adaptive phase space method with application to reflection traveltime tomography*, Inverse Problems **27** (2011), no. 11, 115002, 22, DOI 10.1088/0266-5611/27/11/115002. MR2851908 (2012k:86009)

[10] Eric T. Chung and Jeff C.-F. Wong, *A TV-based iterative regularization method for the solutions of thermal convection problems*, Commun. Comput. Phys. **14** (2013), no. 4, 1120–1146. MR3066967

[11] David Finch, Markus Haltmeier, and Rakesh, *Inversion of spherical means and the wave equation in even dimensions*, SIAM J. Appl. Math. **68** (2007), no. 2, 392–412, DOI 10.1137/070682137. MR2366991 (2008k:35494)

[12] David Finch, Sarah K. Patch, and Rakesh, *Determining a function from its mean values over a family of spheres*, SIAM J. Math. Anal. **35** (2004), no. 5, 1213–1240 (electronic), DOI 10.1137/S0036141002417814. MR2050199 (2005b:35290)

[13] D. Finch and Rakesh. Recovering a function from its spherical mean values in two and three dimensions. *in: Photoacoustic Imaging and Spectroscopy, CRC Press*, 2009.

[14] M. Haltmeier, O. Scherzer, P. Burgholzer, and G. Paltauf, *Thermoacoustic computed tomography with large planar receivers*, Inverse Problems **20** (2004), no. 5, 1663–1673, DOI 10.1088/0266-5611/20/5/021. MR2109142 (2006a:35311)

[15] M. Haltmeier, T. Schuster, and O. Scherzer, *Filtered backprojection for thermoacoustic computed tomography in spherical geometry*, Math. Methods Appl. Sci. **28** (2005), no. 16, 1919–1937, DOI 10.1002/mma.648. MR2170772 (2006d:92023)

[16] Yulia Hristova, *Time reversal in thermoacoustic tomography—an error estimate*, Inverse Problems **25** (2009), no. 5, 055008, 14, DOI 10.1088/0266-5611/25/5/055008. MR2501026 (2010d:78036)

[17] Yulia Hristova, Peter Kuchment, and Linh Nguyen, *Reconstruction and time reversal in thermoacoustic tomography in acoustically homogeneous and inhomogeneous media*, Inverse Problems **24** (2008), no. 5, 055006, 25, DOI 10.1088/0266-5611/24/5/055006. MR2438941 (2010c:65162)

[18] X. Jin and L. V. Wang. Thermoacoustic tomography with correction for acoustic speed variations. *Phys. Med. Biol.*, 51:6437–6448, 2006.

[19] R. A. Kruger, W. L. Kiser, D. R. Reinecke, and G. A. Kruger. Thermoacoustic computed tomography using a conventional linear transducer array. *Med Phys*, 30(5):856–860, May 2003.

[20] R. A. Kruger, D. R. Reinecke, and G. A. Kruger. Thermoacoustic computed tomography–technical considerations. *Med Phys*, 26(9):1832–1837, Sep 1999.
[21] G. Ku, B. Fornage, X. Jin, M. Xu, K. Hunt, and L. V. Wang. Thermoacoustic and photoacoustic tomography of thick biological tissues toward breast imaging. *Tech. Cancer Research and Treatment*, 4:559–565, 2005.
[22] Peter Kuchment and Leonid Kunyansky, *Mathematics of thermoacoustic tomography*, European J. Appl. Math. **19** (2008), no. 2, 191–224, DOI 10.1017/S0956792508007353. MR2400720 (2009c:92026)
[23] S. K. Patch. Thermoacoustic tomography – consistency conditions and the partial scan problem. *Physics in Medicine and Biology*, 49(11):2305–2315, 2004.
[24] Jianliang Qian, Plamen Stefanov, Gunther Uhlmann, and Hongkai Zhao, *An efficient Neumann series-based algorithm for thermoacoustic and photoacoustic tomography with variable sound speed*, SIAM J. Imaging Sci. **4** (2011), no. 3, 850–883, DOI 10.1137/100817280. MR2836390 (2012h:94031)
[25] Plamen Stefanov and Gunther Uhlmann, *Thermoacoustic tomography with variable sound speed*, Inverse Problems **25** (2009), no. 7, 075011, 16, DOI 10.1088/0266-5611/25/7/075011. MR2519863 (2010i:35439)
[26] Plamen Stefanov and Gunther Uhlmann, *Thermoacoustic tomography arising in brain imaging*, Inverse Problems **27** (2011), no. 4, 045004, 26, DOI 10.1088/0266-5611/27/4/045004. MR2781028 (2012f:76113)
[27] L. V. Wang and H.-I. Wu. *Biomedical Optics: Principles and Imaging*. Wiley-Interscience, New Jersey, 2007.
[28] M. Xu and L.-H. V. Wang. Universal back-projection algorithm for photoacoustic computed tomography. *Phys. Rev. E.*, 71:016706, 2005.
[29] M. Xu and L. V. Wang. Photoacoustic imaging in biomedicine. *Review of Scientific Instruments*, 77(4):041101, 2006.
[30] Y. Xu and L. V. Wang. Effects of acoustic heterogeneity in breast thermoacoustic tomography. *IEEE Trans. Ultra. Ferro. and Freq. Cont.*, 50:1134–1146, 2003.
[31] Y. Xu and L. V. Wang. Rhesus monkey brain imaging through intact skull with thermoacoustic tomography. *IEEE Trans. Ultrason., Ferroelectr., Freq. Control*, 53(3):542–548, 2006.
[32] X. Yang and L. V. Wang. Monkey brain cortex imaging by photoacoustic tomography. *J Biomed Opt*, 13(4):044009, 2008.

Department of Mathematics, The Chinese University of Hong Kong, Hong Kong SAR
E-mail address: tschung@math.cuhk.edu.hk

Department of Mathematics, The Chinese University of Hong Kong, Hong Kong SAR
E-mail address: cylam@math.cuhk.edu.hk

Department of Mathematics, Michigan State University, 619 Red Cedar Rd RM C306, East Lansing, Michigan 48824-3429
E-mail address: qian@math.msu.edu

Contemporary Mathematics
Volume **615**, 2014
http://dx.doi.org/10.1090/conm/615/12267

Nonlinear Inversion from Partial EIT Data: Computational Experiments

S. J. Hamilton and S. Siltanen

ABSTRACT. Electrical impedance tomography (EIT) is a non-invasive imaging method in which an unknown physical body is probed with electric currents applied on the boundary, and the internal conductivity distribution is recovered from the measured boundary voltage data. The reconstruction task is a nonlinear and ill-posed inverse problem, whose solution calls for special regularized algorithms, such as D-bar methods which are based on complex geometrical optics solutions (CGOs). In many applications of EIT, such as monitoring the heart and lungs of unconscious intensive care patients or locating the focus of an epileptic seizure, data acquisition on the entire boundary of the body is impractical, restricting the boundary area available for EIT measurements. An extension of the D-bar method to the case when data is collected only on a subset of the boundary is studied by computational simulation. The approach is based on solving a boundary integral equation for the traces of the CGOs using localized basis functions (Haar wavelets). The numerical evidence suggests that the D-bar method can be applied to partial-boundary data in dimension two and that the traces of the partial data CGOs approximate the full data CGO solutions on the available portion of the boundary, for the necessary small k frequencies.

1. Introduction

1.1. EIT and the inverse conductivity problem. Electrical impedance tomography (EIT) is a non-invasive imaging method where an unknown physical body is probed with electric currents, and the internal conductivity distribution is recovered from the measurement data. The reconstruction task is a nonlinear and ill-posed inverse problem, whose solution calls for special regularized algorithms, such as the D-bar method [**KLMS09**]. Applications of EIT include monitoring the heart and lungs of unconscious intensive care patients, industrial process monitoring and underground prospecting.

2010 *Mathematics Subject Classification.* Primary 65N21, 35R30; Secondary 45Q05.

Key words and phrases. Inverse problem, numerical solver, conductivity equation, inverse conductivity problem, complex geometrical optics solution, nonlinear Fourier transform, electrical impedance tomography.

SalWe Research Program for Mind and Body (Tekes - the Finnish Funding Agency for Technology and Innovation grant 1104/10).

Academy of Finland (Finnish Centre of Excellence in Inverse Problems Research 2012–2017, decision number 250215).

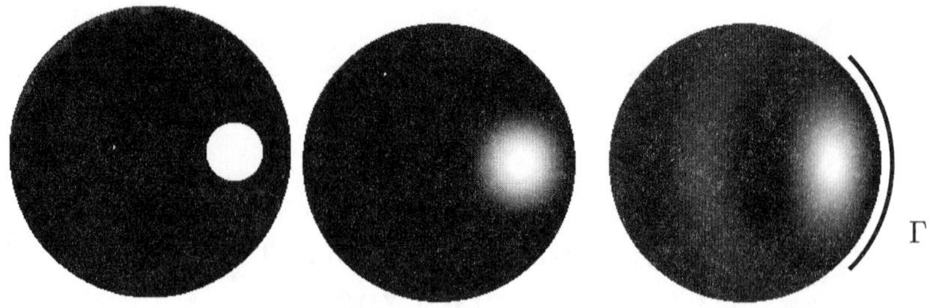

FIGURE 1. Left: original conductivity. Middle: reconstruction from full-boundary data using the D-bar method. Right: reconstruction from partial-boundary data using the proposed method. The subset $\Gamma \subset \partial\Omega$ where the measurements are available is denoted by a black line, which corresponds to 25% of the entire boundary

Practical considerations typically restrict the boundary area available for EIT measurements: for example, it is not sensible to cover a patient completely with electrodes when imaging the heart. In this paper we study a possible extension of the D-bar method to the case when data is collected only on a subset of the boundary. See Figure 1.

The mathematical model of EIT is the inverse conductivity problem introduced by Calderón [**Cal80**]. Let $\Omega \subset \mathbb{R}^n$ be a bounded and simply connected set with a smooth boundary $\partial\Omega$. Let $\sigma : \Omega \to \mathbb{R}$ be an essentially bounded measurable function satisfying $\sigma(x) \geq c > 0$ for almost every $x \in \Omega$. Let $u \in H^1(\Omega)$ be the unique solution to

$$\nabla \cdot \sigma \nabla u = 0 \text{ in } \Omega, \tag{1.1}$$

$$u\big|_{\partial\Omega} = \phi \in H^{1/2}(\partial\Omega). \tag{1.2}$$

The inverse conductivity problem is to recover the conductivity σ from the Dirichlet-to-Neumann (D-N) map defined by

$$\Lambda_\sigma : \phi \mapsto \sigma \frac{\partial u}{\partial \nu}\Big|_{\partial\Omega}.$$

Here $\nu = (\nu_1, \nu_2) = \nu_1 + i\nu_2$ is the unit outward facing normal vector to the boundary. Here ϕ is a voltage distribution applied on the boundary, and $\Lambda_\sigma\phi$ is the resulting current flux through the boundary. Therefore, Λ_σ can be seen as an ideal infinite-precision model of practical voltage-to-current measurements.

Calderón asked two main questions in his seminal article [**Cal80**]:

(i) Is σ uniquely determined by Λ_σ?
(ii) If the answer to (i) is yes, how can one calculate σ from Λ_σ?

In practical EIT imaging only a finite-range and noisy approximate operator Λ_σ^δ is available. In general, Λ_σ^δ is not the D-N map of any conductivity. We usually only know that $\|\Lambda_\sigma^\delta - \Lambda_\sigma\|_Y \leq \delta$. Here Y is an appropriate data space and $\delta > 0$ can be determined from the properties of the measurement device. This leads us to a third question:

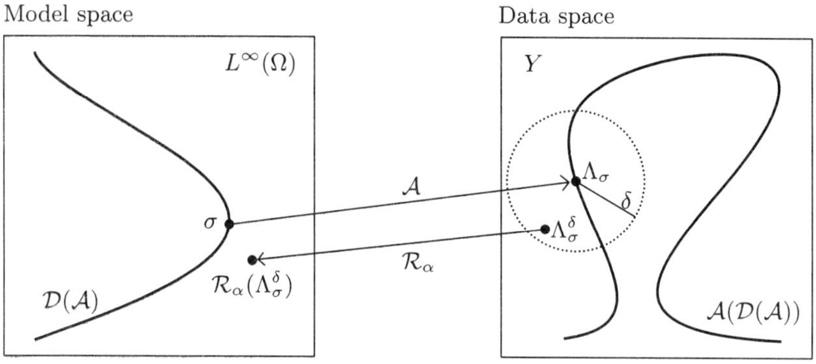

FIGURE 2. Schematic illustration of nonlinear regularization of the EIT problem. Here the forward map is defined as $\mathcal{A}(\sigma) = \Lambda_\sigma$ with the domain of definition denoted by $\mathcal{D}(\mathcal{A}) \subset L^\infty(\Omega)$. The conductivity σ is approximately recovered as $\mathcal{R}_\alpha(\Lambda_\sigma^\delta)$.

(iii) Given Λ_σ^δ and δ, how can one design a continuous map from Y to $L^\infty(\Omega)$ whose output is a useful approximation to σ?

As the inverse conductivity problem is ill-posed, the forward map $\mathcal{A} : \sigma \mapsto \Lambda_\sigma$ does not have a continuous inverse. Therefore, question (iii) needs to be answered by constructing a *regularization strategy* [**EHN96**]. More precisely, a family of continuous mappings $\mathcal{R}_\alpha : Y \to L^\infty(\Omega)$ must be defined, parameterized by $0 < \alpha < \infty$, such that

$$\lim_{\alpha\to 0} \|\mathcal{R}_\alpha(\Lambda_\sigma) - \sigma\|_{L^\infty(\Omega)} = 0, \tag{1.3}$$

for each fixed σ. Note that (1.3) is closely related to question (ii) above. Furthermore, one needs to specify a choice $\alpha = \alpha(\delta)$ for the regularization parameter as a function of the noise level so that $\alpha(\delta) \to 0$ as $\delta \to 0$. Finally, the reconstruction error $\|\mathcal{R}_{\alpha(\delta)}(\Lambda_\sigma^\delta) - \sigma\|_{L^\infty(\Omega)}$ must vanish in the zero noise limit: for any fixed σ we must have

$$\sup_{\Lambda_\sigma^\delta\in Y} \left\{\|\mathcal{R}_{\alpha(\delta)}(\Lambda_\sigma^\delta) - \sigma\|_{L^\infty(\Omega)} \; : \; \|\Lambda_\sigma^\delta - \Lambda_\sigma\|_Y \le \delta\right\} \to 0 \text{ as } \delta \to 0. \tag{1.4}$$

For more details, see Figure 2 and [**KLMS09, MS12**].

1.2. D-bar methods for full-boundary data. From the practical viewpoint, the solution of the inverse conductivity problem is a computational algorithm that implements a regularization strategy $\mathcal{R}_\alpha$ satisfying (1.3) and (1.4). Achieving such a goal is typically a large project involving several milestones, often corresponding to one of the following two types:

(a) A theoretical breakthrough that outlines a computational approach
(b) Successful computational experiments that inspire further theoretical study

Let us review the history of a specific two-dimensional D-bar method for EIT in light of (a) and (b).

1996(a): Nachman showed uniqueness (i) and introduced a infinite-precision reconstruction method (ii) for twice differentiable conductivities in [**Nac96**]. The proof used a nonlinear Fourier transform based on so-called *complex geometrical optics* (CGO) solutions, first defined by Faddeev in 1966 [**Fad66**] and later rediscovered in 1987 by Sylvester and Uhlmann in the context of 3D EIT [**SU87**]. Thus, [**Nac96**] represents a breakthrough in the form of (a) since it is the basis of the first numerical D-bar method [**SMI00, SMI01, MS03**].

2004(b): Isaacson *et al.* demonstrated in [**IMNS04, IMNS06**] that the D-bar method performs well on practical data measured from laboratory phantoms and from human subjects. The mandatory regularization step was provided by low-pass filtering in the nonlinear frequency domain. The need for such filtering is evident from the structure of the experimental nonlinear Fourier transforms: they blow up outside a disc centered at the origin.

2009(a): The numerical evidence from practical imaging experiments [**IMNS04, IMNS06**] inspired a rigorous regularization proof of convergence in the form of (1.4), see [**KLMS09**]. This gave an answer to (iii) and outlined a method for choosing the regularization parameter as the inverse of the nonlinear cutoff frequency. We outline the reconstruction method in Section 2 below.

There is an analogous history for other uniqueness proofs and related algorithms in two-dimensions. We review them briefly below without specifying explicitly the progress steps of types (a) and (b).

Brown and Uhlmann were able to prove uniqueness for real-valued conductivities assuming only one derivative in [**BU97**]. This result was complemented by constructive steps and numerical implementation by Knudsen and Tamasan [**KT04, Knu02, Knu03**]; see also [**KMS04**]. Francini [**Fra00**] extended the uniqueness proof to complex conductivities whose real and imaginary parts are twice differentiable, and her approach was subsequently implemented in [**HHMV12, Ham12, HM13, Her12**]. We outline this reconstruction method in Section 3 below. Both methods involve transforming (1.1) to a first order system of ∂_z and $\overline{\partial}_z$ equations.

Astala and Päivärinta answered Calderón's questions (i) and (ii) in their original smoothness category $\sigma \in L^\infty(\Omega)$, see [**AP06b, AP06a**]. This approach has been implemented numerically as well [**AMPS10, AMP**].

Despite the above developments, some questions still remain open:

- Is it possible to give a regularization analysis (iii) for less smooth conductivities than twice differentiable? There is numerical evidence of type (b) available since all of the above EIT methods produce noise-robust images when applied to data arising from discontinuous conductivities and regularized by nonlinear low-pass filtering [**KLMS08, KLMS07, HHMV12, AMP**].
- Can the D-bar methodology be used in the case of partial-boundary data? We discuss this in Section 1.3 below in the two-dimensional case.

1.3. Extension to partial-boundary data. It is of high practical importance to be able to compute EIT reconstructions from data measured only on a part of the boundary. One possibility for designing such algorithms would be to take one of the recent theoretical breakthroughs, such as [**Knu06, KSU07, NS10, IUY10**], and implement it in the spirit of (a) above. However, we do not discuss such approaches in this paper. We proceed along (b) and produce novel numerical evidence suggesting that it may be possible to use the classical D-bar approach for partial data reconstructions. It is our hope that these computational results inspire further theoretical advances.

Our starting point is the assumption that only a proper subset $\Gamma \subset \partial\Omega$ is available for measurements. We consider voltage-to-current data represented ideally by the restricted D-N map $\widetilde{\Lambda}_\sigma$, defined as follows. Let $\widetilde{\phi} \in H^{1/2}(\partial\Omega)$ satisfy $\text{supp}(\widetilde{\phi}) \subset \Gamma$ and let $u \in H^1(\Omega)$ be the unique solution of the conductivity equation

$$\nabla \cdot \sigma \nabla u = 0 \text{ in } \Omega, \tag{1.5}$$

$$u\big|_{\partial\Omega} = \widetilde{\phi}. \tag{1.6}$$

Our partial D-N map is then defined by

$$\widetilde{\Lambda}_\sigma : \widetilde{\phi} \mapsto \sigma \frac{\partial u}{\partial \nu}\Big|_{\Gamma}. \tag{1.7}$$

The practical data is a finite-range and noisy approximate operator $\widetilde{\Lambda}_\sigma^\delta$ satisfying $\|\widetilde{\Lambda}_\sigma^\delta - \widetilde{\Lambda}_\sigma\|_Y \le \delta$.

Let us briefly explain our approach in the context of the regularized D-bar method [**KLMS09**] based on Nachman's uniqueness proof [**Nac96**]. In the full-boundary data case, it begins by solving this Fredholm integral equation of the second kind for the (approximate) traces of the CGO solutions on $\partial\Omega$:

$$\psi(z,k) = e^{ikz} - \int_{\partial\Omega} G_k(z-\zeta)(\Lambda_\sigma^\delta - \Lambda_1)\psi(\zeta,k)\, dS(\zeta), \quad z \in \partial\Omega, \tag{1.8}$$

where G_k is the Faddeev Green's function [**Fad66**], here defined in the sense of tempered distributions,

$$G_k(z) := e^{ikz} g_k(z), \quad g_k(z) := \frac{1}{(2\pi)^2} \int_{\mathbb{R}^2} \frac{e^{iz\cdot\xi}}{|\xi|^2 + 2k(\xi_1 + i\xi_2)} d\xi. \tag{1.9}$$

In the case of partial-boundary data, we solve the following equation for the unknown functions $\omega(\,\cdot\,,k) : \Gamma \to \mathbb{C}$:

$$\omega(z,k) = e^{ikz} - \int_{\Gamma} G_k(z-\zeta)(\widetilde{\Lambda}_\sigma^\delta - \widetilde{\Lambda}_1)\omega(\zeta,k)\, dS(\zeta), \quad z \in \Gamma \subset \partial\Omega. \tag{1.10}$$

Now the hypothesis is that

$$\psi(z,k)|_\Gamma \approx \omega(z,k), \quad z \in \Gamma \subset \partial\Omega, \quad \text{for some } k \in \mathbb{C}. \tag{1.11}$$

If (1.11) holds, it opens up a variety of extensions of D-bar methods to partial-boundary data applications.

1.4. Focus of this paper. How does one solve (1.10) numerically? Computational solution methods for boundary integral equations (BIEs) of type (1.8), corresponding to the continuum model, have most often been based on representing the unknown CGO solutions in terms of (generalized) trigonometric bases, where the basis functions are essentially supported on the entire boundary [**KLMS09**, **AMP**, **MS12**]. This approach is not directly applicable to partial data problems. In this work we present new numerical experiments on the unit disc (*without loss of generality*) involving the solution of the above-mentioned BIEs using localized basis functions supported only on a subset of the boundary, in this case the Haar wavelets which are naturally applicable to the partial (as well as full) boundary continuum model. See Figure 3.

Let us stress that at present there is no proof available for the solvability of equation (1.10). However, we did not encounter any problems when numerically

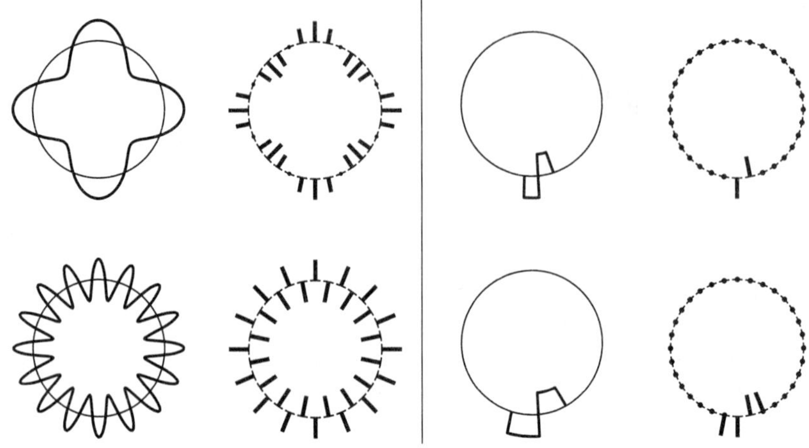

FIGURE 3. Illustration of various basis functions used for solving boundary integral equations. Also shown are voltage patterns applied using 32 electrodes, approximating the basis functions. The right-hand side functions and patterns are localized and therefore may be more suitable for working with partial-boundary data.

solving (1.10), suggesting that it may be possible to prove unique solvability under appropriate assumptions.

We demonstrate that it is possible to recover the traces of the CGO solutions approximately on the part of the boundary available for measurements. In other words, the approximation in (1.11) is quite good in the C^2 and discontinuous conductivity examples we consider. In addition, we show below that these partial traces lead to interesting and useful reconstructions of practically relevant discontinuous conductivities.

Our new results may be useful in extending three-dimensional D-bar reconstructions, such as [**CKS06, BIK, BKM11, DHK0**], to partial-boundary data.

We mention that numerical reconstructions using restricted information about the conductivity have been published in cases of partial-boundary data, see [**MIN99, IIN, IINS10, UW08**]. The present work differs from those in that we aim to recover the full unknown conductivity function instead of inclusions in a known background. Also, there is an alternative methodology for partial-data EIT based on resistor networks [**Mam10, BDMGV10, BDM10**]; our work again represents a very different approach. Finally, we mention that there is a large body of work on iterative solution methods for EIT in cases of partial data; those studies are fundamentally different from our direct (non-iterative) approach.

2. Method 1: The method based on the Schrödinger equation

As mentioned above, the proof by [**Nac96**] transforms the conductivity equation (1.5) to the Schrödinger equation

$$(-\Delta + q(z))\,\psi(z,k) = 0, \quad z \in \Omega \text{ and } k \in \mathbb{C},$$

via the change of variables $\psi(z,k) = \sqrt{\sigma}u(z,k)$, where

$$q(z) = \frac{\Delta\sqrt{\sigma(z)}}{\sqrt{\sigma(z)}},$$

denotes the Schrödinger potential and $\mathbb{R}^2$ is associated with $\mathbb{C}$ via $z = (x,y) = x+iy$. Without loss of generality, the conductivity σ is assumed to be 1 near $\partial\Omega$ and then extended to 1 in all of $\mathbb{C}$. Existence and uniqueness are then studied for the well known Schrödinger equation with CGO solutions ψ that are asymptotic to e^{ikz} and $kz = (k_1 + ik_2)(x+iy)$.

The alternative Lippmann-Schwinger formulation

$$\mu(z,k) = 1 - g_k * (q\mu)\,, \quad z,\ k \in \mathbb{C}, \tag{2.1}$$

of the Schrödinger equation uses the related CGO solutions $\mu(z,k) = e^{-ikz}\psi(z,k)$ where g_k is related to the Faddeev Green's function and defined in (1.9).

The reconstruction method of Nachman [**Nac96**] from infinite-precision data consists of the following two steps:

$$\Lambda_\sigma \xrightarrow{\ 1\ } \mathbf{t}(k) \xrightarrow{\ 2\ } \sigma.$$

Step 1: From boundary measurements Λ_σ to the scattering transform t. For each fixed $k \in \mathbb{C}$, solve in $H^{1/2}(\partial\Omega)$ the integral equation

$$\psi(z,k) = e^{ikz} - \int_{\partial\Omega} G_k(z-\zeta)(\Lambda_\sigma - \Lambda_1)\psi(\zeta,k)\,dS(\zeta), \quad z \in \partial\Omega, \tag{2.2}$$

for the CGO solutions ψ where the D-N map of the homogeneous conductivity 1 is denoted by Λ_1. Then, substitute ψ into the formula for the nonlinear scattering transform $\mathbf{t} : \mathbb{C} \to \mathbb{C}$:

$$\mathbf{t}(k) = \int_{\partial\Omega} e^{i\bar{k}\bar{z}}(\Lambda_\sigma - \Lambda_1)\psi(z,k)\,dS(z), \tag{2.3}$$

where dS denotes arclength measure on $\partial\Omega$.

Step 2: From the scattering transform t to the conductivity σ. Denote $e(z,k) := \exp(i(kz + \overline{k}\overline{z}))$. For each fixed $z \in \Omega$, solve the integral equation

$$\mu(z,k) = 1 + \frac{1}{(2\pi)^2}\int_{\mathbb{R}^2} \frac{\mathbf{t}(k')}{(k-k')\overline{k'}} e(-z,k')\overline{\mu(z,k')}dk_1'dk_2', \tag{2.4}$$

then the conductivity is recovered by $\sigma(z) = \mu(z,0)^2$.

The integral equation (2.4) was obtained from a corresponding partial differential equation, a so-called D-bar equation, which involves the derivative with respect to the complex variable $\overline{k}$. This is where the D-bar method gets its name.

3. Method 2: The method based on a ∂_z and $\overline{\partial}_z$ system

As the existence/uniqueness result by Francini [**Fra00**] holds for complex admittivities $\gamma = \sigma + i\omega\epsilon$, where ω is the frequency of the applied current and ϵ denotes the electrical permittivity, and is an extension of that by Brown and Uhlmann [**BU97**] we will formulate the problem in the complex case. Let $u_1(z,k)$ and $u_2(z,k)$ be two CGO solutions to (1.5) with asymptotic behavior $\frac{e^{ikz}}{ik}$ and $\frac{e^{ik\bar{z}}}{-ik}$, respectively. Introduce a matrix $\Psi(z,k)$ of CGO solutions related to u_1 and u_2 by

$$\begin{pmatrix} \Psi_{11} \\ \Psi_{21} \end{pmatrix} = \gamma^{1/2} \begin{pmatrix} \partial_z u_1 \\ \overline{\partial}_z u_1 \end{pmatrix}, \qquad \begin{pmatrix} \Psi_{12} \\ \Psi_{22} \end{pmatrix} = \gamma^{1/2} \begin{pmatrix} \partial_z u_2 \\ \overline{\partial}_z u_2 \end{pmatrix},$$

for $z \in \Omega$ and $k \in \mathbb{C}$. The transformed system is then

$$D\Psi = Q\Psi, \tag{3.1}$$

where D is a matrix of ∂_z and $\overline{\partial}_z$ partial derivatives and Q represents a matrix potential

$$Q(z) = \begin{pmatrix} 0 & -\frac{1}{2}\partial_z \log\gamma(z) \\ -\frac{1}{2}\bar{\partial}_z \log\gamma(z) & 0 \end{pmatrix}, \qquad D = \begin{pmatrix} \overline{\partial}_z & 0 \\ 0 & \partial_z \end{pmatrix}. \tag{3.2}$$

The admittivity γ is assumed to be 1 near $\partial\Omega$ and is extended to 1 in all of $\mathbb{C}$. Existence and uniqueness of solutions are then studied for (3.1) for $z \in \mathbb{C}$ instead of (1.5). As it is more practical to work with CGOs with finite asymptotic behavior, we often make use of the related matrix of CGO solutions $M(z,k) \sim \begin{pmatrix} 1 & 0 \\ 0 & 1 \end{pmatrix}$ defined by

$$M(z,k) = \Psi(z,k) \begin{pmatrix} e^{-izk} & 0 \\ 0 & e^{i\bar{z}k} \end{pmatrix} = \begin{pmatrix} e^{-izk}\Psi_{11}(z,k) & e^{i\bar{z}k}\Psi_{12}(z,k) \\ e^{-izk}\Psi_{21}(z,k) & e^{i\bar{z}k}\Psi_{22}(z,k) \end{pmatrix}. \tag{3.3}$$

Similarly to Method 1, the full data direct reconstruction algorithm [**HHMV12**] also involves solving Fredholm integral equations of the second kind for CGO solutions using D-N data, evaluating a nonlinear scattering transform $S(k)$, solving a $\overline{\partial}_k$ equation, and using the recovered CGO solutions at $k = 0$ to reconstruct the conductivity and permittivity. The method can be summarized in the following steps:

$$\Lambda_\gamma \xrightarrow{1} S(k) \xrightarrow{2} M(z,0) \xrightarrow{3} \gamma.$$

Step 1: From boundary measurements Λ_γ to the scattering transform S. For fixed $k \in \mathbb{C}\setminus 0$, solve Fredholm integral equations of the second kind on $\partial\Omega$ for the traces of the CGO solutions $u_1(z,k)$ and $u_2(z,k)$:

$$u_1(z,k) = \frac{e^{ikz}}{ik} - \int_{\partial\Omega} G_k(z-\zeta)(\Lambda_\gamma - \Lambda_1)u_1(\zeta,k)\,dS(\zeta), \quad z \in \partial\Omega \tag{3.4}$$

$$u_2(z,k) = \frac{e^{-ik\bar{z}}}{-ik} - \int_{\partial\Omega} G_k(-\bar{z}+\overline{\zeta})(\Lambda_\gamma - \Lambda_1)u_2(\zeta,k)\,dS(\zeta), \quad z \in \partial\Omega. \tag{3.5}$$

Use the traces of u_1 and u_2 to compute the off diagonal entries of the CGO solutions $\Psi(z,k)$ for $z \in \partial\Omega$ from the BIEs

$$\Psi_{12}(z,k) = \int_{\partial\Omega} \frac{e^{i\bar{k}(z-\zeta)}}{4\pi(z-\zeta)} [\Lambda_\gamma - \Lambda_1]\, u_2(\zeta,k)\, dS(\zeta), \quad z \in \partial\Omega \tag{3.6}$$

$$\Psi_{21}(z,k) = \int_{\partial\Omega} \overline{\left[\frac{e^{ik(z-\zeta)}}{4\pi(z-\zeta)}\right]} [\Lambda_\gamma - \Lambda_1]\, u_1(\zeta,k)\, dS(\zeta), \quad z \in \partial\Omega, \tag{3.7}$$

and compute the off-diagonal entries of the scattering matrix $S(k)$

$$S_{12}(k) = \frac{i}{2\pi} \int_{\partial\Omega} e^{-i\bar{k}z} \Psi_{12}(z,k)(\nu_1 + i\nu_2) dS(z), \quad k \in \mathbb{C} \tag{3.8}$$

$$S_{21}(k) = -\frac{i}{2\pi} \int_{\partial\Omega} e^{ik\bar{z}} \Psi_{21}(z,k)(\nu_1 - i\nu_2) dS(z), \quad k \in \mathbb{C}. \tag{3.9}$$

Interpolate the scattering data $S(k)$ to include $k = 0$.

Step 2: From the scattering transform $S(k)$ to CGO solutions $M(z,0)$. Solve the $\bar{\partial}_k$ equation (3.10) for the matrix $M(z,k)$

$$\bar{\partial}_k M(z,k) = M(z,\bar{k}) \begin{pmatrix} e(z,\bar{k}) & 0 \\ 0 & e(z,-k) \end{pmatrix} S(k). \tag{3.10}$$

Step 3: From CGO solutions $M(z,0)$ to the Admittivity $\gamma = \sigma + i\omega\epsilon$. Reconstruct the matrix potential Q from

$$Q_{12}(z) = \frac{\bar{\partial}_z M_+(z,0)}{M_-(z,0)}, \qquad Q_{21}(z) = \frac{\partial_z M_-(z,0)}{M_+(z,0)}, \tag{3.11}$$

where

$$M_+(z,k) = M_{11}(z,k) + e^{-i(kz+\bar{k}\bar{z})} M_{12}(z,k) \tag{3.12}$$

$$M_-(z,k) = M_{22}(z,k) + e^{i(kz+\bar{k}\bar{z})} M_{21}(z,k), \tag{3.13}$$

and use either Q_{12} or Q_{21} to recover γ

$$\gamma(z) = \exp\left\{-\frac{2}{\pi}\int_\Omega \frac{Q_{12}(\zeta)}{\bar{z}-\bar{\zeta}}\, d\mu(\zeta)\right\} = \exp\left\{-\frac{2}{\pi}\int_\Omega \frac{Q_{21}(\zeta)}{z-\zeta}\, d\mu(\zeta)\right\}, \tag{3.14}$$

where the integration takes place over Ω rather than all of $\mathbb{C}$ due to the compact support of the matrix potential Q.

4. Computation of Partial Boundary Data CGO Solutions

In this work we use *localized basis functions* in place of *global basis functions.* As mentioned above, the most commonly used global basis functions for the continuum model are the exponential trigonometric basis functions $e^{in\theta}$. When electrode models (such as the gap, shunt, or complete electrode model) are used, a trigonometric basis of sines and cosines is often used instead [**DM10, IMNS04**]. As mentioned above, the common thread of these global basis functions is that their support is essentially the entire boundary of the domain. By contrast, localized basis functions are supported on a subset of the boundary. Examples of localized basis patterns include the skip patterns and adjacent patterns (see e.g., [**Ham12, Mur07, HM13**]) as well as the Haar wavelets. In this work we use the Haar wavelets as they are localized basis functions that can be naturally used in both the continuum and electrode model cases.

As the boundary integral equations in Methods 1 and 2 are very similar, we will describe, without loss of generality, the computation in detail for Method 1. In order to solve the boundary integral equation (2.2)

$$\psi(z,k) = e^{ikz} - \int_{\partial\Omega} G_k(z-\zeta)\,(\Lambda_\sigma - \Lambda_1)\,\psi(\zeta,k)\,dS(\zeta), \quad z \in \partial\Omega,$$

for the traces of the CGO solutions $\psi(z,k)$, we will need the Dirichlet-to-Neumann (D-N) map, and thus we must first discuss the applied voltage patterns, in this case, the Haar wavelets.

4.1. Description of Haar Wavelets. Let Γ denote a subset of the boundary $\partial\Omega$ and let $|\Gamma| = \mathcal{L}$ denote the length of the subset Γ. The first wavelet is the *scaling function* which we will denote ϕ_1, and is defined as:

$$\begin{aligned} \phi_1(z) &= h_1 \quad z \in \Gamma \subseteq \partial\Omega \\ h_1 &= \sqrt{\tfrac{1}{\mathcal{L}}}. \end{aligned} \tag{4.1}$$

If $z \in \partial\Omega\setminus\Gamma$, the scaling function $\phi_1(z)$ is set to zero, as will be the case for the subsequent Haar wavelets.

For ease of notation, let $d(z)$ be the distance, along the subset Γ of the boundary, a point z is from the beginning point z_0 on Γ (corresponding to the smallest θ value in the traditional counter-clockwise orientation) and $z_{\mathcal{L}}$ the ending point (corresponding to the largest θ value). Thus, $d = 0$ at z_0 and $d = \mathcal{L}$ at $z_{\mathcal{L}}$.

The second wavelet is the so-called *mother wavelet* which we will denote ϕ_2 and is defined as:

$$\phi_2(z) = \begin{cases} h_1, & \{z \in \Gamma | 0 \le d(z) < \frac{\mathcal{L}}{2}\} \\ -h_1, & \{z \in \Gamma | \frac{\mathcal{L}}{2} \le d(z) \le \mathcal{L}\}. \end{cases} \tag{4.2}$$

As the basis functions need to be orthonormal, we require $\langle \phi_m, \phi_n \rangle = \delta_{m,n}$, which the above wavelets satisfy by construction.

The third and fourth Haar wavelets ϕ_3 and ϕ_4 are copies of the mother wavelet ϕ_2, squished into 1/2 the length of the support of ϕ_2 as follows:

$$\begin{aligned} \phi_3(z) &= \begin{cases} h_2, & \{z \in \Gamma | 0 \le d(z) < \frac{\mathcal{L}}{4}\} \\ -h_2, & \{z \in \Gamma | \frac{\mathcal{L}}{4} \le d(z) \le \frac{\mathcal{L}}{2}\}, \end{cases} \\ \phi_4(z) &= \begin{cases} h_2, & \{z \in \Gamma | \frac{\mathcal{L}}{2} \le d(z) < \frac{3\mathcal{L}}{4}\} \\ -h_2, & \{z \in \Gamma | \frac{3\mathcal{L}}{4} \le d(z) \le \mathcal{L}\}, \end{cases} \\ h_2 &= \sqrt{\tfrac{2}{\mathcal{L}}}. \end{aligned} \tag{4.3}$$

Notice that these new wavelets satisfy $\langle \phi_m, \phi_n \rangle = \delta_{m,n}$ for $m, n = 1, \ldots, 4$.

An exact formula for the j-th height function h_j is

$$h_j = \sqrt{\frac{2^{j-1}}{\mathcal{L}}}, \quad j \ge 1,$$

corresponding to Haar wavelets with support width w_j

$$w_j = \frac{\mathcal{L}}{2^{j-1}}, \quad j \geq 2.$$

4.2. Formation of the D-N map Using Haar Wavelets. As the main goal is to use only partial boundary data, thus applying and measuring data only on subset of the domain, it is more natural to apply voltages (rather than currents) and form the Dirichlet-to-Neumann (D-N) map directly. Although in practice currents are frequently applied, and thus the Neumann-to-Dirichlet (N-D) map is formed first (which is done to dampen noise), that approach requires the inversion of the N-D map, which for partial data poses new questions. As a preliminary approach, we proceed with Dirichlet data.

The conductivity equation (1.5) can be solved using the Finite Element Method. For each Haar wavelet, the Dirichlet boundary value problem is solved and the resulting solution u in Ω is used to determine the current flux (Neumann data) at the boundary. This allows the determination of the Neumann data corresponding to the prescribed Dirichlet data and formation the D-N map:

$$\Lambda_\sigma f = \sigma \left.\frac{\partial u}{\partial \nu}\right|_{\partial\Omega}.$$

Note that for the cases in this document, the conductivity on and near the boundary is 1.

The discrete matrix approximation to the D-N map is formed using the following formula for the (m, n)-th entry

$$\Lambda_\sigma^M(m,n) := \langle \Lambda_\sigma \phi_m, \phi_n \rangle = \langle \sigma \nabla \phi_m \cdot \nu, \phi_n \rangle = \langle \nabla \phi_m \cdot \nu, \phi_n \rangle, \tag{4.4}$$

where ϕ_j are the Haar wavelets described in Section 4.1 that now serve as the Dirichlet data and $\langle \cdot, \cdot \rangle$ denotes the L^2 inner product. As ν denotes the outward facing unit normal and Ω is the unit disc, at the boundary point $z = e^{i\theta} = \cos(\theta) + i\sin(\theta)$, we have $\nu = (\cos(\theta), \sin(\theta))$.

4.3. Solution of the Full Data BIE. After forming the D-N map using the Haar wavelets, localized basis functions, we proceed to solving the boundary integral equation for the traces of CGO solutions $\psi(z, k)$. This involves the solution of a Fredholm integral equation of the second kind. Following the approach of [**DM10, HHMV12, HM13, Ham12**], we expand the exponential e^{ikz} and CGOs $\psi(z, k)$ in the Haar wavelet patterns $\{\Phi_j\}_{j=1}^J$. Let z_ℓ be an evaluation point on $\partial\Omega$ and J denote the number of linearly independent Haar wavelet functions used. Then the values of the CGO solution ψ and complex exponential e^{ikz}, for a given complex number k, at position z_ℓ on $\partial\Omega$ are given by

$$\psi(z_\ell, k) \approx \sum_{j=1}^J b_j(k) \Phi_\ell^j, \tag{4.5}$$

and

$$e^{ikz_\ell} \approx \sum_{j=1}^J c_j(k) \Phi_\ell^j, \tag{4.6}$$

where Φ denotes the normalized Haar wavelets such that their ℓ^2 norm is 1, i.e. they are related via

$$\Phi_j = \sqrt{\frac{\mathcal{L}}{L}}\phi_j.$$

Let $\mathbf{b}(k)$ denote the column vector $\mathbf{b}(k) = [b_1(k), \ldots, b_J(k)]^T$, and define $\mathbf{c}(k)$ analogously where T denotes the standard matrix non-conjugate transpose.

Let $E_{\ell'}$ denote the ℓ'-th subdivision of the boundary $\partial\Omega$ ($\ell' = 1, \ldots, L$) centered at the center of the ℓ'-th boundary element $z_{\ell'}$ with length $2\pi/L$. Splitting the integral over $\partial\Omega$ into a sum of integrals over the subsections $E_{\ell'}$

$$\begin{aligned} \psi(z_\ell, k) &\approx e^{ikz_\ell} - \sum_{\ell'=1}^{L} \int_{E_{\ell'}} G_k\left(z_\ell - \zeta\right) \delta\Lambda_\sigma \psi(\zeta_{\ell'}, k)\, dS(\zeta) \\ &= e^{ikz_\ell} - \sum_{\ell'=1}^{L} \int_{E_{\ell'}} G_k\left(z_\ell - \zeta\right)\, dS(\zeta)\, \left[\delta\Lambda_\sigma \psi(\zeta_{\ell'}, k)\right], \end{aligned}$$

where for ease of notation

$$\delta\Lambda_\sigma = \Lambda_\sigma - \Lambda_1.$$

Using the expansions for $\psi(z_\ell, k)$ and e^{ikz_ℓ}, (4.5) and (4.6) respectively, we have

$$\begin{aligned} \sum_{j=1}^{J} b_j(k)\Phi_\ell^j &\approx \sum_{j=1}^{J} c_j(k)\Phi_\ell^j - \sum_{\ell'=1}^{L} \int_{E_{\ell'}} G_k\left(z_\ell - \zeta\right)\, dS(\zeta) \left[\delta\Lambda_\sigma \sum_{j=1}^{J} b_j(k)\Phi_{\ell'}^j\right] \\ &= \sum_{j=1}^{J} c_j(k)\Phi_\ell^j - \sum_{\ell'=1}^{L} \int_{E_{\ell'}} G_k\left(z_\ell - \zeta\right)\, dS(\zeta) \sum_{j=1}^{J} b_j(k) f_j\left(\zeta_{\ell'}\right), \end{aligned}$$

where $f_j\left(\zeta_{\ell'}\right)$ denotes the action of the discretized $\delta\Lambda_\sigma^M$ matrix on the j-th normalized Haar wavelet basis function evaluated at $\zeta_{\ell'}$. Define the matrix approximation to the Faddeev Green's function as

$$\mathbf{G}_k(\ell, \ell') = \begin{cases} G_k\left(z_\ell, \zeta_{\ell'}\right) & \ell \neq \ell' \\ 0 & \ell = \ell', \end{cases} \tag{4.7}$$

removing the singularity at $G_k(0)$. Then

$$\sum_{j=1}^{J} b_j(k)\Phi_\ell^j \approx \sum_{j=1}^{J} c_j(k)\Phi_\ell^j - \frac{2\pi}{L}\sum_{j=1}^{J} b_j(k) \sum_{\ell'=1}^{L} \mathbf{G}_k(\ell, \ell') f_j\left(\zeta_{\ell'}\right). \tag{4.8}$$

Following [**DM10**]

$$f_p(\zeta_{\ell'}) \approx \left(\Phi\delta\Lambda_\sigma^M\right)(\ell', j), \tag{4.9}$$

i.e., the (ℓ', j) entry in the matrix resulting from multiplication of the matrix of normalized basis functions Φ and the discretized difference in D-N maps $\delta\Lambda_\sigma^M$. Using the properties of matrix multiplication, equation (4.8) can be rewritten as

$$\sum_{j=1}^{J} b_j(k)\Phi_\ell^j = \sum_{j=1}^{J} c_j(k)\Phi_\ell^j - \frac{2\pi}{L}\sum_{j=1}^{J} b_j(k) \left(\mathbf{G}_k\Phi\delta\Lambda_\sigma^M\right)(\ell, j),$$

or equivalently,

$$\Phi\mathbf{b} = \Phi\mathbf{c} - \frac{2\pi}{L}\mathbf{G}_k\Phi\delta\Lambda_\sigma^M\mathbf{b},$$

a matrix equation for the unknown coefficients $\mathbf{b}$ which are needed in the normalized Haar wavelet basis expansion of $\psi(z,k)$.

Using the orthonormality of the normalized Haar wavelet basis functions in the matrix Φ, we multiply both sides of the equation by Φ^T, and then solve

$$(I+A)\mathbf{b}=\mathbf{c}, \tag{4.10}$$

where

$$A=\frac{2\pi}{L}\Phi^T\mathbf{G}_k\Phi\delta\Lambda_\sigma^M. \tag{4.11}$$

To reiterate, for each desired value of $k\in\mathbb{C}$, expand e^{ikz} for $z\in\partial\Omega$ in the normalized Haar wavelets Φ to define the vector of coefficients $\mathbf{c}$, and solve the system (4.10) using GMRES for the unknown coefficients $\mathbf{b}$. These coefficients are then used to reconstruct $\psi(z,k)$ for the specified value of k via (4.5).

4.4. Solution of the Partial Data BIE. We now proceed to the problem of interest, namely, the solution of the boundary integral equation (2.2) when only part of the boundary is accessible for data acquisition. The solution method is nearly identical to the full data Haar wavelet case presented above.

Now Γ is a proper subset of the boundary $\partial\Omega$ and the Haar wavelets and D-N map are formed as above. The D-N map now corresponds to data taken only on the proper subset Γ since the applied voltage is 0 off Γ.

Let $\tilde{z}$ denote the boundary values z restricted to Γ and $\tilde{\psi}$ the corresponding partial data CGO solutions. We then expand $e^{ik\tilde{z}}$ and $\tilde{\psi}$ as before and solve the resulting system for each desired value of $k\in\mathbb{C}$:

$$(I+\tilde{A})\tilde{\mathbf{b}}=\tilde{\mathbf{c}},$$

where

$$\tilde{A}=\frac{\mathcal{L}}{L}\tilde{\Phi}^T\tilde{\mathbf{G}}_k\tilde{\Phi}\delta\tilde{\Lambda}_\sigma^M, \tag{4.12}$$

and $\mathcal{L}=|\Gamma|$.

5. Numerical Reconstruction of Conductivities from Partial Data Using Method 2

As stated above, the boundary integral equations (3.4) and (3.5) are nearly identical to the BIE (2.2) described above for Method 1. Thus, their traces on the subset Γ of the boundary can be recovered by solving analogous formulas. The coefficients $\tilde{\mathbf{b}}^{\mathbf{1}}$ for u_1 are determined by solving

$$(I+\tilde{A})\tilde{\mathbf{b}}^{\mathbf{1}}=\tilde{\mathbf{c}}^{\mathbf{1}},$$

where $\tilde{\mathbf{c}}^{\mathbf{1}}$ are the coefficients in the Haar expansion of $\frac{e^{ik\tilde{z}}}{ik}$. Similarly, the coefficients $\tilde{\mathbf{b}}^{\mathbf{2}}$ for u_2 are determined by solving

$$(I+\tilde{A}_2)\tilde{\mathbf{b}}^{\mathbf{2}}=\tilde{\mathbf{c}}^{\mathbf{2}},$$

where $\tilde{\mathbf{c}}^{\mathbf{2}}$ are the coefficients in the Haar expansion of $\frac{e^{-ik\bar{\tilde{z}}}}{-ik}$ and $\tilde{A}_2$ now contains the matrix approximation of $G_k(-\bar{z}+\bar{\zeta})$ instead of $G_k(z-\zeta)$.

A natural question is whether these CGO solutions, which match very well on Γ with their full data counterparts (see Section 6.1), can be used to produce informative reconstructions of the conductivity (and/or permittivity) near the region of the accessible boundary. Our aim was to understand the extent of the impact of the partial data CGO solutions on the remainder of a D-bar algorithm. Therefore, as an initial test, we left the remainder of the algorithm for Method 2 intact which means computing the intermediate CGO solutions Ψ_{12} and Ψ_{21} using the partial D-N map and the partial data CGO solutions u_1 and u_2, computing the scattering transforms S_{12} and S_{21} over Γ, and proceeding with Steps 2-3 as before. The steps of the proposed partial data algorithm are included here for the reader's convenience.

$$\tilde{\Lambda}_\gamma \xrightarrow{1} \tilde{S}(k) \xrightarrow{2} \tilde{M}(z,0) \xrightarrow{3} \tilde{\gamma}.$$

Step 1: From *partial* boundary measurements $\tilde{\Lambda}_\gamma$ to the approx. scattering transform $\tilde{S}(k)$.

For fixed $k \in \mathbb{C}\backslash 0$ such that $|k| < R$ a fixed radius depending on the measured D-N map, solve Fredholm integral equations of the second kind on $\Gamma \subset \partial\Omega$ for the approximate traces of the CGO solutions $\tilde{u}_1(z,k)$ and $\tilde{u}_2(z,k)$ on $\Gamma \subset \partial\Omega$

$$\tilde{u}_1(z,k) = \left.\frac{e^{ikz}}{ik}\right|_\Gamma - \int_\Gamma G_k(z-\zeta)(\tilde{\Lambda}_\gamma - \tilde{\Lambda}_1)\tilde{u}_1(\zeta,k)dS(\zeta), \quad z \in \Gamma \tag{5.1}$$

$$\tilde{u}_2(z,k) = \left.\frac{e^{-ik\bar{z}}}{-ik}\right|_\Gamma - \int_\Gamma G_k(-\bar{z}+\bar{\zeta})(\tilde{\Lambda}_\gamma - \tilde{\Lambda}_1)\tilde{u}_2(\zeta,k)dS(\zeta), \quad z \in \Gamma. \tag{5.2}$$

Use the approximate traces of u_1 and u_2 (namely, $\tilde{u}_1$ and $\tilde{u}_2$) to compute the approximate off diagonal entries of the CGO solutions $\widetilde{\Psi}(z,k)$ for $z \in \Gamma \subset \partial\Omega$ from the BIEs

$$\widetilde{\Psi}_{12}(z,k) = \int_\Gamma \frac{e^{i\bar{k}(z-\zeta)}}{4\pi(z-\zeta)}\left[\tilde{\Lambda}_\gamma - \tilde{\Lambda}_1\right]\tilde{u}_2(\zeta,k)\,dS(\zeta), \quad z \in \Gamma \tag{5.3}$$

$$\widetilde{\Psi}_{21}(z,k) = \int_\Gamma \overline{\left[\frac{e^{ik(z-\zeta)}}{4\pi(z-\zeta)}\right]}\left[\tilde{\Lambda}_\gamma - \tilde{\Lambda}_1\right]\tilde{u}_1(\zeta,k)\,dS(\zeta), \quad z \in \Gamma, \tag{5.4}$$

and compute the off-diagonal entries of the scattering matrix $\tilde{S}(k)$ integrating over $\Gamma \subset \partial\Omega$

$$\tilde{S}_{12}(k) = \frac{i}{2\pi}\int_\Gamma e^{-i\bar{k}z}\widetilde{\Psi}_{12}(z,k)(\nu_1 + i\nu_2)dS(z) \tag{5.5}$$

$$\tilde{S}_{21}(k) = -\frac{i}{2\pi}\int_\Gamma e^{ik\bar{z}}\widetilde{\Psi}_{21}(z,k)(\nu_1 - i\nu_2)dS(z). \tag{5.6}$$

Interpolate the approximate scattering data $\tilde{S}(k)$ to include $k=0$.

Step 2: From the approx. scattering transform $\tilde{S}(k)$ to approx. CGO solutions $\widetilde{M}(z,0)$.

Solve the $\bar{\partial}_k$ equation (3.10) for the matrix of approx. CGO solutions $\widetilde{M}(z,k)$

$$\bar{\partial}_k \widetilde{M}(z,k) = \widetilde{M}(z,\bar{k})\begin{pmatrix} e(z,\bar{k}) & 0 \\ 0 & e(z,-k)\end{pmatrix}\tilde{S}(k). \tag{5.7}$$

Step 3: From the approximate CGO solutions $\widetilde{M}(z,0)$ to approx. admittivity $\tilde{\gamma} = \tilde{\sigma} + i\omega\tilde{\epsilon}$.
Reconstruct the approximate matrix potential $\tilde{Q}$ from

$$\tilde{Q}_{12}(z) = \frac{\bar{\partial}_z \widetilde{M}_+(z,0)}{\widetilde{M}_-(z,0)}, \qquad \tilde{Q}_{21}(z) = \frac{\partial_z \widetilde{M}_-(z,0)}{\widetilde{M}_+(z,0)}, \tag{5.8}$$

where

$$\widetilde{M}_+(z,k) = \widetilde{M}_{11}(z,k) + e^{-i(kz+\bar{k}\bar{z})}\widetilde{M}_{12}(z,k) \tag{5.9}$$

$$\widetilde{M}_-(z,k) = \widetilde{M}_{22}(z,k) + e^{i(kz+\bar{k}\bar{z})}\widetilde{M}_{21}(z,k), \tag{5.10}$$

and use either $\tilde{Q}_{12}$ or $\tilde{Q}_{21}$ to recover the approximation $\tilde{\gamma}$

$$\tilde{\gamma}(z) = \exp\left\{-\frac{2}{\pi}\int_\Omega \frac{\tilde{Q}_{12}(\zeta)}{\bar{z}-\bar{\zeta}}\, d\mu(\zeta)\right\} = \exp\left\{-\frac{2}{\pi}\int_\Omega \frac{\tilde{Q}_{21}(\zeta)}{z-\zeta}\, d\mu(\zeta)\right\}, \tag{5.11}$$

where the integration takes place over Ω rather than all of $\mathbb{C}$ due to the compact support of the matrix potential $\tilde{Q}$.

For the numerical details regarding how to implement Steps 2-3 see [**HHMV12, Ham12, HM13**].

6. Computational Experiments

We considered two test problems. Test 1 aims to determine how a partial data D-N map affects the values of the traces of the CGO solutions on the accessible portion of the boundary. Test 2 aims to determine the effect of the partial data CGO solutions on a D-bar algorithm.

6.1. Test 1: Partial Data Traces of CGO Solutions. For the first test problem we considered the C^2 smooth conductivity given in Figure 4. The conductivity equation was first solved using the Finite Element method with 256 Haar wavelets with essential support on the entire boundary, serving as 256 different Dirichlet boundary conditions. We considered the 3/4, 1/2, and 1/4 data problems with 192, 128, and 64 Haar wavelets respectively. Each of the partial data cases is centered around $z = 1$, i.e. $\theta = 0$.

Using Method 1, we solved the full data matrix formulation (4.10) of the boundary integral equation for the traces of the CGO solutions ψ. The partial data traces of the CGO solutions were recovered by solving (4.12). In order to evaluate how well the reconstructed traces compare to the true traces we also solved the Lippmann-Schwinger equation (2.1) using the twice-differentiable conductivity in Figure 4.

Figures 5 and 6 show the recovered traces of the CGO solutions $\psi(z,k)$ for $k = 0.5$ and $-4i$ respectively, plotted against the true traces produced via the Lippmann-Schwinger computation. Preliminary results suggest that for small magnitude k, the partial data CGO solutions agree with the full data (and true) solutions on the accessible part of the domain. As the magnitude of the frequency parameter k increases, the partial data CGO solutions begin to drift slightly from the full data solutions. However, in the nonlinear CGO approaches typically used in EIT imaging, only low frequency CGO solutions are used and therefore these results are very promising.

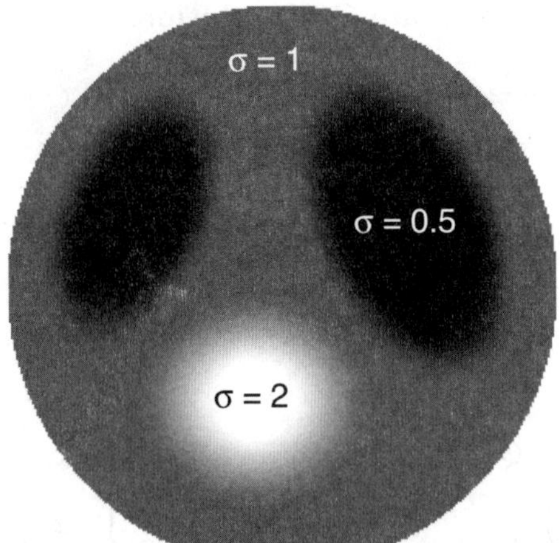

FIGURE 4. The C^2 conductivity used in Test 1.

6.2. Test 2: D-Bar Reconstructions of Conductivities using Partial Data CGO Solutions. In Test 1, we saw that the traces of the CGO solutions computed using the partial D-N data corresponding to an accessible subset Γ of the boundary $\partial\Omega$ match the true, as well as full data, traces very well on Γ for low frequencies k and C^2 smooth conductivities. Real life situations often involve cases where the conductivities are not smooth but instead only bounded. In order to determine if the traces of the partial data CGO solutions still provide useful information when the smoothness is relaxed, we consider the discontinuous test conductivity shown in Figure 7.

This test phantom could represent a saline filled tank containing an object of higher conductivity.

As the smoothness condition for Method 2 is violated (as well as that for Method 1), we cannot compare the full or partial data traces of the CGO solutions to their "true" traces. Precisely, we cannot compute the potential Q (or q) and the associated $\partial_z - \overline{\partial}_z$ system (or Lippmann-Schwinger equation) for the "true" traces of the CGO solutions. Instead, we compare the partial D-N data traces of the CGO solutions to the corresponding full D-N data traces. Figures 8 and 9 shows the reconstructed traces of the CGO solutions $\tilde{u}_1$ and $\tilde{u}_2$, respectively, for $k = 3+3i$ resulting from the conductivity distribution in Figure 7 plotted for full, 3/4, 1/2, and 1/4 D-N data with 256, 192, 128, and 64 Haar wavelets respectively. Clearly the partial D-N data traces of the CGO solutions appear to approximate the full D-N data traces of the CGO solutions in the accessible region Γ of the boundary.

Next we used the partial D-N data traces of the CGO solutions $\tilde{u}_1$ and $\tilde{u}_2$ in the modified D-bar algorithm for Method 2, described above in Section 5, to determine their effect on the algorithm and thus the reconstructed conductivity distribution. Figures 10 and 11 show the reconstructed conductivity from full, 3/4, 1/2, and 1/4 D-N data using scattering data satisfying $|k| \leq 3$ and $|k| \leq 4$, respectively. The range of the reconstructed values decreases with the size of the accessible region of the boundary Γ, however an object of higher conductivity is clearly visible in all cases. As the magnitude of k increases the reconstructed values of the conductivity

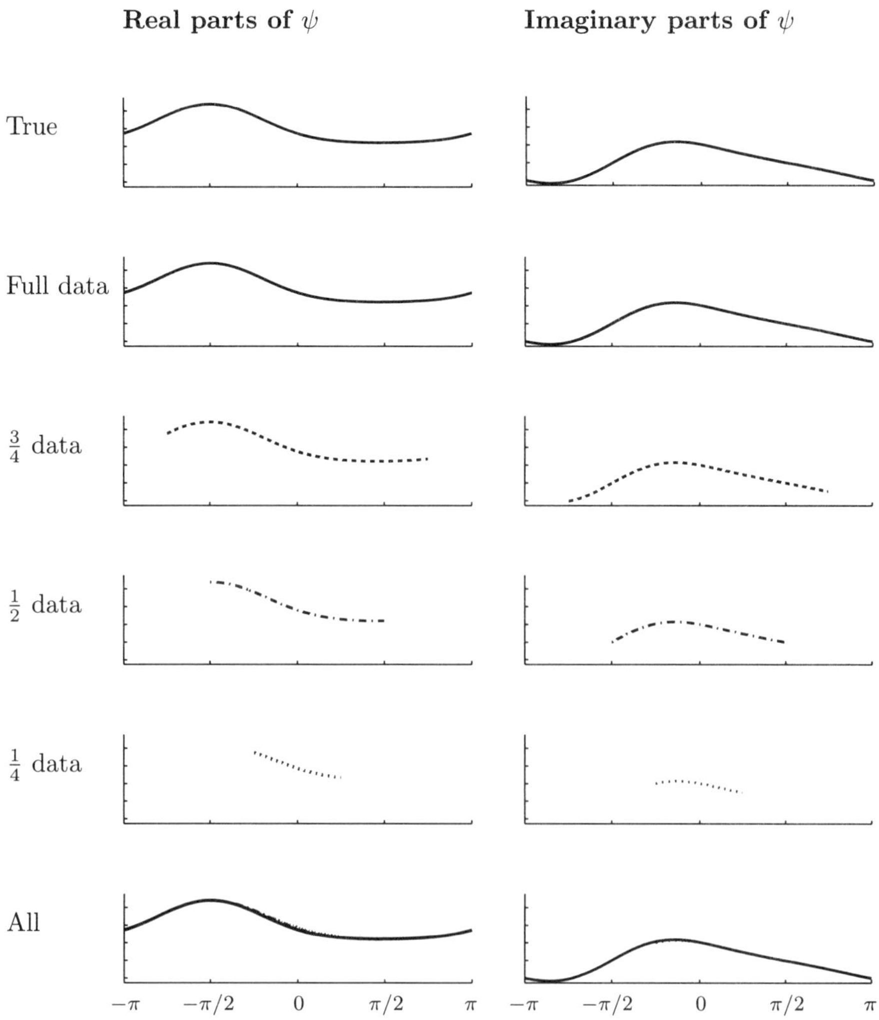

FIGURE 5. Traces of the CGO solutions ψ corresponding to the C^2 conductivity in Figure 4. Here $k = 0.5$.

improve. However, as in the full data D-N case, increasing the scattering radius too much can introduce artifacts into the reconstruction.

Note that the reconstructions of the conductivity shown in Figures 10 and 11 are all plotted on their own scales. In both figures we are clearly able to determine whether the inclusion is more or less conductive than the background, as well as its approximate location, even from as little as 25% D-N data.

Figure 12 shows the real and imaginary parts of the scattering transform $S_{21}(k)$ for Full, 3/4, 1/2, and 1/4 data with $|k| \leq 4$. Note that the scattering data is clearly affected by the loss of information in the D-N map, yet the reconstructions of the conductivity (seen in Figures 10 and 11) continue to contain valuable information. If one were to continue in this direction (using the partial data D-bar algorithm

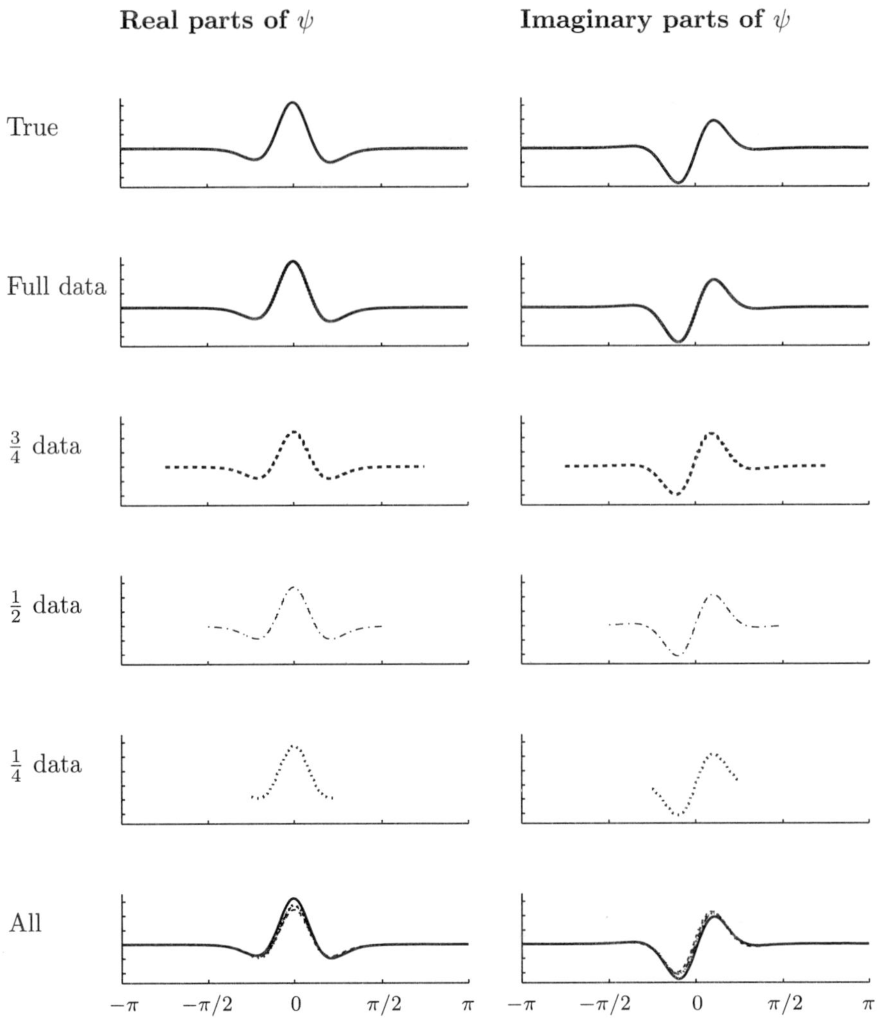

FIGURE 6. Traces of the CGO solutions ψ corresponding to the C^2 conductivity in Figure 4. Here $k = -4i$.

described in Section 5) a more in-depth study to determine which values of k are admissible in the scattering data is recommended.

7. Conclusions

The first step in nonlinear EIT imaging uses the voltage and current boundary data to determine the traces of the CGO solutions at the boundary. This is done by solving a boundary integral equation which is a Fredholm equation of the second kind, e.g., (2.2).

In this work, we used simulated partial boundary data and a wavelet-based integral equation solver to demonstrate that CGO solutions can be approximately

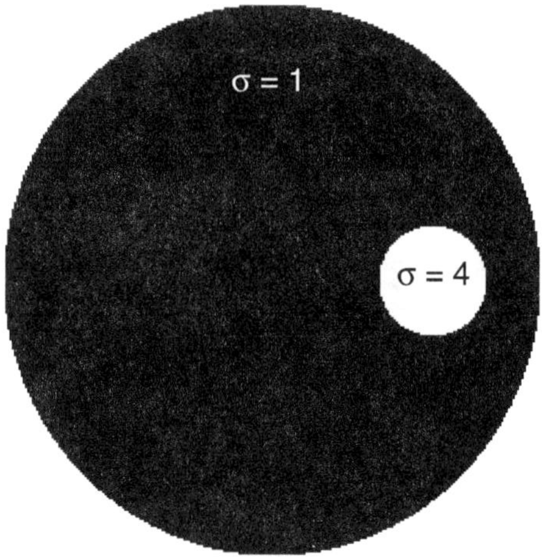

FIGURE 7. The discontinuous conductivity used in Test 2.

recovered from partial data, on the part of the boundary where the data was acquired. This result is clearly seen in Figures 5 and 6 for a C^2 conductivity, and in Figures 8 and 9 for a discontinuous conductivity. In addition, we have demonstrated that such partial data CGO solutions can be used in existing full data D-bar methods to provide useful and informative reconstructions, even in the case of discontinuous conductivities, see Figures 10 and 11.

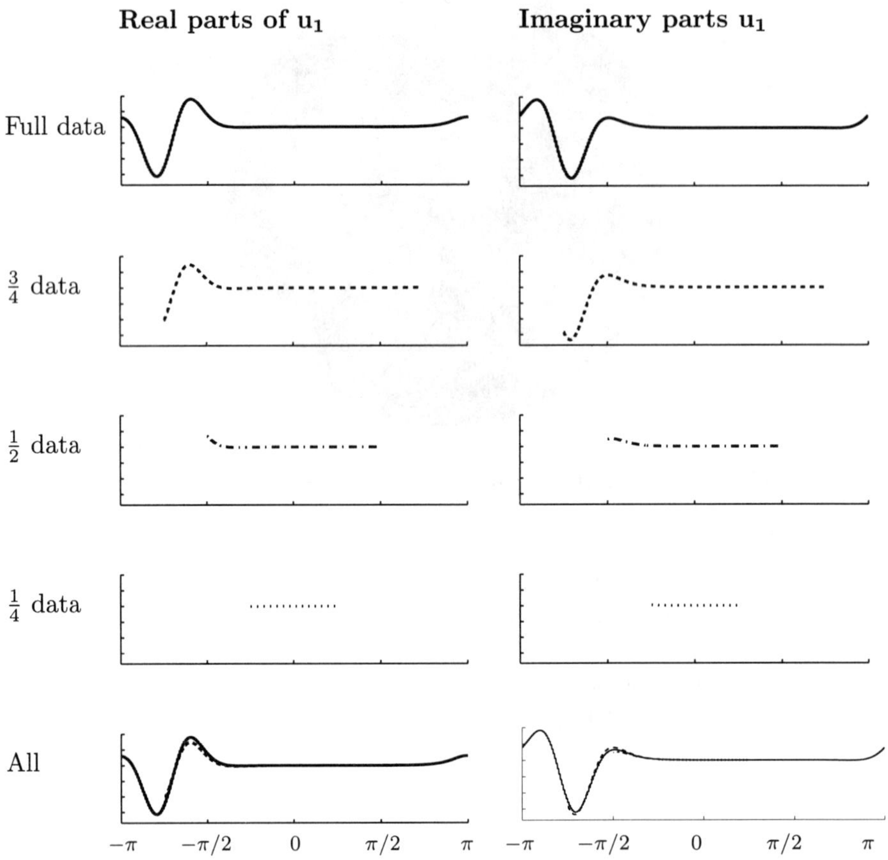

FIGURE 8. Traces of the CGO solutions u_1 corresponding to the discontinuous conductivity in Figure 7. Here $k = 3 + 3i$.

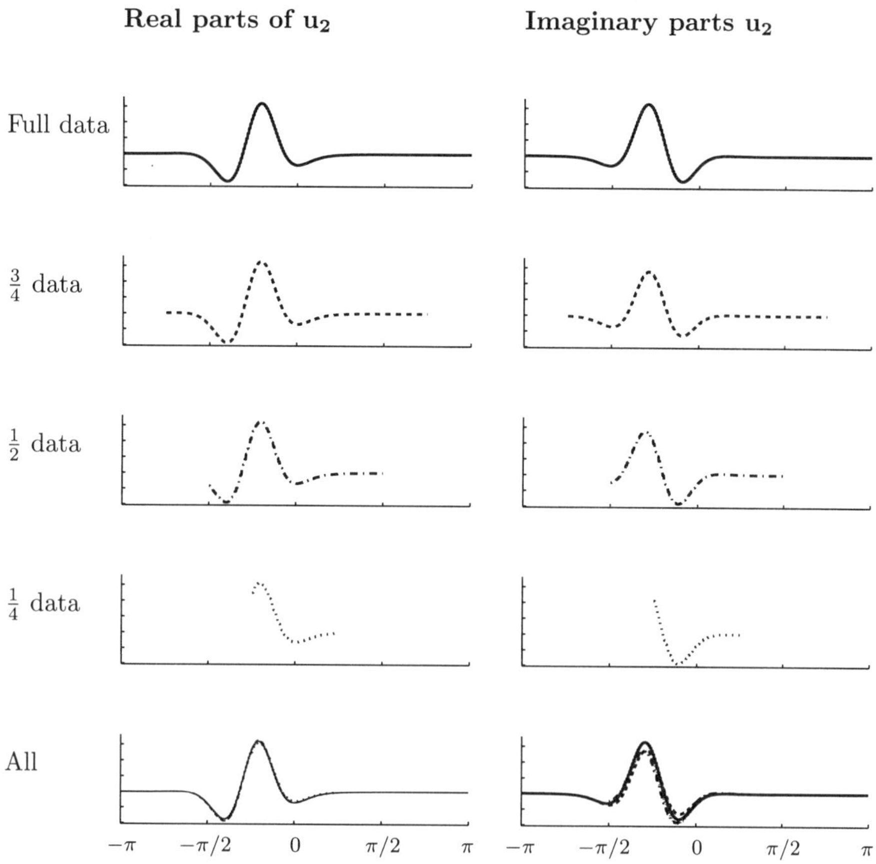

FIGURE 9. Traces of the CGO solutions u_2 corresponding to the discontinuous conductivity in Figure 7. Here $k = 3 + 3i$.

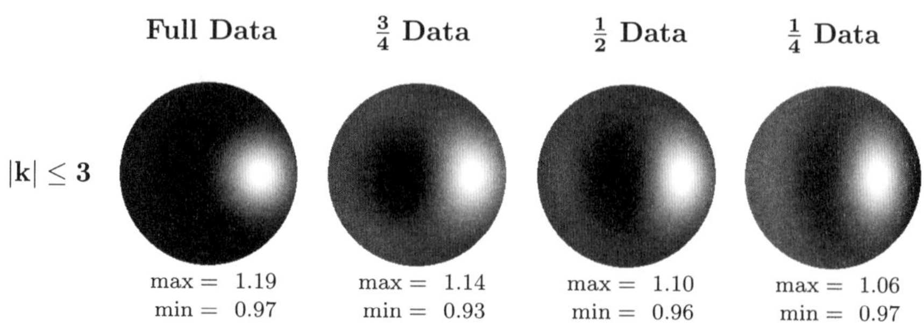

FIGURE 10. Reconstructions of the discontinuous conductivity in Figure 7 produced using scattering data for $|k| \leq 3$ using the method described in Section 5. From left to right, the reconstructions are for Full, 3/4, 1/2, and 1/4 Dirichlet-to-Neumann data.

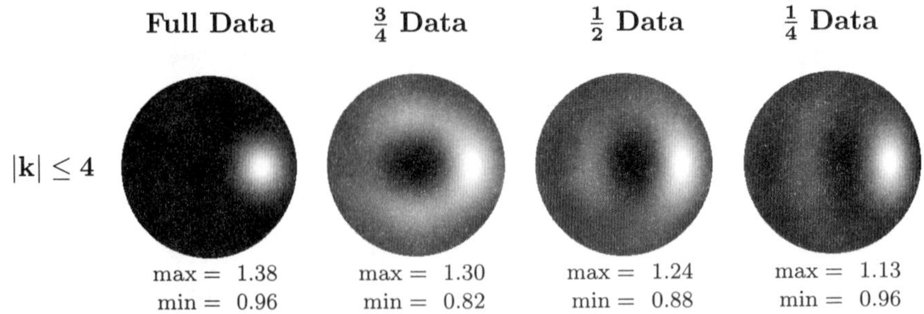

FIGURE 11. Reconstructions of the discontinuous conductivity in Figure 7 produced using scattering data for $|k| \leq 4$ using the method described in Section 5. From left to right, the reconstructions are for Full, 3/4, 1/2, and 1/4 Dirichlet-to-Neumann data.

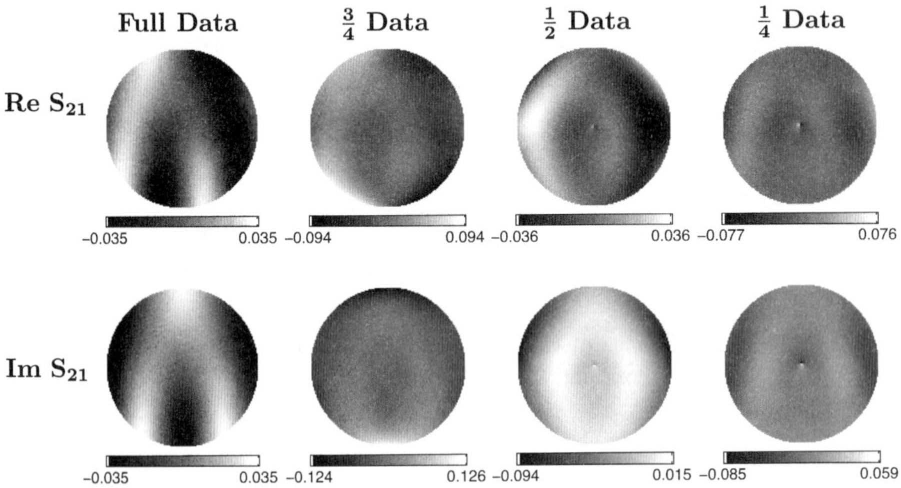

FIGURE 12. Scattering transforms S_{21} for the discontinuous conductivity in Figure 7 for $|k| \leq 4$ using the method described in Section 5. From left to right, the plots are for Full, 3/4, 1/2, and 1/4 Dirichlet-to-Neumann data.

Acknowledgments

The study was supported by the SalWe Research Program for Mind and Body (Tekes - the Finnish Funding Agency for Technology and Innovation grant 1104/10) and by the Academy of Finland (Finnish Centre of Excellence in Inverse Problems Research 2012–2017, decision number 250215).

References

[AMP] Kari Astala, Jennifer L. Mueller, Lassi Päivärinta, Allan Perämäki, and Samuli Siltanen, *Direct electrical impedance tomography for nonsmooth conductivities*, Inverse Probl. Imaging **5** (2011), no. 3, 531–549, DOI 10.3934/ipi.2011.5.531. MR2825727 (2012g:65238)

[AMPS10] Kari Astala, Jennifer L. Mueller, Lassi Päivärinta, and Samuli Siltanen, *Numerical computation of complex geometrical optics solutions to the conductivity equation*, Appl. Comput. Harmon. Anal. **29** (2010), no. 1, 2–17, DOI 10.1016/j.acha.2009.08.001. MR2647008 (2011e:65240)

[AP06a] K. Astala and L. Päivärinta, *A boundary integral equation for Calderón's inverse conductivity problem*, Proc. 7th Internat. Conference on Harmonic Analysis, Collectanea Mathematica, 2006.

[AP06b] Kari Astala and Lassi Päivärinta, *Calderón's inverse conductivity problem in the plane*, Ann. of Math. (2) **163** (2006), no. 1, 265–299, DOI 10.4007/annals.2006.163.265. MR2195135 (2007b:30019)

[BDM10] L. Borcea, V. Druskin, and A. V. Mamonov, *Circular resistor networks for electrical impedance tomography with partial boundary measurements*, Inverse Problems **26** (2010), no. 4, 045010, 30, DOI 10.1088/0266-5611/26/4/045010. MR2608623 (2011f:65239)

[BDMGV10] L. Borcea, V. Druskin, A. V. Mamonov, and F. Guevara Vasquez, *Pyramidal resistor networks for electrical impedance tomography with partial boundary measurements*, Inverse Problems **26** (2010), no. 10, 105009, 36, DOI 10.1088/0266-5611/26/10/105009. MR2719770 (2011f:65238)

[BIK] G. Boverman, D. Isaacson, T.-J. Kao, Saulnier, G. J., and J. C. Newell, *Methods for direct image reconstruction for* EIT *in two and three dimensions*, Proceedings of the 2008 Electrical Impedance Tomography Conference (Dartmouth College, in Hanover, New Hampshire, USA), June 16 to 18 2008.

[BKM11] Jutta Bikowski, Kim Knudsen, and Jennifer L. Mueller, *Direct numerical reconstruction of conductivities in three dimensions using scattering transforms*, Inverse Problems **27** (2011), no. 1, 015002, 19, DOI 10.1088/0266-5611/27/1/015002. MR2746405 (2012a:65301)

[BU97] Russell M. Brown and Gunther A. Uhlmann, *Uniqueness in the inverse conductivity problem for nonsmooth conductivities in two dimensions*, Comm. Partial Differential Equations **22** (1997), no. 5-6, 1009–1027, DOI 10.1080/03605309708821292. MR1452176 (98f:35155)

[Cal80] Alberto-P. Calderón, *On an inverse boundary value problem*, (Rio de Janeiro, 1980), Soc. Brasil. Mat., Rio de Janeiro, 1980, pp. 65–73. MR590275 (81k:35160)

[CKS06] H. Cornean, K. Knudsen, and S. Siltanen, *Towards a d-bar reconstruction method for three-dimensional EIT*, J. Inverse Ill-Posed Probl. **14** (2006), no. 2, 111–134, DOI 10.1163/156939406777571102. MR2242300 (2007f:65046)

[DHK0] Fabrice Delbary, Per Christian Hansen, and Kim Knudsen, *Electrical impedance tomography: 3D reconstructions using scattering transforms*, Applicable Analysis **0** (0), no. 0, 1–19.

[DM10] M. DeAngelo and J. L. Mueller, *2D D-bar reconstructions of human chest and tank data using an improved approximation to the scattering transform*, Physiological Measurement **31** (2010), 221–232.

[EHN96] Heinz W. Engl, Martin Hanke, and Andreas Neubauer, *Regularization of inverse problems*, Mathematics and its Applications, vol. 375, Kluwer Academic Publishers Group, Dordrecht, 1996. MR1408680 (97k:65145)

[Fad66] L. D. Faddeev, *Increasing solutions of the Schrödinger equation*, Soviet Physics Doklady **10** (1966), 1033–1035.

[Fra00] Elisa Francini, *Recovering a complex coefficient in a planar domain from the Dirichlet-to-Neumann map*, Inverse Problems **16** (2000), no. 1, 107–119, DOI 10.1088/0266-5611/16/1/309. MR1741230 (2001b:35292)

[Ham12] S. J. Hamilton, *A Direct D-bar reconstruction algorithm for complex admittivies in $W^{2,\infty}(\Omega)$ for the 2-D EIT problem*, Ph.D. Thesis, Colorado State University, Fort Collins, CO, Summer 2012.

[Her12] C. N. L. Herrera, *Um metodo D-bar para estimar admitividade em 2-D atraves de tomografia por impedancia electrica*, Ph.D. Thesis, University of São Paulo, São Paulo, Brazil, Fall 2012.

[HHMV12] S. J. Hamilton, C. N. L. Herrera, J. L. Mueller, and A. Von Herrmann, *A direct D-bar reconstruction algorithm for recovering a complex conductivity in 2D*, Inverse Problems **28** (2012), no. 9, 095005, 24, DOI 10.1088/0266-5611/28/9/095005. MR2959281

[HM13] S. J. Hamilton and J. L. Mueller, *Direct EIT reconstructions of complex admittivities on a chest-shaped domain in 2-D*, IEEE Transactions on Medical Imaging **32** (2013), no. 4, 757–769.

[IIN] Takanori Ide, Hiroshi Isozaki, Susumu Nakata, Samuli Siltanen, and G. Uhlmann, *Probing for electrical inclusions with complex spherical waves*, Comm. Pure Appl. Math. **60** (2007), no. 10, 1415–1442, DOI 10.1002/cpa.20194. MR2342953 (2008j:35186)

[IINS10] T. Ide, H. Isozaki, S. Nakata, and S. Siltanen, *Local detection of three-dimensional inclusions in electrical impedance tomography*, Inverse Problems **26** (2010), no. 3, 035001, 17, DOI 10.1088/0266-5611/26/3/035001. MR2580544 (2011a:65375)

[IMNS04] D. Isaacson, J. L. Mueller, J. C. Newell, and S. Siltanen, *Reconstructions of chest phantoms by the D-bar method for electrical impedance tomography*, IEEE Transactions on Medical Imaging **23** (2004), 821–828.

[IMNS06] D. Isaacson, J.L. Mueller, J.C. Newell, and S. Siltanen, *Imaging cardiac activity by the D-bar method for electrical impedance tomography*, Physiological Measurement **27** (2006), S43–S50.

[IUY10] Oleg Yu. Imanuvilov, Gunther Uhlmann, and Masahiro Yamamoto, *The Calderón problem with partial data in two dimensions*, J. Amer. Math. Soc. **23** (2010), no. 3, 655–691, DOI 10.1090/S0894-0347-10-00656-9. MR2629983 (2012c:35472)

[KLMS07] Kim Knudsen, Matti Lassas, Jennifer L. Mueller, and Samuli Siltanen, *D-bar method for electrical impedance tomography with discontinuous conductivities*, SIAM J. Appl. Math. **67** (2007), no. 3, 893–913, DOI 10.1137/060656930. MR2300316 (2007m:35276)

[KLMS08] Kim Knudsen, Matti Lassas, Jennifer L. Mueller, and Samuli Siltanen, *Regularized D-bar method for the inverse conductivity problem*, Inverse Probl. Imaging **3** (2009), no. 4, 599–624, DOI 10.3934/ipi.2009.3.599. MR2557921 (2011d:35518)

[KLMS09] Kim Knudsen, Matti Lassas, Jennifer L. Mueller, and Samuli Siltanen, *Regularized D-bar method for the inverse conductivity problem*, Inverse Probl. Imaging **3** (2009), no. 4, 599–624, DOI 10.3934/ipi.2009.3.599. MR2557921 (2011d:35518)

[KMS04] K. Knudsen, J. Mueller, and S. Siltanen, *Numerical solution method for the dbar-equation in the plane*, J. Comput. Phys. **198** (2004), no. 2, 500–517, DOI 10.1016/j.jcp.2004.01.028. MR2062911 (2005g:65185)

[Knu02] K. Knudsen, *On the inverse conductivity problem*, Ph.D. thesis, Department of Mathematical Sciences, Aalborg University, Denmark, 2002.

[Knu03] ______, *A new direct method for reconstructing isotropic conductivities in the plane*, Physiological Measurement **24** (2003), no. 2, 391–403.

[Knu06] Kim Knudsen, *The Calderón problem with partial data for less smooth conductivities*, Comm. Partial Differential Equations **31** (2006), no. 1-3, 57–71, DOI 10.1080/03605300500361610. MR2209749 (2006k:35303)

[KSU07] Carlos E. Kenig, Johannes Sjöstrand, and Gunther Uhlmann, *The Calderón problem with partial data*, Ann. of Math. (2) **165** (2007), no. 2, 567–591, DOI 10.4007/annals.2007.165.567. MR2299741 (2008k:35498)

[KT04] Kim Knudsen and Alexandru Tamasan, *Reconstruction of less regular conductivities in the plane*, Comm. Partial Differential Equations **29** (2004), no. 3-4, 361–381, DOI 10.1081/PDE-120030401. MR2041600 (2005e:35250)

[Mam10] Alexander Vasilyevich Mamonov, *Resistor networks and optimal grids for the numerical solution of electrical impedance tomography with partial boundary measurements*, ProQuest LLC, Ann Arbor, MI, 2010. Thesis (Ph.D.)–Rice University. MR2792941

[MIN99] J.L. Mueller, D. Isaacson, and J. C. Newell, *A reconstruction algorithm for electrical impedance tomography data collected on rectangular electrode arrays*, IEEE Transactions on Biomedical Engineering **49** (1999), 1379–1386.

[MS03] Jennifer L. Mueller and Samuli Siltanen, *Direct reconstructions of conductivities from boundary measurements*, SIAM J. Sci. Comput. **24** (2003), no. 4, 1232–1266, DOI 10.1137/S1064827501394568. MR1976215 (2004c:78036)

[MS12] Jennifer L. Mueller and Samuli Siltanen, *Linear and nonlinear inverse problems with practical applications*, Computational Science & Engineering, vol. 10, Society for Industrial and Applied Mathematics (SIAM), Philadelphia, PA, 2012. MR2986262

[Mur07] E. K. Murphy, *2-D D-Bar conductivity reconstructions on non-circular domains*, Ph.D. Thesis, Colorado State University, Fort Collins, CO, Fall 2007.

[Nac96] Adrian I. Nachman, *Global uniqueness for a two-dimensional inverse boundary value problem*, Ann. of Math. (2) **143** (1996), no. 1, 71–96, DOI 10.2307/2118653. MR1370758 (96k:35189)

[NS10] Adrian Nachman and Brian Street, *Reconstruction in the Calderón problem with partial data*, Comm. Partial Differential Equations **35** (2010), no. 2, 375–390, DOI 10.1080/03605300903296322. MR2748629 (2012b:35368)

[SMI00] Samuli Siltanen, Jennifer Mueller, and David Isaacson, *An implementation of the reconstruction algorithm of A. Nachman for the 2D inverse conductivity problem*, Inverse Problems **16** (2000), no. 3, 681–699, DOI 10.1088/0266-5611/16/3/310. MR1766226 (2001g:35269)

[SMI01] Samuli Siltanen, Jennifer Mueller, and David Isaacson, *Reconstruction of high contrast 2-D conductivites by the algorithm of A. Nachman*, Proceedings of the 2000 conference on Radon transforms and tomography (et. al. ed. Quinto, ed.), vol. 278, 2001, pp. 241–254.

[SU87] John Sylvester and Gunther Uhlmann, *A global uniqueness theorem for an inverse boundary value problem*, Ann. of Math. (2) **125** (1987), no. 1, 153–169, DOI 10.2307/1971291. MR873380 (88b:35205)

[UW08] Gunther Uhlmann and Jenn-Nan Wang, *Reconstructing discontinuities using complex geometrical optics solutions*, SIAM J. Appl. Math. **68** (2008), no. 4, 1026–1044, DOI 10.1137/060676350. MR2390978 (2009d:35347)

University of Helsinki, Department of Mathematics and Statistics, P.O. Box 68, FI-00014 Helsinki, Finland

E-mail address: `sarah.hamilton@helsinki.fi`

University of Helsinki, Department of Mathematics and Statistics, P.O. Box 68, FI-00014 Helsinki, Finland

E-mail address: `samuli.siltanen@helsinki.fi`

Contemporary Mathematics
Volume **615**, 2014
http://dx.doi.org/10.1090/conm/615/12268

Increasing stability of the inverse boundary value problem for the Schrödinger equation

V. Isakov, S. Nagayasu, G. Uhlmann, and J.-N. Wang

Isakov, Nagayasu, and Wang would like to dedicate this work to Gunther Uhlmann for his 60th birthday

ABSTRACT. In this work we study the phenomenon of increasing stability in the inverse boundary value problem for the Schrödinger equation. This problem was previously considered by Isakov in which he discussed the phenomenon in different ranges of the wave number (or energy). The main contribution of this work is to provide a unified and easier approach to the same problem based on the complex geometrical optics solutions.

1. Introduction

Most of inverse problems are known to be severely ill-posed. This weakness makes it extremely difficult to design reliable reconstruction algorithms in practice. However, in some cases, it has been observed numerically that the stability increases with respect to some parameter such as the wave number (or energy) (see, for example, [**6**] for the inverse obstacle scattering problem). Several rigorous justifications of the increasing stability phenomena in different settings were obtained by Isakov *et al* [**3**, **4**, **9**, **11**, **12**]. In particular, in [**12**], Isakov considered the Helmholtz equation with a potential

$$(\Delta + k^2 + q(x))u(x) = 0 \text{ in } \Omega \subset \mathbb{R}^n \tag{1.1}$$

with $n \geq 3$. He obtained stability estimates of determining q by the Dirichlet-to-Neumann map for different ranges of k, which demonstrate the increasing stability phenomena in k. The purpose of this work is to provide a more straightforward way to derive a similar estimate for the inverse boundary value for (1.1). In [**12**], Isakov used *real* geometrical optics solutions for the large wave number k. In this work, by more careful choice of an additional large parameter and a priori constraints we are able to use *complex* geometrical optics (CGO) solutions introduced by Calderón

2010 *Mathematics Subject Classification.* Primary 35R30; Secondary 65N21.

Key words and phrases. Stability estimate, Inverse boundary value problem, Schrödinger equation, Complex geometrical optics solutions.

The first author was in part supported by the NSF grant DMS 10-08902 and by Emylou Keith and Betty Dutcher Distinguished Professorship at WSU.

The second author was partly supported by Grant-in-Aid for Young Scientists (B).

The third author was partly supported by NSF and The Fondation de Sciences Mathématiques de Paris.

The fourth author was partly supported by the NSC grant 99-2115-M-002-006-MY3.

[5] and Sylvester-Uhlmann **[17]** for all $k \geq 1$. This will simplify the proof in **[12]**. Recently similar results were obtained by Isaev and Novikov **[10]** by using less explicit and more complicated methods of scattering theory.

In this work, instead of considering the Dirichlet-to-Neumann map, we define the boundary measurements to be the Cauchy data corresponding to (1.1)

$$\mathcal{C}_q = \left\{ \left(u|_{\partial\Omega}, \frac{\partial u}{\partial \nu}\bigg|_{\partial\Omega} \right), \text{ where } u \text{ is a solution to (1.1)} \right\}.$$

Hereafter, ν is the unit outer normal vector of $\partial\Omega$. Assume that $\mathcal{C}_{q_1}$ and $\mathcal{C}_{q_2}$ are two Cauchy data associated with refraction indices q_1 and q_2, respectively. To measure the distance between two Cauchy data, we define

$$\operatorname{dist}(\mathcal{C}_{q_1}, \mathcal{C}_{q_2}) = \max\left\{ \max_{(f,g)\in\mathcal{C}_{q_1}} \min_{(\widetilde{f},\widetilde{g})\in\mathcal{C}_{q_2}} \frac{\|(f,g)-(\widetilde{f},\widetilde{g})\|_{H^{1/2}\oplus H^{-1/2}}}{\|(f,g)\|_{H^{1/2}\oplus H^{-1/2}}}, \right.$$
$$\left. \max_{(f,g)\in\mathcal{C}_{q_2}} \min_{(\widetilde{f},\widetilde{g})\in\mathcal{C}_{q_1}} \frac{\|(f,g)-(\widetilde{f},\widetilde{g})\|_{H^{1/2}\oplus H^{-1/2}}}{\|(f,g)\|_{H^{1/2}\oplus H^{-1/2}}} \right\},$$

where

$$\|(f,g)\|_{H^{1/2}\oplus H^{-1/2}} = \left(\|f\|^2_{H^{1/2}(\partial\Omega)} + \|g\|^2_{H^{-1/2}(\partial\Omega)} \right)^{1/2}.$$

Note that $\operatorname{dist}(\mathcal{C}_{q_1}, \mathcal{C}_{q_2})$ is a case of distance between two subspaces in a Hilbert space. This notion has a long tradition and still attracts attentions, for instance, see the book **[1**, Ch. III] and the recent literature **[14]**.

Our main theorem is stated as follows.

THEOREM 1.1. *Let $n \geq 3$. Assume $\mathcal{C}_{q_1}$ and $\mathcal{C}_{q_2}$ are Cauchy data corresponding to $q_1(x)$ and $q_2(x)$, respectively. Let $s > n/2$ and $M > 0$. Assume $\|q_l\|_{H^s(\Omega)} \leq M$ $(l = 1, 2)$ and* $\operatorname{supp}(q_1 - q_2) \subset \Omega$. *Denote $\widetilde{q}$ the zero extension of $q_1 - q_2$. Then for $k \geq 1$ and* $\operatorname{dist}(\mathcal{C}_{q_1}, \mathcal{C}_{q_2}) \leq 1/e$ *we have the following stability estimate:*

$$\|\widetilde{q}\|_{H^{-s}(\mathbb{R}^n)} \leq Ck^4 \operatorname{dist}(\mathcal{C}_{q_1}, \mathcal{C}_{q_2}) + C\left(k + \log \frac{1}{\operatorname{dist}(\mathcal{C}_{q_1}, \mathcal{C}_{q_2})}\right)^{-(2s-n)}, \tag{1.2}$$

where $C > 0$ depends only on n, s, Ω, M and $\operatorname{supp}(q_1 - q_2)$.

From estimate (1.2), we can see that the influence of the logarithmic part becomes small when k is large. Indeed, in view of the right hand side of (1.2), we can check that

$$k^4\varepsilon \geq \left(k + \log\frac{1}{\varepsilon}\right)^{-(2s-n)}$$

provided $k \geq k(\varepsilon)$, where $k(\varepsilon)$ solves

$$k(\varepsilon)^4\varepsilon = (k(\varepsilon) + \log\frac{1}{\varepsilon})^{-(2s-n)}.$$

We can see that $k(\varepsilon)$ is a decreasing function of ε and $k(\varepsilon) \approx \varepsilon^{-1/(2s-n+4)}$. In other words, given

$$\operatorname{dist}(\mathcal{C}_{q_1}, \mathcal{C}_{q_2}) \leq \varepsilon, \tag{1.3}$$

the stability behaves more like Lipschitz when k is sufficient large, precisely, $k \geq k(\varepsilon)$ (see Figure 1). On the other hand, if k is confined in a finite interval, then for small ε, the stability estimate is more or less of a logarithmic type.

We also want to point out that if (1.3) is true for all $k \neq 0$, then by substituting $k = k(\varepsilon)$ into the right hand side of (1.2), we obtain a Hölder type estimate in terms of the Cauchy data $\mathcal{C}_q$ at all wave numbers. To facilitate further discussions, we denote $\mathcal{C}_q$ at wave number k^2 as $\mathcal{C}_{q,k^2}$. It is clear that the information of $\mathcal{C}_{q,k^2}$ is contained in that of the Dirichlet-to-Neumann map $\Lambda_{q,z}$, defined by

$$\Lambda_{q,z} f = \frac{\partial u}{\partial \nu}|_{\partial\Omega}$$

with u satisfying $(-\Delta - q)u = zu$ in Ω and $u|_{\partial\Omega} = f$, for all $\mathbb{C} \ni z \notin \sigma(-\Delta - q)$, where $\sigma(-\Delta - q)$ is the set of Dirichlet eigenvalues of $-\Delta - q$ on Ω. Note that $\Lambda_{q,z}$ can be extended to a meromorphic function in $\mathbb{C}$ with poles at $\sigma(-\Delta - q)$. $\Lambda_{q,z}$ is closely to the *boundary spectral data* $\{\lambda_k, \partial\phi_k/\partial\nu|_{\partial\Omega}\}_{k=1}^{\infty}$, where $\lambda_k \in \sigma(-\Delta - q)$ and ϕ_k is the corresponding normalized eigenfunction, and the *Hyperbolic Dirichlet-to-Neumann map* given by

$$\Lambda_q^H : f \mapsto \frac{\partial v}{\partial \nu}|_{\partial\Omega\times[0,T]},$$

where v satisfies

$$\begin{cases} \partial_{tt} v - \Delta v - qv = 0 \quad \text{in} \quad \Omega \times [0, T], \\ v = f \quad \text{on} \quad \partial\Omega \times [0, T], \\ v(x, 0) = \partial_t v(x, 0) = 0 \quad x \in \Omega. \end{cases}$$

In fact, these three data are equivalent (see, for example, [**13**]). Hölder type stability estimates of recovering q using the boundary spectral data and the hyperbolic Dirichlet-to-Neumann map were first derived by Alessandrini, Sylvester [**2**] and by Sun [**16**], respectively.

Finally, we would like to point out that unlike in the acoustic case where the constant associated with the Lipschitz estimate grows exponentially in k [**15**], the constant here grows only polynomially in k. Similarly, the corresponding constant obtained in [**12**] (see estimate (8) there) also grows polynomially in k.

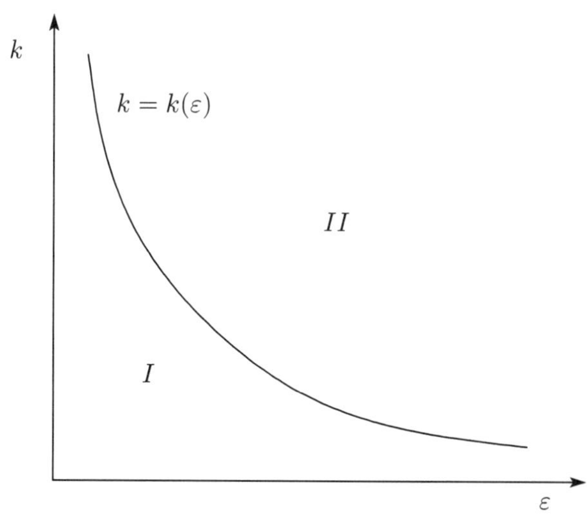

FIGURE 1. The stability estimate is Logarithmic dominated in Region I and is Lipschitz dominated in Region II.

The paper is organized as follows. In Section 2, we will collect some known results about the CGO solutions and an estimate for the difference of potentials, which are essential tools in the proof. In Section 3, we present a detailed proof of Theorem 1.1.

2. Preliminaries

To begin, we state the existence of CGO solutions for (1.1). These special solutions are first constructed by Sylvester and Uhlmann [**17**]. Another construction based on the Fourier series is given by Hähner [**8**].

LEMMA 2.1. *Let $s > n/2$. Assume that $\zeta = \eta + i\xi$ $(\eta, \xi \in \mathbb{R}^n)$ satisfies*

$$|\eta|^2 = k^2 + |\xi|^2 \quad \text{and} \quad \eta \cdot \xi = 0,$$

i.e., $\zeta \cdot \zeta = k^2$. Then there exist constants C_* and $C > 0$, which are independent of k, such that if $|\xi| > C_*\|q\|_{H^s(\Omega)}$ then there exists a solution u to the equation (1.1) *of the form*

$$u(x) = e^{i\zeta\cdot x}\big(1 + \psi(x)\big), \tag{2.1}$$

where ψ has the estimate

$$\|\psi\|_{H^s(\Omega)} \le \frac{C}{|\xi|}\|q\|_{H^s(\Omega)}.$$

REMARK 2.2. *Note that the correction term ψ decays in $Im\,\zeta$. This property is crucial in obtaining that the constant associated with the Lipschiz estimate grows only polynomially in k.*

Next inequality follows from the weak formulation of the equation (1.1). We refer to [**7**] for the proof.

PROPOSITION 2.3. *Let u_l and $\mathcal{C}_{q_l}$ be solution and Cauchy data to the equation* (1.1) *with $q = q_l$, respectively $(l = 1, 2)$. Then the following estimate holds:*

$$\begin{aligned}&\left|\int_\Omega (q_2 - q_1)u_1u_2\,dx\right|\\ &\le \left\|\left(u_1, \frac{\partial u_1}{\partial\nu}\right)\right\|_{H^{1/2}\oplus H^{-1/2}} \left\|\left(u_2, \frac{\partial u_2}{\partial\nu}\right)\right\|_{H^{1/2}\oplus H^{-1/2}} \mathrm{dist}(\mathcal{C}_{q_1}, \mathcal{C}_{q_1}).\end{aligned}$$

3. Proof of main theorem

To prove Theorem 1.1, we first derive two lemmas.

LEMMA 3.1. *Under the assumptions in Theorem* 1.1,

$$|\mathcal{F}\widetilde{q}(r\omega)| \le Ck^4e^{Ca}\,\mathrm{dist}(\mathcal{C}_{q_1}, \mathcal{C}_{q_2}) + \frac{C}{a}\|\widetilde{q}\|_{H^{-s}(\mathbb{R}^n)} \tag{3.1}$$

holds for $k \ge 1$, $r \ge 0$, $\omega \in \mathbb{R}^n$ with $|\omega| = 1$ and $a > C_*M$ with $k^2 + a^2 > r^2/4$, where $C > 0$ depends only on n, s, M, Ω and $\mathrm{supp}\,(q_1 - q_2)$ and C_* is the constant given in Lemma 2.1.

PROOF. We will use CGO solutions (2.1) with appropriately chosen parameter ζ. Let us denote $\zeta_l = \eta_l + i\xi_l$, $l = 1, 2$. We can choose $\omega^\perp, \widetilde{\omega}^\perp \in \mathbb{R}^n$ satisfying

$$\omega\cdot\omega^\perp = \omega\cdot\widetilde{\omega}^\perp = \omega^\perp\cdot\widetilde{\omega}^\perp = 0 \quad \text{and} \quad |\omega^\perp| = |\widetilde{\omega}^\perp| = 1.$$

Now we set

$$\xi_1 = a\omega^\perp, \quad \eta_1 = -\frac{r}{2}\omega + \sqrt{k^2 + a^2 - \frac{r^2}{4}}\,\widetilde{\omega}^\perp,$$
$$\xi_2 = -\xi_1 \quad \text{and} \quad \eta_2 = -r\omega - \eta_1,$$

and thus

$$\xi_l \cdot \eta_l = 0, \quad |\eta_l|^2 = k^2 + |\xi_l|^2$$

and $|\xi_l| = a \geq C_* M \geq C_* \|q_\ell\|_{H^s(\Omega)}$. From Lemma 2.1, there exist CGO solutions

$$u_l(x) = e^{i\zeta_l x}\big(1 + \psi_l(x)\big)$$

to equation (1.1) with $q = q_l$, where ψ_l satisfies

$$\|\psi_l\|_{H^s(\Omega)} \leq \frac{C}{|\xi_l|}\|q_l\|_{H^s(\Omega)}.$$

Note that ψ_l also satisfies the estimate

$$\|\psi_l\|_{H^s(\Omega)} \leq \frac{C}{|\xi_l|}\|q_l\|_{H^s(\Omega)} \leq \frac{CM}{a} < \frac{CM}{C_* M} = \frac{C}{C_*}. \tag{3.2}$$

Now, by Proposition 2.3 and using the relation $-r\omega = \zeta_1 + \zeta_2$, we have that

$$\left|\int_\Omega \widetilde{q}(x) e^{-ir\omega\cdot x}(1+\psi_1)(1+\psi_2)\,dx\right| = \left|\int_\Omega (q_2 - q_1) u_1 u_2\,dx\right|$$
$$\leq \left\|\left(u_1, \frac{\partial u_1}{\partial \nu}\right)\right\|_{H^{1/2}\oplus H^{-1/2}} \left\|\left(u_2, \frac{\partial u_2}{\partial \nu}\right)\right\|_{H^{1/2}\oplus H^{-1/2}} \operatorname{dist}(\mathcal{C}_{q_1}, \mathcal{C}_{q_1}).$$

Subsequently, we obtain

$$\begin{aligned} |\mathcal{F}\widetilde{q}(r\omega)| &= \left|\int_\Omega \widetilde{q}(x) e^{-ir\omega\cdot x}\,dx\right| \\ &\leq \left|\int_\Omega \widetilde{q}(x) e^{-ir\omega\cdot x}(1+\psi_1)(1+\psi_2)\,dx\right| \\ &\qquad + \left|\int_\Omega \widetilde{q}(x) e^{-ir\omega\cdot x}(\psi_1 + \psi_2 + \psi_1\psi_2)\,dx\right| \\ &\leq \left\|\left(u_1, \frac{\partial u_1}{\partial \nu}\right)\right\|_{H^{1/2}\oplus H^{-1/2}} \left\|\left(u_2, \frac{\partial u_2}{\partial \nu}\right)\right\|_{H^{1/2}\oplus H^{-1/2}} \operatorname{dist}(\mathcal{C}_{q_1}, \mathcal{C}_{q_1}) \\ &\qquad + \left|\int_\Omega \widetilde{q}(x) e^{-ir\omega\cdot x}(\psi_1 + \psi_2 + \psi_1\psi_2)\,dx\right|. \end{aligned} \tag{3.3}$$

In view of (3.3), we want to estimate $\big\|(u_l, \partial u_l/\partial\nu)\big\|_{H^{1/2}\oplus H^{-1/2}}$. Recall that u_l solves (1.1) with $q = q_l$. Using assumptions $\|q_l\|_{H^s(\Omega)} \leq M$, and $s > n/2$, and $k \geq 1$, we have that

$$\left\|\frac{\partial u_l}{\partial \nu}\right\|_{H^{-1/2}(\partial\Omega)} \leq Ck^2\|u_l\|_{L^2(\Omega)} + C\|\nabla u_l\|_{L^2(\Omega)}$$

and thus

$$\left\|\left(u_l, \frac{\partial u_l}{\partial \nu}\right)\right\|_{H^{1/2}\oplus H^{-1/2}} \leq Ck^2\|u_l\|_{L^2(\Omega)} + C\|\nabla u_l\|_{L^2(\Omega)}.$$

We now choose $R_0 > 0$ large enough such that $\Omega \subset B_{R_0}(0)$. Then we have

$$|u_l(x)| \leq e^{-\xi_l\cdot x}\big(1 + |\psi_l(x)|\big) \leq Ce^{|\xi_l|R_0} = Ce^{aR_0}$$

since

$$|\psi_l(x)| \le \|\psi_l\|_{L^\infty(\Omega)} \le C\|\psi_l\|_{H^s(\Omega)} \le C$$

by $s > n/2$ and (3.2). It follows that

$$\|u_l\|_{L^2(\Omega)} \le Ce^{aR_0}.$$

On the other hand, in view of $\|\nabla\psi_l\|_{L^2(\Omega)} \le \|\psi_l\|_{H^s(\Omega)} \le C$ ($s > n/2 \ge 3/2 > 1$) and (3.2), we can estimate

$$\begin{aligned}\|\nabla u_l\|_{L^2(\Omega)} &= \left\|iu_l\zeta_l + e^{i\zeta_l\cdot\bullet}\nabla\psi_l\right\|_{L^2(\Omega)} \le |\zeta_l|\|u_l\|_{L^2(\Omega)} + e^{|\xi_l|R_0}\|\nabla\psi_l\|_{L^2(\Omega)}\\ &\le C(k+|\xi_l|)e^{aR_0} + Ce^{|\xi_l|R_0} = C(k+a)e^{aR_0} + Ce^{aR_0} \le Cke^{Ca}.\end{aligned}$$

Summing up, we obtain

$$\begin{aligned}\left\|\left(u_l, \frac{\partial u_l}{\partial\nu}\right)\right\|_{H^{1/2}\oplus H^{-1/2}} &\le Ck^2\|u_l\|_{L^2(\Omega)} + C\|\nabla u_l\|_{L^2(\Omega)}\\ &\le Ck^2e^{Ca} + Cke^{Ca} \le Ck^2e^{Ca}.\end{aligned}\tag{3.4}$$

Note that here C depends on n, s, M, and the diameter of Ω.

Let $\chi \in C_0^\infty(\Omega)$ be a cut-off function satisfying $\chi \equiv 1$ near $\operatorname{supp}(q_1 - q_2)$, then we have

$$\begin{aligned}&\left|\int_\Omega \widetilde{q}(x)e^{-ir\omega\cdot x}(\psi_1+\psi_2+\psi_1\psi_2)\,dx\right|\\ &\qquad= \left|\int_\Omega \widetilde{q}(x)\chi(x)e^{-ir\omega\cdot x}(\psi_1+\psi_2+\psi_1\psi_2)\,dx\right|\\ &\qquad\le \int_\Omega |\widetilde{q}(x)||\chi(\psi_1+\psi_2+\psi_1\psi_2)|\,dx\\ &\qquad\le \|\widetilde{q}\|_{H^{-s}(\Omega)}\|\chi(\psi_1+\psi_2+\psi_1\psi_2)\|_{H^s(\Omega)}.\end{aligned}\tag{3.5}$$

Since $s > n/2$ and (3.2), we can estimate

$$\begin{aligned}&\|\chi(\psi_1+\psi_2+\psi_1\psi_2)\|_{H^s(\Omega)}\\ &\le \|\chi\|_{H^s(\Omega)}\big(\|\psi_1\|_{H^s(\Omega)} + \|\psi_2\|_{H^s(\Omega)} + \|\psi_1\|_{H^s(\Omega)}\|\psi_2\|_{H^s(\Omega)}\big)\\ &\le \|\chi\|_{H^s(\Omega)}\left(\frac{CM}{a} + \frac{CM}{a} + \frac{C}{C_*}\cdot\frac{CM}{a}\right) \le \frac{C}{a}.\end{aligned}\tag{3.6}$$

Finally, (3.1) follows from (3.3), (3.4), (3.5), and (3.6). □

The following lemma is an easy corollary of Lemma 3.1.

LEMMA 3.2. *Suppose that the assumptions in Theorem 1.1 hold. Let $R > C_*M$ with C_* being the constant given in Lemma 2.1. Then for $k \ge 1$, $r \ge 0$ and $\omega \in \mathbb{R}^n$ with $|\omega| = 1$, the following estimates hold true: if $0 \le r \le k + R$ then*

$$|\mathcal{F}\widetilde{q}(r\omega)| \le Ck^4e^{CR}\operatorname{dist}(\mathcal{C}_{q_1}, \mathcal{C}_{q_2}) + \frac{C}{R}\|\widetilde{q}\|_{H^{-s}(\mathbb{R}^n)};\tag{3.7}$$

if $r \ge k + R$ then

$$|\mathcal{F}\widetilde{q}(r\omega)| \le Ck^4e^{Cr}\operatorname{dist}(\mathcal{C}_{q_1}, \mathcal{C}_{q_2}) + \frac{C}{r}\|\widetilde{q}\|_{H^{-s}(\mathbb{R}^n)}.\tag{3.8}$$

PROOF. It is enough to take $a = R$ when $0 \le r \le k + R$, and take $a = r$ when $r \ge k + R$ in Lemma 3.1. □

Now we prove our main theorem.

PROOF OF THEOREM 1.1. Written in polar coordinates, we have that

$$
\begin{aligned}
(3.9)\qquad \|\widetilde{q}\|^2_{H^{-s}(\mathbb{R}^n)} &= C\int_0^\infty \int_{|\omega|=1} |\mathcal{F}\widetilde{q}(r\omega)|^2(1+r^2)^{-s} r^{n-1}\, d\omega dr \\
&= C\Big(\int_0^{k+R}\int_{|\omega|=1} |\mathcal{F}\widetilde{q}(r\omega)|^2(1+r^2)^{-s} r^{n-1}\, d\omega dr \\
&\quad + \int_{k+R}^{T}\int_{|\omega|=1} |\mathcal{F}\widetilde{q}(r\omega)|^2(1+r^2)^{-s} r^{n-1}\, d\omega dr \\
&\quad + \int_{T}^{\infty}\int_{|\omega|=1} |\mathcal{F}\widetilde{q}(r\omega)|^2(1+r^2)^{-s} r^{n-1}\, d\omega dr\Big) \\
&=: C(I_1+I_2+I_3),
\end{aligned}
$$

where $R > C_* M$ and $T \geq k+R$ are parameters which will be chosen later.

Our task now is to estimate each integral separately. We begin with I_3. Since $|\mathcal{F}\widetilde{q}(r\omega)| \leq C\|q_1-q_2\|_{L^2(\Omega)}$, $q_1-q_2 \in H^s_0(\Omega)$ and $s > n/2$, we get

$$
\begin{aligned}
(3.10)\qquad I_3 &\leq C\int_T^\infty \|q_1-q_2\|^2_{L^2(\Omega)}(1+r^2)^{-s} r^{n-1}\, dr \leq CT^{-m}\|q_1-q_2\|^2_{L^2(\Omega)} \\
&\leq CT^{-m}\left(\varepsilon\|q_1-q_2\|^2_{H^{-s}(\Omega)} + \frac{1}{\varepsilon}\|q_1-q_2\|^2_{H^s(\Omega)}\right) \\
&\leq CT^{-m}\left(\varepsilon\|\widetilde{q}\|^2_{H^{-s}(\mathbb{R}^n)} + \frac{1}{\varepsilon}\right)
\end{aligned}
$$

for $\varepsilon > 0$, where $m := 2s-n$.

On the other hand, by estimate (3.7), we can obtain

$$
\begin{aligned}
(3.11)\quad I_1 &\leq \int_0^{k+R}\left(Ck^4 e^{CR}\operatorname{dist}(\mathcal{C}_{q_1},\mathcal{C}_{q_2}) + \frac{C}{R}\|\widetilde{q}\|_{H^{-s}(\mathbb{R}^n)}\right)^2 (1+r^2)^{-s} r^{n-1}\, dr \\
&\leq C\left(k^8 e^{CR}\operatorname{dist}(\mathcal{C}_{q_1},\mathcal{C}_{q_2})^2 + \frac{1}{R^2}\|\widetilde{q}\|^2_{H^{-s}(\mathbb{R}^n)}\right)\int_0^\infty (1+r^2)^{-s} r^{n-1}\, dr \\
&= C\left(k^8 e^{CR}\operatorname{dist}(\mathcal{C}_{q_1},\mathcal{C}_{q_2})^2 + \frac{1}{R^2}\|\widetilde{q}\|^2_{H^{-s}(\mathbb{R}^n)}\right).
\end{aligned}
$$

In the same way, using estimate (3.8), we have

$$
\begin{aligned}
(3.12)\quad I_2 &\leq C\int_{k+R}^{T}\left(Ck^4 e^{Cr}\operatorname{dist}(\mathcal{C}_{q_1},\mathcal{C}_{q_2}) + \frac{C}{r}\|\widetilde{q}\|_{H^{-s}(\mathbb{R}^n)}\right)^2 (1+r^2)^{-s} r^{n-1}\, dr \\
&\leq Ck^8\operatorname{dist}(\mathcal{C}_{q_1},\mathcal{C}_{q_2})^2\int_{k+R}^{T} e^{Cr}(1+r^2)^{-s} r^{n-1}\, dr \\
&\qquad + C\|\widetilde{q}\|^2_{H^{-s}(\mathbb{R}^n)}\int_{k+R}^{T}(1+r^2)^{-s} r^{n-1}\, dr \\
&\leq C\left(k^8 e^{CT}\operatorname{dist}(\mathcal{C}_{q_1},\mathcal{C}_{q_2})^2 + \frac{1}{R^2}\|\widetilde{q}\|^2_{H^{-s}(\mathbb{R}^n)}\right),
\end{aligned}
$$

where we have used

$$\int_{k+R}^{T} e^{Cr}(1+r^2)^{-s}r^{n-1}\,dr \le e^{CT}\int_{k+R}^{T}(1+r^2)^{-s}r^{n-1}\,dr$$
$$\le e^{Ct}\int_{0}^{\infty}(1+r^2)^{-s}r^{n-1}\,dr = Ce^{CT},$$

$$\int_{k+R}^{T}(1+r^2)^{-s}r^{n-1}\,dr \le \int_{k+R}^{T} r^{-2s+n-1}\,dr$$
$$\le \frac{1}{2s-n+2}\frac{1}{(k+R)^{2s-n+2}} \le \frac{C}{(k+R)^2} \le \frac{C}{R^2},$$

and $s > n/2$, $k \ge 1$. Combining (3.9)–(3.12) gives

$$\begin{aligned}
\|\widetilde{q}\|^2_{H^{-s}(\mathbb{R}^n)} &\le C(I_1+I_2+I_3)\\
&\le C\left(k^8e^{CR}\operatorname{dist}(\mathcal{C}_{q_1},\mathcal{C}_{q_2})^2+\frac{1}{R^2}\|\widetilde{q}\|^2_{H^{-s}(\mathbb{R}^n)}\right)\\
&\quad +C\left(k^8e^{CT}\operatorname{dist}(\mathcal{C}_{q_1},\mathcal{C}_{q_2})^2+\frac{1}{R^2}\|\widetilde{q}\|^2_{H^{-s}(\mathbb{R}^n)}\right)\\
&\quad +CT^{-m}\left(\varepsilon\|\widetilde{q}\|^2_{H^{-s}(\mathbb{R}^n)}+\frac{1}{\varepsilon}\right)\\
&\le C\left(\frac{2}{R^2}+\varepsilon T^{-m}\right)\|\widetilde{q}\|^2_{H^{-s}(\mathbb{R}^n)}+Ck^8e^{CR}\operatorname{dist}(\mathcal{C}_{q_1},\mathcal{C}_{q_2})^2\\
&\quad +Ck^8e^{CT}\operatorname{dist}(\mathcal{C}_{q_1},\mathcal{C}_{q_2})^2+\frac{CT^{-m}}{\varepsilon}.
\end{aligned} \tag{3.13}$$

To continue, we consider the following two cases:

$$\text{(i) } k+R\le p\log\frac{1}{A} \quad\text{and}\quad \text{(ii) } k+R\ge p\log\frac{1}{A},$$

where $R > C_*M$ and $p > 0$ are constants which will be determined later. We begin with the first case (i). Taking

$$R > 2\sqrt{C} \tag{3.14}$$

and $\varepsilon = cT^m$ ($c \ll 1$), we deduce that

$$\|\widetilde{q}\|^2_{H^{-s}(\mathbb{R}^n)} \le Ck^8A + Ck^8e^{CT}A + CT^{-2m} \tag{3.15}$$

for any $T \ge k+R$ by (3.13), where $A = \operatorname{dist}(\mathcal{C}_{q_1},\mathcal{C}_{q_2})^2$.

Now we choose $T = p\log(1/A)$, which is greater than or equal to $k+R$ by the condition (i). Our current aim is to show that there exists $C_1 > 0$ such that

$$k^8e^{CT}A \le C_1\left(k+\log\frac{1}{A}\right)^{-2m} \tag{3.16}$$

and

$$T^{-2m} \le C_1\left(k+\log\frac{1}{A}\right)^{-2m}. \tag{3.17}$$

Substituting (3.16) and (3.17) into (3.15) clearly implies (1.2). We remark that (3.17) is equivalent to

$$C_1^{-1/2m}\left(k+\log\frac{1}{A}\right)\leq p\log\frac{1}{A}. \tag{3.18}$$

Since we have

$$k+\log\frac{1}{A}\leq(k+R)+\log\frac{1}{A}\leq(p+1)\log\frac{1}{A}$$

by (i), condition (3.18) (i.e. (3.17)) holds whenever

$$C_1^{-1/2m}\leq\frac{p}{p+1}. \tag{3.19}$$

On the other hand, condition (3.16) is equivalent to

$$8\log k+(Cp-1)\log\frac{1}{A}+2m\log\left(k+\log\frac{1}{A}\right)\leq\log C_1. \tag{3.20}$$

Using (i), we can bound the left-hand side of (3.20) by

$$(\text{LHS of }(3.20))\leq 8\log p+2m\log(p+1)+(Cp-1)\log\frac{1}{A}+2(m+4)\log\log\frac{1}{A}.$$

Choosing

$$p\leq\frac{1}{2C}, \tag{3.21}$$

we can see that

$$\begin{aligned}
&(\text{LHS of }(3.20))\\
&\leq 8\log\frac{1}{2C}+2m\log\left(\frac{1}{2C}+1\right)-\frac{1}{2}\log\frac{1}{A}+2(m+4)\log\log\frac{1}{A}\\
&\leq 8\log\frac{1}{2C}+2m\log\left(\frac{1}{2C}+1\right)+\max_{z\geq 2}\left(-\frac{1}{2}z+2(m+4)\log z\right)\\
&= 8\log\frac{1}{2C}+2m\log\left(\frac{1}{2C}+1\right)+2(m+4)\big(\log(4m+16)-1\big).
\end{aligned}$$

Therefore, condition (3.20) (i.e. (3.16)) is satisfied provided

$$8\log\frac{1}{2C}+2m\log\left(\frac{1}{2C}+1\right)+2(m+4)\big(\log(4m+16)-1\big)\leq\log C_1. \tag{3.22}$$

Next we consider case (ii). We choose $T=k+R$ and observe that the term I_2 in (3.9) does not appear in this case. Hence, instead of (3.13), we have

$$\begin{aligned}
&\|\widetilde{q}\|^2_{H^{-s}(\mathbb{R}^n)}\\
&\leq C\left(\frac{1}{R^2}+\varepsilon T^{-m}\right)\|\widetilde{q}\|^2_{H^{-s}(\mathbb{R}^n)}+Ck^8e^{CR}\operatorname{dist}(\mathcal{C}_{q_1},\mathcal{C}_{q_2})^2+\frac{CT^{-m}}{\varepsilon}
\end{aligned}$$

Setting $\varepsilon=T^m/R^2$ implies that

$$\|\widetilde{q}\|^2_{H^{-s}(\mathbb{R}^n)}\leq\frac{2C}{R^2}\|\widetilde{q}\|^2_{H^{-s}(\mathbb{R}^n)}+Ck^8e^{CR}A+CR^2(k+R)^{-2m}.$$

Now we choose

$$R>2\sqrt{C} \tag{3.23}$$

and obtain that

$$\|\widetilde{q}\|^2_{H^{-s}(\mathbb{R}^n)}\leq Ck^8A+C(k+R)^{-2m},$$

which implies the desired estimate (1.2) since from condition (ii) we have

$$k+R \geq \frac{k}{2}+\frac{k+R}{2} \geq \frac{k}{2}+\frac{p}{2}\log\frac{1}{A} \geq \frac{\min\{p,1\}}{2}\left(k+\log\frac{1}{A}\right).$$

As the last step, we choose appropriate R, p, and C_1 to complete the proof. We first pick $R > C_* M$ sufficiently large satisfying (3.14) and (3.23) and then choose p small enough satisfying (3.21). Finally, we take C_1 large enough satisfying (3.19) and (3.22). □

4. Conclusion

We think that increasing stability is an important feature of the inverse boundary problem for the Schrödinger potential which should lead to higher resolution of numerical algorithms. It is important to collect numerical evidence of this phenomenon. Our method is based on the CGO solutions constructed in [**8**] where the constants in Lemma 2.1 are explicit. So most likely one can give explicit constants in Theorem 1.1 at least for particular domains Ω like balls. Contrary to the acoustic case [**15**], the constants in the estimate (1.2) depend only polynomially on k. It is an important and challenging question to determine whether the exponential dependence on k of the estimates in [**15**] is indeed generic if there are no assumptions on rays.

Acknowledgements

The authors would like to thank the referee for many insightful comments and for bringing several related papers to our attention.

References

[1] N. I. Akhiezer and I. M. Glazman, *Theory of linear operators in Hilbert space*, Dover Publications Inc., New York, 1993. Translated from the Russian and with a preface by Merlynd Nestell; Reprint of the 1961 and 1963 translations; Two volumes bound as one. MR1255973 (94i:47001)

[2] Giovanni Alessandrini and John Sylvester, *Stability for a multidimensional inverse spectral theorem*, Comm. Partial Differential Equations **15** (1990), no. 5, 711–736, DOI 10.1080/03605309908820705. MR1070844 (91i:35198)

[3] Deepak Aralumallige Subbarayappa and Victor Isakov, *On increased stability in the continuation of the Helmholtz equation*, Inverse Problems **23** (2007), no. 4, 1689–1697, DOI 10.1088/0266-5611/23/4/019. MR2348729 (2008i:35038)

[4] Deepak Aralumallige Subbarayappa and Victor Isakov, *Increasing stability of the continuation for the Maxwell system*, Inverse Problems **26** (2010), no. 7, 074005, 14, DOI 10.1088/0266-5611/26/7/074005. MR2644032 (2011a:35511)

[5] Alberto-P. Calderón, *On an inverse boundary value problem*, (Rio de Janeiro, 1980), Soc. Brasil. Mat., Rio de Janeiro, 1980, pp. 65–73. MR590275 (81k:35160)

[6] David Colton, Houssem Haddar, and Michele Piana, *The linear sampling method in inverse electromagnetic scattering theory*, Inverse Problems **19** (2003), no. 6, S105–S137, DOI 10.1088/0266-5611/19/6/057. Special section on imaging. MR2036524 (2004m:35270)

[7] J. Feldman, M. Salo and G. Uhlmann, *Calderón's problem: An inroduction to inverse problems.* Preliminary notes on the book in preparation. http://www.math.ubc.ca/ feldman/ibook/

[8] Peter Hähner, *A periodic Faddeev-type solution operator*, J. Differential Equations **128** (1996), no. 1, 300–308, DOI 10.1006/jdeq.1996.0096. MR1392403 (97d:35021)

[9] Tomasz Hrycak and Victor Isakov, *Increased stability in the continuation of solutions to the Helmholtz equation*, Inverse Problems **20** (2004), no. 3, 697–712, DOI 10.1088/0266-5611/20/3/004. MR2067496 (2005f:35051)

[10] Mikhail I. Isaev and Roman G. Novikov, *Energy and regularity dependent stability estimates for the Gel'fand inverse problem in multidimensions*, J. Inverse Ill-Posed Probl. **20** (2012), no. 3, 313–325, DOI 10.1515/jip-2012-0024. MR2984491

[11] Victor Isakov, *Increased stability in the continuation for the Helmholtz equation with variable coefficient*, Control methods in PDE-dynamical systems, Contemp. Math., vol. 426, Amer. Math. Soc., Providence, RI, 2007, pp. 255–267, DOI 10.1090/conm/426/08192. MR2311529 (2008f:35056)

[12] Victor Isakov, *Increasing stability for the Schrödinger potential from the Dirichlet-to-Neumann map*, Discrete Contin. Dyn. Syst. Ser. S **4** (2011), no. 3, 631–640, DOI 10.3934/dcdss.2011.4.631. MR2746425 (2011m:35419)

[13] Alexander Katchalov, Yaroslav Kurylev, and Matti Lassas, *Inverse boundary spectral problems*, Chapman & Hall/CRC Monographs and Surveys in Pure and Applied Mathematics, vol. 123, Chapman & Hall/CRC, Boca Raton, FL, 2001. MR1889089 (2003e:58045)

[14] Andrew Knyazev, Abram Jujunashvili, and Merico Argentati, *Angles between infinite dimensional subspaces with applications to the Rayleigh-Ritz and alternating projectors methods*, J. Funct. Anal. **259** (2010), no. 6, 1323–1345, DOI 10.1016/j.jfa.2010.05.018. MR2659762 (2011j:47068)

[15] Sei Nagayasu, Gunther Uhlmann, and Jenn-Nan Wang, *Increasing stability in an inverse problem for the acoustic equation*, Inverse Problems **29** (2013), no. 2, 025012, 11, DOI 10.1088/0266-5611/29/2/025012. MR3020433

[16] Zi Qi Sun, *On continuous dependence for an inverse initial-boundary value problem for the wave equation*, J. Math. Anal. Appl. **150** (1990), no. 1, 188–204, DOI 10.1016/0022-247X(90)90207-V. MR1059582 (91i:35024)

[17] John Sylvester and Gunther Uhlmann, *A global uniqueness theorem for an inverse boundary value problem*, Ann. of Math. (2) **125** (1987), no. 1, 153–169, DOI 10.2307/1971291. MR873380 (88b:35205)

Department of Mathematics and Statistics, Wichita State University, Kansas 67260-0033

E-mail address: victor.isakov@wichita.edu

Department of Mathematical Sciences, University of Hyogo, Himeji 671-2201, Japan

E-mail address: sei@sci.u-hyogo.ac.jp

Department of Mathematics, University of Washington, Box 354305, Seattle, Washington 98195-4350 & Fondation de Sciences Mathématiques de Paris

E-mail address: gunther@math.washington.edu

Institute of Applied Mathematical Sciences, NCTS (Taipei), National Taiwan University, Taipei 106, Taiwan.

E-mail address: jnwang@math.ntu.edu.tw

Contemporary Mathematics
Volume **615**, 2014
http://dx.doi.org/10.1090/conm/615/12290

Recent progress of inverse scattering theory on non-compact manifolds

Hiroshi Isozaki, Yaroslav Kurylev, and Matti Lassas

Dedicated to Gunther Uhlmann on the occasion of his 60th birthday

ABSTRACT. We give a brief survey for the recent development of inverse scattering theory on non-compact Riemannian manifolds. The main theme is the reconstruction of the manifold and the metric from the scattering matrix.

1. Introduction

Scattering theory for waves in classical or quantum physics has a long history, and nowadays there is an abundance of works devoted to the forward and inverse problems of potential scattering for Schrödinger equations, and obstacle scattering for wave equations. Much less is known, however, about the inverse scattering on manifolds, where the main theme is the reconstruction of the manifold itself and its Riemannian metric from the scattering matrix associated with the Laplace-Beltrami operator. We have been working on this subject by choosing typical examples possessing characteristic features of the problem. The aim of this paper is to present a birds-eye view based on the results we have obtained so far.

In §2, we discuss a basic framework of our issue. Scattering of waves is a time-dependent phenomenon, however, it is a consequence of properties of the continuous spectrum of the underlying Laplacian. Therefore, we formulate the problem mainly from the stationary view point with emphasis on the solution space of the *Helmholtz equation*. In §3, we explain the outline of our theory for the forward and inverse problems ignoring the detailed assumptions of the metric. In §4, we pick up four metrics we have already studied. They are well-known standard examples, and, viewed from the growth order of volume near infinity, range from an exponentially growing case to an exponentially decaying case. In §5, we devote ourselves to the case of 2-dim. arithmetic surface and discuss the inverse scattering from cusp.

In §6, we mention some recent results for metrics with intermediate behavior.

Due to the lack of space, we cannot talk about all of the important works. For example, [**12**], [**11**], [**13**], [**50**], [**54**], [**55**] are dealing with problems related to ours from different view points.

1991 *Mathematics Subject Classification.* Primary 35R30; Secondary 81U40.

2. Review of forward and inverse problems

2.1. Time-dependent scattering problem. Let us consider $\mathbb{R}^n$, $n > 1$, with a Riemannian metric $G = g_{ij}dx^i dx^j$, where we use the Einstein summation convention of omitting sum signs. We assume that asymptotically, when $|x| \to \infty$, $g_{ij}(x) \simeq \delta_{ij}$,, see e.g. (4.1). How can one recover the metric tensor g_{ij} from some "physical" observations? One way is to consider wave motions associated with this metric:

$$\partial_t^2 v(t,x) = \Delta_G v(t,x), \tag{2.1}$$

where Δ_G is the (negative) Laplace-Beltrami operator associated with G. In the remost past (i.e. $t \to -\infty$) and in the remote future (i.e. $t \to \infty$), the solution to (2.1) behaves like $v_\pm$ satisfying $\partial_t^2 v_\pm = \Delta v_\pm$, Δ being the standard Euclidean Laplacian on $\mathbb{R}^n$. The mapping assigning the far-field pattern of v_- to that of v_+ is called the *scattering operator.* It is a common belief that, under natural situations, this scattering operator determines the original physical system, i.e. the metric, at least up to changes of coordinates. In other words, from the observation of all far-field patterns of waves at infinity, one can reconstruct the metric G. For elliptic inverse boundary value problems on compact or non-compact Riemannian manifolds there are also non-uniqueness results; one can use the fact that the measurements are invariant in diffeomorphisms of the manifold to create counterexamples for the uniqueness of inverse problems and even invisibility results, see [**18, 19, 35, 44**] and the references in [**20, 21**]

Another way is to use the wave motion of quantum mechanical particles, in which case we use the Schrödinger equation

$$i\partial_t v(t,x) = \Delta_G v(t,x), \tag{2.2}$$

instead of (2.1). (See e.g. [**56**] or [**61**]). Both of these time-dependent scattering problems are reformulated in the same stationary (time-independent) picture.

2.2. Stationary scattering theory. Consider a time-periodic motion of (2.1), i.e. put $v(t,x) = e^{-ikt}u(x)$ in (2.1). Then we get

$$(-\Delta_G - k^2)u(x) = 0, \quad k > 0. \tag{2.3}$$

Since all solution of the Helmholtz equation (2.3) with unperturbed Laplacian, Δ, can be written as superpositions of plane waves and taking into account that $g_{ij} \simeq \delta_{ij}$, we seek a solution of (2.3) admitting the following asymptotic expansion

$$u(x) \simeq e^{ik\omega\cdot x} + \frac{e^{ikr}}{r^{(n-1)/2}} a(\hat{x},\omega;k), \quad \hat{x} = x/r, \quad \text{as} \quad r = |x| \to \infty, \tag{2.4}$$

where $\omega \in S^{n-1}$. The 2nd term of the right-hand side represents the scattered spherical wave. In the case of quantum mechanics, for a given beam of particles with initial direction ω, $|a(\theta,\omega;k)|^2$ is proportional to the number of particles scattered to the θ-direction. This is the physical quantity observed in the experiment. Let $A(k) : L^2(S^{n-1}) \to L^2(S^{n-1})$ be the integral operator with the kernel $a(\hat{x},\omega;k)$. By a suitable choice of the constant $C(k)$, $S(k) = I - C(k)A(k)$ is a unitary operator on $L^2(S^{n-1})$, which is the well-known *Heisenberg's S-matrix.* The above mentioned scattering operator in the time-dependent formulation is written in terms this S-matrix.

2.3. Geometric scattering theory. Let us look at the stationary scattering theory from a geometrical view point. Consider the restriction of the Fourier transform on S^{n-1}:

$$\left(\mathcal{F}_0(k)^*\varphi\right)(x) = \int_{S^{n-1}} e^{ik\omega\cdot x}\varphi(\omega)d\omega, \quad k>0.$$

This is sometimes called the *Herglotz integral* or the *Poisson integral.* It is the eigenoperator of $-\Delta$ in the sense that it satisfies the equation

$$(-\Delta - k^2)\mathcal{F}_0(k)^*\varphi = 0.$$

By the stationary phase method, it admits the asymptotic expansion:

$$(2.5)\quad (\mathcal{F}_0(k)^*\varphi)(x) \simeq C_+(k)\frac{e^{ikr}}{r^{(n-1)/2}}\varphi(\hat{x}) + C_-(k)\frac{e^{-ikr}}{r^{(n-1)/2}}\varphi(-\hat{x}), \quad r=|x|\to\infty.$$

Let us note that the spherical waves $e^{\pm ikr}/r^{(n-1)/2}$ appear in the asymptotic expansion of the Green functions $G_0^{(\pm)}(x,y;k)$ of $-\Delta - k^2$ in $\mathbb{R}^n$:

$$\int_{\mathbb{R}^n} G_0^{(\pm)}(x,y;k)f(y)dy \simeq \widetilde{C}_\pm(k)\frac{e^{\pm ikr}}{r^{(n-1)/2}}\widehat{f}(\pm k\hat{x}),$$

where $\widehat{f}(\xi)$ is the Fourier transform of $f(y)$. The unit sphere S^{n-1} can be regarded as a boundary at infinity of $\mathbb{R}^n$. Therefore, for $\mathbb{R}^n$ with Euclidean metric, one can associate the *manifold at infinity,* S^{n-1}, and the integral transform on it, $\mathcal{F}_0(k)^*$. It is an eigenoperator of $-\Delta$ and has the asymptotic expansion (2.5).

The stationary scattering theory asserts that these properties are transferred to $\mathbb{R}^n$ with metric G. Namely, one can associate a generalized eigenoperator $\mathcal{F}(k)^*$ on $L^2(S^{n-1})$ satisfying

$$(-\Delta_G - k^2)\mathcal{F}(k)^*\varphi = 0, \quad \text{for all } k^2 \in \sigma_c(-\Delta_G),$$

and admitting the asymptotic expansion (2.5) with $\mathcal{F}_0(k)^*$ replaced by $\mathcal{F}(k)^*$. Moreover, by introducing a Banach space $\mathcal{B}^*$ by

$$(2.6)\quad \mathcal{B}^* \ni u \Longleftrightarrow \sup_{R>1}\frac{1}{R}\int_{|x|<R}|u(x)|^2dx < \infty,$$

one can show that

$$\mathcal{N}(k) := \{u\in\mathcal{B}^*\,;(-\Delta_G - k^2)u = 0\} = \mathcal{F}(k)^*\left(L^2(S^{n-1})\right),$$

and that for any $\varphi_- \in L^2(S^{n-1})$, there exist unique $u\in\mathcal{N}(k)$ and $\varphi_+\in L^2(S^{n-1})$ satisfying

$$u(x) \simeq C_+(k)\frac{e^{ikr}}{r^{(n-1)/2}}\varphi_+(\hat{x}) + C_-(k)\frac{e^{-ikr}}{r^{(n-1)/2}}\varphi_-(\hat{x}), \quad r=|x|\to\infty.$$

Here, by $f\simeq g$, we mean

$$\lim_{R\to\infty}\frac{1}{R}\int_{|x|<R}|f(x)-g(x)|^2dx = 0.$$

The mapping

$$S_{geo}(k) : L^2(S^{n-1}) \ni \varphi_- \to \varphi_+ \in L^2(S^{n-1})$$

is unitary. In fact, it is related with the above S-matrix as follows:

$$(2.7)\quad S_{geo}(k) = JS(k),$$

where $J\phi(\omega) = \phi(-\omega)$.

Motivated by the above, let us give an overview how a geometric scattering theory on a Riemannian manifold can be formulated. We are given a non-compact Riemannian manifold $\mathcal{M}$ having the boundary at infinity M, which is a compact Riemannian manifold of dimension $n-1$. We then construct an operator $\mathcal{F}(k)^*$: $L^2(M) \to L^2_{loc}(\mathcal{M})$ such that

$$(-\Delta_G - k^2)\mathcal{F}(k)^*\varphi = 0, \quad \text{for all } k^2 \in \sigma_c(-\Delta_G),$$

with the property that, by a suitable choice of the Banach space $\mathcal{B}^*$ of functions in $\mathcal{M}$,

$$\mathcal{N}(k) := \{u \in \mathcal{B}^* \,;\, (-\Delta_G - k^2)u = 0\} = \mathcal{F}(k)^* \left(L^2(M)\right).$$

Moreover, there exist functions $w_\pm(k) \in \mathcal{B}^*$ such that for any $\varphi_- \in L^2(M)$ there exist unique $u \in \mathcal{N}(k)$ and $\varphi_+ \in L^2(M)$ admitting the asymptotic expansion

$$u \simeq w_+(k)\varphi_+ + w_-(k)\varphi_-,$$

and the mapping

$$S_{geo}(k) : L^2(M) \ni \varphi_- \to \varphi_+ \in L^2(M)$$

is unitary. We then formulate our inverse problem as follows : *Reconstruct the manifold $\mathcal{M}$ and its metric from the knowledge of the S-matrix $S_{geo}(k)$.*

2.4. Inverse scattering. The inverse potential scattering on the half-line was solved in 1950's by Gel'fand-Levitan and Marchenko [**17**], [**46**]. The multi-dimensional extension was proposed by the seminal work of Faddeev [**16**]. We do not enter into the detailed exposition here (see e.g. [**30**]), however, let us emphasize that the so called *CGO solution* of the Schrödinger equation introduced by Calderón and implemented into a powerfull tool by Sylvester and Uhlmann [**58**] (in fact, it was done independently of Faddeev) stimulated the works of Nachman [**53**] and Khenkin-Novikov [**36**] to bring the progress in the multi-dimensional inverse problem.

As for our inverse scattering problem, we use the *boundary control* (BC) method. It is now regarded as an established theory for the inverse spectral problem, however, a long time was necessary to settle the whole idea. The precursor of the BC method appeared in the work of M. G. Krein on the 1-dimensional wave equation [**37**], [**38**]. Although there is a similarity of Krein's theory to that of Gel'fand-Levitan-Marchenko, his idea is based on the hyperbolic nature of the wave equation, i.e. finite velocity of the wave propagation and the notion of the domain of influence. However, by passing to the Fourier transform in t, he used it in the form of the analyticity properties with respect to the spectral parameter. Blagovestcenskii [**9**] analyzed Krein's idea, using the finite velocity and controllability of the filled domain, to derive a Volterra-type equation for unknown functions. In [**10**], he also found a crucial idea of evaluating the inner product of waves with given data on the boundary by the spectral data. These ideas were then extended to the multi-dimensional case by the works of Belishev and Belishev-Blagovestcenskii [**3**], [**5**]. Belishev and Kurylev [**7**] used the BC method to solve the Gel'fand inverse problem on compact Riemannian manifolds. We must also mention that the step of controllability was completed by Tataru [**59**], which gave a final form for Holmgren's uniqueness theorem. The BC method can also be used to reconstruct non-smooth manifolds [**2**], [**42**] or solve inverse problems for the heat and Schrödinger equations on Riemannian manifolds [**33**]. For the more detailed history of the BC method and the related reconstruction methods, see [**4, 8, 32, 39**].

3. Bird's-eye view of the inverse scattering problem for the metric

We shall discuss here "typical" results which we expect to hold. Therefore, we state Theorems and Lemmas "formally" without specfying the detailed assumptions.

3.1. General framework. We consider an n-dimensional, non-compact, connected, Riemannian manifold $\mathcal{M}$ with the following properties :

$$\mathcal{M} = \mathcal{K} \cup \mathcal{M}_1 \cup \cdots \cup \mathcal{M}_N, \tag{3.1}$$

where $\mathcal{K}$ is relatively compact, and $\mathcal{M}_i$'s (called *ends*) are non-compact. We allow $\mathcal{K}$ to have an arbitrarily metric, however, each $\mathcal{M}_i$ is assumed to be diffeomorphic to $(1,\infty) \times M_i$, where M_i is a compact $(n-1)$-dimensional manifold endowed with a metric $h_i = (h_i)_{\alpha\beta}(y)dy^\alpha dy^\beta$. Let us denote points of $(1,\infty) \times M_i$ by (r,y). We also assume that on each end $\mathcal{M}_i$, the metric G is asymptotically equal to the warped product:

$$G \sim (dr)^2 + \rho_i(r)(h_i)_{\alpha\beta}(y)dy^\alpha dy^\beta, \quad \text{as} \quad r \to \infty, \tag{3.2}$$

with some positive function $\rho_i(r) \in C^\infty(\mathbb{R})$. We assume $\mathcal{M}$ to be complete, hence $-\Delta_G\big|_{C_0^\infty(\mathcal{M})}$ is essentially self-adjoint. For the sake of simplicity, assume that $\sigma_c(-\Delta_G) = [E_0,\infty)$, and put

$$H = -\Delta_G - E_0.$$

The properties of the continuous spectrum of H are determined by those of ends $\mathcal{M}_i$. Therefore, we reduce our analysis to a suitable model space, whose choice depends largely on the nature of $\rho_i(r)$, and shall be discussed separately.

3.2. Analysis on the model space. To fix our assumptions, we consider the case when the boundary $\partial\mathcal{M}$ is empty. Note that, however, one could consider, in a similar manner, also the case when $\partial\mathcal{M} \neq \emptyset$ with Neumann, Dirichlet or other self-adjoint boundary conditions. Given $\rho(r) \in C^\infty(\mathbb{R})$, we take $\mathcal{M}_0 = I_0 \times M_0$, where $I_0 = (c_0,\infty)$ is an open interval such that $c_0 < 1$, equipped with the metric $G_0 = (dr)^2 + \rho(r)h_0$. By imposing the Neumann boundary condition at $r = c_0$, we assume that the Laplace-Beltrami operator $H_0 = -\Delta_{G_0}$ on $\mathcal{M}_0$ is self-adjoint in $L^2(I_0, L^2(M_0); \rho(r)^{(n-1)/2}dr)$. Assume that $\sigma_c(-\Delta_{G_0}) = [E_0,\infty)$, and put $H_0 = -\Delta_{G_0} - E_0$, $R_0(z) = (H_0 - z)^{-1}$. Let us call H_0 the *free operator* on $\mathcal{M}_0$. Our first task is to prove the limiting absorption principle, which is going to be explained below.

Let $A = \int_{\mathbb{R}} \lambda dE_A(\lambda)$ be a self-adjoint operator on a Hilbert space $\mathcal{H}$. Then for $\lambda \in \sigma(A)$, $(A-\lambda)^{-1}$ does not exist. However, for $\lambda \in \sigma_c(A)$, sometimes one can prove the existence of the limit

$$(A - \lambda \mp i0)^{-1} = \lim_{\epsilon\to 0}(A - \lambda \mp i\epsilon)^{-1} \in \mathbf{B}(\mathcal{H}_+;\mathcal{H}_-),$$

where $\mathbf{B}(X;Y)$ means the set of all bounded operators from a Banach space X to a Banach space Y, and $\mathcal{H}_\pm$ are suitable Banach spaces skirting $\mathcal{H}$:

$$\mathcal{H}_+ \subset \mathcal{H} \subset \mathcal{H}_-,$$

with inclusions dense and continuous, and $\mathcal{H}_- = (\mathcal{H}_+)^*$ through the inner product of $\mathcal{H}$. This is called the *limiting absorption principle*, abbreviated as LAP. If the

LAP is proved on an interval $I \subset \sigma_c(A)$, by Stone's formula, we have

$$E_A(I)\mathcal{H} \subset \mathcal{H}_{ac}(A),$$

where the absolutely continuous subspace $\mathcal{H}_{ac}(A)$ is defined by

$$\mathcal{H}_{ac}(A) \ni u \Longleftrightarrow (E_A(\lambda)u, u) \text{ is absolutely continuous w.r.t. } d\lambda.$$

In all examples we have already treated, the space $\mathcal{H}_-$ is defined as follows.

$$\mathcal{H}_- \ni u \Longleftrightarrow \sup_{R>1} \frac{1}{R} \int_{c_0<r<R} \|u(r)\|^2_{L^2(M_0)} \rho(r)^{(n-1)/2} dr < \infty. \tag{3.3}$$

This is a dual space of some Banach space $\mathcal{B}$ using a dyadic decomposition of the manifold $\mathcal{M}_0$, and is denoted by $\mathcal{B}^*$ below. To prove LAP, usually we need to avoid the end points of $\sigma_c(H)$, the eigenvalues embedded in the continuous spectrum, and thresholds (the energies at which the nature of scattering changes). Let us call the set of these points the exceptional set, and denote it by $\mathcal{E}_0$. This is shown to be discrete with possible accumulation points 0 and on thresholds.

Our next task is to analyse the asymptotic behavior of the resolvent $R_0(\lambda \pm i0)$ as $r \to \infty$. For $f, g \in \mathcal{H}_-$ we use the notation $f \simeq g$ to denote

$$f \simeq g \Longleftrightarrow \lim_{R\to\infty} \frac{1}{R} \int_{c_0<r<R} \|f(r) - g(r)\|^2_{L^2(M_0)} \rho(r)^{(n-1)/2} dr = 0.$$

Then for any $k > 0$ such that $k^2 \notin \mathcal{E}_0$, there exist bounded operators $\mathcal{F}_0^{(\pm)}(k) \in \mathbf{B}(\mathcal{B}; L^2(M_0))$ and functions $w_\pm(k) \in \mathcal{B}^*$ such that for $f \in \mathcal{B}$,

$$R_0(k^2 \pm i0)f \simeq w_\pm(k)\mathcal{F}_0^{(\pm)}(k)f. \tag{3.4}$$

3.3. Gluing the estimates for ends. Having established the LAP and the asymptotic expansion (3.4) on each end, we need to glue them together to obtain global results. Let $\{\chi_j\}_{j=0}^N$ be a partition of unity on $\mathcal{M}$ such that $\sum_{j=0}^N \chi_j = 1$, and, for $j = 1, \cdots, N$, $\mathrm{supp}(\chi_j) \subset \mathcal{M}_j$, $\chi_j(r) = 1$ if $r > 2$. We take $\widetilde{\chi}_j \in C^\infty(\mathcal{M}_j)$ such that $\mathrm{supp}\,(\widetilde{\chi}_j) \subset \mathcal{M}_j$ and $\widetilde{\chi}_j = 1$ on $\mathrm{supp}\,\chi_j$. Let $\mathcal{H}_{free(j)}$ be the free operator on $\mathcal{M}_j$ and assume, for simplicity, that for all $j = 1, \ldots, N$, we have $\sigma_c(\mathcal{H}_{free(j)}) = [E_0, \infty)$. Then we put $R_{free(j)}(z) = (H_{free(j)} - z)^{-1}$ and

$$A(z) = \sum_{j=1}^N A_j(z)\widetilde{\chi}_j,$$

$$A_j(z) = [H, \chi_j]R_{free(j)}(z) + \chi_j(H - H_{free(j)})\widetilde{\chi}_j R_{free(j)}(z).$$

Then we have

$$R(z) = \sum_{j=1}^N \chi_j R_{free(j)}(z)\widetilde{\chi}_j + R(z)(\chi_0 - A(z)). \tag{3.5}$$

This formula and the perturbation technique enable us to prove LAP for H. Let $\mathcal{E} = \big(\cup_{j=1}^N \mathcal{E}_j\big) \cup \big(\sigma_p(H) \cap [0, \infty)\big)$, $\mathcal{E}_j$ being the exceptional set for $H_{free(j)}$. One can show that $\mathcal{E}$ is discrete with possible accumulation points 0 and on thresholds of $H_{free(j)}$. We define the space $\mathcal{B}^*$ by gluing $\mathcal{B}_j{}^*$ using the partition of unity, namely,

$$\mathcal{B}^* = \{u \in L^2_{loc}(\mathcal{M}) : \chi_j u \in \mathcal{B}_j^*\}.$$

THEOREM 3.1. *For $\lambda \in [0, \infty) \setminus \mathcal{E}$, there exists a limit $R(\lambda \pm i0) \in \mathbf{B}(\mathcal{B}; \mathcal{B}^*)$.*

By taking the adjoint in (3.5), we can derive the asymptotic expansion of the resolvent on each end:

$$R(k^2 \pm i0)f \simeq w_j^{(\pm)}(k)\mathcal{F}_j^{(\pm)}(k)f, \quad f \in \mathcal{B}. \tag{3.6}$$

We now introduce the L^2-space over the manifold at infinity of $\mathcal{M}$:

$$\mathbf{h}_\infty = \oplus_{j=1}^N L^2(M_j). \tag{3.7}$$

We also put

$$\mathcal{F}^{(\pm)}(k) = (\mathcal{F}_1^{(\pm)}(k), \cdots, \mathcal{F}_N^{(\pm)}k)) \in \mathbf{B}(\mathcal{B}\,;\,\mathbf{h}_\infty), \tag{3.8}$$

$$\widehat{\mathcal{H}} = L^2((0,\infty), \mathbf{h}_\infty;\, dk). \tag{3.9}$$

Here by $L^2(I, \mathbf{h}; dk)$ with given auxiliary Hilbert space $\mathbf{h}$, we mean the Hilbert space of all $\mathbf{h}$-valued L^2-functions on an interval I with respect to the measure dk.

THEOREM 3.2. *(1) The operator $\mathcal{F}^{(\pm)}$ defined for $f \in \mathcal{B}$ by $(\mathcal{F}^{(\pm)}f)(k) = \mathcal{F}^{(\pm)}(k)f$ is uniquely extended to a partial isometry on $L^2(\mathcal{M})$ with initial set $\mathcal{H}_{ac}(H)$ and final set $\widehat{\mathcal{H}}$.*
(2) For $f \in D(H)$, we have

$$(\mathcal{F}^{(\pm)}Hf)(k) = k^2(\mathcal{F}^{(\pm)}f)(k).$$

(3) $\mathcal{F}^{(\pm)}(k)^*$ is an eigenoperator in the sense that

$$(H - k^2)\mathcal{F}^{(\pm)}(k)^*\varphi = 0, \quad \textit{for all } \varphi \in \mathbf{h}_\infty.$$

(4) For any $f \in \mathcal{H}_{ac}(H)$, the inversion formula holds:

$$f = \int_0^\infty \mathcal{F}^{(\pm)}(k)^*(\mathcal{F}^{(\pm)}f)(k)dk.$$

We have thus constructed a spectral representation (generalized Fourier transformation) for H.

LEMMA 3.3. *For any $k^2 \in (0,\infty) \setminus \mathcal{E}$,*

$$\mathcal{N}(k) := \{u \in \mathcal{B}^*\,;\, (-\Delta_G - E_0 - k^2)u = 0\} = \mathcal{F}^{(\pm)}(k)^*\mathbf{h}_\infty.$$

THEOREM 3.4. *For any $\varphi^{(-)} = (\varphi_1^{(-)}, \cdots, \varphi_N^{(-)}) \in \mathbf{h}_\infty$, there exist unique $u \in \mathcal{N}(k)$ and $\varphi^{(+)} = (\varphi_1^{(+)}, \cdots, \varphi_N^{(+)}) \in \mathbf{h}_\infty$ such that on each end $\mathcal{M}_j$, the asymptotic expansion*

$$u \simeq w_j^{(-)}(k)\varphi_j^{(-)} + w_j^{(+)}(k)\varphi_j^{(+)}$$

holds. Moreover the operator

$$S_{geo}(k) : \mathbf{h}_\infty \to \mathbf{h}_\infty,$$
$$S_{geo}(k)\varphi^{(-)} = \varphi^{(+)}$$

is unitary.

3.4. Rellich's theorem, Radiation condition. In the Euclidean case, that is, for the manifold $\mathbb{R}^n$, there are two types of boundary conditions at infinity. A solution to the equation $(-\Delta_G - k^2)u = f$, $k > 0$, in $\mathbb{R}^n$ is said to satisfy the outgoing radiation condition, if $(\frac{\partial}{\partial r} - ik)u = o(|x|^{-(n-1)/2})$. Similarly, u is said to satisfy the incoming radiation condition, if $(\frac{\partial}{\partial r} + ik)u = o(|x|^{-(n-1)/2})$. Using the space $\mathcal{B}^*$, we can rewrite these conditions as

$$\lim_{R\to\infty} \frac{1}{R} \int_{1<r<R} \left\|(\frac{\partial}{\partial r} \mp ik)u(r)\right\|^2_{L^2(S^{n-1})} r^{(n-1)} dr = 0 \tag{3.10}$$

Moreover, the classical Rellich theorem says that if $u \in \mathcal{B}^*(\mathbb{R}^n)$ solves a homogeneous Helmholtz equation and satisfies an incoming or outgoing radiation condition (3.10), then $u = 0$.

These results can be extended to a more general case. Namely, for any $k > 0$ such that $k^2 \notin \mathcal{E}$, it is possible to define an incoming and outgoing radiation conditions for $u \in \mathcal{B}^*$. They are defined by

$$\lim_{R\to\infty} \frac{1}{R} \sum_{j=1}^{N} \left(\int_{1<r<R} \left\|(\frac{\partial}{\partial r} - iE_j(k))u(\cdot, r)\right\|^2_{L^2(M_j)} \rho_j(r)^{(n-1)/2} dr \right) = 0, \tag{3.11}$$

and

$$\lim_{R\to\infty} \frac{1}{R} \sum_{j=1}^{N} \left(\int_{1<r<R} \left\|(\frac{\partial}{\partial r} + iE_j(k))u(\cdot, r)\right\|^2_{L^2(M_j)} \rho_j(r)^{(n-1)/2} dr \right) = 0, \tag{3.12}$$

corrrespondingly. Here $E_j(k)$ is some operator in $L^2(M_j)$.

In typical situations, the space $\mathcal{B}^*$ enjoys the following property.

THEOREM 3.5. *Let $u \in \mathcal{B}^*$ satisfies $(-\Delta_G - E_0 - \lambda)u = 0$ for $\lambda \in (0, \infty) \setminus \mathcal{E}$ together with an outgoing* (3.11)*, or incoming* (3.12) *radiation condition. Then $u = 0$.*

Namely, as for the decay at infinity, $\mathcal{B}^*$ is the smallest space for non-trivial solutions to the Helmholtz equation $(-\Delta_G - E_0 - \lambda)u = 0$. This is a generalization of the classical Rellich's theorem.

THEOREM 3.6. *A solution $u \in \mathcal{B}^*$ of the equation $(-\Delta_G - E_0 - k^2)u = f \in \mathcal{B}$ satisfying either the outgoing radiation condition, or the incoming radiation condition is unique and is given by $u = R(k^2 + i0)f$, or $u = R(k^2 - i0)f$, respectively.*

3.5. Inverse problem. Having completed the forward problem, we can now enter into the inverse problem. Since $\mathcal{M}$ has N number of ends, the S-matrix $S_{geo}(k)$ is an $N \times N$ matrix-valued unitary operator, $S_{geo}(k) = \big(S_{ij}(k)\big)_{i,j=1}^{N}$. We pick up the entry $S_{11}(k)$, and try to reconstruct the whole manifold $\mathcal{M}$ and the Riemannain metric from $S_{11}(k)$.

Take the end $\mathcal{M}_1$ and split $\mathcal{M}$ into 2 parts:

$$\mathcal{M} = \mathcal{M}_{ext} \cup \mathcal{M}_{int}, \quad \mathcal{M}_{ext} = \mathcal{M}_1 \cap \{r > 2\}, \quad \mathcal{M}_{int} = \mathcal{M} \setminus \mathcal{M}_{ext}. \tag{3.13}$$

Let H_{ext} be the self-adjoint unbounded operator $-\Delta_G$ in $L^2(\mathcal{M}_{ext})$ identifed by the Neumann boundary condition at $r = 2$. Let H_{int} be the similarly defined self-adjoint unbounded operator $-\Delta_G$ in $L^2(\mathcal{M}_{int})$ with Neumann boundary condition at $r = 2$. One can then solve the boundary value problem on $\mathcal{M}_{ext}$ and $\mathcal{M}_{int}$:

$$\begin{cases} (-\Delta_G - E_0 - k^2)u = 0, & \text{in} \quad \mathcal{M}_{ext}, \\ \dfrac{\partial u}{\partial r} = f, & \text{on} \quad \partial\mathcal{M}_1 = \{r = 2\}, \\ u \text{ satisfies the outgoing radiation condition.} \end{cases} \tag{3.14}$$

$$\begin{cases} (-\Delta_G - E_0 - k^2)u = 0, & \text{in} \quad \mathcal{M}_{int}, \\ \dfrac{\partial u}{\partial r} = f, & \text{on} \quad \partial\mathcal{M}_1 = \{r = 2\}, \\ u \text{ satisfies the outgoing radiation condition if } \mathrm{N} \geq 2. \end{cases} \tag{3.15}$$

Let $\Lambda_{ext}(k)$ and $\Lambda_{int}(k)$ be the associated Neumann to Dirichlet maps:

$$\Lambda_{ext}(k) : f \to u_{ext}\big|_{\{r=2\}}, \tag{3.16}$$

$$\Lambda_{int}(k) : f \to u_{int}\big|_{\{r=2\}}, \tag{3.17}$$

u_{ext} and u_{int} being a unique solution to (3.14) and (3.15), respectively. Here and below, we assume that k^2 is not on the exceptional sets associated with $\mathcal{H}_{ext}$ and $\mathcal{H}_{int}$ or an eigenvalue of H_{int} if $\mathcal{M}_{int}$ is a bounded region.

The following lemma is a bridge between the scattering problem and the *interior* boundary value problem.

LEMMA 3.7. *Given $\Lambda_{ext}(k)$ for all $k > 0$, the S-matrix $S_{11}(k)$ and the ND map $\Lambda_{int}(k)$ determine each other for all $k > 0, k^2 \notin \mathcal{E}$.*

This is a well-known fact. The proof for the case of the Euclidean space is given in [**25**], and the hyperbolic space case is deal with in [**26**]. In the proof of this lemma, Rellich's uniqueness theorem plays a crucial role.

With the aid of Lemma 3.7, assuming that we know the end $\mathcal{M}_1$, we can obtain $\Lambda_{int}(k)$ for all $k \in \mathbb{C}$ from the $(1,1)$-component of the S-matrix $S_{11}(k)$ for $k > 0$. One can then apply the BC method to $\mathcal{M}_{int}$, namely we have

THEOREM 3.8. *Suppose we are given two (possibly non-compact) Riemannian manifolds $\mathcal{M}^{(1)}_{int}$, $\mathcal{M}^{(2)}_{int}$ such that there is a diffeomorphism $\psi : \partial\mathcal{M}^{(1)}_{int} \to \partial\mathcal{M}^{(2)}_{int}$ and the Dirichlet-to-Neumann maps satisfy $\psi^*(\Lambda^{(2)}_{int}(k)f) = \Lambda^{(1)}_{int}(k)\psi^* f$ for all $k \in (a,b)$, $b > a > 0$ and $f \in C_0^\infty(\partial\mathcal{M}^{(2)}_{int})$. Then $\mathcal{M}^{(1)}$ and $\mathcal{M}^{(2)}$ are isometric.*

The detailed proof of this theorem it is too long to review here but we explain below the sketch of the main steps of the proof. The details are seen in [**7**,**32**] for the case of compact manifolds and in [**27**,**43**] for non-compact manifolds. The first step needed to prove Theorem 3.8 is to consider the hyperbolic problem corresponding to (3.15)

$$\begin{cases} (\partial_t^2 - \Delta_G)u^F(x,t) = 0, & \text{in} \quad \mathcal{M}_{int} \times \mathbb{R}_+, \\ \dfrac{\partial u^F}{\partial \nu} = F(x,t), & \text{on} \quad \partial\mathcal{M}_{int} \times \mathbb{R}_+, \\ u^F(x,t)|_{t=0} = 0, & \partial_t u^F(x,t)|_{t=0} = 0. \end{cases} \tag{3.18}$$

Assume next that we are given $\partial\mathcal{M}_{int}$ and the operator $\Lambda_{int}(k)$ for $k \in (a,b)$. Since $\Lambda_{int}(k)$ is meromorphic operator depending on $k \in \mathbb{C}$, this data determine $\Lambda_{int}(k)$ for all $k \in \mathbb{C}$. Since the Fourier transform $(\mathcal{F}_{t\to\omega}u^F)(x,\omega)$ of $u^F(x,t)$ with respect to t satisfies (3.15) with $k^2 + E_0 = \omega^2$, we can determine the hyperbolic Dirichlet-to-Neumann map $\Lambda^{(h)} : F \mapsto u^F|_{\partial\mathcal{M}_{int}\times\mathbb{R}_+}$. This map can be used to compute inner products of waves.

LEMMA 3.9 (Blagovestchenskii identity). *Let* $F, H \in C_0^\infty(\partial\mathcal{M}_{int} \times \mathbb{R}_+)$ *and* $T > 0$. *Then*

$$\int_{\mathcal{M}_{int}} u^F(x,T)u^H(x,T)\, dV_G(x) = \frac{1}{2}\int_L \int_{\partial\mathcal{M}_{int}} (F(x,t)(\Lambda^{(h)}H)(x,s) - (\Lambda^{(h)}H)(x,t)h(x,s))\, dS_G(x)dtds, \tag{3.19}$$

where

$$L = \{(s,t) : \ 0 \le t + s \le 2T, \ t < s, \ t, s > 0\}.$$

PROOF. Let

$$w(t,s) = \int_{\mathcal{M}_{int}} u^F(x,t)u^H(x,s)\, dV_G(x).$$

Then, by integration by parts, we see that

$$\begin{aligned}(\partial_t^2 - \partial_s^2)w(t,s) &= \int_{\mathcal{M}_{int}} [\Delta_G u^F(x,t)u^H(x,s) - u^F(x,t)\Delta_G u^H(x,s)]\, dV_G(x)\\ &= \int_{\partial\mathcal{M}_{int}} [\partial_\nu u^F(t)u^H(s) - u^F(t)\partial_\nu u^H(s)]\, dS_G(x)\\ &= \int_{\partial\mathcal{M}_{int}} [\Lambda^{(h)}F(x,t)u^H(x,s) - u^F(x,t)\Lambda^{(h)}H(x,s)]\, dS_G(x).\end{aligned}$$

Moreover,

$$w|_{t=0} = w|_{s=0} = 0, \quad \partial_t w|_{t=0} = \partial_s w|_{s=0} = 0.$$

Thus, w is the solution of the initial-boundary value problem for the one-dimensional wave equation in the domain $(t,s) \in [0,2T] \times [0,2T]$ with a known source term and zero initial and boundary conditions. Solving this problem, we determine $w(t,s)$ in the domain where $t + s \le 2T$ and $t < s$. In particular, $w(T,T)$ gives the assertion. □

The other result is based on the following fundamental theorem by D. Tataru [**59**, **60**].

THEOREM 3.10. *Let* $u = u(x,t)$ *solve the wave equation* $u_{tt} - \Delta_G u = 0$ *in* $\mathcal{M}_{int} \times \mathbb{R}$ *and* $u|_{\Gamma\times(0,2T_1)} = \partial_\nu u|_{\Gamma\times(0,2T_1)} = 0$, *where* $\emptyset \ne \Gamma \subset \partial\mathcal{M}_{int}$ *is open. Then*

$$u(x,t) = 0 \text{ in } K_{\Gamma,T_1}, \tag{3.20}$$

where

$$K_{\Gamma,T_1} = \{(x,t) \in \mathcal{M}_{int} \times \mathbb{R} : d(x,\Gamma) < T_1 - |t - T_1|\}$$

is the double cone of influence.

(The proof of this theorem, in full generality, is in [**59**]. A simplified proof for the considered case is in [**32**].)

The observability Theorem 3.10 gives rise to the following approximate controllability:

COROLLARY 3.11. For any open $\Gamma \subset \partial M$ and $T_1 > 0$,

$$\mathrm{cl}_{L^2(\mathcal{M}_{int})}\{u^F(\cdot, T_1) : F \in C_0^\infty(\Gamma \times (0, T_1))\} = L^2(\mathcal{M}_{int}(\Gamma, T_1)).$$

Here

$$\mathcal{M}_{int}(\Gamma, T_1) = \{x \in \mathcal{M}_{int} : \ d(x, \Gamma) < T_1\} = K_{\Gamma, T_1} \cap \{t = T_1\}$$

is the domain of influence of Γ at time T_1 and $L^2(\mathcal{M}_{int}(\Gamma, T_1)) = \{a \in L^2(\mathcal{M}_{int}) : \mathrm{supp}(a) \subset \mathcal{M}_{int}(\Gamma, T_1)\}$.

LEMMA 3.12. *Let $T > 0$ and $\Gamma_j \subset \partial\mathcal{M}_{int}$, $j = 1, \dots, J$, be non-empty, relatively compact open sets, $0 \le T_j^- < T_j^+ \le T$. Assume we are given $\partial\mathcal{M}_{int}$ and the response operator $\Lambda^{(h)}$. This data determines the inner product*

$$J_{\mathcal{N}}^T(F_1, F_2) = \int_{\mathcal{N}} u^{F_1}(x, t) u^{F_2}(x, t) \, dV_G(x)$$

for given $t > 0$ and $F_1, F_2 \in C_0^\infty(\partial\mathcal{M}_{int} \times \mathbb{R}_+)$, where

$$\mathcal{N} = \bigcap_{j=1}^{J} (\mathcal{M}_{int}(\Gamma_j, T_j^+) \setminus \mathcal{M}_{int}(\Gamma_j, T_j^-)) \subset \mathcal{M}_{int}. \tag{3.21}$$

We give the proof of Lemma 3.12 only for the case when $J = 1$ and $T_1^- = 0$ so that $\mathcal{N} = \mathcal{M}_{int}(\Gamma_1, T_1^+)$. We note that when the manifold $\partial\mathcal{M}_{int}$ and the response operator $\Lambda^{(h)}$ are given, we can determine the metric $G_{\partial\mathcal{M}_{int}}$ of the boundary $\partial\mathcal{M}_{int}$. We say that the manifold $\partial\mathcal{M}_{int}$ with its metric and the response operator $\Lambda^{(h)}$ form the *boundary data.*

Consider first the case $F_1 = F_2 = F$. Let $B = \Gamma_1 \times [T - T_1^+, T])$. For all $H \in C_0^\infty(B)$ it holds that $\mathrm{supp}(u^H(\cdot, T)) \subset \mathcal{N}$, and thus

$$\begin{aligned}&\|u^F(T) - u^H(T)\|^2_{L^2(\mathcal{M}_{int})} \\ &\quad = \int_{\mathcal{N}} (u^F(x, T) - u^H(x, T))^2 dV_G(x) + \int_{\mathcal{M}_{int} \setminus \mathcal{N}} (u^F(x, T))^2 dV_G(x).\end{aligned}$$

Let $\chi_{\mathcal{N}}(x)$ be the characteristic function of the set $\mathcal{N}$. By Corollary 3.11, there is $H \in C_0^\infty(B)$ such that the norm $\|\chi_{\mathcal{N}} u^F(T) - u^H(T)\|_{L^2(M, dV_\mu)}$ is arbitrarily small. This shows that $J_{\mathcal{N}}^T(F_1, F_2)$ can be found by

$$J_{\mathcal{N}}^T(F, F) = \|u^F(T)\|^2_{L^2(\mathcal{M}_{int})} - \inf_{H \in C_0^\infty(B)} \mathcal{F}(H), \tag{3.22}$$

where

$$\mathcal{F}(H) = \|u^F(T) - u^H(T)\|^2_{L^2(\mathcal{M}_{int})}.$$

Since $\mathcal{F}(H)$ can be computed when the boundary data is given by Lemma 3.9, it follows that we can determine $J_{\mathcal{N}}^T(F, F)$ for any $F \in C_0^\infty(\partial M \times \mathbb{R}_+)$. Now, since

$$J_{\mathcal{N}}^T(F_1, F_2) = \frac{1}{4}(J_{\mathcal{N}}^T(F_1 + F_2, F_1 + F_2) - J_{\mathcal{N}}^T(F_1 - F_2, F_1 - F_2)),$$

Lemma 3.12 follows in the case when $J = 1$ and $T_1^- = 0$.

To reconstruct $(\mathcal{M}_{int}, G)$, we use a special representation, the *boundary distance representation*, $R(\mathcal{M}_{int})$, of $\mathcal{M}_{int}$ and show that the boundary data determine $R(\mathcal{M}_{int})$. Consider a map $R : \mathcal{M}_{int} \to C(\partial\mathcal{M}_{int})$,

$$R(x) = r_x(\cdot); \quad r_x(z) = d_g(x, z), \ z \in \partial\mathcal{M}_{int}, \tag{3.23}$$

i.e., $r_x(\cdot)$ is the *distance function* from $x \in \mathcal{M}_{int}$ to the points on $\partial\mathcal{M}_{int}$. The image $R(\mathcal{M}_{int}) \subset C(\partial\mathcal{M}_{int})$ of R is called the boundary distance representation of $\mathcal{M}_{int}$.

The set $R(\mathcal{M}_{int})$ is a metric space with the distance inherited from $C(\partial\mathcal{M}_{int})$ with the standard norm $\|f\|_\infty = \max_{z\in\partial\mathcal{M}_{int}} |f(z)|$ of $C(\partial\mathcal{M}_{int})$. We denote this distance by d_∞. The map R, due to the triangular inequality, is Lipschitz,

$$d_\infty(r_x, r_y) \le d_g(x, y). \tag{3.24}$$

Note that if $\partial\mathcal{M}_{int}$ is unbounded, which corresponds to the possible case of unbounded M_i, $r_x(\cdot)$ becomes unbounded and, instead of the distance $d_\infty(r_x, r_y)$, it is convenient to use a modified one

$$d_b(r_x, r_y) = \sup_{z\in\partial\mathcal{M}_{int}} \left(\frac{|r_x(z) - r_y(z)|}{1 + |r_x(z) - r_y(z)|}\right).$$

When x, y run over a compact in $\mathcal{M}_{int}$, $d_b(r_x, r_y)$ is equivalent to $d_\infty(r_x, r_y)$.

The map $R : \mathcal{M}_{int} \to R(\mathcal{M}_{int}) \subset C(\partial\mathcal{M}_{int})$ is an embedding. Many results of differential geometry, such as Whitney or Nash embedding theorems, concern the question how an abstract manifold can be embedded to some simple space, such as a higher dimensional Euclidean space. In the inverse problem we need to construct a "copy" of the unknown manifold in some known space, and as we assume that the boundary is given, we do this by embedding the manifold $\mathcal{M}_{int}$ to the known, although infinite dimensional function space $C(\partial\mathcal{M}_{int})$.

The basic observation needed to construct $R(\mathcal{M}_{int})$ is that the set $\mathcal{N}$ given in (3.21) has non-zero measure if and only if

$$P(\{(\Gamma_j, T_j^+, T_j^-) :\ j = 1, \dots, J\}) = \sup_f J_N^T(f, f),$$

is non-zero.

THEOREM 3.13. *Let $\{z_n\}_{n=1}^\infty$ be a dense set on $\partial\mathcal{M}_{int}$ and $r(\cdot) \in C(\partial\mathcal{M}_{int})$ be an arbitrary continuous function. Then $r \in R(\mathcal{M}_{int})$ if and only if for all $L \in \mathbb{N}_+$, it holds that*

$$P\Big(\Big\{(z_j, r(z_n) + \frac{1}{L}, r(z_n) - \frac{1}{L}); j = 1, \dots, L\Big\}\Big) > 0. \tag{3.25}$$

Moreover, condition (3.25) can be verified when the manifold $\partial\mathcal{M}_{int}$ and the response operator $\Lambda^{(h)}$ are given. Hence the pair $(\partial\mathcal{M}_{int}, \Lambda^{(h)})$ determines uniquely the boundary distance representation $R(\mathcal{M}_{int})$ of $(\mathcal{M}_{int}, G)$.

The idea of to proof Theorem 3.13 is that we see that the set $\{x \in \mathcal{M}_{int};\ |d(z_j, x) - r(z_n)| < \frac{1}{L}$ for all $j = 1, \dots, L\}$ contains a point x_0 for all L if and only if $r_{x_0}(z) = r(z)$ for all $z \in \partial\mathcal{M}_{int}$.

Next let us consider a compact manifold $(\mathcal{M}_{int}, G)$ that is geodesically regular manifold, i.e.,

i) For any $x, y \in \mathcal{M}_{int}$ there is a unique geodesic γ joining these points.
ii) Any geodesic $\gamma([a, b])$ can be continued to a geodesic $\gamma([a', b'])$ whose endpoints lie on the boundary.

Consider $R(\mathcal{M}_{int})$ as a metric space $(R(\mathcal{M}_{int}), d_\infty)$ with the distance inherited from $L^\infty(\partial\mathcal{M}_{int})$, $d_\infty(r_x, r_y) = \|r_x - r_y\|_{L^\infty(\partial\mathcal{M}_{int})}$. Then the mapping R is an isometry, i.e.

$$d_\infty(r_x, r_y) = d(x, y).$$

Indeed, let $\gamma([a, b])$ be the shortest geodesic from y to x. Continue this geodesic to the shortest geodesic $\gamma([a', b])$, where $z = \gamma(a') \in \partial\mathcal{M}_{int}$. Then

$$r_x(z) - r_y(z) = |\gamma([a', b])| - |\gamma([a', a])| = b - a = d(x, y).$$

Hence,

$$d(x,y) \le d_\infty(r_x, r_y).$$

The opposite inequality is valid by the triangular inequality, yielding that $d(x,y) = d_\infty(r_x, r_y)$. This means that, if we know *a priori* that $(\mathcal{M}_{int}, G)$ is geodesically regular, then $(R(\mathcal{M}_{int}), d_\infty))$ is isometric to $(\mathcal{M}_{int}, G)$ and the problem of the reconstruction of $(\mathcal{M}_{int}, G)$ from the data $(\partial\mathcal{M}_{int}, \Lambda^{(h)})$ is solved. The above ideas can be generalized to prove Theorem 3.8.

Let us note that only one diagonal entry $S_{11}^{(2)}(k)$ is necessary to recover the whole manifold using Lemma 3.7 and Theorem 3.8. Therefore, our final inverse procedure is as follows.

THEOREM 3.14. *Suppose we are given two non-compact Riemannian manifolds $\mathcal{M}^{(1)}$, $\mathcal{M}^{(2)}$ having N_1 and N_2 numbers of ends, respectively. Let $\mathcal{M}_1^{(1)}$ and $\mathcal{M}_1^{(2)}$ be their first ends, and suppose that there exists $r_0 > 1$ such that $\mathcal{M}_1^{(1)} \cap \{r > r_0\}$ and $\mathcal{M}_1^{(2)} \cap \{r > r_0\}$ are isometric. Assume furthermore that their $(1,1)$-entries of S-matrices coincide : $S_{11}^{(1)}(k) = S_{11}^{(2)}(k)$ for all $k > 0$, $k^2 \notin \mathcal{E}^{(1)} \cup \mathcal{E}^{(2)}$. Then $\mathcal{M}^{(1)}$ and $\mathcal{M}^{(2)}$ are isometric.*

4. Examples

4.1. Model metric. Spectral properties of $-\Delta_G$ depends largely on the growth of the manifold $\mathcal{M}$ at infinity. Looking at an end $(1,\infty) \times M$, we pick up here 4 basic examples, and examine what is going on. We have so far studied the following cases:

(1) $G = (dr)^2 + e^{2r}h$,

(2) $G = (dr)^2 + r^2 h$,

(3) $G = (dr)^2 + h$,

(4) $G = (dr)^2 + e^{-2r}h$,

where h is a metric on M.

These are well-known classical examples, and have some distinguished properties.

4.2. Tools for the resolvent estimates. Before going into the details, we explain here the method for resolvent estimates. There are several ways of proving LAP for Laplacians on manifolds. One is E. Mourre's abstract commutator theory [**51**], another is Melrose' theory of scattering metric [**48**], [**49**]. Our method is different from both of them. We employ the classical Eidus' approach of integration by parts [**14**]. Given the equation $(-\Delta_{G_0} - E_0 - z)u = f$ on the model space $\mathcal{M}_0 = I_0 \times M_0$, we expand u by the eigenvectors of $-\Delta_{M_0}$. The problem is then reduced to the 1-dimensional case. The main step is to multiply this 1-dimensional equation by (the derivative of) solution and integrate by parts to obtain some identities. Standard arguments from perturbation theory then prove the necessary resolvent estimates. See e.g. [**28**], Lemmas 2.4, 2.5, Theorem 2.7, or [**27**], Lemmas 2.4 $\sim$ 2.8. This method is no less powerful than the above two machineries, and gives us results by elementary computation. All of the above 4 examples are treated by this method.

4.3. Euclidean metric. Let us start with the Euclidean metric (2). In this case, we take $\mathcal{M}_0 = \mathbb{R}^n$, and the perturbed metric $g_{ij}dx^i dx^j$ is assumed to satisfy

$$\partial_x^\alpha (g_{ij}(x) - \delta_{ij}) = O(|x|^{-1-|\alpha|-\epsilon_0}), \quad \text{for all } \alpha \text{ and some } \epsilon_0 > 0. \tag{4.1}$$

There are plenty of papers dealing with scattering in the asymptotically Euclidean space, and all the requisites are prepared in e.g. [**27**] or [**61**]. Consulting them, one can see that the arguments in §3 work well, and Theorem 3.14 holds. In this case, there are no embedded eigenvalues in $(0, \infty)$. The radiation condition is the standard one:

$$\lim_{R\to\infty} \frac{1}{R} \int_{|x|<R} \left|\left(\frac{\partial}{\partial r} \mp ik\right)u\right|^2 dx = 0.$$

4.4. Cylindrical ends. Let us consider (3). In the daily life example, it arises as the problem of waveguide. As a model space, we take M_0 to be a compact $(n-1)$-dimensional manifold without boundary. Let $\mathcal{M}_0 = (1, \infty) \times (M_0, h)$, and $\Delta_0 = \Delta_{(M_0,h)}$. The perturbed metric is assumed to satisfy

$$|\partial_X^\alpha (g_{ij}(X) - h_{ij}(x))| \leq C_\alpha (1 + r)^{-1-\epsilon}, \quad \text{for all } \alpha \text{ and some } \epsilon > 0$$

where x denotes the variable in M_0, and $X = (r, x)$.

The waveguide has many features different from the conical Euclidean metric (2). In the first place, it may have eigenvalues embedded in the continuous spectrum. The 2nd feature is that the scattering in the waveguide has many channels. Let $\{\lambda_m\}_{m=1}^\infty$ be the eigenvalues of $-\Delta_0$. Then, since $-\Delta_G$ has the form

$$-\Delta_G = -\left(\frac{\partial}{\partial r}\right)^2 - \Delta_0,$$

when the total system has energy $\lambda > 0$, only the states with $\lambda_m \leq \lambda$ occur in the scattering phenomena. The radiation condition is thus defined by

$$\lim_{R\to\infty} \frac{1}{R} \int_{1<r<R} \left\|\left(\frac{\partial}{\partial r} \mp i\sqrt{\lambda + \Delta_0}\right)u(r, \cdot)\right\|_{L^2(M_0)}^2 dr = 0.$$

Because of this channel property, the scattering matrix changes its size according to the energy, higher the energy, bigger the size of the S-matrix.

To discuss the inverse scattering, we assume that one end is strictly cylindrical. Namely, the end $\mathcal{M}_1$ is equal to $[1, \infty) \times M_0$ with metric $(dr)^2 + h$. We take an artificial boundary $r = 2$ in $\mathcal{M}_1$, and split $\mathcal{M}$ into the exterior and interior domains. As in §3, we derive interior N-D map from the knowledge of $(1,1)$-entry of the S-matrix. However, the physical S-matrix is not sufficient to determine the interior N-D map. The key observation here is that the physical S-matrix $S_{11}(k)$ admits an analytic continuation into the *upper* (physical) half plane, and, down to the real axis, determines non-physical scattering matrix, which is defined using the exponentially growing solutions of the reduced wave equation instead of the physical plane wave. Therefore Theorem 3.14 holds in this case.

The exponentially growing solution is the crucial idea found by Faddeev in his multi-dimensional inverse scattering theory, as well as Calderón's inverse boundary value problem. What is interesting is that this apparently artificial exponentially growing solutions appear naturally in the waveguide problem. The details are given in [**28**].

We can also allow the ends of type (2) and (3) at the same time. Namely, if our ends $\mathcal{M}_1, \cdots, \mathcal{M}_N$ are composed of two parts: $\mathcal{M}_1, \cdots, \mathcal{M}_\mu$, which are Euclidean ends, and $\mathcal{M}_{\mu+1}, \cdots, \mathcal{M}_N$, which are cylindrical ends, the results in subsections 4.3, 4.4 are applied as well.

4.5. Hyperbolic space. We turn to the case (1). The problem here is that in dimensions ≥ 3, the infinity of hyperbolic manifold may have a complicated structure. Therefore, we restrict ourselves to the simple case, and as a model space, we take $\mathcal{M}_0 = M_0 \times (0, \infty)$ equipped with the metric $((dy)^2 + h(x, dx))/y^2$, where M_0 is a compact manifold equipped with metric $h(x, dx)$. This model has two infinities, $y = 0$ corresponding to the infinite volume, which we call *regular end*, and $y = \infty$ corresponding to *cusp*. We take $H_0 = -\Delta_{\mathcal{M}_0} - (n-1)^2/4$. Then $\sigma(H_0) = [0, \infty)$. We split $\mathcal{H}_0$ into two parts : $\{0 < y < 1\}$ and $\{y > 1\}$. By the change of variable $y = e^{-r}$, the former becomes a model space for (1). The change of variable $y = e^r$ makes the latter to a model space for (4).

The perturbed metric is assumed to have the form

$$ds^2 = y^{-2}\big((dy)^2 + h(x, dx) + A(x, y, dx, dy)\big),$$

$$A(x, y, dx, dy) = \sum_{i,j=1}^{n-1} a_{ij}(x, y) dx^i dx^j + 2 \sum_{i=1}^{n-1} a_{in}(x, y) dx^i dy + a_{nn}(x, y)(dy)^2,$$

where $a_{ij}(x, y)$ satisfies, for some $\epsilon_0 > 0$

$$|\widetilde{D}_x^\alpha D_y^\beta\, a(x, y)| \leq C_{\alpha\beta}(1 + |\log y|)^{-\min(|\alpha|+\beta, 1) - 1 - \epsilon_0}, \quad \text{for all} \quad \alpha, \beta.$$

Here, $D_y = y\partial y$, $\widetilde{D}_x = \widetilde{y}(y)\partial_x$, $\widetilde{y}(y) \in C^\infty((0, \infty))$, $\widetilde{y}(y) = y$ for $y > 2$, $\widetilde{y}(y) = 1$ for $0 < y < 1$.

If one of the ends is a regular infinity, there are no embedded eigenvalues. If all the ends are cusps, there may be embedded eigenvalues. The radiation condition is formulated as

$$\lim_{R\to\infty} \frac{1}{\log R} \int_{1/R<y<1} \big\|\big(D_y - i\sigma_\pm(k)\big)u(y)\big\|^2_{L^2(M_0)} \frac{dy}{y^n} = 0,$$

$$\lim_{R\to\infty} \frac{1}{\log R} \int_{1<y<R} \big\|\big(D_y - i\sigma_\mp(k)\big)u(y)\big\|^2_{L^2(M_0)} \frac{dy}{y^n} = 0,$$

$$\sigma_\pm(k) = \frac{n-1}{2} \mp ik.$$

Note that these are standard radiation conditions
(3.11) and (3.12) since $r = |\log y|$.

The inverse scattering from regular ends works well as in §3. Namely, suppose we are given two such asymptotically hyperbolic manifolds $\mathcal{M}^{(1)}$, $\mathcal{M}^{(2)}$ whose regular ends $\mathcal{M}_1^{(1)}$ and $\mathcal{M}_1^{(2)}$ are isometric, and the $(1, 1)$ components of the S-matrix coincide for all k. Then $\mathcal{M}^{(1)}$ and $\mathcal{M}^{(2)}$ are isometric. This is proven in [**27**].

We remark that Sá Barreto [**57**], using the framework of scattering metric due to Melrose, proved the existence of isometry between two asymptotically hyperbolic manifolds having the same total scattering matrices without the assumption that $\mathcal{M}_1^{(1)}$ and $\mathcal{M}_1^{(2)}$ are isometric, although the decay assumption at infinity are different from ours. In [**22**], it is extended to asymptotically hyperbolic complex manifolds.

4.6. Poisson integral and the space $\mathcal{B}^*$. Before entering into the problem of cusp, we briefly look at some aspects of Poisson integral. Let us take the most basic example of Poincaré disc : $D = \{z \in \mathbb{C}\,;\, |z| < 1\}$. It is well-known that the Poisson integral

$$u(z) = \frac{1}{2\pi}\int_0^{2\pi}\left(\frac{1-|z|^2}{|e^{i\theta}-z|^2}\right)^s f(\theta)d\theta, \tag{4.2}$$

$f(\theta)$ being a function on the boundary $\partial D = S^1$, is a solution to the equation

$$(-\Delta_G - 4s(1-s))u = 0 \tag{4.3}$$

for all $s \in \mathbb{C}$. Our space $\mathcal{B}^*$, for which $f(\theta) \in L^2(S^1)$, has the following special feature : $\{u \in \mathcal{B}^*\,;\, (-\Delta_G - \frac{1}{4} - k^2)u = 0\}$ is the smallest space of solutions as regard to decay at infinity. In fact, the Rellich type theorem says that the solution of (4.3) decaying faster than the elements of $\mathcal{B}^*$ vanishes identically. Although this space is the smallest, it contains sufficiently many solutions for the inverse scattering. The largest solution space was found by Helgason [**23**], namely all solutions of (4.3) are represented as (4.2), where $f(\theta)$ is Sato's hyperfunction. This theorem was extended to the general symmetric spaces by Kashiwara-Kowata-Minemura-Okamoto-Oshima-Tanaka [**34**]. The theorem of Helgason suggests that one may control the solution space of the equation $(-\Delta_G - \frac{1}{4} - k^2)u = 0$ through function spaces on the boundary at infinity. This enables us to extend the notion of S-matrix. We utilize this idea for the cusp.

5. Arithmetic surface and generalized S-matrix

We restrict ourselves here to the 2-dimensional case. Lots of examples of hyperbolic surfaces are given by the action of discrete groups on the upper half plane. If we consider the geometrically finite case, which implies that the associated quotient space is a polygon with sides of finite number of geodesics, the infinity consists of a finite number of cusp and funnel (the latter being a slightly perturbed regular infinity discussed in 4.5). However, if the Fuchsian group contains elliptic elements, the quotient space $\mathcal{M}$ has singularities. Note that $\mathcal{M}$ itself is an analytic manifold without singularities, however, the hyperbolic metric induced from $\mathbb{C}_+$ becomes singular at elliptic fixed points. Thus they are orbifolds, moreover, many examples of quotient spaces appearing in number theory have only cusp at infinity. They are non-compact, but has a finite volume. The result of the previous section cannot be applied to this case. As a matter of fact, the S-matrix at a cusp end does not have enough information to reconstruct the metric, since the cusp gives only one-dimensional contribution to the continuous spectrum. A remedy lies in generalizing the notion of S-matrix.

Assume that $\mathcal{M}_1 = (1,\infty) \times (-1/2, 1/2)$ with two sides identified. Take any solution u of the equation $(H - k^2)u = 0$, and expand it into the Fourier series on $\mathcal{M}_1$:

$$u(x,y) = \sum_{n\in\mathbf{Z}} e^{2\pi nx} u_n(y).$$

Then u_n satisfies

$$y^2\big(-\partial_y^2 + (2ny)^2\big)u_n(y) - \frac{1}{4}u_n(y) = k^2 u_n(y),$$

hence is written as

$$u_n(y) = \begin{cases} \widetilde{a}_n\, y^{1/2} I_{-ik}(2\pi|n|y) + \widetilde{b}_n\, y^{1/2} K_{ik}(2\pi|n|y), & n \neq 0, \\ a_0\, y^{1/2-ik} + b_0\, y^{1/2+ik}, & n = 0. \end{cases}$$

where I_ν and K_ν are modified Bessel functions. We then see that *all* solutions of the equation $(H - k^2)u = 0$ behave like

$$u(x,y) \sim a_0 y^{1/2-ik} + \sum_{n\neq 0} a_n e^{inx+|n|y} + b_0 y^{1/2+ik} + \sum_{n\neq 0} b_n e^{inx-|n|y}, \tag{5.1}$$

as $y \to \infty$. Given an exponentially growing wave represented by the first two terms in (5.1), one can uniquely construct a solution of $(H - k^2)u = 0$ belonging to $\mathcal{B}^*$ on ends $\mathcal{M}_j$, $j \neq 1$, and behaving like (5.1) in $\mathcal{M}_1$. We call the mapping

$$\mathcal{S}_{11}(k) : \{a_n\}_{n\in\mathbf{Z}} \to \{b_n\}_{n\in\mathbf{Z}} \tag{5.2}$$

the *generalized S-matrix*. Passing to the Fourier series, we see that to define (5.2), we are using a class of analytic functionals bigger than Sato's hyperfunction. Note that $\mathcal{S}_{11}(k)$ is an infinite matrix, and the usual S-matrix $S_{11}(k)$ is its $(0,0)$-entry. This generalized S-matrix has enough information for the inverse scattering. We make an artificial boundary in $\mathcal{M}_1$ and consider the boundary value problem in the interior domain. Then the knowledge of $\mathcal{S}_{11}(k)$ for all k enables us to determine the N-D map $\Lambda_{int}(k)$, hence to apply the BC method.

In the interior domain $\mathcal{M}_{int}$, there is a finite number singular points. Since they have a special structure, we can deal with $\mathcal{M}_{int}$ as a manifold with *conical singularities*, to which we can apply the BC method as well. We can thus prove the following theorem.

THEOREM 5.1. *Suppose we are given two 2-dimensional asymptotically hyperbolic surfaces with conical singularities $\mathcal{M}^{(1)}, \mathcal{M}^{(2)}$. Suppose they have pure cusp ends $\mathcal{M}_1^{(1)}, \mathcal{M}_1^{(2)}$, i.e. with metric $((dy)^2 + (dx)^2)/y^2$, and their generalized S-matrices coincide : $\mathcal{S}_{11}^{(1)}(k) = \mathcal{S}_{11}^{(2)}(k)$, for all k. Then $\mathcal{M}^{(1)}$ and $\mathcal{M}^{(2)}$ are isometric.*

In particular, if $\mathcal{M}^{(i)} = \Gamma^{(i)}/\mathbf{H}^2$ with geometrically finite Fuchsian group $\Gamma^{(i)}$, and their generalized S-matrix coincide, $\Gamma^{(1)}$ and $\Gamma^{(2)}$ are conjugate. See [**29**] for details. Also note that the results in subsections 4.5 and 5.1 also hold when $\mathcal{M}$ has both regular ends and cusps.

As in the previous cases, the key idea for the proof of Theorem 5.1 is the use of exponentially growing solutions of the Helmholtz equation, which appears naturally here because of the form of hyperbolic metric. For cylindrical ends, we have encountered a similar situation, in which case, however, the non-physical scattering matrix is obtained by the analytic continuation of the physical scattering matrix. This is not true for the cusp. In fact, Zelditch [**62**] constructed an example of non-isometric hyperbolic surfaces with the same physical scattering matrix.

6. Works in progress

6.1. Higher dimensional asymptotically hyperbolic orbifolds. In higer dimensions, we have new phenomenon in the problem in §5. Consider the case of $n = 3$, and let the Picard group $\Gamma = SL(2, \mathbf{Z}+i\mathbf{Z})$ act on $\mathbf{H}^3$ through quarternions. The resulting quotient space is a 3-dimensional hyperbolic manifold with singularities, i.e. an orbifold. Unlike the 2-dimensional case, the singularities are not confined in a compact set. They form an unbounded curve, and the manifold at infinity is not a smooth manifold, but a 2-dimensional orbifold (see [**15**]). However, the results for the forward problem also hold in this case with no essential change, and we are also expecting the inverse scattering results as well.

The structure of the manifold at infinity of general hyperbolic manifold is complicated (see e.g. [**24**]). The forward and inverse scattering for this general case are challenging problems.

6.2. Intermediate metrics. We have worked on the problems (1)–(4), however, there is no reason that we must restrict ourselves to these cases. There is a wide area of problems for the intermediate metrics between (1) and (4), and those outside. For example, [**40**], [**31**] are dealing with non-existence of embedded eigenvalues, and [**41**] studies the LAP. We expect that the inverse scattering theory can also be developed in these cases.

References

[1] Shmuel Agmon, *Spectral theory of Schrödinger operators on Euclidean and on non-Euclidean spaces*, Comm. Pure Appl. Math. **39** (1986), no. S, suppl., S3–S16, DOI 10.1002/cpa.3160390703. Frontiers of the mathematical sciences: 1985 (New York, 1985). MR861480 (88c:35113)

[2] Michael Anderson, Atsushi Katsuda, Yaroslav Kurylev, Matti Lassas, and Michael Taylor, *Boundary regularity for the Ricci equation, geometric convergence, and Gel′fand's inverse boundary problem*, Invent. Math. **158** (2004), no. 2, 261–321, DOI 10.1007/s00222-004-0371-6. MR2096795 (2005h:53051)

[3] M. I. Belishev, *An approach to multidimensional inverse problems for the wave equation* (Russian), Dokl. Akad. Nauk SSSR **297** (1987), no. 3, 524–527; English transl., Soviet Math. Dokl. **36** (1988), no. 3, 481–484. MR924687 (89c:35152)

[4] M. I. Belishev, *Boundary control in reconstruction of manifolds and metrics (the BC method)*, Inverse Problems **13** (1997), no. 5, R1–R45, DOI 10.1088/0266-5611/13/5/002. MR1474359 (98k:58073)

[5] M. I. Belishev and A. S. Blagovestcenskii, *Multidimensional analogs of the Gel'fand-Levitan-Krein equations in inverse problems for the wave equations*, Ill-posed problems of Mathematical Physics and Analysis, Nobosibirsk : Nauk (1992), 50-63.

[6] M. I. Belishev and Ya. V. Kurylëv, *A nonstationary inverse problem for the multidimensional wave equation "in the large"* (Russian), Zap. Nauchn. Sem. Leningrad. Otdel. Mat. Inst. Steklov. (LOMI) **165** (1987), no. Mat. Vopr. Teor. Rasprostr. Voln. 17, 21–30, 189, DOI 10.1007/BF01095575; English transl., J. Soviet Math. **50** (1990), no. 6, 1944–1951. MR918955 (89e:35143)

[7] Michael I. Belishev and Yaroslav V. Kurylev, *To the reconstruction of a Riemannian manifold via its spectral data (BC-method)*, Comm. Partial Differential Equations **17** (1992), no. 5-6, 767–804, DOI 10.1080/03605309208820863. MR1177292 (94a:58199)

[8] Kenrick Bingham, Yaroslav Kurylev, Matti Lassas, and Samuli Siltanen, *Iterative time-reversal control for inverse problems*, Inverse Probl. Imaging **2** (2008), no. 1, 63–81, DOI 10.3934/ipi.2008.2.63. MR2375323 (2009c:35470)

[9] A. S. Blagoveščenskiĭ, *The local method of solution of the nonstationary inverse problem for an inhomogeneous string* (Russian), Trudy Mat. Inst. Steklov. **115** (1971), 28–38. MR0307558 (46 #6678)

[10] A. S. Blagoveščenskiĭ, *The nonselfadjoint inverse matrix boundary problem for a hyperbolic differential equation* (Russian), Problems of mathematical physics, No. 5: spectral theory (Russian), Izdat. Leningrad. Univ., Leningrad, 1971, pp. 38–62. MR0303124 (46 #2262)

[11] David Borthwick, *Spectral theory of infinite-area hyperbolic surfaces*, Progress in Mathematics, vol. 256, Birkhäuser Boston Inc., Boston, MA, 2007. MR2344504 (2008h:58056)

[12] David Borthwick, Chris Judge, and Peter A. Perry, *Selberg's zeta function and the spectral geometry of geometrically finite hyperbolic surfaces*, Comment. Math. Helv. **80** (2005), no. 3, 483–515, DOI 10.4171/CMH/23. MR2165200 (2006f:58038)

[13] T. J. Christiansen, *Sojourn times, manifolds with infinite cylindrical ends, and an inverse problem for planar waveguides*, J. Anal. Math. **107** (2009), 79–106, DOI 10.1007/s11854-009-0004-5. MR2496400 (2010k:35527)

[14] D. M. Èĭdus, *The principle of limiting amplitude* (Russian), Uspehi Mat. Nauk **24** (1969), no. 3(147), 91–156. MR0601072 (58 #29156)

[15] J. Elstrodt, F. Grunewald, and J. Mennicke, *Groups acting on hyperbolic space*, Springer Monographs in Mathematics, Springer-Verlag, Berlin, 1998. Harmonic analysis and number theory. MR1483315 (98g:11058)

[16] L. D. Faddeev, *Inverse problem of quantum scattering theory*, J. Sov. Math. **5** (1976), 334-396.

[17] I. M. Gel′fand and B. M. Levitan, *On the determination of a differential equation from its spectral function* (Russian), Izvestiya Akad. Nauk SSSR. Ser. Mat. **15** (1951), 309–360. MR0045281 (13,558f)

[18] Allan Greenleaf, Matti Lassas, and Gunther Uhlmann, *On nonuniqueness for Calderón's inverse problem*, Math. Res. Lett. **10** (2003), no. 5-6, 685–693, DOI 10.4310/MRL.2003.v10.n5.a11. MR2024725 (2005f:35316)

[19] Allan Greenleaf, Yaroslav Kurylev, Matti Lassas, and Gunther Uhlmann, *Full-wave invisibility of active devices at all frequencies*, Comm. Math. Phys. **275** (2007), no. 3, 749–789, DOI 10.1007/s00220-007-0311-6. MR2336363 (2008g:78016)

[20] Allan Greenleaf, Yaroslav Kurylev, Matti Lassas, and Gunther Uhlmann, *Invisibility and inverse problems*, Bull. Amer. Math. Soc. (N.S.) **46** (2009), no. 1, 55–97, DOI 10.1090/S0273-0979-08-01232-9. MR2457072 (2010d:35399)

[21] Allan Greenleaf, Yaroslav Kurylev, Matti Lassas, and Gunther Uhlmann, *Cloaking devices, electromagnetic wormholes, and transformation optics*, SIAM Rev. **51** (2009), no. 1, 3–33, DOI 10.1137/080716827. MR2481110 (2010b:35484)

[22] Colin Guillarmou and Antônio Sá Barreto, *Scattering and inverse scattering on ACH manifolds*, J. Reine Angew. Math. **622** (2008), 1–55, DOI 10.1515/CRELLE.2008.064. MR2433611 (2010i:58028)

[23] Sigurđur Helgason, *A duality for symmetric spaces with applications to group representations*, Advances in Math. **5** (1970), 1–154 (1970). MR0263988 (41 #8587)

[24] Peter D. Hislop, *The geometry and spectra of hyperbolic manifolds*, Proc. Indian Acad. Sci. Math. Sci. **104** (1994), no. 4, 715–776. Spectral and inverse spectral theory (Bangalore, 1993). MR1319884 (96b:58117)

[25] Victor Isakov and Adrian I. Nachman, *Global uniqueness for a two-dimensional semilinear elliptic inverse problem*, Trans. Amer. Math. Soc. **347** (1995), no. 9, 3375–3390, DOI 10.2307/2155015. MR1311909 (95m:35202)

[26] Hiroshi Isozaki, *Inverse spectral problems on hyperbolic manifolds and their applications to inverse boundary value problems in Euclidean space*, Amer. J. Math. **126** (2004), no. 6, 1261–1313. MR2102396 (2005m:35311)

[27] Hiroshi Isozaki and Yaroslav Kurylev, *Spectral theory and inverse problems on asymptotically hyperbolic manifolds*, Spectral and scattering theory and related topics, RIMS Kôkyûroku Bessatsu, B16, Res. Inst. Math. Sci. (RIMS), Kyoto, 2010, pp. 29–73. MR2743480 (2012b:58047)

[28] Hiroshi Isozaki, Yaroslav Kurylev, and Matti Lassas, *Forward and inverse scattering on manifolds with asymptotically cylindrical ends*, J. Funct. Anal. **258** (2010), no. 6, 2060–2118, DOI 10.1016/j.jfa.2009.11.009. MR2578464 (2010k:58047)

[29] H. Isozaki, Y. Kurylev and M. Lassas, *Conic singularities, generalized scattering matrix and inverse scattering on asymptotically hyperbolic surfaces*, arXiv:1108.1577.

[30] Hiroshi Isozaki, *Inverse spectral theory*, Topics in the theory of Schrödinger operators, World Sci. Publ., River Edge, NJ, 2004, pp. 93–143, DOI 10.1142/9789812562470_0004. MR2117108

[31] K. Ito and E. Skibsted, *Absence of embedded eigenvalues for Riemannian Laplacians*, Adv. Math. **248** (2013), 945–962, DOI 10.1016/j.aim.2013.08.023. MR3107533

[32] Alexander Katchalov, Yaroslav Kurylev, and Matti Lassas, *Inverse boundary spectral problems*, Chapman & Hall/CRC Monographs and Surveys in Pure and Applied Mathematics, vol. 123, Chapman & Hall/CRC, Boca Raton, FL, 2001. MR1889089 (2003e:58045)

[33] A. Katchalov, Y. Kurylev, M. Lassas, and N. Mandache, *Equivalence of time-domain inverse problems and boundary spectral problems*, Inverse Problems **20** (2004), no. 2, 419–436, DOI 10.1088/0266-5611/20/2/007. MR2065431 (2005d:35270)

[34] M. Kashiwara, A. Kowata, K. Minemura, K. Okamoto, T. Ōshima, and M. Tanaka, *Eigenfunctions of invariant differential operators on a symmetric space*, Ann. of Math. (2) **107** (1978), no. 1, 1–39, DOI 10.2307/1971253. MR485861 (81f:43013)

[35] M. Kerker, *Invisible bodies*, J. Opt. Soc. Am. **65** (1975), 376.

[36] R. G. Novikov and G. M. Khenkin, *The $\bar{\partial}$-equation in the multidimensional inverse scattering problem* (Russian), Uspekhi Mat. Nauk **42** (1987), no. 3(255), 93–152, 255. MR896879 (89d:35043)

[37] M. G. Kreĭn, *Determination of the density of a nonhomogeneous symmetric cord by its frequency spectrum* (Russian), Doklady Akad. Nauk SSSR (N.S.) **76** (1951), 345–348. MR0042045 (13,43d)

[38] M. G. Kreĭn, *On inverse problems for a nonhomogeneous cord* (Russian), Doklady Akad. Nauk SSSR (N.S.) **82** (1952), 669–672. MR0052636 (14,649f)

[39] Katsiaryna Krupchyk, Yaroslav Kurylev, and Matti Lassas, *Inverse spectral problems on a closed manifold* (English, with English and French summaries), J. Math. Pures Appl. (9) **90** (2008), no. 1, 42–59, DOI 10.1016/j.matpur.2008.02.009. MR2435213 (2010h:58052)

[40] Hironori Kumura, *The radial curvature of an end that makes eigenvalues vanish in the essential spectrum. I*, Math. Ann. **346** (2010), no. 4, 795–828, DOI 10.1007/s00208-009-0410-0. MR2587093 (2011a:53055)

[41] H. Kumura, *Limiting absorption principle on manifolds having ends with various measure growth rate limits*, arXiv. math DG/0606125.

[42] Y. Kurylev, M. Lassas, T. Yamaguchi. *Uniqueness and stability in inverse spectral problems for collapsing manifolds,* arxiv.math: 1209.587v2

[43] M. Lassas, M. Salo, G. Uhlmann, *Wave imaging.* Chapter in Handbook of Mathematical Methods in Imaging, O. Scherzer (Ed.), Springer-Verlag, 2011, 50 pp

[44] Matti Lassas, Michael Taylor, and Gunther Uhlmann, *The Dirichlet-to-Neumann map for complete Riemannian manifolds with boundary*, Comm. Anal. Geom. **11** (2003), no. 2, 207–221, DOI 10.4310/CAG.2003.v11.n2.a2. MR2014876 (2004h:58033)

[45] Matti Lassas and Gunther Uhlmann, *On determining a Riemannian manifold from the Dirichlet-to-Neumann map* (English, with English and French summaries), Ann. Sci. École Norm. Sup. (4) **34** (2001), no. 5, 771–787, DOI 10.1016/S0012-9593(01)01076-X. MR1862026 (2003e:58037)

[46] V. A. Marčenko, *On reconstruction of the potential energy from phases of the scattered waves* (Russian), Dokl. Akad. Nauk SSSR (N.S.) **104** (1955), 695–698. MR0075402 (17,740e)

[47] Rafe R. Mazzeo and Richard B. Melrose, *Meromorphic extension of the resolvent on complete spaces with asymptotically constant negative curvature*, J. Funct. Anal. **75** (1987), no. 2, 260–310, DOI 10.1016/0022-1236(87)90097-8. MR916753 (89c:58133)

[48] Rafe Mazzeo and András Vasy, *Analytic continuation of the resolvent of the Laplacian on symmetric spaces of noncompact type*, J. Funct. Anal. **228** (2005), no. 2, 311–368, DOI 10.1016/j.jfa.2004.10.003. MR2175410 (2006m:58047)

[49] Richard B. Melrose, *Geometric scattering theory*, Stanford Lectures, Cambridge University Press, Cambridge, 1995. MR1350074 (96k:35129)

[50] Richard Melrose and Gunther Uhlmann, *Generalized backscattering and the Lax-Phillips transform*, Serdica Math. J. **34** (2008), no. 1, 355–372. MR2414425 (2010a:35179)

[51] E. Mourre, *Absence of singular continuous spectrum for certain selfadjoint operators*, Comm. Math. Phys. **78** (1980/81), no. 3, 391–408. MR603501 (82c:47030)

[52] W. Müller, *Manifolds with cusp of rank one, spectral theory and L^2-index theorem*, Lecture Notes in Math. **1224**, Springer (1987).

[53] Adrian I. Nachman, *Reconstructions from boundary measurements*, Ann. of Math. (2) **128** (1988), no. 3, 531–576, DOI 10.2307/1971435. MR970610 (90i:35283)

[54] Lassi Päivärinta, Mikko Salo, and Gunther Uhlmann, *Inverse scattering for the magnetic Schrödinger operator*, J. Funct. Anal. **259** (2010), no. 7, 1771–1798, DOI 10.1016/j.jfa.2010.06.002. MR2665410 (2011f:35363)

[55] Peter Perry, *The spectral geometry of geometrically finite hyperbolic manifolds*, Spectral theory and mathematical physics: a Festschrift in honor of Barry Simon's 60th birthday, Proc. Sympos. Pure Math., vol. 76, Amer. Math. Soc., Providence, RI, 2007, pp. 289–327. MR2310208 (2008i:58030)

[56] Michael Reed and Barry Simon, *Methods of modern mathematical physics. III*, Academic Press [Harcourt Brace Jovanovich Publishers], New York, 1979. Scattering theory. MR529429 (80m:81085)
[57] Antônio Sá Barreto, *Radiation fields, scattering, and inverse scattering on asymptotically hyperbolic manifolds*, Duke Math. J. **129** (2005), no. 3, 407–480, DOI 10.1215/S0012-7094-05-12931-3. MR2169870 (2006g:58058)
[58] John Sylvester and Gunther Uhlmann, *A global uniqueness theorem for an inverse boundary value problem*, Ann. of Math. (2) **125** (1987), no. 1, 153–169, DOI 10.2307/1971291. MR873380 (88b:35205)
[59] Daniel Tataru, *Unique continuation for solutions to PDE's; between Hörmander's theorem and Holmgren's theorem*, Comm. Partial Differential Equations **20** (1995), no. 5-6, 855–884, DOI 10.1080/03605309508821117. MR1326909 (96e:35019)
[60] Daniel Tataru, *On the regularity of boundary traces for the wave equation*, Ann. Scuola Norm. Sup. Pisa Cl. Sci. (4) **26** (1998), no. 1, 185–206. MR1633000 (99e:35129)
[61] D. R. Yafaev, *Mathematical scattering theory*, Mathematical Surveys and Monographs, vol. 158, American Mathematical Society, Providence, RI, 2010. Analytic theory. MR2598115 (2012d:47033)
[62] Steven Zelditch, *Kuznecov sum formulae and Szegö limit formulae on manifolds*, Comm. Partial Differential Equations **17** (1992), no. 1-2, 221–260, DOI 10.1080/03605309208820840. MR1151262 (93e:58184)

Institute of Mathematics, University of Tsukuba, Tsukuba, 305-8571, Japan
E-mail address: isozakih@math.tsukuba.ac.jp

Department of Mathematics, University College of London, United Kingdom
E-mail address: y.kurylev@ucl.ac.uk

Department of Mathematics and Statistics, University of Helsinki, Finland
E-mail address: mjl@rni.helsinki.fi

Contemporary Mathematics
Volume **615**, 2014
http://dx.doi.org/10.1090/conm/615/12260

On an inverse problem for the Steklov spectrum of a Riemannian surface

Alexandre Jollivet and Vladimir Sharafutdinov

Abstract. The paper is devoted to the inverse problem of reconstruction, up to a natural gauge transform, of a smooth simply connected Riemannian surface with nonempty boundary from its Steklov spectrum. We demonstrate that the problem has two other equivalent forms: (1) the problem of recovering a positive function on the unit circle from the eigenvalue spectrum of some operator and (2) the problem of recovering an immersion of the unit disk to the Euclidean plane from the corresponding Steklov spectrum. The latter problem is a natural generalization of the classical problem of recovering a planar domain from its Steklov spectrum. We also give qualitative statements on the Steklov spectrum for two classes of Riemannian surfaces.

1. Introduction

We start with posing the problem in an arbitrary dimension, although the two-dimensional case is discussed only in the main part of the paper.

Throughout the paper, the term "smooth" is used as a synonym of "C^∞-smooth". Let (M,g) be a compact connected smooth Riemannian manifold of dimension n with the nonempty boundary ∂M. *The Laplace – Beltrami operator* is defined in local coordinates by

$$\Delta_g = (\det g)^{-1/2} \sum_{i,j=1}^{n} \frac{\partial}{\partial x_i}\left((\det g)^{1/2} g^{ij} \frac{\partial}{\partial x_j}\right),$$

where $(g^{ij}) = (g_{ij})^{-1}$ and $\det g = \det(g_{ij})$. *The Dirichlet-to-Neumann operator* (the DN map)

$$\Lambda_g : C^\infty(\partial M) \to C^\infty(\partial M)$$

is defined by $\Lambda_g f = \frac{\partial u}{\partial \nu}\big|_{\partial M}$, where ν is the unit outward normal to the boundary and u is the solution to the Dirichlet problem

$$\begin{cases} \Delta_g u = 0 \\ u|_{\partial M} = f \end{cases}.$$

2000 *Mathematics Subject Classification.* Primary 35R30; Secondary 58J32.
Key words and phrases. Steklov spectrum, Riemannian surface, Inverse spectral problem.
The work was started when the second author visited Université de Cergy–Pontoise. The work was accomplished when the second author stayed at Institut Mittag-Leffler in the scope of the program "Inverse Problems". The second author is grateful to both the institutes for the support and hospitality.

As well known, Λ_g is a non-negative self-dual pseudodifferential operator of order 1. Therefore it has a discrete eigenvalue spectrum

$$\mathfrak{S}(M,\Lambda_g)=\{0=\lambda_0<\lambda_1\le\lambda_2\le\dots\le\lambda_k\to\infty\},$$

each eigenvalue is repeated according to its multiplicity, which is called *the Steklov spectrum* of (M,g). Steklov [**14**] introduced the spectrum in the case when M is a domain in $\mathbb{R}^n$ and g is the Euclidean metric.

For isoperimetric inequalities on the Steklov eigenvalues and for bounds on their multiplicities we refer the reader to the recent papers [**2**, **4**, **5**, **7**] and references therein.

The inverse problem for the Steklov spectrum is posed as follows:

PROBLEM 1.1. To which extent is a compact connected Riemannian manifold (M,g) with a nonempty boundary determined by the Steklov spectrum $\mathfrak{S}(M,\Lambda_g)$?

Isometric Riemannian manifolds have coincident Steklov spectra. This is the first obvious ambiguity in Problem 1.1: one has to recover a Riemannian manifold up to an isometry.

Let us use the term "a Riemannian surface" for a two-dimensional Riemannian manifold. The Laplace – Beltrami operator on a Riemannian surface (M,g) possesses the following conformal invariance: $\Delta_{\rho^2g}=\rho^{-2}\Delta_g$ for any function $0<\rho\in C^\infty(M)$. This implies $\Lambda_{\rho^2g}=\Lambda_g$ if $\rho|_{\partial M}=1$. In particular, $\mathfrak{S}(M,\Lambda_{\rho^2g})=\mathfrak{S}(M,\Lambda_g)$ for every function $0<\rho\in C^\infty(M)$ satisfying $\rho|_{\partial M}=1$.

In the present paper, we study Problem 1.1 for a compact and simply connected Riemannian surface (M,g) with a nonempty boundary. Such a surface is diffeomorphic to the disc

$$D=\{(x,y)\mid x^2+y^2\le 1\}\subset\mathbb{R}^2=\mathbb{C}.$$

Without loss of generality, we can assume $M=D$. For a Riemannian metric g on D, we abbreviate the notation $\mathfrak{S}(D,\Lambda_g)$ to $\mathfrak{S}(\Lambda_g)$, where Λ_g is the pseudodifferential operator on the unit circle

$$\gamma=\partial D=\{e^{i\theta}\mid\theta\in\mathbb{R}\}. \tag{1.1}$$

Given a Riemannian metric g on D and diffeomorphism $\Phi:D\to D$, the metric $g'=\Phi^*g$ is defined by $g'_p(v,w)=g_{\Phi(p)}((d_p\Phi)v,(d_p\Phi)w)$ for a point $p\in D$ and vectors v and w belonging to the tangent space T_pD, where $d_p\Phi:T_pD\to T_{\Phi(p)}D$ is the differential of Φ. The statement "the metrics g and g' on D are isometric" is equivalent to the statement "there exists a diffeomorphism $\Phi:D\to D$ such that $g'=\Phi^*g$".

We suggest that two mentioned ambiguities, the isometry invariance and conformal invariance, exhaust the non-uniqueness in Problem 1.1 for simply connected Riemannian surfaces. In other words, we have the following

CONJECTURE 1.2. Two Riemannian metrics g_j $(j=1,2)$ on D satisfy

$$\mathfrak{S}(\Lambda_{g_1})=\mathfrak{S}(\Lambda_{g_2}) \tag{1.2}$$

if and only if there exist a diffeomorphism $\Phi:D\to D$ and function $0<\rho\in C^\infty(D)$ such that $\rho|_\gamma=1$ and

$$g_2=\rho^2\Phi^*g_1. \tag{1.3}$$

The question becomes much easier if the isospectral hypothesis (1.2) is replaced by the assumption on intertwining the operators Λ_{g_1} and Λ_{g_2}. Let us recall the definition.

For a manifold N, two linear operators $A_j : C^\infty(N) \to C^\infty(N)$ $(j = 1, 2)$ are *intertwined* if there exists a diffeomorphism $\varphi : N \to N$ such that $A_2 = \varphi^* \circ A_1 \circ \varphi^{*-1}$, where $\varphi^* : C^\infty(N) \to C^\infty(N)$ is defined by $\varphi^* f = f \circ \varphi$.

THEOREM 1.3. *For two Riemannian metrics g_j $(j = 1, 2)$ on D, the operators Λ_{g_1} and Λ_{g_2} are intertwined if and only if there exist a diffeomorphism $\Phi : D \to D$ and function $0 < \rho \in C^\infty(D)$ such that $\rho|_\gamma = 1$ and $g_2 = \rho^2 \Phi^* g_1$.*

If two linear operators are intertwined, they are isospectral. Is the converse statement true for some classes of operators? To authors' knowledge, the question is open even in the one-dimensional case. Nevertheless, by virtue of Theorem 1.3, Conjecture 1.2 is equivalent to the statement: For two Riemannian metrics g_j $(j = 1, 2)$ on D, the operators Λ_{g_1} and Λ_{g_2} are intertwined if $\mathfrak{S}(\Lambda_{g_1}) = \mathfrak{S}(\Lambda_{g_2})$.

The paper is organized as follows. In Section 2, after recalling some known facts on the Dirichlet-to-Neumann operator, we prove Theorem 1.3.

In Section 3, we demonstrate that Conjecture 1.2 has an equivalent form in terms of the problem of recovering a positive function a on the unit circle from the eigenvalue spectrum of the operator $a^{-1}\Lambda_e$, where e is the Euclidean metric. We prove an analogous of Theorem 1.3 for the latter problem.

In Section 4, we consider the problem of recovering an immersion $\mathfrak{I} : D \to \mathbb{R}^2$ from the Steklov spectrum $\mathfrak{S}(\mathfrak{I}^* e)$. In the particular case of an embedding $\mathfrak{I}$, this is equivalent to the classical problem [**3**] of recovering a planar domain $\Omega = \mathfrak{I}(D)$ from the Steklov spectrum $\mathfrak{S}(\Omega, \Lambda_e)$. We again give an equivalent version of Conjecture 1.2 in terms of immersions and prove an analogous of Theorem 1.3.

In Section 5, we recall qualitative properties of the Steklov spectrum and describe the eigenvalue spectrum of $a_m^{-1}\Lambda_e$ for functions a_m of the form $a_m^{-1}(e^{i\theta}) = b_0 + 2b_1 \cos(m\theta)$. Our main results are Theorem 5.3 and Corollary 5.4 that describe the eigenvalue spectrum for the functions a_2.

In Section 6, we interpret the eigenvalue problem for the operator $a^{-1}\Lambda_e$ as a scalar Riemann – Hilbert problem in order to prove Theorem 5.3.

The appendix contains proofs of Theorems 5.1 and 5.2 and of Lemma 6.2.

2. The problem of determining a metric on the disc

A Riemannian metric g on the disc D induces the metric g_∂ on the circle $\gamma = \partial D$. The latter metric can be written in the form $g_\partial = ds_g^2$, where ds_g is a smooth one-form on γ which does not vanish at any point. The form is uniquely determined if it is assumed to be a *positive* form, i.e., $ds_g = c(\theta)\, d\theta$, where $0 < c \in C^\infty(\gamma)$ and the cyclic coordinate θ on γ is defined by (1.1). We call ds_g *the arc-length form* of the metric g. If, for a diffeomorphism $\Phi : D \to D$, we set $\varphi = \Phi|_\gamma : \gamma \to \gamma$, then $ds_{\Phi^* g} = \pm\varphi^*(ds_g)$, where the sign $(+)$ should be chosen if Φ preserves orientation, and $(-)$ otherwise.

As mentioned in the Introduction, $\Lambda_{\rho^2 g} = \Lambda_g$ for $0 < \rho \in C^\infty(D)$ if $\rho|_\gamma = 1$. We will need the following generalization.

LEMMA 2.1. *Given a Riemannian metric g on D and function $0 < \rho \in C^\infty(D)$, let $a = \rho|_\gamma \in C^\infty(\gamma)$. Then*

$$\Lambda_{\rho^2 g} = a^{-1}\Lambda_g \tag{2.1}$$

and

$$ds_{\rho^2 g} = a\, ds_g. \tag{2.2}$$

In particular,

$$\Lambda_{\rho^2 g} = \Lambda_g \quad \textit{and} \quad ds_{\rho^2 g} = ds_g \quad \textit{for} \quad 0 < \rho \in C^\infty(D),\ \rho|_\gamma = 1.$$

PROOF. Given $f \in C^\infty(\gamma)$, let $u \in C^\infty(D)$ be the solution to the Dirichlet problem

$$\Delta_g u = 0 \quad \text{in} \quad D, \quad u|_\gamma = f.$$

It also solves the problem

$$\Delta_{\rho^2 g} u = 0 \quad \text{in} \quad D, \quad u|_\gamma = f.$$

Therefore

$$\Lambda_g f = \frac{\partial u}{\partial \nu}\Big|_\gamma, \quad \Lambda_{\rho^2 g} f = \frac{\partial u}{\partial \nu'}\Big|_\gamma,$$

where ν and ν' are unit outward normal vectors to γ in metrics g and $\rho^2 g$ respectively. The vectors are related by $\nu' = a^{-1}\nu$. Therefore

$$\Lambda_{\rho^2 g} f = \frac{\partial u}{\partial \nu'}\Big|_\gamma = \frac{\partial u}{\partial (a^{-1}\nu)}\Big|_\gamma = a^{-1}\frac{\partial u}{\partial \nu}\Big|_\gamma = a^{-1}\Lambda_g f.$$

Equality (2.2) is obvious. □

REMARK. Since we will refer to [**13**] several times, we emphasize the following difference between our notations and that of [**13**]. We systematically use the equality $g' = \rho^2 g$ for conformally equivalent metrics while the formula $g' = \rho g$ is used in [**13**]. In particular, (2.1) and (2.2) take the form $\Lambda_{\rho g} = \frac{1}{\sqrt{a}}\Lambda_g$ and $ds_{\rho g} = \sqrt{a}\, ds_g$ in notations of [**13**]. Here, we use the notation $\rho^2 g$ just to avoid the appearance of the square root in (2.1) and (2.2).

The arc-length form ds_g is uniquely determined by the DN map Λ_g. This follows from the statement: the full symbol of the operator Λ_g is equal to $|\xi|$, where ξ is the Fourier-dual variable of the arc-length s_g. The statement is proved in [**9**] and [**3**]. Taking this fact into account, Theorem 1.2 of [**13**] can be stated as follows:

PROPOSITION 2.2. *If $\Lambda_{g_1} = \Lambda_{g_2}$ for two Riemannian metrics on D, then there exist a diffeomorphism $\Phi : D \to D$ and function $0 < \rho \in C^\infty(D)$ such that $\Phi|_\gamma = I$ (= the identity), $\rho|_\gamma = 1$, and $g_2 = \rho^2\Phi^* g_1$.*

Proposition 2.2 differs from [**13**, Theorem 2.1] by the absence of the hypothesis $ds_{g_1} = ds_{g_2}$. But the latter equality follows from $\Lambda_{g_1} = \Lambda_{g_2}$ as we have just mentioned.

We also refer the reader to [**8**] for the unique determination of a two-dimensional compact connected Riemannian manifold M with a nonempty boundary ∂M from its DN map up to a conformal factor and up to an isometry that leaves invariant ∂M.

PROOF OF THEOREM 1.3. If $\rho|_\gamma = 1$, then (1.3) implies

$$\Lambda_{g_2} = \Lambda_{\Phi^* g_1} = \varphi^* \circ \Lambda_{g_1} \circ \varphi^{*-1}$$

for $\varphi = \Phi|_\gamma$.

Conversely, let two metrics g_j $(j = 1, 2)$ on D satisfy

$$\Lambda_{g_2} = \varphi^* \circ \Lambda_{g_1} \circ \varphi^{*-1} \tag{2.3}$$

for some diffeomorphism $\varphi : \gamma \to \gamma$. Extend φ to a diffeomorphism $\Phi : D \to D$ and set

$$\widetilde{g}_2 = \Phi^* g_1. \tag{2.4}$$

Then

$$\Lambda_{\widetilde{g}_2} = \varphi^* \circ \Lambda_{g_1} \circ \varphi^{*-1}. \tag{2.5}$$

Comparing (2.3) and (2.5), we have

$$\Lambda_{\widetilde{g}_2} = \Lambda_{g_2}.$$

By Proposition 2.2, there exist a diffeomorphism $\Psi : D \to D$ and function $0 < \rho \in C^\infty(D)$ such that $\rho|_\gamma = 1$ and

$$g_2 = \rho^2 \Psi^* \widetilde{g}_2. \tag{2.6}$$

From (2.4) and (2.6),

$$g_2 = \rho^2 \Psi^* \Phi^* g_1 = \rho^2 (\Phi \circ \Psi)^* g_1.$$

□

3. The problem of determining a positive function on the circle

For a smooth map $\varphi : \gamma \to \gamma$, the derivative $\frac{d\varphi}{d\theta} \in C^\infty(\gamma)$ is defined by $\varphi^*(d\theta) = \frac{d\varphi}{d\theta}\, d\theta$, where θ is the cyclic coordinate defined by (1.1).

Let e be the Euclidean metric on $\mathbb{R}^2$. The following proposition is an extended version of [**13**, Theorem 1.3].

Proposition 3.1. Given a Riemannian metric g on D, there exist an orientation preserving diffeomorphism $\Phi : D \to D$ and two positive functions $\rho, \mu \in C^\infty(D)$ satisfying

$$\rho|_\gamma = 1, \quad \Delta_e(\ln \mu) = 0$$

such that

(a) $\Phi^* g = \rho^2 \mu^2 e$.

(b) There exists a smooth immersion $\mathfrak{I} : D \to \mathbb{R}^2$ such that $\mathfrak{I}^* e = \widetilde{\rho}^2 g$, where $\widetilde{\rho} = (\rho \circ \Phi^{-1})^{-1}$. In particular, $h = \widetilde{\rho}^2 g$ is a flat metric, i.e., its Gaussian curvature is identically equal to zero.

(c) If $\varphi = \Phi|_\gamma$ and $a = \mu|_\gamma$, then

$$\varphi^* \circ \Lambda_g \circ \varphi^{*-1} = a^{-1} \Lambda_e \tag{3.1}$$

and

$$\varphi^*(ds_g) = a\, d\theta. \tag{3.2}$$

Only statement (c) is explicitly formulated in [**13**, Theorem 1.3] although statements (a) and (b) also participate in the proof. Let us briefly recall our arguments for the proof.

Sketch of the proof of Proposition 3.1. Given a Riemannian metric g on D, we can find a function $0 < \widetilde{\rho} \in C^\infty(D)$ such that $\widetilde{\rho}|_\gamma = 1$ and $h = \widetilde{\rho}^2 g$ is a flat metric, see [**13**, Lemma 2.1]. Extend h to a flat metric on the open disc $D_\varepsilon = \{(x,y) \mid x^2+y^2 < 1+\varepsilon\}$ for some $\varepsilon > 0$ and denote the extension by h again. The flat metric h is locally isometric to the Euclidean metric e, i.e., for every point $p \in D_\varepsilon$, there exist a neighborhood $U \subset D_\varepsilon$ and orientation preserving isometric embedding $(U,h) \to (\mathbb{R}^2, e)$ which is defined uniquely up to the composition of a rotation and parallel translation. Continuing such embeddings along curves and using the monodromy principle, we obtain an immersion $\mathfrak{I} : D \to \mathbb{R}^2$ such that $\mathfrak{I}^* e = h$. This proves statement (b).

The flat metric h determines the complex structure $\mathcal{C}_h$ on D_ε, see [**13**, Lemma 2.5]. We consider D as a closed domain in the complex manifold $(D_\varepsilon, \mathcal{C}_h)$. On the other hand, D is a closed domain in $(\mathbb{C}, \mathcal{C}_e)$, where $\mathcal{C}_e$ is the standard complex structure. By the Riemann theorem on the existence of a conformal map between two simply connected domains, there exists a biholomorphism of closed domains $\Phi : (D, \mathcal{C}_e|_{\text{int}\, D}) \to (D, \mathcal{C}_h|_{\text{int}\, D})$. By statement (iii) of [**13**, Lemma 2.5], $\Phi^* h = \mu^2 e$ for some function $0 < \mu \in C^\infty(D)$ such that $\ln \mu$ is an e-harmonic function.

From equalities $h = \widetilde{\rho}^2 g$ and $\Phi^* h = \mu^2 e$, we derive

$$\rho^2 \mu^2 e = \rho^2 \Phi^* h = \rho^2 \Phi^*(\widetilde{\rho}^2 g) = \rho^2 (\widetilde{\rho} \circ \Phi)^2 \Phi^* g = \Phi^* g$$

if $\rho = (\widetilde{\rho} \circ \Phi)^{-1}$. This proves statement (a).

Set $\varphi = \Phi|_\gamma$. Repeating the arguments from the beginning of [**13**, Section 3], we prove that

$$\varphi^* \circ \Lambda_h \circ \varphi^{*-1} = a^{-1} \Lambda_e, \tag{3.3}$$

where the function $a \in C^\infty(\gamma)$ is defined by

$$\varphi^*(ds_h) = a\, d\theta. \tag{3.4}$$

Equalities $h = \widetilde{\rho}^2 g$ and $\widetilde{\rho}|_\gamma = 1$ imply that $\Lambda_g = \Lambda_h$ and $ds_g = ds_h$. Together with (3.3) and (3.4), this gives (3.1) and (3.2). Finally, statement (a) together with the boundary condition $\rho|_\gamma = 1$ implies $\varphi^*(ds_g) = \mu|_\gamma\, d\theta$. Comparing the last equality with (3.2), we obtain $a = \mu|_\gamma$. □

Statement (c) of Proposition 3.1 relates the inverse problem of recovering a metric on D to the problem of finding a positive function $a \in C^\infty(\gamma)$ from the eigenvalue spectrum $\mathfrak{S}(a^{-1}\Lambda_e)$ of the operator $a^{-1}\Lambda_e$.

Statement (b) of Proposition 3.1 has an important corollary. Choosing a biholomorphism of $(D, \mathcal{C}_h|_{\text{int}\, D})$ onto the upper half-plane, we can repeat all arguments from the proof of [**3**, Theorem 1] to obtain the following:

Proposition 3.2. For a Riemannian metric g on D, let L be the length of γ in g and let

$$\mathfrak{S}(\Lambda_g) = \{0 = \lambda_0 < \lambda_1 \le \lambda_2 \le \dots \le \lambda_k \le \dots\}$$

be the Steklov spectrum. Then

$$\lambda_k = \frac{2\pi}{L}\Big[\frac{k+1}{2}\Big] + O(k^{-N}) \quad \text{as} \quad k \to \infty$$

for every $N > 0$, where $[x]$ is the integer part of x. In particular, L is uniquely determined by $\mathfrak{S}(\Lambda_g)$.

The same statement is valid for the asymptotics of the spectrum $\mathfrak{S}(a^{-1}\Lambda_e)$, where the constant L is defined by $L = \int_0^{2\pi} a(\theta)\,d\theta$.

LEMMA 3.3. *Let $\Phi : D \to D$ be a smooth map such that $\Phi(\gamma) \subset \gamma$ and $\Phi|_{int D} : int D \to \mathbb{C}$ is either a holomorphic or antiholomorphic function. Set $\varphi = \Phi|_\gamma : \gamma \to \gamma$. Assume the derivative $d\varphi/d\theta$ does not vanish at any point. Then*

$$\Lambda_e \circ \varphi^* = \left|\frac{d\varphi}{d\theta}\right| \varphi^* \circ \Lambda_e. \tag{3.5}$$

If two functions $0 < a_j \in C^\infty(\gamma)$ $(j = 1, 2)$ are related by the equation

$$a_2 = \left|\frac{d\varphi}{d\theta}\right| a_1 \circ \varphi, \tag{3.6}$$

then

$$(a_2^{-1}\Lambda_e) \circ \varphi^* = \varphi^* \circ (a_1^{-1}\Lambda_e) \tag{3.7}$$

and

$$\mathfrak{S}(a_1^{-1}\Lambda_e) \subset \mathfrak{S}(a_2^{-1}\Lambda_e). \tag{3.8}$$

PROOF. Let, for definiteness, $\Phi|_{\text{int } D}$ be a holomorphic function. Given a function $f \in C^\infty(\gamma)$, let $u \in C^\infty(D)$ be the solution to the Dirichlet problem

$$\Delta_e u = 0 \quad \text{in} \quad D, \quad u|_\gamma = f.$$

The function $\widetilde{u} = u \circ \Phi$ solves the problem

$$\Delta_e \widetilde{u} = 0 \quad \text{in} \quad D, \quad \widetilde{u}|_\gamma = \varphi^* f.$$

Therefore

$$\Lambda_e f = \left.\frac{\partial u}{\partial \nu}\right|_\gamma, \quad \Lambda_e(\varphi^* f) = \left.\frac{\partial \widetilde{u}}{\partial \nu}\right|_\gamma. \tag{3.9}$$

We are going to relate the right-hand sides of (3.9). For $z \in \gamma$, let ν_z be the unit outward vector normal to γ at the point z. Then

$$\frac{\partial \widetilde{u}}{\partial \nu}(z) = (d_z(u \circ \Phi))\nu_z = (d_{\varphi(z)} u \circ d_z\Phi)\nu_z. \tag{3.10}$$

The differential

$$d_z\Phi : T_z D = \mathbb{C} \to \mathbb{C} = T_{\Phi(z)} D$$

is just the multiplication by $\Phi'(z)$. In particular, the differential commutes with the multiplication by the imaginary unit. Since $\Phi(\gamma) \subset \gamma$ and the vector $i\nu_z$ is tangent to γ, the vector $i(d_z\Phi)\nu_z$ is also tangent to γ. This implies

$$(d_z\Phi)\nu_z = |\Phi'(z)|\, \nu_{\Phi(z)} = \left|\frac{d\varphi}{d\theta}(z)\right| \nu_{\varphi(z)}.$$

With the help of this, formula (3.10) becomes

$$\frac{\partial \widetilde{u}}{\partial \nu}(z) = \left|\frac{d\varphi}{d\theta}(z)\right| (d_{\varphi(z)} u)\nu_{\varphi(z)} = \left|\frac{d\varphi}{d\theta}(z)\right| \frac{\partial u}{\partial \nu}(\varphi(z)).$$

Comparing this with (3.9), we obtain

$$(\Lambda_e \circ \varphi^*) f = \Big(\left|\frac{d\varphi}{d\theta}\right| \varphi^* \circ \Lambda_e\Big) f.$$

This proves (3.5).

Let two functions $0 < a_j \in C^\infty(\gamma)$ $(j = 1, 2)$ satisfy (3.6). Multiply (3.5) by a_2^{-1} from the left

$$(a_2^{-1}\Lambda_e) \circ \varphi^* = a_2^{-1} \left|\frac{d\varphi}{d\theta}\right| \varphi^* \circ (a_1 a_1^{-1}\Lambda_e) = a_2^{-1} \left|\frac{d\varphi}{d\theta}\right| (a_1 \circ \varphi)\, \varphi^* \circ (a_1^{-1}\Lambda_e) = \varphi^* \circ (a_1^{-1}\Lambda_e).$$

This proves (3.7). Inclusion (3.8) obviously follows from (3.7). □

DEFINITION 3.4. Two functions $0 < a_j \in C^\infty(\gamma)$ $(j = 1, 2)$ are said to be e-conformally equivalent if there exists an e-conformal (or e-anticonformal) transformation $\Phi : D \to D$ such that (3.6) holds for $\varphi = \Phi|_\gamma$.

THEOREM 3.5. *Two functions $0 < a_j \in C^\infty(\gamma)$ $(j = 1, 2)$ are e-conformally equivalent if and only if the operators $a_1^{-1}\Lambda_e$ and $a_2^{-1}\Lambda_e$ are intertwined.*

PROOF. The "only if" statement follows from Lemma 3.3. Indeed, in our case φ^* is invertible and (3.7) can be written as

$$a_2^{-1}\Lambda_e = \varphi^* \circ a_1^{-1}\Lambda_e \circ \varphi^{*-1},$$

i.e., φ intertwines $a_1^{-1}\Lambda_e$ and $a_2^{-1}\Lambda_e$.

Conversely, let two functions $0 < a_j \in C^\infty(\gamma)$ $(j = 1, 2)$ be such that the operators $a_1^{-1}\Lambda_e$ and $a_2^{-1}\Lambda_e$ are intertwined, i.e.,

$$a_2^{-1}\Lambda_e = \psi^* \circ a_1^{-1}\Lambda_e \circ \psi^{*-1} \tag{3.11}$$

for some diffeomorphism $\psi : \gamma \to \gamma$. For each $j = 1, 2$, we extend a_j to a function $0 < \rho_j \in C^\infty(D)$,

$$\rho_j|_\gamma = a_j$$

and define the metric g_j on D by

$$g_j = \rho_j^2 e. \tag{3.12}$$

Hence,

$$ds_j = a_j\, d\theta. \tag{3.13}$$

By Lemma 2.1,

$$\Lambda_{g_j} = a_j^{-1}\Lambda_e. \tag{3.14}$$

From (3.11) and (3.14),

$$\Lambda_{g_2} = \psi^* \circ \Lambda_{g_1} \circ \psi^{*-1}. \tag{3.15}$$

By Theorem 1.3, (3.15) implies the existence of a diffeomorphism $\Phi : D \to D$ and of a function $0 < \rho \in C^\infty(D)$ such that $\rho|_\gamma = 1$ and

$$g_2 = \rho^2 \Phi^* g_1. \tag{3.16}$$

Hence

$$ds_{g_2} = \pm\varphi^*(ds_{g_1}), \quad \text{where} \quad \varphi = \Phi|_\gamma. \tag{3.17}$$

From (3.12) and (3.16),

$$\rho_2^2 e = g_2 = \rho^2 \Phi^* g_1 = \rho^2 \Phi^*(\rho_1^2 e) = \rho^2 (\rho_1 \circ \Phi)^2 \Phi^* e,$$

i.e.,

$$\Phi^* e = \left(\frac{\rho_2}{\rho(\rho_1 \circ \Phi)}\right)^2 e.$$

This means that Φ is an e-conformal (or e-anticonformal) transformation of the disc D.

Using (3.13) and (3.17), we obtain

$$a_2\, d\theta = ds_{g_2} = \pm\varphi^*(ds_{g_1}) = \pm\varphi^*(a_1\, d\theta) = (a_1 \circ \varphi)\left|\frac{d\varphi}{d\theta}\right| d\theta,$$

i.e.,

$$a_2 = (a_1 \circ \varphi)\left|\frac{d\varphi}{d\theta}\right|. \tag{3.18}$$

Since Φ is an e-conformal (or e-anticonformal) transformation and $\varphi = \Phi|_\gamma$, (3.18) means the e-conformal equivalence of a_1 and a_2. □

Conjecture 1.2 has the following form in terms of e-conformally equivalent functions.

CONJECTURE 3.6. For two functions $0 < a_j \in C^\infty(\gamma)$ $(j = 1, 2)$, the equality

$$\mathfrak{S}(a_1^{-1}\Lambda_e) = \mathfrak{S}(a_2^{-1}\Lambda_e)$$

holds if and only if these functions are e-conformally equivalent.

A wrong version of the conjecture was supposed in [**13**, Problem 3.2].

PROOF OF THE EQUIVALENCE OF CONJECTURES 1.2 AND 3.6. First of all, the "if" statement of Conjecture 3.6 follows from Theorem 3.5 since intertwined operators have coincident spectra. The implication Conjecture 1.2 $\longrightarrow$ Conjecture 3.6 is proved by the same arguments we have used in the proof of Theorem 3.5. The reverse implication is a little bit more tricky.

Let Riemannian metrics g_j $(j = 1, 2)$ on D satisfy (1.2). Applying Proposition 3.1 to each of the metrics, we find an orientation preserving diffeomorphism $\Phi_j : D \to D$ and positive functions $\rho_j, \mu_j \in C^\infty(D)$ such that $\rho_j|_\gamma = 1$, $\Delta_e(\ln \mu_j) = 0$,

$$\Phi_j^* g_j = \rho_j^2 \mu_j^2 e, \tag{3.19}$$

and for $\varphi_j = \Phi_j|_\gamma$, $a_j = \mu_j|_\gamma$,

$$\varphi_j^* \circ \Lambda_{g_j} \circ \varphi_j^{*-1} = a_j^{-1}\Lambda_e. \tag{3.20}$$

From (1.2) and (3.20), $\mathfrak{S}(a_1^{-1}\Lambda_e) = \mathfrak{S}(a_2^{-1}\Lambda_e)$. Assuming Conjecture 3.6 to be true, we find an e-conformal (or e-anticonformal) transformation Φ such that

$$a_2 = (a_1 \circ \varphi)\left|\frac{d\varphi}{d\theta}\right|, \quad \text{where} \quad \varphi = \Phi|_\gamma. \tag{3.21}$$

We remember also that

$$\Phi^* e = |\Phi'|^2 e, \tag{3.22}$$

where $\Phi' = d\Phi/dz$ $(\Phi' = d\Phi/d\bar{z})$ for the holomorphic (antiholomorphic) function Φ.

From (3.19) and (3.22), we derive

$$g_2 = \Phi_2^{*-1}(\rho_2^2\mu_2^2 e) = \Phi_2^{*-1}(\rho_2^2\mu_2^2|\Phi'|^{-2}\Phi^* e).$$

Substitute the value $e = \rho_1^{-2}\mu_1^{-2}\Phi_1^* g_1$, that follows from (3.19), into the right-hand side of the latter formula

$$g_2 = \Phi_2^{*-1}\Big(\rho_2^2\mu_2^2|\Phi'|^{-2}\Phi^*(\rho_1^{-2}\mu_1^{-2}\Phi_1^* g_1)\Big)$$
$$= \Phi_2^{*-1}\Big(\rho_2^2\mu_2^2|\Phi'|^{-2}(\rho_1\circ\Phi)^{-2}(\mu_1\circ\Phi)^{-2}\Phi^*\Phi_1^* g_1\Big).$$

Rewrite this in the form

$$g_2 = \Phi_2^{*-1}\left\{\left(\frac{\rho_2}{\rho_1\circ\Phi}\right)^2\left[\frac{\mu_2}{(\mu_1\circ\Phi)|\Phi'|}\right]^2\Phi^*\Phi_1^* g_1\right\}. \tag{3.23}$$

Let us demonstrate that the function in brackets on the right-hand side of (3.23) is identically unit, i.e.,

$$\mu_2 = (\mu_1\circ\Phi)|\Phi'|. \tag{3.24}$$

Indeed, both

$$\ln\mu_2 \quad \text{and} \quad \ln(\mu_1\circ\Phi) + \ln|\Phi'|$$

are e-harmonic functions in D. Therefore, to prove (3.24), it suffices to show that this equality holds on γ. But on γ, (3.24) coincides with (3.21) since $\mu_j|_\gamma = a_j$ and $|\Phi'|_\gamma = |d\varphi/d\theta|$.

By (3.24), formula (3.23) simplifies to the following one:

$$g_2 = \Phi_2^{*-1}(\widetilde{\rho}^2\Phi^*\Phi_1^* g_1) = (\widetilde{\rho}\circ\Phi_2^{-1})^2\Phi_2^{*-1}\Phi^*\Phi_1^* g_1,$$

where $\widetilde{\rho} = \rho_2/\rho_1\circ\Phi$. Setting $\rho = \widetilde{\rho}\circ\Phi_2^{-1}$ and $\Psi = \Phi_1\circ\Phi\circ\Phi_2^{-1}$, we obtain

$$g_2 = \rho^2\Psi^* g_1, \quad \rho|_\gamma = 1.$$

□

Theorem 3.5 can be generalized. We first generalize Definition 3.4 as follows.

Definition 3.7. Let g be a Riemannian metric on D. Two functions $0 < a_j \in C^\infty(\gamma)$ $(j = 1, 2)$ are said to be g-conformally equivalent if there exists a g-conformal (or g-anticonformal) transformation

$$\Phi : D \to D, \quad \Phi^* g = \lambda^2 g$$

such that

$$a_2 = c\,(a_1\circ\varphi)$$

for $\varphi = \Phi|_\gamma$ and $c = \lambda|_\gamma$.

Theorem 3.8. *Let g be a Riemannian metric on D. Two functions $0 < a_j \in C^\infty(\gamma)$ $(j = 1, 2)$ are g-conformally equivalent if and only if the operators $a_1^{-1}\Lambda_g$ and $a_2^{-1}\Lambda_g$ are intertwined.*

Proof. The proof of the "if" statement repeats, with obvious changes, the corresponding part of the proof of Theorem 3.5. So, we present the proof of the statement "only if".

Assume functions $0 < a_j \in C^\infty(\gamma)$ to be such that the operators $a_1^{-1}\Lambda_g$ and $a_2^{-1}\Lambda_g$ are intertwined. On using Proposition 3.1, we find an orientation preserving diffeomorphism $\Phi : D \to D$ and two functions $0 < \rho, \mu \in C^\infty(D)$ such that

$$\Phi^* g = \rho^2\mu^2 e, \quad \rho|_\gamma = 1, \quad \Delta_e(\ln\mu) = 0 \tag{3.25}$$

and

$$\varphi^* \circ \Lambda_g \circ \varphi^{*-1} = a^{-1}\Lambda_e \tag{3.26}$$

for $a = \mu|_\gamma$ and $\varphi = \Phi|_\gamma$.

We set

$$b_j = a\,(a_j \circ \varphi) \quad (j = 1, 2). \tag{3.27}$$

From this, we derive with the help of (3.26)

$$b_j^{-1}\Lambda_e = (a_j \circ \varphi)^{-1}a^{-1}\Lambda_e = (a_j \circ \varphi)^{-1}\varphi^* \circ \Lambda_g \circ \varphi^{*-1} = \varphi^* \circ a_j^{-1}\Lambda_g \circ \varphi^{*-1}.$$

This means that the operators $b_j^{-1}\Lambda_e$ and $a_j^{-1}\Lambda_g$ are intertwined. Since $a_1^{-1}\Lambda_g$ and $a_2^{-1}\Lambda_g$ are assumed to be intertwined, we conclude: $b_1^{-1}\Lambda_e$ and $b_2^{-1}\Lambda_e$ are intertwined.

By Theorem 3.5, b_1 and b_2 are e-conformally equivalent, i.e., there exists an e-conformal (or e-anticonformal) $\Psi : D \to D$ such that

$$b_2 = \left|\frac{d\psi}{d\theta}\right| b_1 \circ \psi \quad \text{for} \quad \psi = \Psi|_\gamma. \tag{3.28}$$

From (3.27) and (3.28), we find the relationship between a_1 and a_2

$$a_2 \circ \varphi = a^{-1}b_2 = a^{-1}\left|\frac{d\psi}{d\theta}\right| b_1 \circ \psi = a^{-1}\left|\frac{d\psi}{d\theta}\right| (a\,a_1 \circ \varphi)\circ\psi = a^{-1}\left|\frac{d\psi}{d\theta}\right| (a\circ\psi)(a_1\circ\varphi\circ\psi),$$

i.e.,

$$a_2 = \frac{a \circ \psi \circ \varphi^{-1}}{a \circ \varphi^{-1}}\left|\frac{d\psi}{d\theta} \circ \varphi^{-1}\right| (a_1 \circ \varphi \circ \psi \circ \varphi^{-1}).$$

Introducing the notations

$$\widetilde{\Psi} = \Phi \circ \Psi \circ \Phi^{-1}, \quad \widetilde{\psi} = \widetilde{\Psi}|_\gamma,$$

we write the result as

$$a_2 = \frac{a \circ \psi \circ \varphi^{-1}}{a \circ \varphi^{-1}}\left|\frac{d\psi}{d\theta} \circ \varphi^{-1}\right| (a_1 \circ \widetilde{\psi}). \tag{3.29}$$

Let us demonstrate that $\widetilde{\Psi}$ is a g-conformal (or g-anticonformal) transformation. Indeed, by twice usage of (3.25), we deduce

$$\begin{aligned}
\widetilde{\Psi}^* g &= \Phi^{*-1}\Psi^*\Phi^* g = \Phi^{*-1}\Psi^*(\rho^2\mu^2 e) \\
&= (\rho \circ \Psi \circ \Phi^{-1})^2(\mu \circ \Psi \circ \Phi^{-1})^2\Phi^{*-1}\Psi^* e \\
&= (\rho \circ \Psi \circ \Phi^{-1})^2(\mu \circ \Psi \circ \Phi^{-1})^2\Phi^{*-1}(|\Psi'|^2 e) \\
&= (\rho \circ \Psi \circ \Phi^{-1})^2(\mu \circ \Psi \circ \Phi^{-1})^2|\Psi' \circ \Phi^{-1}|^2\Phi^{*-1} e \\
&= (\rho \circ \Psi \circ \Phi^{-1})^2(\mu \circ \Psi \circ \Phi^{-1})^2|\Psi' \circ \Phi^{-1}|^2\Phi^{*-1}(\rho^{-2}\mu^{-2}\Phi^* g) \\
&= \Big(\frac{\rho \circ \Psi \circ \Phi^{-1}}{\rho \circ \Phi^{-1}}\Big)^2\Big(\frac{\mu \circ \Psi \circ \Phi^{-1}}{\mu \circ \Phi^{-1}}\Big)^2|\Psi' \circ \Phi^{-1}|^2 g.
\end{aligned}$$

We write the result in the form

$$\widetilde{\Psi}^* g = \widetilde{\lambda}^2 g, \tag{3.30}$$

where

$$\widetilde{\lambda} = \frac{\rho \circ \Psi \circ \Phi^{-1}}{\rho \circ \Phi^{-1}}\frac{\mu \circ \Psi \circ \Phi^{-1}}{\mu \circ \Phi^{-1}}|\Psi' \circ \Phi^{-1}|.$$

In particular, (3.30) means that $\widetilde{\Psi}$ is a g-conformal (or g-anticonformal) transformation. Restricting the last formula to γ and using $\rho|_\gamma = 1$, $\mu|_\gamma = a$, and $\Psi'|_\gamma = d\psi/d\theta$, we get

$$\widetilde{\lambda}|_\gamma = \frac{a \circ \psi \circ \varphi^{-1}}{a \circ \varphi^{-1}} \left| \frac{d\psi}{d\theta} \circ \varphi^{-1} \right|.$$

With the help of this, formula (3.29) takes the form

$$a_2 = \widetilde{\lambda}|_\gamma (a_1 \circ \widetilde{\psi}). \tag{3.31}$$

Formulas (3.30) and (3.31) mean that a_1 and a_2 are g-conformally equivalent. □

Finally, Conjecture 3.6 is equivalent to the following more general statement. The equivalence is proved by the same arguments as we have used in the proof of Theorem 3.8.

CONJECTURE 3.9. Let g be a Riemannian metric on D. For two functions $0 < a_j \in C^\infty(\gamma)$, the equality

$$\mathfrak{S}(a_1^{-1}\Lambda_g) = \mathfrak{S}(a_2^{-1}\Lambda_g)$$

holds if and only if these functions are g-conformally equivalent.

4. The problem of recovering a planar domain

Let $\Omega \subset \mathbb{R}^2$ be a simply connected domain bounded by a smooth closed curve. How far is such a domain determined by the Steklov spectrum $\mathfrak{S}(\Omega, \Lambda_e)$, where e is the Euclidean metric? The problem was considered in [**3**]. We are going to generalize the problem a little bit by involving multisheet domains. A typical example of such a domain is shown on the picture.

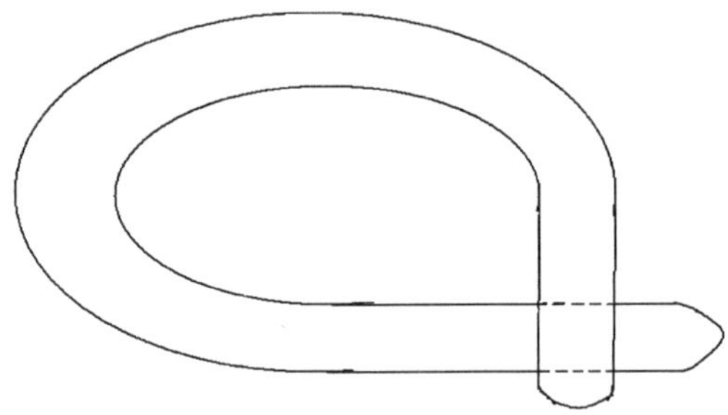

FIGURE 1

How can one define rigorously a (smooth simply connected) multisheet domain? One of possible definitions is as follows: this is a domain diffeomorphic to the disc D and situated on the Riemann surface of a multivalued analytic function. However, the following definition is more appropriate for our purposes.

DEFINITION 4.1. Let $\mathcal{I}$ be the set of all smooth immersions $\mathfrak{J} : D \to \mathbb{R}^2$. Introduce the equivalence relation on $\mathcal{I}$ as follows: $\mathfrak{J}_1 \sim \mathfrak{J}_2$ if there exists a diffeomorphism $\Phi : D \to D$ such that $\mathfrak{J}_2 = \mathfrak{J}_1 \circ \Phi$. Elements of the factor-set $\mathcal{I}/\sim$ are called (smooth, closed, bounded, simply connected) multisheet domains in $\mathbb{R}^2$.

At first sight, the definition can seem too complicated. Nevertheless, in our opinion, every reasonable definition of a multisheet domain should be equivalent to this one. In particular, if an immersion $\mathfrak{I} : D \to \mathbb{R}^2$ is injective, then its equivalence class is uniquely determined by the image $\Omega = \mathfrak{I}(D)$.

The natural conjecture is that a smooth bounded simply connected domain $\Omega \subset \mathbb{R}^2$ is determined by the Steklov spectrum $\mathfrak{S}(\Omega, \Lambda_e)$ uniquely up to an isometry. For multisheet domains, the conjecture sounds as follows:

CONJECTURE 4.2. For two smooth immersions $\mathfrak{I}_j : D \to \mathbb{R}^2$ $(j = 1, 2)$, the equality

$$\mathfrak{S}(\Lambda_{\mathfrak{I}_1^* e}) = \mathfrak{S}(\Lambda_{\mathfrak{I}_2^* e}) \tag{4.1}$$

holds if and only if there exist a diffeomorphism $\Phi : D \to D$ and isometry $I : (\mathbb{R}^2, e) \to (\mathbb{R}^2, e)$ such that

$$\mathfrak{I}_2 = I \circ \mathfrak{I}_1 \circ \Phi. \tag{4.2}$$

We are going to demonstrate the conjecture is equivalent to our previous Conjectures 1.2 and 3.6. But first we will prove the following:

THEOREM 4.3. *For two smooth immersions $\mathfrak{I}_j : D \to \mathbb{R}^2$ $(j = 1, 2)$, the operators $\Lambda_{\mathfrak{I}_1^* e}$ and $\Lambda_{\mathfrak{I}_2^* e}$ are intertwined if and only if there exist a diffeomorphism $\Phi : D \to D$ and isometry $I : (\mathbb{R}^2, e) \to (\mathbb{R}^2, e)$ such that $\mathfrak{I}_2 = I \circ \mathfrak{I}_1 \circ \Phi$.*

PROOF. The "if" statement is obvious. Indeed, since $I^* e = e$, (4.2) implies $\mathfrak{I}_2^* e = \Phi^*(\mathfrak{I}_1^* e)$ and the operators $\Lambda_{\mathfrak{I}_1^* e}$ and $\Lambda_{\mathfrak{I}_2^* e}$ are intertwined by $\varphi = \Phi|_\gamma$.

Now, we prove the statement "only if". Let the operators $\Lambda_{\mathfrak{I}_1^* e}$ and $\Lambda_{\mathfrak{I}_2^* e}$ be intertwined. By Theorem 1.3, there exist a diffeomorphism $\Phi : D \to D$ and function $0 < \rho \in C^\infty(D)$ such that

$$\mathfrak{I}_2^* e = \rho^2 \Phi^* \mathfrak{I}_1^* e, \quad \rho|_\gamma = 1.$$

We rewrite this in the form

$$h_2 = \rho^2 h_1, \quad \text{where} \quad h_1 = \Phi^* \mathfrak{I}_1^* e,\ h_2 = \mathfrak{I}_2^* e.$$

Both h_1 and h_2 are flat metrics. For flat metrics, equality $h_2 = \rho^2 h_1$ holds if and only if $\Delta_{h_1}(\ln \rho) = 0$. Together with the boundary condition $\rho|_\gamma = 1$, this gives $\rho \equiv 1$. Thus,

$$\mathfrak{I}_2^* e = (\mathfrak{I}_1 \circ \Phi)^* e. \tag{4.3}$$

Two immersions $\widetilde{\mathfrak{I}}_1 = \mathfrak{I}_1 \circ \Phi$ and $\mathfrak{I}_2$ satisfy $\widetilde{\mathfrak{I}}_1^* e = \mathfrak{I}_2^* e$ if and only if $\mathfrak{I}_2 = I \circ \widetilde{\mathfrak{I}}_1$ for some isometry $I : (\mathbb{R}^2, e) \to (\mathbb{R}^2, e)$. Therefore (4.3) implies (4.2). □

PROOF OF THE EQUIVALENCE OF CONJECTURES 1.2 AND 4.2. The implication Conjecture 1.2 $\longrightarrow$ Conjecture 4.2 is proved by the same arguments as we have used in the proof of Theorem 4.3. Let us prove the reverse implication.

Assume metrics g_j $(j = 1, 2)$ on D to satisfy (1.2). By Proposition 3.1, there exist a function $0 < \rho_j \in C^\infty(D)$, $\rho_j|_\gamma = 1$, and immersion $\mathfrak{I}_j : D \to \mathbb{R}^2$ such that

$$h_j = \rho_j^2 g_j = \mathfrak{I}_j^* e.$$

Hence,

$$\mathfrak{S}(\Lambda_{\mathfrak{I}_j^* e}) = \mathfrak{S}(\Lambda_{g_j}).$$

Together with (1.2), this gives (4.1). Assuming Conjecture 4.2 to be true, we obtain (4.2). From (4.2),

$$h_2 = \mathfrak{I}_2^* e = (I \circ \mathfrak{I}_1 \circ \Phi)^* e = \Phi^* \mathfrak{I}_1^* I^* e = \Phi^* \mathfrak{I}_1^* e = \Phi^* h_1,$$

i.e.,

$$h_2 = \Phi^* h_1.$$

Now,

$$g_2 = \rho_2^{-2} h_2 = \rho_2^{-2} \Phi^* h_1 = \rho_2^{-2} \Phi^* (\rho_1^2 g_1) = \Big(\frac{\rho_1 \circ \Phi}{\rho_2}\Big)^2 \Phi^* g_1.$$

Setting $\rho = \rho_1 \circ \Phi / \rho_2$, we have

$$g_2 = \rho^2 \Phi^* g_1, \quad \rho|_\gamma = 1.$$

This is the statement of Conjecture 1.2. □

The following statement was proved by Weinstock [**16**]:

If $\mathfrak{S}(\Omega, \Lambda_e) = \{0 = \lambda_0 < \lambda_1 \le \dots\}$ is the Steklov spectrum of a smooth bounded simply connected domain $\Omega \subset \mathbb{R}^2$ and L is the length of $\partial\Omega$, then

$$\lambda_1 \le \frac{2\pi}{L} \tag{4.4}$$

and the equality holds if and only if Ω is a disc.

Weinstock's proof works for multisheet domains as well and the statement can be presented in the form:

PROPOSITION 4.4. For a smooth immersion $\mathfrak{I} : D \to \mathbb{R}^2$, let $\mathfrak{S}(\Lambda_{\mathfrak{I}^* e}) = \{0 = \lambda_0 < \lambda_1 \le \dots\}$ be the Steklov spectrum of the metric $\mathfrak{I}^* e$ and L be the length of γ in the metric. Then (4.4) is valid. The equality in (4.4) holds if and only if there exist a diffeomorphism $\Phi : D \to D$ and isometry $I : (\mathbb{R}^2, e) \to (\mathbb{R}^2, e)$ such that $\mathfrak{I} = I \circ \mathfrak{I}_L \circ \Phi$, where $\mathfrak{I}_L(z) = \frac{L}{2\pi} z$.

On using the equivalence of Conjecture 4.2 to Conjectures 1.2 and 3.6, the latter statement can be transformed to two other forms.

PROPOSITION 4.5. For a function $0 < a \in C^\infty(\gamma)$, let $\mathfrak{S}(a^{-1}\Lambda_e) = \{0 = \lambda_0 < \lambda_1 \le \dots\}$ and $L = \int_0^{2\pi} a(\theta)\, d\theta$. Then (4.4) is valid. The equality in (4.4) holds if and only if the function a is e-conformally equivalent to the constant $L/2\pi$.

PROPOSITION 4.6. For a Riemannian metric g on D, let $\mathfrak{S}(\Lambda_g) = \{0 = \lambda_0 < \lambda_1 \le \dots\}$ be the Steklov spectrum and L be the length of γ in the metric g. Then (4.4) is valid. The equality in (4.4) holds if and only if there exist a diffeomorphism $\Phi : D \to D$ and function $0 < \rho \in C^\infty(D)$ such that $\rho|_\gamma = 1$ and $g = (\frac{L}{2\pi}\rho)^2 \Phi^* e$.

5. Two classes of Riemannian metrics

In the rest of the text we use the following notations. $L^2(\gamma)$ is the space of square integrable complex functions on γ with the scalar product and norm defined by

$$\langle f, g \rangle = \frac{1}{2\pi} \int_0^{2\pi} f(e^{i\theta}) \bar{g}(e^{i\theta})\, d\theta, \quad \|f\|^2 = \langle f, f \rangle.$$

For $f \in L^2(\gamma)$, we denote by f_n the n-th Fourier coefficient,

$$f_n = \frac{1}{2\pi} \int_0^{2\pi} f(e^{i\theta}) e^{-in\theta}\, d\theta.$$

For $1 \leq m \in \mathbb{N}$, we denote by $H^m(\gamma)$ the Hilbert space

$$H^m(\gamma) = \Big\{ f \in L^2(\gamma) \mid \sum_{n\in\mathbb{Z}} |n|^{2m} |f_n|^2 < \infty \Big\}$$

endowed with the scalar product defined by

$$\langle f, g\rangle_{H^m} = \sum_{n\in\mathbb{Z}} (1+n^2)^m f_n \bar{g}_n \quad \text{for} \quad f, g \in H^m(\gamma).$$

Then $\Lambda_e(e^{in\theta}) = |n| e^{in\theta}$ for $n \in \mathbb{Z}$, and $\mathfrak{S}(\Lambda_e) = \mathbb{N}$ where 0 is a simple eigenvalue while others eigenvalues are double. We will denote by e_n the function $e_n(e^{i\theta}) = e^{in\theta}$ for $n \in \mathbb{Z}$.

5.1. General statements on the eigenvalue spectrum. Given a function $0 < a \in C^\infty(\gamma)$, observe that $a^{-1}\Lambda_e = a^{-\frac{1}{2}}(a^{-\frac{1}{2}}\Lambda_e a^{-\frac{1}{2}})a^{\frac{1}{2}}$, where $a^{-\frac{1}{2}}\Lambda_e a^{-\frac{1}{2}} : H^1(\gamma) \to L^2(\gamma)$ is a positive self-adjoint operator, i.e., $\langle a^{-\frac{1}{2}}\Lambda_e a^{-\frac{1}{2}}u, u\rangle \geq 0$ for $u \in H^1(\gamma)$. Then $\mathfrak{S}(a^{-1}\Lambda_e) = \mathfrak{S}(a^{-\frac{1}{2}}\Lambda_e a^{-\frac{1}{2}})$ consists of an increasing sequence of eigenvalues λ_n $(n = 0, 1, 2, \dots)$ counted with multiplicities, where $\lambda_0 = 0$ is a simple eigenvalue and every λ_n is of a finite multiplicity. The multiplicity of λ_1 is not greater than 2, see [**5, 7**]. The fact that $\lambda_0 = 0$ is simple follows from the equality $\Lambda_e 1 = 0$ and $\|a^{-1}\Lambda_e u\| \geq \frac{\|\Lambda_e u\|}{\max a} \geq \frac{\|u-u_0\|}{\max a}$ for $u \in H^1(\gamma)$.

Every eigenspace $\ker(a^{-1}\Lambda_e - \lambda_n)$ consists of smooth functions and can be spanned by real-valued functions. Indeed, if $u \in C^\infty(\gamma)$ is an eigenvector for $a^{-1}\Lambda_e$, then the complex conjugate function $\bar{u}$ is also an eigenvector, and $a^{-1}\Lambda_e(\Re u) = \lambda \Re u$ and $a^{-1}\Lambda_e(\Im u) = \lambda \Im u$.

The spectrum of $a^{-1}\Lambda_e$ is generically simple (see [**1, 15**]) which means, in particular, that there exists a dense subset $\mathcal{S}$ of $\{0 < a \in C^\infty(\gamma)\}$ endowed with its usual topology such that all eigenvalues of $a^{-1}\Lambda_e$ are simple for any $a \in \mathcal{S}$.

We will now describe the eigenvalue spectrum of $a^{-1}\Lambda_e$ for two classes of functions a, first when $a(e^{i\theta}) = a_0 + 2a_1\cos\theta$ for $(a_0, a_1) \in \mathbb{R}^2$, $0 < 2|a_1| < a_0$, and then when $a(e^{i\theta})^{-1} = b_0 + 2b_1\cos(m\theta)$ for $(b_0, b_1, m) \in \mathbb{R}^2 \times \mathbb{N}$, $0 < 2|b_1| < b_0$ and $m \geq 2$.

When $a(e^{i\theta})^{-1} = b_0 + 2b_1\cos\theta$ for $(b_0, b_1) \in \mathbb{R}^2$, $0 < 2|b_1| < b_0$, then a is e-conformally equivalent to the constant-valued function a_0:

$$a_0 \left| \frac{d\varphi}{d\theta} \right| = a, \tag{5.1}$$

where $\varphi(e^{i\theta}) = \exp\left(i a_0^{-1} \int_0^\theta a(e^{is})ds\right) = \frac{z-r}{1-rz}\Big|_{z=e^{i\theta}}$ and $r \in (-1, 1)$ is a root of the polynomial $b_1X^2 + b_0X + b_1$ (see also [**12**, Remark 3]). Hence in that case $\mathfrak{S}(a^{-1}\Lambda_e) = \mathfrak{S}(a_0^{-1}\mathbb{N}) = a_0^{-1}\mathbb{N}$, where 0 is a simple eigenvalue and all others are double eigenvalues.

5.2. Simple eigenvalues. First we exhibit an example of a class of functions a that belongs to $\mathcal{S}$ defined above. We have the following result.

THEOREM 5.1. *All eigenvalues of $a^{-1}\Lambda_e$ are simple when $a(e^{i\theta}) = a_0 + 2a_1\cos\theta$ for $(a_0, a_1) \in \mathbb{R}^2$, $0 < 2|a_1| < a_0$.*

The proof of Theorem 5.1 is given in the appendix. It is based on the following reformulation of the eigenvalue problem for the operator $a^{-1}\Lambda_e$. For $\lambda \in (0, +\infty)$ and $u \in C^\infty(\gamma)$, the equation $\Lambda_e u = \lambda a u$ holds if and only if the Fourier coefficients

$(u_n)_{n\in\mathbb{Z}}$ of the functions u satisfy the following linear second order recurrence with non-constant coefficients

$$-a_1u_{n+1} + (\frac{|n|}{\lambda} - a_0)u_n - a_1u_{n-1} = 0 \quad \text{for} \quad n \in \mathbb{Z}.$$

We introduce some notations before studying the second class of functions a. Given $0 < a \in C^\infty(\gamma)$ and $2 \le m \in \mathbb{N}$, define the holomorphic function $\Phi(z) = z^m$ on $\operatorname{int} D$ and set $\varphi = \Phi|_\gamma$. Then $|\Phi'|\big|_\gamma = m$ and Lemma 3.3 implies

$$(5.2) \qquad (a\circ\varphi)^{-1}\Lambda_e\varphi^*f = m\varphi^*\big(a^{-1}\Lambda_e f\big) \quad \text{for} \quad f \in C^\infty(\gamma).$$

Let $t_m(a)$ denote $a\circ\varphi$. For $j = 0,\dots,m-1$, we denote by $V_{j,m}$ (resp. $\tilde V_{j,m}$) the closure in $H^1(\gamma)$ (resp. $L^2(\gamma)$) of the vector space spanned by the vectors e_{j+km} ($k \in \mathbb{Z}$). For $j = 1,\dots,[\frac{m}{2}]$, we denote by $W_{j,m}$ (resp. $\tilde W_{j,m}$) the vector space $V_{j,m} \bigoplus V_{m-j,m}$ (resp. $\tilde V_{j,m} \bigoplus \tilde V_{m-j,m}$). Note that $W_{j,m}$ is invariant under the complex conjugation. If m is even, then $W_{\frac{m}{2},m} = V_{\frac{m}{2},m}$.

Given a bounded linear operator $B : H^1(\gamma) \to L^2(\gamma)$, we denote by $B|_{V_{j,m}}$ (resp. $B|_{W_{j,m}}$) the restriction of B to $V_{j,m}$ (resp. $W_{j,m}$) for every $0 \le j \le m-1$. The operator Λ_e maps $V_{j,m}$ to $\tilde V_{j,m}$ and the multiplication operators $t_m(a)^{-\frac12}$ and $t_m(a)^{\frac12}$ map $V_{j,m}$ to $V_{j,m}$. Therefore $\Lambda_e|_{V_{j,m}}$, $t_m(a)^{-\frac12}\Lambda_e t_m(a)^{-\frac12}|_{V_{j,m}}$, and $t_m(a)^{-1}\Lambda_e|_{V_{j,m}}$ define unbounded operators in $\tilde V_{j,m}$ where their domain is $V_{j,m}$. The same statement is valid when $V_{j,m}$ and $\tilde V_{j,m}$ are replaced by $W_{j,m}$ and $\tilde W_{j,m}$ respectively. We also have

$$\mathfrak{S}((t_m(a))^{-1}\Lambda_e) = \bigcup_{j=0}^{m-1} \mathfrak{S}((t_m(a))^{-1}\Lambda_e|_{V_{j,m}})$$

$$= \mathfrak{S}((t_m(a))^{-1}\Lambda_e|_{V_{0,m}}) \bigcup_{j=1}^{[\frac{m}{2}]} \mathfrak{S}((t_m(a))^{-1}\Lambda_e|_{W_{j,m}}).$$

The following statement is proved along the same lines as Theorem 5.1. The proof is presented in the appendix.

THEOREM 5.2. *Given integers $m \ge 2$ and $1 \le j \le m-1$, let $a(e^{i\theta}) := (b_0 + 2b_1\cos\theta)^{-1}$ for some $(b_0, b_1) \in \mathbb{R}^2$ satisfying $b_0 > 2|b_1| > 0$. All eigenvalues of $(t_m(a))^{-1}\Lambda_e|_{V_{j,m}}$ are simple.*

On using the invariance of $t_m(a)^{-1}\Lambda_e$ with respect to the complex conjugation, we obtain the corollary: in notations of Theorem 5.2, positive eigenvalues of the operator $(t_m(a))^{-1}\Lambda_e|_{W_{j,m}}$ ($1 \le j < m/2$) are double, and these eigenvalues of the operator $(t_m(a))^{-1}\Lambda_e$ are at least double.

5.3. Double eigenvalues.

THEOREM 5.3. *For $(b_0, b_1) \in \mathbb{R}^2$ satisfying $b_0 > 2|b_1| > 0$, let $a(e^{i\theta}) := (b_0 + 2b_1\cos\theta)^{-1}$. Then*

(1) $\mathfrak{S}((t_m(a))^{-1}\Lambda_e|_{V_{0,m}}) = ma_0^{-1}\mathbb{N}$ *for every integer $m \ge 2$.*

(2) *For integers $m \ge 2$ and $k > 0$, mka_0^{-1} is a double eigenvalue of the operator $(t_m(a))^{-1}\Lambda_e$.*

The first statement of the theorem follows immediately from (5.2) and (5.1). The second statement will be proved in Section 6.2. The proof is based on the reformulation of the eigenvalue problem for the operator $(t_m(a))^{-1}\Lambda_e$ as a scalar Riemann – Hilbert problem, see Section 6.

In the case of $m = 2$, we have the following more precise statement:

COROLLARY 5.4. In notations of Theorem 5.3,

$$\mathfrak{S}((t_2(a))^{-1}\Lambda_e) = \bigcup_{k=1}^{\infty}\{\lambda_{k,-},\lambda_{k,+}\}\bigcup(2a_0^{-1}\mathbb{N}).$$

For $k > 0$, $2ka_0^{-1}$ is a double eigenvalue while $\lambda_{k,-}$ and $\lambda_{k,+}$ are simple eigenvalues of the operator $(t_2(a))^{-1}\Lambda_e$. The following inequalities hold

$$(2k-2)a_0^{-1} < \lambda_{k,-} < \lambda_{k,+} < 2ka_0^{-1} \quad \text{for} \quad k > 0. \tag{5.3}$$

Corollary 5.4 follows from Theorems 5.2 and 5.3 taking into account the continuity of eigenvalues with respect to a continuous perturbation of $(t_2(a))^{-1}\Lambda_e$, see [**6**]. More precisely, let $a(\varepsilon, e^{i\theta}) = a_0\sqrt{b_0^2 - 4\varepsilon^2 b_1^2}\,(b_0 + 2\varepsilon b_1\cos(2\theta))^{-1}$ for $\varepsilon \in (0,1]$. Then $a_0(\varepsilon,.) = a_0$ and eigenvalues of $(a(\varepsilon,.)^{-1}\Lambda_e)|_{V_{1,2}}$ are simple by Theorem 5.2. Let $\{\lambda_{\pm,\varepsilon,k} \mid k \in \mathbb{N}\}$ be the set of positive eigenvalues of the latter operator ordered so that $\lambda_{-,\varepsilon,k} < \lambda_{+,\varepsilon,k} < \lambda_{-,\varepsilon,k+1}$. Then, by continuity of eigenvalues with respect to ε, we have $\lambda_{\pm,\varepsilon,k} \to (2k+1)a_0^{-1}$ as $\varepsilon \to 0^+$. With the help of Theorem 5.3, this implies (5.3).

Quite similarly, for an integer $m \geq 3$, we can prove: mka_0^{-1} is a double eigenvalue of the operator $(t_m(a))^{-1}\Lambda_e$ for every $0 < k \in \mathbb{N}$ and

$$\mathfrak{S}((t_m(a))^{-1}\Lambda_e) \setminus (ma_0^{-1}\mathbb{N}) = \bigcup_{k=1}^{\infty}\bigcup_{j=1}^{2m-2}\{\lambda_{k,j}\}, \quad \text{where} \quad ma_0^{-1}(k-1) < \lambda_{k,j} < ma_0^{-1}k.$$

We also mention a possible direction for a generalization of Corollary 5.4 at the end of Section 6.3.

To authors' knowledge, no bounded simply connected domain is known with a Steklov eigenvalue of multiplicity greater than 2.

5.4. The Hilbert transform and an integral equation. Observe that (see for example [**11**])

$$\Lambda_e = \sqrt{-\frac{d^2}{d\theta^2}} = H \circ \frac{d}{d\theta} = \frac{d}{d\theta} \circ H,$$

where H is the Hilbert transform on $L^2(\gamma)$ defined by
(5.4)

$$(Hu)(e^{i\theta}) = i^{-1}\sum_{n\in\mathbb{Z},\ n\neq 0}\operatorname{sgn}(n)u_n e^{in\theta} = (2\pi)^{-1}\text{p.v.}\int_0^{2\pi}\frac{1+\cos(\theta-t)}{\sin(\theta-t)}u(e^{it})\,dt.$$

In particular, $H(\chi_\theta)(e^{it}) = \frac{1}{\pi}(-\ln)\left(\left|\sin\left(\frac{\theta-t}{2}\right)\right|/\left|\sin\left(\frac{t}{2}\right)\right|\right)$ for $(t,\theta) \in \mathbb{R}\times(0,2\pi)$, where $\chi_\theta(e^{it}) = 1$ for $t \in (0,\theta)$ and $\chi_\theta(e^{it}) = 0$ for $t \in [\theta, 2\pi]$. Hence the eigenvalue problem for $a^{-1}\Lambda_e$ can be transformed into an integral equation as follows:

$$(a^{-1}\Lambda_e u = \lambda u) \Leftrightarrow u(e^{i\theta}) = u(1) + \frac{\lambda}{\pi}\int_0^{2\pi}(-\ln)\left(\left|\frac{\sin\left(\frac{\theta-t}{2}\right)}{\sin\left(\frac{t}{2}\right)}\right|\right)a(e^{it})u(e^{it})\,dt$$

for $u \in C^\infty(\gamma)$ and $\lambda \in (0,+\infty)$. We do not use the integral equation in the rest of the text. However, the Fredholm determinant of the square of the self-adjoint compact operator in $L^2([0,2\pi])$ with the nonnegative integral kernel $a^{1/2}(e^{i\theta})(-$

$\ln)\,(|\sin(\frac{\theta-t}{2})|)\,a^{1/2}(e^{it})$ may be of some interest for the Steklov inverse spectral problem.

It is also of some interest to mention that the Steklov eigenvalues asymptotics presented in Proposition 3.2 can be derived from Proposition 3.1 on using the following fact: the commutator $[H,b]=Hb-bH$ is a smoothing operator for any $b\in C^\infty(\gamma)$. Indeed, as follows from (5.4), the kernel of $[H,b]$ is the smooth function $\mathcal{K}$ on $\gamma\times\gamma$ defined by

$$\mathcal{K}(e^{i\theta},e^{it})=(2\pi)^{-1}\cot\big(\frac{\theta-t}{2}\big)(b(e^{it})-b(e^{i\theta}))\quad\text{for}\quad(\theta,t)\in[0,2\pi]\times[0,2\pi].$$

Hence

$$(5.5)\qquad (a^{-1/2}\Lambda_e a^{-1/2})^2=-\Big(a^{-1/2}\frac{d}{d\theta}a^{-1}\frac{d}{d\theta}a^{-1/2}\Big)+a^{-1/2}\frac{d}{d\theta}[H,a^{-1}]\Lambda_e a^{-1/2}$$

and $a^{-1/2}\frac{d}{d\theta}[H,a^{-1}]\Lambda_e a^{-1/2}$ is a smoothing operator.

Using (5.5) and the min-max principle, one obtains the following asymptotics for eigenvalues and eigenvectors of $a^{-1/2}\Lambda_e a^{-1/2}$. Define

$$v_{\pm,n}(e^{i\theta})=\Big(\frac{a(e^{i\theta})}{a_0}\Big)^{1/2}\exp\left(\pm i\left[\frac{n+1}{2}\right]\int_0^\theta\frac{a(e^{is})}{a_0}\,ds\right)$$

and let $(u_n)_{n\in\mathbb{N}}$ be an orthonormal basis of $L^2(\mathbb{S}^1)$ such that $a^{-1/2}\Lambda_e a^{-1/2}u_n=\lambda_n u_n$. Then

$$\lambda_n=a_0^{-1}\left[\frac{n+1}{2}\right]+O(n^{-N})\quad\text{as}\quad n\to+\infty,$$
$$u_n-\langle u_n,v_{+,n}\rangle v_{+,n}-\langle u_n,v_{-,n}\rangle v_{-,n}=O(n^{-N})\quad\text{as}\quad n\to+\infty$$

for every $N>0$. The multiplicity of the eigenvalue λ_n is at most 2 for n large enough, see [**7**, Proposition 1.4.2].

In the case of $a_0=1$, the operator $a^{-1/2}\Lambda_e a^{-1/2}-h^*\Lambda_e(h^{-1})^*$ is a smoothing operator, where the diffeomorphism $h:\gamma\to\gamma$ is defined by $h(e^{i\theta})=\exp\Big(i\int_0^\theta a(e^{is})\,ds\Big)$, see [**3**].

In the next section, we relate the eigenvalue problem for $a^{-1}\Lambda_e$ to a scalar Riemann – Hilbert problem. The relationship will be used in our proof of Theorem 5.3, but it is also of some independent interest. We also refer the reader to [**10**, pp. 182] for a connection between Fredholm operators and Riemann – Hilbert problems.

6. A scalar Riemann – Hilbert problem

6.1. Preliminaries. In this paragraph we obtain equivalent formulations of the eigenvalue problem for $a^{-1}\Lambda_e$ as a scalar Riemann-Hilbert problem.

Let $\mathcal{H}(O)$ be the space of holomorphic functions on an open set $O\subset\mathbb{C}$. Given $u\in C^\infty(\gamma)$, define the functions $u_+\in\mathcal{H}(\operatorname{int}D)\cap C^\infty(D)$ and $u_-\in\mathcal{H}(\mathbb{C}\backslash D)\cap C^\infty(\mathbb{C}\backslash\operatorname{int}D)$ by $u_+(z)=u_0/2+\sum_{n=1}^{+\infty}u_nz^n$ and $u_-(z)=u_0/2+\sum_{n=1}^{+\infty}u_{-n}z^{-n}$ respectively. Then, for $z=e^{i\theta}$, $u(z)=u_+(z)+u_-(z)$ and

$$(6.1)\qquad \Lambda_e u(z)=z\frac{d(u_+-u_-)}{dz}(z)=i^{-1}\frac{du_+}{d\theta}(e^{i\theta})-i^{-1}\frac{du_-}{d\theta}(e^{i\theta}).$$

If u is a real function, then $u_n=\bar u_{-n}$ and $u_-(z)=\overline{u_+(\bar z^{-1})}$.

On using (6.1) and the complex conjugation invariance of $a^{-1}\Lambda_e$, we obtain for a real valued function $u \in C^\infty(\gamma)$ and $\lambda \in \mathbb{R}$

$$(6.2) \qquad (\Lambda_e u = \lambda a u) \Leftrightarrow \left(\Re\left(z\frac{du_+}{dz}(z) - \lambda a(z)u_+(z)\right) = 0 \quad for \quad z \in \gamma\right).$$

REMARK 6.1. Statement (6.2) can be written in two other equivalent forms. First, for a complex function $u \in C^\infty(\gamma)$,

$$(\Lambda_e u = \lambda a u) \Leftrightarrow \left(z\frac{du_+}{dz}(z) - \lambda a(z)u_+(z) = z\frac{du_-}{dz}(z) + \lambda a(z)u_-(z) \quad for \quad z \in \gamma\right).$$

Second, given a complex function $u \in C^\infty(\gamma)$, let $v_\pm$ be the 2π-periodic and bounded function on the closed half-plane $\{x + iy \mid \pm y \geq 0\}$ defined by $v_\pm(z) = u_0/2 + \sum_{n=1}^{+\infty} u_{\pm n}e^{inz}$. Then

$$(\Lambda_e u = \lambda a u) \Leftrightarrow \left(\frac{dv_+}{dx}(x) - i\lambda a(x)v_+(x) = \frac{dv_-}{dx}(x) + i\lambda a(x)v_-(x) \quad for \quad x \in \mathbb{R}\right).$$

6.2. The case of a rational function a. We are going to reformulate (6.2) when a is a rational function of $z \in \gamma$. We will need the following

LEMMA 6.2. *Let $f \in \mathcal{H}(int D) \cap C(D)$ and $n > 0$. Then $\Re(f(z)/z^n) = 0$ for all $z \in \gamma$ if and only if there exists a sequence $(\alpha_0, \dots, \alpha_n) \in \mathbb{R} \times \mathbb{C}^n$ such that $f(z) = i\alpha_0 z^n + \sum_{j=1}^n (\alpha_j z^{n+j} - \bar{\alpha}_j z^{n-j})$ for $z \in D$.*

For the sake of consistency, the proof of Lemma 6.2 is postponed to the appendix.

Let $(\zeta_1, \dots, \zeta_M)$ and $(\eta_1, \dots, \eta_N)$ be two sequences of points from $\operatorname{int} D\backslash\{0\}$ such that $\zeta_j \neq \eta_l$ for all j and l. Define functions p and q on $\mathbb{C}\backslash\{0\}$ by

$$(6.3)\ p(z) = c_1 z^{-M}\Pi_{j=1}^M (z - \zeta_j)(z - \bar{\zeta}_j^{-1}), \quad q(z) = c_2 z^{-N}\Pi_{j=1}^N (z - \eta_j)(z - \bar{\eta}_j^{-1}),$$

where the constants $c_k \neq 0$ $(k = 1, 2)$ are such that $\bar{c}_1 = c_1\Pi_{j=1}^M(\zeta_j/\bar{\zeta}_j)$ and $\bar{c}_2 = c_2\Pi_{j=1}^N(\eta_j/\bar{\eta}_j)$. Then $\overline{p(\bar{z}^{-1})} = p(z)$ and $\overline{q(\bar{z}^{-1})} = q(z)$ for $z \in \mathbb{C}\backslash\{0\}$. The functions p and q are real-valued on γ and do not vanish at any point of γ. Next, set

$$(6.4) \qquad a(z) = \frac{p(z)}{q(z)} \quad \text{for} \quad z \in \mathbb{C}\backslash(Z \cup \{0\}),$$

where Z is the set of zeros of q. For so chosen a, (6.2) takes the form

$$(\Lambda_e u = \lambda a u) \Leftrightarrow \left(\Re\left(zq(z)\frac{d}{dz}u_+(z) - \lambda p(z)u_+(z)\right) = 0 \quad for \quad z \in \gamma\right).$$

With the help of Lemma 6.2, this implies the statement: λ is an eigenvalue of $a^{-1}\Lambda_e$ if and only if there exist a function $u_+ \in \mathcal{H}(\operatorname{int} D) \cap C^\infty(D)$ and sequence $(\alpha_0, \dots, \alpha_s) \in \mathbb{R} \times \mathbb{C}^s$ such that

$$(6.5) \quad zq(z)\frac{du_+}{dz}(z) - \lambda p(z)u_+(z) = i\alpha_0 + \sum_{j=1}^s (\alpha_j z^j - \bar{\alpha}_j z^{-j}) \quad \text{for} \quad z \in D\backslash\{0\},$$

where $s = \max(M, N - 1)$. This is a linear first order differential equation with rational coefficients and rational right-hand side. Therefore the solution u_+ admits a holomorphic continuation to $\mathbb{C}\backslash Z_-$, where Z_- denotes the set of zeros of q which are outside D. If $u_+ \in \mathcal{H}(\mathbb{C}\backslash Z_-)$ solves equation (6.5) and c is a real constant, then $u_+ + ic$ solves a similar equation that differs from (6.5) by values of parameters $(\alpha_0,\dots,\alpha_s)\in\mathbb{R}\times\mathbb{C}^s$ only. Since the integral $p_0=(2\pi^{-1})\int_0^{2\pi}p(e^{i\theta})\,d\theta$ is real and nonzero, we can choose the constant c such that $\alpha_0=0$ in the corresponding equation. We have thus proved

LEMMA 6.3. *Let the function $a\in C^\infty(\gamma)$ be defined by (6.3) and (6.4). A real function $u\in C^\infty(\gamma)$ is an eigenvector of $a^{-1}\Lambda_e$ with eigenvalue λ if and only if there exists $(\alpha_1,\dots,\alpha_s)\in\mathbb{C}^s$ such that*

$$(6.6)\quad zq(z)\frac{du_+}{dz}(z)-\lambda p(z)u_+(z)=\sum_{j=1}^{s}(\alpha_j z^j-\bar\alpha_j z^{-j})\quad \textit{for}\quad z\in\mathbb{C}\backslash(Z_-\cup\{0\}),$$

where $s=\max(M,N-1)$ and Z_- is the set of zeros of q which are outside D.

REMARK 6.4. If $N=0$, then q is a nonzero constant function, $Z_-=\emptyset$, and u_+ in (6.6) is an entire function. If $s=0$, then the right-hand side of (6.6) is zero.

6.3. Proof of Theorem 5.3. Let $a(e^{i\theta})=(b_0+2b_1\cos(m\theta))^{-1}$ for some $(b_0,b_1)\in\mathbb{R}^2$ satisfying $b_0>2|b_1|>0$ and $m\in\mathbb{N}$. We can set $b_0=1$ and replace b_1 by some $b\in(0,\frac12)$ without loss of generality. Then a admits a holomorphic continuation given by

$$(6.7)\qquad a(z)=\frac{z^m}{b+z^m+bz^{2m}}\quad\text{for}\quad z\in\mathbb{C}\backslash Z_m,$$

where $Z_m:=\{\omega e^{i\frac{2j\pi}{m}},\ \bar\omega^{-1}e^{i\frac{2j\pi}{m}}\mid 1\le j\le m\}$ and ω is a root in D of the polynomial $b+X^m+bX^{2m}$.

Formula (6.7) is the partial case of (6.4) for $p(z)=1$ and $q(z)=bz^{-m}+1+bz^m$. We denote by $Z_{m,-}$ the set $\{\bar\omega^{-1}e^{i\frac{2j\pi}{m}}\mid 1\le j\le m\}$. Lemma 6.3 implies: a real $u\in C^\infty(\gamma)$ satisfies $a^{-1}\Lambda_e u=\lambda u$ if and only if there exists $(\alpha_1,\dots,\alpha_{m-1})\in\mathbb{C}^{m-1}$ such that

(6.8)

$$(b+z^m+bz^{2m})\frac{du_+}{dz}(z)-\lambda z^{m-1}u_+(z)=\sum_{l=1}^{m-1}(\alpha_l z^{m-1+l}-\bar\alpha_l z^{m-1-l})\text{ for }z\in\mathbb{C}\backslash Z_{m,-}.$$

REMARK 6.5. Since $W_{j,m}$ $(1\le j<m/2)$ is spanned by $e_{\pm j+nm}$ for $n\in\mathbb{Z}$, we obtain: if λ is an eigenvalue for $a^{-1}\Lambda_e|_{W_{j,m}}$, we can set $\alpha_l=0$ for $l\notin\{j,m-j\}$ on the right-hand side of (6.8). If λ is an eigenvalue for $a^{-1}\Lambda_{e|W_{m/2,m}}$ and m is even, we can set $\alpha_l=0$ for $l\ne m/2$ on the right-hand side of (6.8).

Let $r\in(-1,0)$ be a root of the polynomial $b+X+bX^2$ (we recall that $a_0^{-1}=(1-4b^2)^{1/2}$). When $m=1$ and $\lambda=ka_0^{-1}$ for some positive integer k, equation (6.8) becomes $du_+(z)/dz-k\big(a_0b(z-r)(z-r^{-1})\big)^{-1}u_+(z)=0$ or

$$\frac{1}{i}\frac{d}{d\theta}u_+(e^{i\theta})-\frac{k}{a_0(1+2b\cos\theta)}u_+(e^{i\theta})=0.$$

Integrating the equation and using $\big(a_0b(z-r)(z-r^{-1})\big)^{-1}=(z-r)^{-1}-(z-r^{-1})^{-1}$, we obtain

$$u_+(z)=C\left(\frac{z-r}{1-rz}\right)^k\quad\text{for}\quad z\ne r^{-1},\quad u_+(e^{i\theta})=C\exp\Big(\frac{ik}{a_0}\int_0^\theta\frac{ds}{1+2b\cos s}\Big)$$

with a complex constant C. This implies for u being a real eigenfunction of $a^{-1}\Lambda_e|_{V_{0,m}}$ related to mka_0^{-1}

$$u_+(z) = C\left(\frac{z^m - r}{z^m - r^{-1}}\right)^k \text{ for } z \in \mathbb{C}\backslash Z_{m,-}. \tag{6.9}$$

Given $(\alpha_1, \ldots, \alpha_{m-1}) \in \mathbb{C}^{m-1}$, let us consider the first order differential equation

(6.10)

$$(b + z^m + bz^{2m})\frac{dv_+}{dz}(z) - kma_0^{-1}z^{m-1}v_+(z) = \sum_{j=1}^{m-1}(\alpha_j z^{m-1+j} - \bar{\alpha}_j z^{m-1-j}), \quad z \in \mathbb{C}\backslash Z_{m,-}.$$

We are going to prove that the existence of a holomorphic solution v_+ on $\mathbb{C}\backslash Z_{m,-}$ implies that $\alpha_l = 0$ $(1 \le l \le m-1)$. Together with (6.9), this will give us the statement of Theorem 5.3.

Assume the existence of a solution $v_+ \in \mathcal{H}(\mathbb{C}\backslash Z_{m,-})$ to equation (6.10). We use the variation of constants. Define the function C by

$$C(z) = u_+(z)^{-1}v_+(z) = \left(\frac{z^m - r^{-1}}{z^m - r}\right)^k v_+(z) \text{ for } z \in \mathbb{C}\backslash Z_m, \tag{6.11}$$

where $u_+(z) = \left(\frac{z^m - r}{z^m - r^{-1}}\right)^k$. The function C is holomorphic on $\mathbb{C}\backslash Z_m$. Hence using (6.10) and (6.11) we obtain

$$(b+z^m+bz^{2m})u_+(z)\frac{d}{dz}C(z) = \sum_{j=1}^{m-1}(\alpha_j z^{m-1+j} - \bar{\alpha}_j z^{m-1-j}) \text{ for } z \in \mathbb{C}\backslash Z_m. \tag{6.12}$$

We set

(6.13)

$$R(z) = \frac{d}{dz}C(z) = b^{-1}\frac{(z^m - r^{-1})^{k-1}}{(z^m - r)^{k+1}}\sum_{j=1}^{m-1}(\alpha_j z^{m-1+j} - \bar{\alpha}_j z^{m-1-j}) \text{ for } z \in \mathbb{C}\backslash Z_m.$$

As follows from (6.13), there exists a constant δ such that $|R(z)| \le \delta|z|^{-2}$ for $z \in \mathbb{C}$, $|z| \ge 2(-r)^{1/m}$.

Since the rational function R is the complex derivative of a holomorphic function on $\mathbb{C}\backslash Z_m$, the residue of R at each point of Z_m is equal to zero. Therefore R integrates to zero over the curve $\gamma_l : \mathbb{R} \to \mathbb{C}$ $(0 \le l \le m-1)$ defined by $\gamma_l(t) = -te^{i\frac{2\pi l}{m}}$ for $t \le 0$ and $\gamma_l(t) = te^{\frac{i2\pi(l+1)}{m}}$ for $t > 0$:

$$\int_0^{+\infty} R\big(te^{i\frac{2\pi l}{m}}\big)\, dt - e^{i\frac{2\pi}{m}}\int_0^{+\infty} R\big(te^{i\frac{2\pi(l+1)}{m}}\big)\, dt = 0. \tag{6.14}$$

We remind that $-r > 0$ and $z^m - r \ne 0$ for any point z of γ_l. As is seen from (6.13), R is integrable on γ_l and $\rho\int_0^{2\pi} R(\rho e^{i\theta})d\theta = O(\rho^{-1})$ as $\rho \to +\infty$.

Using (6.12) and (6.14), we obtain the following linear system of equation on $(\alpha_1, \ldots, \alpha_{m-1})$:

(6.15)

$$\sum_{j=1}^{m-1}\Big(\alpha_j e^{i\frac{2\pi l(j-1)}{m}}(1 - e^{i\frac{2\pi j}{m}})I_{m-1+j}(-r) - \bar{\alpha}_j e^{i\frac{2\pi l(-j-1)}{m}}(1-e^{-i\frac{2\pi j}{m}})I_{m-1-j}(-r)\Big) = 0$$

for $0 \le l \le m-1$, where

$$(6.16)\quad I_{l'}(\sigma) = \int_0^{+\infty} \frac{(\sigma t^m+1)^{k-1}}{(t^m+\sigma)^{k+1}} t^{l'}\, dt \text{ for } (\sigma, l') \in (0,1)\times \mathbb{N},\ 0 \le l' \le 2m-2.$$

From (6.15) we obtain

$$\sum_{j=1}^{m-1} (\beta^{l(j-1)} - \beta^{lj})\,(\alpha_j I_{m-1+j}(-r) - \bar{\alpha}_{m-j} I_{j-1}(-r)) = 0 \text{ for } 0 \le l \le m-1,$$

where $\beta = e^{i\frac{2\pi}{m}}$. Since the Van der Monde matrix $(\beta^{j(l-1)})_{1\le j\le m, 1\le l\le m}$ is invertible, we obtain $\alpha_j I_{m-1+j}(-r) - \bar{\alpha}_{m-j} I_{j-1}(-r) = 0$ for $j = 1 \dots m-1$. In other words,

$$(6.17)\quad \alpha_j\big(I_{m-1+j}(-r) I_{2m-j-1}(-r) - I_{j-1}(-r) I_{m-j-1}(-r)\big) = 0 \text{ for } 1 \le j \le m-1.$$

We use the following

LEMMA 6.6. *For $(\sigma, l) \in (0,1)\times\mathbb{N}$ satisfying $1 \le l \le m-1$, the inequalities $0 < I_{2m-l-1}(\sigma) < I_{l-1}(\sigma)$ hold.*

By Lemma 6.6, $I_{m-1+j}(-r) I_{2m-j-1}(-r) - I_{j-1}(-r) I_{m-j-1}(-r) < 0$ for $1 \le j \le m-1$. Therefore (6.17) implies $\alpha_j = 0$ for $j = 1\dots m-1$. This finishes the proof of Theorem 5.3. □

PROOF OF LEMMA 6.6. Given $\sigma \in (0,1)$ and $l \in \mathbb{N}$ satisfying $1 \le l \le m-1$, we set $\alpha = l/m \in (0,1)$. We change the integration variable $s = t^m$ on (6.16) to obtain

$$I_{2m-l-1}(\sigma) = \frac{1}{m}\int_0^{+\infty} \frac{(\sigma s+1)^{k-1}}{(s+\sigma)^{k+1}} s^{1-\alpha}\, ds,$$
$$I_{l-1}(\sigma) = \frac{1}{m}\int_0^{+\infty} \frac{(\sigma s+1)^{k-1}}{(s+\sigma)^{k+1}} s^{\alpha-1}\, ds.$$

The integration variable change $s := s^{-1}$ gives

$$\int_1^{+\infty} \frac{(\sigma s+1)^{k-1}}{(s+\sigma)^{k+1}} s^{1-\alpha}\, ds = \int_0^1 \frac{(\sigma+s)^{k-1}}{(1+\sigma s)^{k+1}} s^{\alpha-1}\, ds.$$

Therefore

$$I_{2m-l-1}(\sigma) = \frac{1}{m}\int_0^1 \left(\frac{(\sigma s+1)^{k-1}}{(s+\sigma)^{k+1}} s^{1-\alpha} + \frac{(\sigma+s)^{k-1}}{(1+\sigma s)^{k+1}} s^{\alpha-1}\right) ds,$$
$$I_{l-1}(\sigma) = \frac{1}{m}\int_0^1 \left(\frac{(\sigma s+1)^{k-1}}{(s+\sigma)^{k+1}} s^{\alpha-1} + \frac{(\sigma+s)^{k-1}}{(1+\sigma s)^{k+1}} s^{1-\alpha}\right) ds.$$

Two last formulas imply

$$(6.18)\quad m(I_{l-1}(\sigma) - I_{2m-l-1}(\sigma)) = \int_0^1 s^{\alpha-1}(1-s^{2(1-\alpha)})\left(\frac{(\sigma s+1)^{k-1}}{(s+\sigma)^{k+1}} - \frac{(\sigma+s)^{k-1}}{(1+\sigma s)^{k+1}}\right) ds.$$

Then note that

$$\frac{(\sigma s+1)^{k-1}}{(s+\sigma)^{k+1}} > \frac{(\sigma+s)^{k-1}}{(1+\sigma s)^{k+1}} \text{ and } (1-s^{2(1-\alpha)}) > 0 \text{ for } s \in (0,1) \text{ since } \alpha \in (0,1).$$

Thus the integrand on the right-hand side of (6.18) is positive and the inequality $I_{l-1}(\sigma) - I_{2m-l-1}(\sigma) > 0$ holds. □

We indicate a possible direction for a generalization of Corollary 5.4. Let $(\lambda, u) \in (0, +\infty) \times C^\infty(\gamma)$ and $1 \le j \le [m/2]$ so that $u \in W_{j,m}$. Hence $u_0 = 0$ and $u_+(0) = 0$, where u_+ is defined at the beginning of Section 6. Assume $\lambda \notin ma_0^{-1}\mathbb{N}$ and $a^{-1}\Lambda_e u = \lambda u$ for $a(e^{i\theta}) = (1 + 2b\cos(m\theta))^{-1}$. Then, using Remark 6.5, we see that $u_+ \in \mathcal{H}(\mathbb{C}\backslash Z_{m,-})$ and

$$(b+z^m+bz^{2m})\frac{d}{dz}u_+(z)-\lambda z^{m-1}u_+(z) = \alpha_{m-j}z^{2m-1-j}+\alpha_j z^{m-1+j}-\bar{\alpha}_{m-j}z^{j-1}-\bar{\alpha}_j z^{m-1-j}$$

for $z \in \mathbb{C}\backslash Z_{m,-}$ with some $(\alpha_j, \alpha_{m-j}) \in \mathbb{C}^2$. Integrate the equation to obtain
(6.19)

$$\begin{aligned} u_+(z) = za_0 \int_0^1 &\left(\frac{(tz)^m - r^{-1}}{(tz)^m - r}\right)^{\frac{\lambda a_0}{m}} \left(\frac{1}{(tz)^m - r^{-1}} - \frac{1}{(tz)^m - r}\right) \\ &\times \left(\alpha_{m-j}(tz)^{2m-1-j} + \alpha_j(tz)^{m-1+j} - \bar{\alpha}_{m-j}(tz)^{j-1} - \bar{\alpha}_j(tz)^{m-1-j}\right)dt \end{aligned}$$

for $z \in \mathbb{C}\backslash K$, where $K = \bigcup_{l=1}^{m}\left(e^{i\pi/m+2\pi(l-1)/m}\left[(-r)^{1/m}, (-r)^{-1/m}\right]\right)$. In (6.19), we have actually used the following branch of the logarithm: $\ln(z) = \ln|z| + i\arg z$ for $z \in \mathbb{C}\backslash(-\infty, 0]$ and $\arg z \in (-\pi, \pi)$. As is seen from (6.19), $u_+ \in \mathcal{H}(\mathbb{C}\backslash K)$. The condition $u_+ \in \mathcal{H}(\mathbb{C}\backslash Z_{m,-})$ gives
(6.20)

$$\begin{aligned} 0 &= u_+(s^{\frac{1}{m}}\omega_l e^{i0^+}) - u_+(s^{\frac{1}{m}}\omega_l e^{i0^-}) \\ &= \frac{2}{\lambda}\sin\left(\frac{\lambda a_0 \pi}{m}\right)\left(\frac{s-1}{r^{-2}-s}\right)^{\frac{\lambda a_0}{m}} \Big[\alpha_{m-j}\omega_l^{m-j}c_k(1-\frac{j}{m}, \mu) \\ &\quad + \alpha_j\omega_l^j c_k(\frac{j}{m}, \mu) - \bar{\alpha}_{m-j}\omega_l^{j-m}c_k(\frac{j}{m}-1, \mu) - \bar{\alpha}_j\omega_l^{-j}c_k(-\frac{j}{m}, \mu)\Big] \end{aligned}$$

for $s \in (1, r^{-2})$ and for $0 \le l \le m-1$, where $\omega_l = e^{i\pi/m+i2\pi l/m}(-r)^{1/m}$, $k = [\lambda a_0/m]$, and $\mu = \lambda - ma_0^{-1}k$. The function c_k is smooth on $(-1,1) \times (0,1)$ and

$$c_0(\alpha, \mu) = r^{-2\mu} + \alpha\int_0^1 \left(\left(\frac{r^{-2}-t}{1-t}\right)^\mu - r^{-2\mu}\right)t^{\alpha-1}dt, \quad (\alpha, \mu) \in (-1,1) \times (0,1).$$

We also define for $(\alpha, \mu) \in (0, \frac{1}{2}] \times (0,1)$

$$F_k(\alpha, \mu) = c_k(\alpha-1, \mu)c_k(-\alpha, \mu) - r^2 c_k(1-\alpha, \mu)c_k(\alpha, \mu).$$

For fixed $k \in \mathbb{N}$ and $\mu \in (0,1)$, the linear system of equations (6.20) has a nontrivial solution $(\alpha_j, \alpha_{m-j}) \in \mathbb{C}^2$ if and only if $F_k(\alpha, \mu) = 0$. We therefore obtain, for $\mu \in (0,1)$ and $(j,k) \in \mathbb{N}^2$ $(1 \le j \le [m/2])$,

$$mka_0^{-1} + \mu \in \mathfrak{S}(a^{-1}\Lambda_{e|W_{j,m}}) \Leftrightarrow F_k(\frac{j}{m}, \mu) = 0.$$

Finally, studying growth properties of functions $F_k(., \mu)$ at fixed $\mu \in (0,1)$, one may get some knowledge of multiplicities of eigenvalues of $a^{-1}\Lambda_e$ which belong to the interval $(mka_0^{-1}, m(k+1)a_0^{-1})$. More precisely, if the inequality $\frac{\partial F_0}{\partial \alpha}(\alpha, \mu) > 0$ was proved, then one would obtain the statement: the eigenvalues of $a^{-1}\Lambda_e|_{W_{j,m}}$ ($1 \le j < m/2$) belonging to $(0, ma_0^{-1})$ are double eigenvalues of the operator $a^{-1}\Lambda_e$; and for even m, the eigenvalues of $a^{-1}\Lambda_{e|W_{\frac{m}{2},m}}$ belonging $(0, ma_0^{-1})$ are simple eigenvalues of $a^{-1}\Lambda_e$.

Appendix A. Proofs of Theorems 5.1 and 5.2 and of Lemma 6.2

To prove Theorem 5.1, we need the following lemma that will be proved at the end of the section.

LEMMA A.1. *Given $(\lambda, a_0, a_1) \in (0, +\infty)^2 \times \mathbb{C}$ satisfying $|a_1| > 0$, let G_λ denote the two-dimensional complex vector space of complex-valued sequences $(v_n)_{n\in\mathbb{N}}$ satisfying*

$$-\lambda a_1 v_{k+1} + (k - \lambda a_0)v_k - \lambda a_1 v_{k-1} = 0 \tag{A.1}$$

for $k \geq 1$. Let also $\tilde{G}_\lambda$ be the subspace of G_λ consisting of sequences satisfying $\sum_{k=0}^{+\infty}(1 + k^2)|v_k|^2 < \infty$. Then the dimension of the vector space $\tilde{G}_\lambda$ is at most one.

PROOF OF THEOREM 5.1. We remind that 0 is a simple eigenvalue of $a^{-1}\Lambda_e$. If $v \in C^\infty(\gamma)$ solves the equation $\Lambda_e v = \lambda a v$ with $\lambda > 0$, Fourier coefficients of v satisfy (A.1) for all $k \in \mathbb{Z}$. Hence $v_{k-1} = \lambda^{-1}a_1^{-1}(-\lambda a_1 v_{k+1} + (k - \lambda a_0)v_k)$ $(k \leq 1)$; and $v \equiv 0$ if $v_n = 0$ for all $n \geq 1$. We have thus the embedding $f : \ker(a^{-1}\Lambda_e - \lambda) \to \tilde{G}_\lambda$ defined by $f(v) = (v_n)_{n\in\mathbb{N}}$. This implies with the help of Lemma A.1 $\dim_\mathbb{C}(\ker(a^{-1}\Lambda_e - \lambda)) \leq \dim_\mathbb{C}(\tilde{G}_\lambda) \leq 1$. □

To prove Theorem 5.2, we need the following lemma that will be proved at the end of the section.

LEMMA A.2. *Let $(\lambda, b_0, b_1, m) \in [0, +\infty) \times (0, +\infty) \times \mathbb{C} \times \mathbb{N}$ satisfy $m \geq 2$ and $b_0 > 2|b_1| > 0$. We denote by $F_{\lambda,j}$ $(1 \leq j \leq m - 1)$ the two-dimensional complex vector space of complex-valued sequences $(u_n)_{n\in\mathbb{N}}$ satisfying*

$$(mk + m + j)b_1 u_{k+1} + ((mk + j)b_0 - \lambda)u_k + (mk - m + j)b_1 u_{k-1} = 0 \tag{A.2}$$

for $k \geq 1$. Let also $S_{\lambda,j}$ be the subspace of $F_{\lambda,j}$ consisting of sequences satisfying $\sum_{k=0}^{+\infty}(1 + k^2)|u_k|^2 < \infty$. Then the dimension of the complex vector space $S_{\lambda,j}$ is at most one.

PROOF OF THEOREM 5.2. We mimic the proof of Theorem 5.1. If $u \in C^\infty(\gamma)$ solves the equation $\Lambda_e u = \lambda t_m(a)u$ and $u \in V_{j,m}$, then the Fourier coefficients $(u_{mk+j})_{k\in\mathbb{Z}}$ of u satisfy (A.2) for every $k \in \mathbb{Z}$. Hence $u_{mk-m+j} = -((mk - m + j)b_1)^{-1}((mk + m + j)b_1 u_{mk+m+j} + ((mk + j)b_0 - \lambda)u_{mk+j})$ $(k \leq 1)$; and $u \equiv 0$ if $u_{mn+j} = 0$ for all $n \geq 1$. We have thus the embedding $f : \ker((t_m(a))^{-1}\Lambda_e|_{V_{j,m}} - \lambda) \to S_{\lambda,j}$ defined by $f(u) = (u_{mn+j})_{n\in\mathbb{N}}$. This implies with the help of Lemma A.2 $\dim_\mathbb{C}(\ker((t_m(a))^{-1}\Lambda_e|_{V_{j,m}} - \lambda)) \leq 1$. □

PROOF OF LEMMA A.1. For $(u_n)_{n\in\mathbb{N}} \in G_\lambda$,

$$\begin{pmatrix} u_k \\ u_{k+1} \end{pmatrix} = \tilde{M}_k \begin{pmatrix} u_{k-1} \\ u_k \end{pmatrix} \quad \text{for} \quad k \geq 1,$$

where $\tilde{M}_k$ is the invertible 2×2-matrix

$$\tilde{M}_k = \begin{pmatrix} 0 & 1 \\ -1 & \frac{k-\lambda a_0}{\lambda a_1} \end{pmatrix}. \tag{A.3}$$

Let $\tilde{k}_\lambda \geq 2$ be such that

$$k - \lambda a_0 \geq 2\lambda|a_1| \quad \text{for} \quad k \geq \tilde{k}_\lambda. \tag{A.4}$$

Then using (A.3) we have for any $(u_n)_{n\in\mathbb{Z}}$

$$\begin{pmatrix} u_{\tilde{k}_\lambda-1} \\ u_{\tilde{k}_\lambda} \end{pmatrix} = \Pi_{i=1}^{\tilde{k}_\lambda-1}\tilde{M}_i \begin{pmatrix} u_0 \\ u_1 \end{pmatrix}.$$

Since $\Pi_{i=1}^{\tilde{k}_\lambda-1}\tilde{M}_i$ is invertible, we can consider the sequence $(v_n)_{n\in\mathbb{N}} \in G_\lambda$ such that $(v_{\tilde{k}_\lambda-1}, v_{\tilde{k}_\lambda}) = (0,1)$. We will prove that

$$|v_k| \geq |v_{k-1}| \geq 1 \quad \text{for} \quad k \geq \tilde{k}_\lambda. \tag{A.5}$$

This will prove that $\sum_{k=0}^{\infty}|v_k|^2 = \infty$, and hence $(v_n)_{n\in\mathbb{N}} \notin \tilde{G}_\lambda$. Therefore we will obtain $\tilde{G}_\lambda \neq G_\lambda$. Since $\tilde{G}_\lambda \subset G_\lambda$ and G_λ is two-dimensional, we will obtain that $\tilde{G}_\lambda$ is at most one-dimensional. This will prove Lemma A.1.

We prove (A.5) by induction. The inequalities (A.5) are trivially satisfied for $k = \tilde{k}_\lambda$. Assume (A.5) to hold for some $k \geq \tilde{k}_\lambda$. Then from (A.1)

$$|v_{k+1}| \geq \frac{|k - a_0\lambda|}{\lambda|a_1|}|v_k| - |v_{k-1}|.$$

Using the induction hypothesis (A.5) and (A.4), we obtain

$$|v_{k+1}| \geq \frac{|k-\lambda a_0| - \lambda|a_1|}{\lambda|a_1|}|v_k| \geq |v_k| \geq 1$$

and the induction step is done. □

REMARK A.3. The solution $(v_k)_{k\in\mathbb{N}}$ of (A.1) used in the proof of Lemma A.1 grows at a rate faster than the exponential one. Indeed, taking (A.1) and (A.5) into account,

$$\frac{k-\lambda a_0 - \lambda|a_1|}{\lambda|a_1|} \leq \frac{|v_{k+1}|}{|v_k|} \leq \frac{k-\lambda a_0 + \lambda|a_1|}{\lambda|a_1|} \quad \text{for} \quad k \geq \tilde{k}_\lambda$$

and

$$\frac{k-\lambda a_0 - \lambda|a_1|}{\lambda|a_1|} \to +\infty \quad \text{as} \quad k \to +\infty,$$

where $\tilde{k}_\lambda$ is defined by (A.4).

PROOF OF LEMMA 6.2. First we reduce the question to the case of $n = 0$. Indeed, given $f \in \mathcal{H}(\text{int}D) \cap C(D)$, let $\alpha_{n-j} = \frac{1}{j!}\frac{d^j}{dz^j}f(0)$ $(0 \leq j \leq n-1)$ for some $n > 0$. Since $\Re(\frac{\sum_{j=1}^{n}(\alpha_j z^{n-j} - \bar{\alpha}_j z^{n+j})}{z^n}) = 0$ for $z \in \gamma$, we have $(\Re\tilde{f})|_\gamma = 0$, where the function $\tilde{f}(z) = \frac{f(z) - \sum_{j=1}^{n}(\alpha_j z^{n-j} - \bar{\alpha}_j z^{n+j})}{z^n}$ is continuous in D and holomorphic in int$D$. The condition $(\Re\tilde{f})|_\gamma = 0$ is equivalent to $\tilde{f}|_D = i\alpha_0$ for some $\alpha_0 \in \mathbb{R}$, which proves the Lemma modulo the case of $n = 0$.

Now, assume f to satisfy $f|_D = i\alpha_0$ for some $\alpha_0 \in \mathbb{R}$. Then obviously $\Re(f(z)) = 0$ for $z \in \gamma$. Conversely assume that $\Re(f)|_\gamma = 0$. This implies, by maximum principle, $\Re f|_D = 0$ since $\Re f$ is a harmonic function in int$D$. Since f is holomorphic in D, it must be identically equal to some pure imaginary constant. □

PROOF OF LEMMA A.2. For $(u_n)_{n\in\mathbb{N}} \in F_{\lambda,j}$,

$$\begin{pmatrix} u_k \\ u_{k+1} \end{pmatrix} = M_k \begin{pmatrix} u_{k-1} \\ u_k \end{pmatrix} \quad \text{for} \quad k \geq 1,$$

where M_k is the invertible 2×2-matrix

$$M_k = \begin{pmatrix} 0 & 1 \\ -\frac{mk-m+j}{mk+m+j} & -\frac{(mk+j)b_0-\lambda}{(mk+m+j)b_1} \end{pmatrix}. \tag{A.6}$$

Let $k_\lambda \geq 2$ be such that

$$b_0(mk+j) - \lambda \geq 2(mk+m+j)|b_1| \quad \text{for} \quad k \geq k_\lambda \tag{A.7}$$

(such a k_λ exists since $b_0 > 2|b_1| > 0$). Then using (A.6) we have for any $(u_n)_{n\in\mathbb{Z}}$

$$\begin{pmatrix} u_{k_\lambda-1} \\ u_{k_\lambda} \end{pmatrix} = \Pi_{i=1}^{k_\lambda-1} M_i \begin{pmatrix} u_0 \\ u_1 \end{pmatrix}.$$

Since $\Pi_{i=1}^{k_\lambda-1} M_i$ is invertible, we can consider the sequence $(v_n)_{n\in\mathbb{N}} \in F_{\lambda,j}$ such that $(v_{k_\lambda-1}, v_{k_\lambda}) = (0,1)$. We are going to demonstrate that

$$|v_k| \geq |v_{k-1}| \geq 1, \text{ for } k \geq k_\lambda. \tag{A.8}$$

This will prove $\sum_{k=0}^{\infty} |v_k|^2 = \infty$, and hence $(v_n)_{n\in\mathbb{N}} \notin S_{\lambda,j}$. Therefore we will obtain $S_{\lambda,j} \neq F_{\lambda,j}$. Since $S_{\lambda,j} \subset F_{\lambda,j}$ and $F_{\lambda,j}$ is two-dimensional, we will obtain that $S_{\lambda,j}$ is at most one-dimensional. This will prove Lemma A.2.

We prove (A.8) by induction. The inequalities (A.8) are trivially satisfied for $k = k_\lambda$. Assume (A.8) to hold for some $k \geq k_\lambda$. Then from (A.2),

$$|v_{k+1}| \geq \frac{|(mk+j)b_0 - \lambda|}{(mk+m+j)|b_1|}|v_k| - \frac{mk-m+j}{mk+m+j}|v_{k-1}|.$$

Using the induction hypothesis (A.8) and (A.7), we obtain

$$|v_{k+1}| \geq \frac{|(mk+j)b_0 - \lambda| - (mk-m+j)|b_1|}{(mk+m+j)|b_1|}|v_k| \geq |v_k| \geq 1$$

and the induction step is done. □

Note that similarly to Remark A.3 the solution $(v_k)_{k\in\mathbb{N}}$ of (A.2) has an exponential increase.

References

[1] J. H. Albert, *Genericity of simple eigenvalues for elliptic PDE's*, Proc. Amer. Math. Soc. **48** (1975), 413–418. MR0385934 (52 #6793)

[2] Bruno Colbois, Ahmad El Soufi, and Alexandre Girouard, *Isoperimetric control of the Steklov spectrum*, J. Funct. Anal. **261** (2011), no. 5, 1384–1399, DOI 10.1016/j.jfa.2011.05.006. MR2807105 (2012m:35328)

[3] Julian Edward, *An inverse spectral result for the Neumann operator on planar domains*, J. Funct. Anal. **111** (1993), no. 2, 312–322, DOI 10.1006/jfan.1993.1015. MR1203456 (94c:58217)

[4] Ailana Fraser and Richard Schoen, *The first Steklov eigenvalue, conformal geometry, and minimal surfaces*, Adv. Math. **226** (2011), no. 5, 4011–4030, DOI 10.1016/j.aim.2010.11.007. MR2770439 (2012f:58054)

[5] P. Jammes, *Prescription du spectre de Steklov dans une classe conforme*, preprint 2012, arXiv:1209.4571.

[6] Tosio Kato, *Perturbation theory for linear operators*, Classics in Mathematics, Springer-Verlag, Berlin, 1995. Reprint of the 1980 edition. MR1335452 (96a:47025)

[7] M. Karpukhin, G. Kokarev and I. Polterovich, *Multiplicity bounds for Steklov eigenvalues on Riemannian surfaces*, preprint 2012, arXiv:1209.4869.

[8] Matti Lassas and Gunther Uhlmann, *On determining a Riemannian manifold from the Dirichlet-to-Neumann map* (English, with English and French summaries), Ann. Sci. École Norm. Sup. (4) **34** (2001), no. 5, 771–787, DOI 10.1016/S0012-9593(01)01076-X. MR1862026 (2003e:58037)

[9] John M. Lee and Gunther Uhlmann, *Determining anisotropic real-analytic conductivities by boundary measurements*, Comm. Pure Appl. Math. **42** (1989), no. 8, 1097–1112, DOI 10.1002/cpa.3160420804. MR1029119 (91a:35166)

[10] Solomon Grigorevich Mikhlin and Siegfried Prössdorf, *Singular integral operators*, Mathematische Lehrbücher und Monographien, II. Abteilung: Mathematische Monographien [Mathematical Textbooks and Monographs, Part II: Mathematical Monographs], vol. 68, Akademie-Verlag, Berlin, 1986. Translated from the German by Albrecht Böttcher and Reinhard Lehmann. MR867687 (87k:45001)

[11] Leonid Pestov and Gunther Uhlmann, *Two dimensional compact simple Riemannian manifolds are boundary distance rigid*, Ann. of Math. (2) **161** (2005), no. 2, 1093–1110, DOI 10.4007/annals.2005.161.1093. MR2153407 (2006c:53038)

[12] I. Kh. Sabitov, *Locally Euclidean metrics with a given geodesic curvature of the boundary* (Russian, with Russian summary), Tr. Mat. Inst. Steklova **266** (2009), no. Geometriya, Topologiya i Matematicheskaya Fizika. II, 218–226, DOI 10.1134/S0081543809030122; English transl., Proc. Steklov Inst. Math. **266** (2009), no. 1, 210–218. MR2603269 (2011b:53026)

[13] V. A. Sharafutdinov, *The geometric problem of electrical impedance tomography in the disk* (Russian, with Russian summary), Sibirsk. Mat. Zh. **52** (2011), no. 1, 223–238, DOI 10.1134/S0037446606010198; English transl., Sib. Math. J. **52** (2011), no. 1, 178–190. MR2810264 (2012c:35481)

[14] W. Stekloff, *Sur les problèmes fondamentaux de la physique mathématique (suite et fin)* (French), Ann. Sci. École Norm. Sup. (3) **19** (1902), 455–490. MR1509018

[15] K. Uhlenbeck, *Generic properties of eigenfunctions*, Amer. J. Math. **98** (1976), no. 4, 1059–1078. MR0464332 (57 #4264)

[16] Robert Weinstock, *Inequalities for a classical eigenvalue problem*, J. Rational Mech. Anal. **3** (1954), 745–753. MR0064989 (16,368c)

Laboratoire de Physique Théorique et Modélisation, CNRS UMR 8089/Université de Cergy-Pontoise, 95302 Cergy-Pontoise, France

E-mail address: alexandre.jollivet@u-cergy.fr

Sobolev Institute of Mathematics and Novosibirsk State University, Russia

E-mail address: sharaf@math.nsc.ru

Contemporary Mathematics
Volume **615**, 2014
http://dx.doi.org/10.1090/conm/615/12245

Recent progress in the Calderón problem with partial data

Carlos Kenig and Mikko Salo

ABSTRACT. We survey recent results on Calderón's inverse problem with partial data, focusing on three and higher dimensions.

CONTENTS

1. Introduction

The Calderón problem with partial data asks to determine the electrical conductivity of a body from electrical measurements made on certain subsets of the boundary. This is a fundamental inverse problem, also mentioned as an open question in Gunther Uhlmann's ICM address [**Uh98**]. Subsequent years have seen several advances in partial data problems, many of them due to Gunther Uhlmann and his collaborators, and it is a pleasure for us to survey some of these developments in this volume in honor of Gunther's remarkable career.

Let us give the formal statement of the problem. If $\Omega \subset \mathbb{R}^n$, $n \geq 2$, is a bounded domain with C^∞ boundary, if $\gamma \in L^\infty(\Omega)$ is a positive function (the electrical conductivity of the medium), and if Γ_D, Γ_N are open subsets of $\partial\Omega$, we consider the partial Cauchy data set

$$C_\gamma^{\Gamma_D,\Gamma_N} = \{(u|_{\Gamma_D}, \gamma\partial_\nu u|_{\Gamma_N})\,;\ \mathrm{div}(\gamma\nabla u) = 0 \text{ in } \Omega,\ u \in H^1(\Omega),$$
$$\mathrm{supp}(u|_{\partial\Omega}) \subset \Gamma_D\}.$$

2010 *Mathematics Subject Classification.* Primary 35R30.
The first author was partly supported by NSF.
The second author was supported in part by the Academy of Finland and an ERC Starting Grant.

In the Calderón problem with partial data, the objective is to determine the conductivity γ from the knowledge of $C_\gamma^{\Gamma_D,\Gamma_N}$ for given (possibly very small) sets Γ_D, Γ_N. Here ∂_ν is the normal derivative, and the conormal derivative $\gamma \partial_\nu u|_{\partial\Omega}$ is interpreted in the weak sense as an element of $H^{-1/2}(\partial\Omega)$. Thus we have

$$C_\gamma^{\Gamma_D,\Gamma_N} \subset H^{1/2}(\Gamma_D) \times H^{-1/2}(\Gamma_N).$$

It is very useful to consider the related inverse boundary value problem for the Schrödinger equation with partial data. There, the objective is to determine a potential $q \in L^\infty(\Omega)$ from the partial Cauchy data set

$$C_q^{\Gamma_D,\Gamma_N} = \{(u|_{\Gamma_D}, \partial_\nu u|_{\Gamma_N})\,;\, (-\Delta + q) = 0 \text{ in } \Omega,\ u \in H_\Delta(\Omega), \\ \operatorname{supp}(u|_{\partial\Omega}) \subset \Gamma_D\}.$$

Here, the space $H_\Delta(\Omega)$ is defined by

$$H_\Delta(\Omega) = \{u \in L^2(\Omega)\,;\, \Delta u \in L^2(\Omega)\}.$$

It is known (see [**BU01**]) that for u in this space, the trace $u|_{\partial\Omega}$ and normal derivative $\partial_\nu u|_{\partial\Omega}$ are in $H^{-1/2}(\partial\Omega)$ and $H^{-3/2}(\partial\Omega)$, respectively. Therefore,

$$C_q^{\Gamma_D,\Gamma_N} \subset H^{-1/2}(\Gamma_D) \times H^{-3/2}(\Gamma_N).$$

We formulate the relevant partial data problems as follows:

QUESTION 1. *Let Γ_D, Γ_N be open subsets of $\partial\Omega$ and let $\gamma_1, \gamma_2 \in L^\infty(\Omega)$. If*

$$C_{\gamma_1}^{\Gamma_D,\Gamma_N} = C_{\gamma_2}^{\Gamma_D,\Gamma_N},$$

is it true that $\gamma_1 = \gamma_2$?

QUESTION 2. *Let Γ_D, Γ_N be open subsets of $\partial\Omega$ and let $q_1, q_2 \in L^\infty(\Omega)$. If*

$$C_{q_1}^{\Gamma_D,\Gamma_N} = C_{q_2}^{\Gamma_D,\Gamma_N},$$

is it true that $q_1 = q_2$?

In both problems above, we think of $u|_{\partial\Omega}$ as Dirichlet data (voltage) prescribed only on Γ_D, and we assume that one can measure the Neumann data of the corresponding solution (outgoing current) on Γ_N. The Cauchy data set is determined by the Dirichlet-to-Neumann map (DN map) $\Lambda_\gamma : H^{1/2}(\partial\Omega) \to H^{-1/2}(\partial\Omega)$, defined by

$$\Lambda_\gamma : u|_{\partial\Omega} \mapsto \gamma \partial_\nu u|_{\partial\Omega} \quad \text{where } u \in H^1(\Omega) \text{ solves } \operatorname{div}(\gamma \nabla u) = 0 \text{ in } \Omega.$$

The partial Cauchy data set is obtained from the graph of Λ_γ as

$$C_\gamma^{\Gamma_D,\Gamma_N} = \{(f|_{\Gamma_D}, \Lambda_\gamma f|_{\Gamma_N})\,;\, f \in H^{1/2}(\partial\Omega),\ \operatorname{supp}(f) \subset \Gamma_D\}.$$

Also $C_q^{\Gamma_D,\Gamma_N}$ is obtained by restricting the graph of the DN map $\Lambda_q : u|_{\partial\Omega} \mapsto \partial_\nu u|_{\partial\Omega}$, where $(-\Delta + q)u = 0$ in Ω, provided that 0 is not a Dirichlet eigenvalue of $-\Delta + q$ in Ω. We also note that the problems above concern the case where the potential is grounded on $\partial\Omega \setminus \Gamma_D$. Similarly one could think that part of the boundary is insulated, which would lead to partial data problems for the Neumann-to-Dirichlet map (ND map).

One can think of three subcases of the above problems:

- *Full data*: $\Gamma_D = \Gamma_N = \partial\Omega$.
- *Local data*: $\Gamma_D = \Gamma_N = \Gamma$, where Γ can be any nonempty open subset of $\partial\Omega$.
- *Data on disjoint sets*: Γ_D and Γ_N are disjoint open sets.

The most complete results are of course available for the full data case. Moreover, virtually all known results involve some version of the method of complex geometrical optics (CGO) solutions. This method has its origin in the works [**Fa66**], [**Ca80**], and major results for full data inverse boundary value problems based on the CGO method are [**SU87**], [**HT13**] in dimensions $n \geq 3$ and [**Na96**], [**AP06**], [**Bu08**] in the case $n = 2$. In particular, it has been proved that the set $C_\gamma^{\partial\Omega,\partial\Omega}$ determines uniquely a conductivity $\gamma \in C^1(\overline{\Omega})$ if $n \geq 3$ and a conductivity $\gamma \in L^\infty(\Omega)$ if $n = 2$. See the survey [**Uh09**] for further references.

For the partial data question where the sets Γ_D or Γ_N may not be the whole boundary, we mention here four main approaches. Formulated in terms of the Schrödinger problem, it is known that $C_q^{\Gamma_D,\Gamma_N}$ determines q in Ω in the following cases:

(1) $n \geq 3$, the set Γ_D is possibly very small, and Γ_N is slightly larger than $\partial\Omega \setminus \Gamma_D$ (Kenig, Sjöstrand, and Uhlmann [**KSU07**])

(2) $n \geq 3$ and $\Gamma_D = \Gamma_N = \Gamma$, and $\partial\Omega \setminus \Gamma$ is either part of a hyperplane or part of a sphere (Isakov [**Is07**])

(3) $n = 2$ and $\Gamma_D = \Gamma_N = \Gamma$, where Γ can be an arbitrary open subset of $\partial\Omega$ (Imanuvilov, Uhlmann, and Yamamoto [**IUY10**])

(4) $n \geq 2$, linearized partial data problem, $\Gamma_D = \Gamma_N = \Gamma$ where Γ can be an arbitrary open subset of $\partial\Omega$ (Dos Santos, Kenig, Sjöstrand, and Uhlmann [**DKSjU09**])

Here approach (1) requires roughly complementary sets Γ_D and Γ_N, whereas approaches (2)–(4) deal with the local data problem. Approaches (1)–(3) also give a partial data result for the conductivity equation with the same assumptions on the dimension and the sets Γ_D, Γ_N. In (4), the linearized partial data problem asks to show injectivity of the Fréchet derivative of Λ_q at $q = 0$ instead of injectivity of the full map $q \mapsto \Lambda_q$, when restricted to the sets Γ_D and Γ_N.

It is interesting that each of the four approaches is based on a version of CGO solutions, but still the approaches are distinct in the sense that none of the above results is contained in any of the others. Approach (1) is based on Carleman estimates with boundary terms for limiting Carleman weights, whereas approach (2) is based on reflection arguments and the full data methods of [**SU87**]. Approach (3) uses limiting Carleman weights with critical points and stationary phase, but involves complex analysis and is therefore restricted to two dimensions. Approach (4) is based on analytic microlocal analysis but so far only works for the linearized case.

In a recent work [**KS12**], we unified the approaches (1) and (2) and extended both of them, giving new partial data results in dimensions $n \geq 3$ also on certain Riemannian manifolds. Below, we will explain the approaches (1) and (2) from the new perspective obtained from [**KS12**], and we will also give detailed proofs of certain partial data results in [**KS12**] restricting to the special case of Euclidean domains (and mostly to linear Carleman weights as in [**BU01**]). We hope that the present treatment will be more accessible to readers familiar with Euclidean analysis than the geometric paper [**KS12**]. Approaches (3) and (4) give rather final results in the local data problem for $n = 2$ and for the linearized case. We refer to the recent surveys [**GT13**], [**IY13b**] on two-dimensional partial data problems for more details on approach (3). Approach (4) will be discussed in Section 7, and some open problems are listed in Section 8.

We list here some further references for partial data results, first for the case $n \geq 3$. The Carleman estimate approach was initiated in [**BU01**] and [**KSU07**]. Based on this approach, there are low regularity results [**Kn06**], [**Zh12**], results for other scalar equations [**DKSjU07**], [**KS07**], [**Ch11**], [**Ch12**] and systems [**ST10**], stability results [**HW06**], [**HW07**], [**Tz08**], [**CDR12**], and reconstruction results [**NS10**]. The reflection approach was introduced in [**Is07**], and has been employed for the Maxwell system [**COS09**], [**Ca11**]. Partial data results for slab geometries are given in [**LiU10**], [**KLU12**]. A result analogous to Theorem 2.7 was independently obtained in [**IY13a**].

In two dimensions, the main partial data result is [**IUY10**] which has been extended to more general equations [**IUY11a**], combinations of measurements on disjoint sets [**IUY11b**], less regular coefficients [**IY12a**], and some systems [**IY12b**]. An earlier result is in [**ALP05**]. In the case of Riemann surfaces with boundary, corresponding partial data results are given in [**GT11a**], [**GT11b**], [**AGTU11**].

For piecewise analytic conductivities, uniqueness in the local data problem follows from boundary determination results [**KV84**], [**KV85**]. See [**Ge08**] for a related construction. An early result for the linearized problem in an annular domain in $\mathbb{R}^2$ with no measurements on the inner boundary is in [**Hä98**]. In the case when the conductivity is known near the boundary, the partial data problem can be reduced to the full data problem [**Is88**], [**AU04**], [**Fa07**], [**Be09**], [**AK12**], [**HPS12**]. Also, we remark that in the corresponding problem for the wave equation, it has been known for a long time (see [**KKL01**]) that measuring the Dirichlet and Neumann data of waves on an arbitrary open subset of the boundary is sufficient to determine the coefficients uniquely up to natural gauge transforms. Recent partial results for the case where Dirichlet and Neumann data are measured on disjoint sets are in [**LO10**], [**LO12**].

The paper is structured as follows. Section 1 is the introduction. In Section 2 we give precise statements for the various partial data results in the literature, concentrating on the Schrödinger problem. Section 3 describes the main strategy for proving most of these results, by reducing to a density statement for products of solutions that vanish on parts of the boundary. A Carleman estimate with boundary terms relevant for partial data results is proved in Section 4. This is a special case of the corresponding estimate in [**KS12**], but we give self-contained proofs in the Euclidean case. Section 5 discusses the construction of CGO solutions vanishing on parts of the boundary, and Section 6 explains the corresponding uniqueness results relying on the injectivity of a mixed Fourier transform/local attenuated geodesic ray transform of the unknown coefficient. The linearized problem is considered in Section 7, and Section 8 lists open questions related to the partial data problem.

2. Partial data results

In this section we give precise statements of the partial data results mentioned in the introduction. We will only consider the Schrödinger problem, since the conductivity problem can be reduced to that case by the well known relation

$$\Lambda_{q_\gamma} f = \gamma^{-1/2}\Lambda_\gamma(\gamma^{-1/2} f) + \frac{1}{2}\gamma^{-1}(\partial_\nu \gamma) f, \qquad q_\gamma = \frac{\Delta(\gamma^{1/2})}{\gamma^{1/2}}.$$

This reduction works if the conductivities are sufficiently regular and have the same boundary values up to first order on part of the boundary (if $\Gamma_D = \Gamma_N$ the latter condition follows from boundary determination).

It will be useful to introduce some notation. In all results below, we assume that $\Omega \subset \mathbb{R}^n$ is a bounded open connected set with C^∞ boundary. If $\varphi \in C^\infty(\overline{\Omega})$ is a real valued function, we can write the boundary of Ω as the disjoint union

$$\partial\Omega = \partial\Omega_+ \cup \partial\Omega_- \cup \partial\Omega_{\mathrm{tan}}$$

where

$$\begin{aligned}
\partial\Omega_\pm &= \partial\Omega_\pm(\varphi) = \{x \in \partial\Omega\,;\, \pm\nabla\varphi(x)\cdot\nu(x) > 0\},\\
\partial\Omega_{\mathrm{tan}} &= \partial\Omega_{\mathrm{tan}}(\varphi) = \{x \in \partial\Omega\,;\, \nabla\varphi(x)\cdot\nu(x) = 0\}.
\end{aligned}$$

Here ν is the outer unit normal of $\partial\Omega$.

The functions φ that can be used in partial data results are typically *limiting Carleman weights*. This is a special class of functions, introduced in [**KSU07**], that coincides with the set of harmonic functions (with some restriction on their critical points) if $n = 2$, and if $n \geq 3$ it consists of the following six functions up to translation and scaling:

$$\begin{gathered}
\alpha\cdot x, \quad \log|x|, \quad \frac{\alpha\cdot x}{|x|^2}, \quad \arg((\alpha + i\beta)\cdot x),\\
\arg(e^{i\theta}(x+i\xi)^2), \quad \log\frac{|x+\xi|^2}{|x-\xi|^2}.
\end{gathered}$$

Here α and β are orthogonal unit vectors, $\theta \in [0, 2\pi)$, $\xi \in \mathbb{R}^n \setminus \{0\}$, and the argument function is defined by

$$\arg(z) = 2\arctan\frac{\mathrm{Im}(z)}{|z| + \mathrm{Re}(z)}, \qquad z \in \mathbb{C}\setminus(-\infty, 0].$$

We refer to [**DKSaU09**] for more information and a thorough analysis of limiting Carleman weights that have no critical points.

It is suggested by the methods of [**KS12**] that for a fixed limiting Carleman weight φ, measuring Neumann data on $\partial\Omega_+(\varphi)$ for Dirichlet data input of $\partial\Omega_-(\varphi)$, with no measurements required on $\partial\Omega_{\mathrm{tan}}(\varphi)$, might be sufficient for determining the unknown coefficients (and this should also hold with $\partial\Omega_+(\varphi)$ and $\partial\Omega_-(\varphi)$ interchanged). All results described below can be understood in light of this idea, but most of them require measurements on $\partial\Omega_\pm(\varphi)$ for a large family of different φ's instead of just one φ.

2.1. **The result of [KSU07].** This result is stated in terms of the front and back faces of $\partial\Omega$ with respect to some point x_0 which is outside the convex hull $\mathrm{ch}(\overline{\Omega})$ of $\overline{\Omega}$. Note that if Ω is strictly convex, the front face can be made arbitrarily small by placing x_0 close to the boundary, but in this case the back face will be very large.

THEOREM 2.1. *Let $\Omega \subset \mathbb{R}^n$, $n \geq 3$. If $x_0 \in \mathbb{R}^n \setminus \mathrm{ch}(\overline{\Omega})$, define the front and back face of $\partial\Omega$ by*

$$F(x_0) = \{x \in \partial\Omega\,;\, (x - x_0) \cdot \nu(x) \leq 0\},$$
$$B(x_0) = \{x \in \partial\Omega\,;\, (x - x_0) \cdot \nu(x) \geq 0\}.$$

Let Γ_D, Γ_N be open subsets of $\partial\Omega$ with $F(x_0) \subset \Gamma_D$ and $B(x_0) \subset \Gamma_N$. If $q_1, q_2 \in L^\infty(\Omega)$ and if

$$C_{q_1}^{\Gamma_D,\Gamma_N} = C_{q_2}^{\Gamma_D,\Gamma_N},$$

then $q_1 = q_2$.

We also state another partial data result, which follows from the previous theorem by placing x_0 sufficiently far from Ω. This is close to the earlier result of [**BU01**].

THEOREM 2.2. *Let $\Omega \subset \mathbb{R}^n$, $n \geq 3$. If $\alpha \in \mathbb{R}^n$ is a unit vector, define the front and back face of $\partial\Omega$ by*

$$F(\alpha) = \{x \in \partial\Omega\,;\, \alpha \cdot \nu(x) \leq 0\},$$
$$B(\alpha) = \{x \in \partial\Omega\,;\, \alpha \cdot \nu(x) \geq 0\}.$$

Let Γ_D, Γ_N be open subsets of $\partial\Omega$ with $F(\alpha) \subset \Gamma_D$ and $B(\alpha) \subset \Gamma_N$. If $q_1, q_2 \in L^\infty(\Omega)$ and if

$$C_{q_1}^{\Gamma_D,\Gamma_N} = C_{q_2}^{\Gamma_D,\Gamma_N},$$

then $q_1 = q_2$.

Both results can be understood as follows: partial measurements are sufficient for determining the potential provided that Γ_D and Γ_N satisfy

$$\partial\Omega_-(\varphi) \cup \partial\Omega_{\tan}(\varphi) \subset \Gamma_D, \qquad \partial\Omega_+(\varphi) \cup \partial\Omega_{\tan}(\varphi) \subset \Gamma_N$$

for a suitable family of limiting Carleman weights. In the first theorem $\varphi(x) = \log|x - x_0'|$ where x_0' ranges over a small neighborhood of x_0, and in the second theorem $\varphi(x) = \alpha' \cdot x$ where α' ranges over a small neighborhood of α in the unit sphere.

2.2. **The result of [Is07].** This is a local data result where one measures both Dirichlet and Neumann data on the same set $\Gamma = \Gamma_D = \Gamma_N$, but where the inaccessible part $\Gamma_0 = \partial\Omega \setminus \Gamma$ has to be part of a hyperplane.

THEOREM 2.3. *Let $\Omega \subset \mathbb{R}^n$, $n \geq 3$. Assume that $\Omega \subset \{x_n > 0\}$, let Γ be a nonempty open subset of $\partial\Omega$, and assume that $\Gamma_0 = \partial\Omega \setminus \Gamma$ satisfies $\Gamma_0 \subset \{x_n = 0\}$. If $q_1, q_2 \in L^\infty(\Omega)$ and if*

$$C_{q_1}^{\Gamma,\Gamma} = C_{q_2}^{\Gamma,\Gamma},$$

then $q_1 = q_2$.

By applying the Kelvin transform $K(x) = x/|x|^2$, this theorem implies a similar result (also proved in [**Is07**]) where the inaccessible part of the boundary has to be part of a sphere.

THEOREM 2.4. *Let $\Omega \subset \mathbb{R}^n$, $n \geq 3$. Assume that Ω is a strict subset of some ball B, let Γ be a nonempty open subset of $\partial\Omega$, and assume that $\Gamma_0 = \partial\Omega \setminus \Gamma$ satisfies $\Gamma_0 \subset \partial B$. If $q_1, q_2 \in L^\infty(\Omega)$ and if*

$$C_{q_1}^{\Gamma,\Gamma} = C_{q_2}^{\Gamma,\Gamma},$$

then $q_1 = q_2$.

These results can be understood so that measurements on Γ are sufficient to determine the potential if the inaccessible part Γ_0 satisfies certain geometric conditions related to limiting Carleman weights. In the first theorem, it holds that

$$\Gamma_0 \subset \bigcap_{\substack{\alpha \in \mathbb{R}^n, |\alpha|=1 \\ \alpha \cdot e_n = 0}} \partial\Omega_{\tan}(\varphi_\alpha)$$

where $\varphi_\alpha(x) = \alpha \cdot x$ and e_n is the nth coordinate vector. Taking complements, this condition means that Γ should be sufficiently large in the sense that

$$\bigcup_{\substack{\alpha \in \mathbb{R}^n, |\alpha|=1 \\ \alpha \cdot e_n = 0}} (\partial\Omega_+(\varphi_\alpha) \cup \partial\Omega_-(\varphi_\alpha)) \subset \Gamma.$$

In the second theorem, if the coordinates are normalized in such a way that $B = B(\frac{1}{2}e_n, \frac{1}{2})$, one has instead

$$\Gamma_0 \subset \bigcap_{\substack{\alpha \in \mathbb{R}^n, |\alpha|=1 \\ \alpha \cdot e_n = 0}} \partial\Omega_{\tan}(K^*\varphi_\alpha)$$

where K is the Kelvin transform and $K^*\varphi_\alpha$ is the limiting Carleman weight $K^*\varphi_\alpha(x) = \alpha \cdot x/|x|^2$.

2.3. The result of [IUY10]. This is a general local data result that is valid for two-dimensional domains. (It was extended to potentials with $W^{1,p}$, $p > 2$, regularity in [**IY12a**].)

THEOREM 2.5. *Let $\Omega \subset \mathbb{R}^2$ be a bounded open set with C^∞ boundary, and let Γ be a nonempty open subset of $\partial\Omega$. If $q_1, q_2 \in C^{4,\alpha}(\overline{\Omega})$ for some $\alpha > 0$ and if*

$$C_{q_1}^{\Gamma,\Gamma} = C_{q_2}^{\Gamma,\Gamma},$$

then $q_1 = q_2$.

The result is related to limiting Carleman weights as follows. Let Γ be any open subset of $\partial\Omega$, and let $\Gamma_0 = \partial\Omega \setminus \Gamma$ be the inaccessible part of the boundary. The proof of [**IUY10**] begins by showing that one can find a family of harmonic functions (limiting Carleman weights in 2D) $\{\varphi_p\}_{p \in S}$, where S is a dense subset of Ω, such that

$$\Gamma_0 \subset \bigcap_{p \in S} \partial\Omega_{\tan}(\varphi_p)$$

and each φ_p is a Morse function (its critical points are nondegenerate) having a critical point at p. We refer to [**IUY10**] and the surveys [**GT13**], [**IY13b**] for more details about partial data problems in two dimensions.

2.4. The result of [DKSjU09]. As it is explained in Lemma 3.1 below, if $\Gamma_D = \Gamma_N = \Gamma$, where Γ is a nonempty open subset of $\partial\Omega$, and if $C_{q_1}^{\Gamma,\Gamma} = C_{q_2}^{\Gamma,\Gamma}$, then

$$\int_\Omega (q_1 - q_2) u_1 u_2 \, dx = 0$$

for any $u_j \in H_\Delta(\Omega)$ with $(-\Delta + q_j)u_j = 0$ in Ω and $\mathrm{supp}(u_j|_{\partial\Omega}) \subset \Gamma$. In [**DKSjU09**], the authors consider a linearization of this assumption at $q_1 = q_2 = 0$. The main result of [**DKSjU09**] is:

THEOREM 2.6. *Let Ω be a bounded connected open subset of $\mathbb{R}^n$, $n \geq 2$, with connected C^∞ boundary. The set of products of harmonic functions in $C^\infty(\overline{\Omega})$ which vanish on a closed proper subset $F \subset \partial\Omega$ is dense in $L^1(\Omega)$.*

While the proof of this theorem is quite different from the ones of the other results mentioned here, it does depend on the linear limiting Carleman weights $\varphi(x) = \alpha \cdot x$ and the associated complex geometrical optics solutions $e^{\zeta \cdot x}$, where $\zeta \in \mathbb{C}^n$ satisfies $\zeta \cdot \zeta = 0$, or equivalently $\zeta = \tau(\alpha + i\beta)$ where $\alpha, \beta \in \mathbb{R}^n$ are unit vectors satisfying $\alpha \cdot \beta = 0$. See Section 7 for a sketch of the proof of this result.

2.5. **The results of [KS12].** The partial data results in [**KS12**] are valid on a class of Riemannian manifolds of dimension $n \geq 3$, but here we will only mention some consequences for domains in $\mathbb{R}^3$. In all of the results below $\Omega \subset \mathbb{R}^3$ is a bounded open connected set with C^∞ boundary, and we consider the decomposition

$$\partial\Omega = \partial\Omega_+(\varphi) \cup \partial\Omega_-(\varphi) \cup \partial\Omega_{\mathrm{tan}}(\varphi)$$

with respect to a fixed limiting Carleman weight φ. We will also assume that $\partial\Omega_{\mathrm{tan}}$ is decomposed as

$$\partial\Omega_{\mathrm{tan}} = \Gamma_a \cup \Gamma_i$$

where Γ_a is an open subset of $\partial\Omega_{\mathrm{tan}}$ that is accessible to boundary measurements, and Γ_i is the inaccessible part. Further, we will assume that Γ_D and Γ_N are nonempty open sets in $\partial\Omega$ with $\partial\Omega_- \cup \Gamma_a \subset \Gamma_D$ and $\partial\Omega_+ \cup \Gamma_a \subset \Gamma_N$.

The content of the results below is that if the inaccessible part Γ_i satisfies a geometric condition, then Dirichlet measurements on $\partial\Omega_- \cup \Gamma_a$ and Neumann measurements on $\partial\Omega_+ \cup \Gamma_a$ are sufficient to determine the unknown coefficient locally away from Γ_i.

The first theorem corresponds to the case $\varphi(x) = x_1$ and the case where the inaccessible part Γ_i is part of a cylindrical set. This result generalizes Theorem 2.2 (if one chooses $E = \partial\Omega_0$ and $\Gamma_i = \emptyset$) and Theorem 2.3 (if one chooses Ω_0 and E so that $\Omega \cap \{x_3 = 0\} \subset \mathbb{R} \times (\partial\Omega_0 \setminus E)$).

THEOREM 2.7. *Suppose that $\Omega \subset \mathbb{R} \times \Omega_0$ where Ω_0 is a bounded open set with C^∞ boundary in $\mathbb{R}^2$. Let $\varphi(x) = x_1$, and let E be an open subset of $\partial\Omega_0$ such that Γ_i satisfies*

$$\Gamma_i \subset \mathbb{R} \times (\partial\Omega_0 \setminus E).$$

If $q_1, q_2 \in C(\overline{\Omega})$ and if

$$C_{q_1}^{\Gamma_D,\Gamma_N} = C_{q_2}^{\Gamma_D,\Gamma_N},$$

then $q_1 = q_2$ in $\overline{\Omega} \cap (\mathbb{R} \times O)$ where O is the intersection of $\overline{\Omega}_0$ with the union of all lines in $\mathbb{R}^2$ that have $\partial\Omega_0 \setminus E$ on one side.

Note that if Ω_0 is convex and E is connected, then the set O in the previous theorem is just the convex hull of E in $\overline{\Omega}_0$ and one recovers the potential in $\overline{\Omega} \cap (\mathbb{R} \times \mathrm{ch}_{\overline{\Omega}_0}(E))$. The next theorem is related to the logarithmic weight and generalizes Theorem 2.1 (choose $E = \partial\Omega_0$ and $\Gamma_i = \emptyset$). The inaccessible part of the boundary must now be part of a conical set.

THEOREM 2.8. *Let $\overline{\Omega} \subset \{x_3 > 0\}$ and let $\varphi(x) = \log|x|$. Consider the hemisphere $S^2_+ = \{\omega \in S^2 \,;\, \omega_3 > 0\}$, and let M_0 be a compact subdomain of S^2_+ with C^∞ boundary. Let E be an open subset of ∂M_0 such that Γ_i satisfies*

$$\Gamma_i \subset \{r\omega \,;\, r > 0, \omega \in \partial M_0 \setminus E\}.$$

If $q_1, q_2 \in C(\overline{\Omega})$ and if

$$C_{q_1}^{\Gamma_D,\Gamma_N} = C_{q_2}^{\Gamma_D,\Gamma_N},$$

then $q_1 = q_2$ in $\overline{\Omega} \cap \{r\omega\,;\, r > 0, \omega \in O\}$ where O is the union of all great circle segments in S^2_+ such that $\partial M_0 \setminus E$ is on one side of the hyperplane containing the great circle segment.

The final theorem involves the weight $\varphi(x) = \arg(x_1 + ix_2)$ and does not have a counterpart in the previous results. It corresponds to a case where the inaccessible part of the boundary is part of a surface of revolution.

THEOREM 2.9. *Let $\Omega \subset \{x_1 > 0\}$ and let $\varphi(x) = \arg(x_1 + ix_2)$. Let $S = \{(x_1, 0, x_3)\,;\, x_1 > 0\}$ be a half-plane with hyperbolic geodesics given by the half-circles (with $R > 0$ and $\alpha \in \mathbb{R}$)*

$$(x_1(t), x_2(t), x_3(t)) = (R\sin t, 0, R\cos t + \alpha), \quad t \in (0, \pi)$$

or the lines

$$(x_1(t), x_2(t), x_3(t)) = (t, 0, \alpha), \quad t > 0.$$

Let M_0 be a compact subdomain of S with smooth boundary, and let E be an open subset of ∂M_0 such that Γ_i satisfies

$$\Gamma_i \subset \{R_\theta(\partial M_0 \setminus E)\,;\, \theta \in (-\pi, \pi)\}$$

where $R_\theta x = (\tilde{R}_\theta(x_1, x_2), x_3)$ and $\tilde{R}_\theta$ rotates vectors in $\mathbb{R}^2$ by angle θ counterclockwise. That is, we assume that the inaccessible part Γ_i is contained in a surface of revolution obtained by rotating the boundary curve $\partial M_0 \setminus E$ about the x_3-axis.

If $q_1, q_2 \in C(\overline{\Omega})$ and if

$$C_{q_1}^{\Gamma_D,\Gamma_N} = C_{q_2}^{\Gamma_D,\Gamma_N},$$

then $q_1 = q_2$ in $\overline{\Omega} \cap \{R_\theta(O)\,;\, \theta \in (-\pi, \pi)\}$ where O is the union of all geodesics in S that have $\partial M_0 \setminus E$ on one side.

Note that the above results are local results that allow to determine the unknown potential locally away from the inaccessible part Γ_i. The paper [**KS12**] also contains results where one obtains information on the potential near the inaccessible part Γ_i, but this information comes in the form of integrals along broken geodesic rays in the transversal manifold. In the Euclidean case, a continuous curve $\gamma : [0, L] \to \overline{\Omega}_0$ is called a *nontangential broken ray* if γ is obtained by following straight lines that are reflected in the standard way (angle of incidence = angle of reflection) whenever they hit $\partial\Omega_0$, all reflections are nontangential, and all reflection points are distinct. One also needs a somewhat stronger assumption on the sets Γ_D and Γ_N (this assumption was made for simplicity and could be removed in many cases).

Let us state a counterpart of Theorem 2.7 which involves the broken ray transform with exponential attenuation. Note that in the absence of the stronger assumption on Γ_D and Γ_N, this result would generalize Theorem 2.7.

THEOREM 2.10. *Suppose that $\Omega \subset \mathbb{R} \times \Omega_0$ where Ω_0 is a bounded open set with C^∞ boundary in $\mathbb{R}^2$. Let $\varphi(x) = x_1$, and let E be an open subset of $\partial\Omega_0$ such that Γ_i satisfies*

$$\Gamma_i \subset \mathbb{R} \times (\partial\Omega_0 \setminus E).$$

Let $q_1, q_2 \in C(\overline{\Omega})$, let Γ be a neighborhood of $\overline{\partial\Omega_+ \cup \partial\Omega_- \cup \Gamma_a}$, and assume that

$$C_{q_1}^{\Gamma,\Gamma} = C_{q_2}^{\Gamma,\Gamma}.$$

Then for any nontangential broken ray $\gamma : [0, L] \to \overline{\Omega}_0$ *with endpoints on* E*, and given any real number* λ*, one has*

$$\int_0^L e^{-2\lambda t}(q_1 - q_2)\hat{\ }(2\lambda, \gamma(t))\, dt = 0.$$

Here $(\,\cdot\,)\hat{\ }$ *is the Fourier transform in the* x_1 *variable, and* $q_1 - q_2$ *is extended by zero to* $\mathbb{R}^3 \setminus \overline{\Omega}$.

It follows from this result that one could recover the unknown potential also near Γ_i if one knew how to invert the broken ray transform, that is, to recover a function in $\overline{\Omega}_0$ from its (attenuated) integrals over nontangential broken rays with endpoints on a given open subset $E \subset \partial\Omega_0$. This is a question of independent interest and there seem to be only partial results in this direction, see [**Es04**], [**Hu13**], [**Il12**], [**Il13**].

Finally, we mention that the method for proving the above theorems also allows to reduce the overlap for Γ_D and Γ_N needed in [**KSU07**]. An example of such a result is the following (a similar result was also proved in [**IY13b**]).

THEOREM 2.11. *Let* $q_1, q_2 \in C(M)$. *If* φ *is a linear function and if* $\partial\Omega_{tan}$ *has zero measure in* $\partial\Omega$*, then*

$$C_{q_1}^{\partial\Omega_-, \partial\Omega_+} = C_{q_2}^{\partial\Omega_-, \partial\Omega_+} \implies q_1 = q_2.$$

3. Strategy of proof

In this section we give an outline of the proof of the partial data results described above. The proofs proceed in three steps:

1. Via an integral identity, the partial data uniqueness question is reduced to showing that products of solutions of Schrödinger equations, which vanish on suitable parts of the boundary, are dense in $L^1(\Omega)$.
2. Construction of a family of special complex geometrical optics solutions to Schrödinger equations, and showing that certain transforms of any function that is L^2-orthogonal to products of complex geometrical optics solutions must vanish.
3. Showing that the transforms arising in Step 2 are injective, which completes the proof that products of solutions are dense and also the uniqueness proof.

This general outline roughly applies to all of the results in Section 2, but the particulars (the choice of complex geometrical optics solutions, and the transforms in the final step) vary from case to case.

3.1. Reduction to the density of products of solutions.

This step is achieved by the following basic integral identity.

LEMMA 3.1. *If* $\Gamma_D, \Gamma_N \subset \partial\Omega$ *are open and if* $C_{q_1}^{\Gamma_D, \Gamma_N} = C_{q_2}^{\Gamma_D, \Gamma_N}$*, then*

$$\int_\Omega (q_1 - q_2) u_1 u_2 \, dx = 0$$

for any $u_j \in H_\Delta(\Omega)$ *satisfying* $(-\Delta + q_j)u_j = 0$ *in* Ω *and*

$$\operatorname{supp}(u_1|_{\partial\Omega}) \subset \Gamma_D, \quad \operatorname{supp}(u_2|_{\partial\Omega}) \subset \Gamma_N.$$

PROOF. Let u_j be as stated. Since $C_{q_1}^{\Gamma_D,\Gamma_N} = C_{q_2}^{\Gamma_D,\Gamma_N}$, there is a function $\tilde{u}_2 \in H_\Delta(\Omega)$ with $(-\Delta + q_2)\tilde{u}_2 = 0$ in Ω, $\mathrm{supp}(\tilde{u}_2|_{\partial\Omega}) \subset \Gamma_D$, and

$$(u_1|_{\Gamma_D}, \partial_\nu u_1|_{\Gamma_N}) = (\tilde{u}_2|_{\Gamma_D}, \partial_\nu \tilde{u}_2|_{\Gamma_N}).$$

Using that u_1, u_2 and $\tilde{u}_2$ are solutions, we have

$$\begin{aligned}\int_\Omega (q_1 - q_2)u_1 u_2 \, dx &= \int_\Omega [(\Delta u_1)u_2 - u_1(\Delta u_2)] \, dx \\ &= \int_\Omega [(\Delta(u_1 - \tilde{u}_2))u_2 - (u_1 - \tilde{u}_2)(\Delta u_2)] \, dx.\end{aligned}$$

Now $u_1 - \tilde{u}_2|_{\partial\Omega} = 0$, so in fact $u_1 - \tilde{u}_2 \in H^2(\Omega)$ by the properties of the space $H_\Delta(\Omega)$, see [**BU01**]. Recall also that $C^\infty(\overline{\Omega})$ is dense in $H_\Delta(\Omega)$ and that $u_2|_{\partial\Omega} \in H^{-1/2}(\partial\Omega)$ and $\partial_\nu u_2|_{\partial\Omega} \in H^{-3/2}(\partial\Omega)$. These facts make it possible to integrate by parts, and we obtain that

$$\int_\Omega (q_1 - q_2)u_1 u_2 \, dx = \int_{\partial\Omega} [(\partial_\nu(u_1 - \tilde{u}_2))u_2 - (u_1 - \tilde{u}_2)(\partial_\nu u_2)] \, dS$$

in the weak sense. The last expression vanishes since $\partial_\nu(u_1 - \tilde{u}_2)|_{\Gamma_N} = 0$ and $\mathrm{supp}(u_2|_{\partial\Omega}) \subset \Gamma_N$. □

The next statement makes precise the reduction of the partial data problem to density of products.

COROLLARY 3.2. *Let $\Omega \subset \mathbb{R}^n$ be a bounded open set with C^∞ boundary, let $q_1, q_2 \in L^\infty(\Omega)$, and let Γ_D and Γ_N be nonempty open subsets of $\partial\Omega$. If the set*

$$\begin{aligned}\{u_1 u_2 \; ; \; u_j \in H_\Delta(\Omega), \quad (-\Delta + q_j)u_j = 0 \text{ in } \Omega, \\ \mathrm{supp}(u_1|_{\partial\Omega}) \subset \Gamma_D, \quad \mathrm{supp}(u_2|_{\partial\Omega}) \subset \Gamma_N\}\end{aligned}$$

is dense in $L^1(\Omega)$, then the condition $C_{q_1}^{\Gamma_D,\Gamma_N} = C_{q_2}^{\Gamma_D,\Gamma_N}$ implies $q_1 = q_2$.

3.2. **Complex geometrical optics solutions.** To show the density claim in Corollary 3.2, one constructs special solutions $u_j \in H_\Delta(\Omega)$ of the equations $(-\Delta + q_j)u_j = 0$ in Ω with the required boundary conditions. This will involve complex geometrical optics solutions having the form

$$\begin{aligned}u_1 &= e^{-\tau\varphi}(m_1 + r_1), \\ u_2 &= e^{\tau\varphi}(m_2 + r_2)\end{aligned}$$

where $\tau > 0$ is a large parameter, φ is a real valued weight function, m_1 and m_2 are complex amplitudes, and r_1 and r_2 are correction terms that are small when τ is large. The solutions u_1 and u_2 are chosen with opposite signs for the exponential ($e^{-\tau\varphi}$ and $e^{\tau\varphi}$) to make sure that these exponentials go away in the product, allowing one to consider the limit of $u_1 u_2$ as $\tau \to \infty$.

The point is to choose the amplitudes m_1, m_2 so that $\Delta(e^{-\tau\varphi}m_1) \approx 0$ and $\Delta(e^{\tau\varphi}m_2) \approx 0$ in a suitable sense, ensuring that the approximate solutions $e^{-\tau\varphi}m_1$ and $e^{\tau\varphi}m_2$ can be corrected to exact solutions of $(-\Delta + q_j)u_j = 0$. The complex amplitudes often involve phase factors $e^{i\tau\psi_j}$, so the solutions contain complex exponentials $e^{\tau(-\varphi+i\psi_1)}$ and $e^{\tau(\varphi+i\psi_2)}$. This explains why these are called complex geometrical optics solutions.

The choice of the weight function φ is crucial in this process. The correction terms r_1 and r_2 are obtained by solving the equations

$$e^{\tau\varphi}(-\Delta + q_1)(e^{-\tau\varphi} r_1) = f_1 \quad \text{in } \Omega,$$
$$e^{-\tau\varphi}(-\Delta + q_2)(e^{\tau\varphi} r_2) = f_2 \quad \text{in } \Omega$$

where

$$f_1 = -e^{\tau\varphi}(-\Delta + q_1)(e^{-\tau\varphi} m_1),$$
$$f_2 = -e^{-\tau\varphi}(-\Delta + q_2)(e^{\tau\varphi} m_2).$$

Thus one would like to have solvability results for the conjugated operators $e^{\tau\varphi}(-\Delta + q_1)e^{-\tau\varphi}$ and $e^{-\tau\varphi}(-\Delta + q_2)e^{\tau\varphi}$ that come with estimates such as

$$\|r_j\|_{L^2(\Omega)} \leq C\tau^{-1}\|f_j\|_{L^2(\Omega)},$$

for a constant C uniform over all sufficiently large τ. Additionally, one needs some control of the boundary values of r_1 and r_2 on parts of the boundary.

The above procedure can be carried out if φ is a *limiting Carleman weight.* We refer to [**KSU07**], [**DKSaU09**] for more information about the microlocal condition characterizing these weights that ensures the right kind of solvability results. However, as discussed in Section 2, limiting Carleman weights are just harmonic functions (with some restriction on their critical points) if $n = 2$, and in $\mathbb{R}^n$ with $n \geq 3$ there are only six possibilities up to translation and scaling. Moreover, as will be explained in Section 4 below, one can expect relatively good control of the boundary values of r_1 on $\partial\Omega_+(\varphi)$ and weak control on $\partial\Omega_{\mathrm{tan}}(\varphi)$, and similarly relatively good control of the boundary values of r_2 on $\partial\Omega_-(\varphi)$ and weak control on $\partial\Omega_{\mathrm{tan}}(\varphi)$ (this is due to the sign in the exponential $e^{\mp\tau\varphi}$). This will follow from a Carleman estimate (a weighted L^2 estimate with exponential weights depending on the large parameter τ) with boundary terms.

It remains to describe the construction of the amplitudes m_1 and m_2. The following three conditions are typical ones that one might like the amplitudes to satisfy:

1. $\|e^{\tau\varphi}(-\Delta + q_1)(e^{-\tau\varphi} m_1)\|_{L^2(\Omega)} = \|e^{-\tau\varphi}(-\Delta + q_2)(e^{\tau\varphi} m_2)\|_{L^2(\Omega)} = O(1)$ as $\tau \to \infty$,
2. $\|m_1\|_{L^2(\partial\Omega_+(\varphi))} = \|m_2\|_{L^2(\partial\Omega_-(\varphi))} = O(1)$, and $\|m_j\|_{L^2(\partial\Omega_{\mathrm{tan}}(\varphi))} = o(1)$ as $\tau \to \infty$,
3. the set of limits $\{\lim_{\tau\to\infty} m_1 m_2\}$ for all such m_1 and m_2 is a dense set in $L^1(\Omega)$.

The first condition means that $e^{-\tau\varphi} m_1$ and $e^{\tau\varphi} m_2$ are approximate solutions in a suitable sense, allowing to find correction terms r_j with $\|r_j\|_{L^2(\Omega)} = o(1)$ as $\tau \to \infty$. The second condition comes from the fact that one wants roughly that

$$u_1|_{\partial\Omega_+ \cup \partial\Omega_{\mathrm{tan}}} = 0, \qquad u_2|_{\partial\Omega_- \cup \partial\Omega_{\mathrm{tan}}} = 0,$$

or equivalently

$$r_1|_{\partial\Omega_+ \cup \partial\Omega_{\mathrm{tan}}} = -m_1|_{\partial\Omega_+ \cup \partial\Omega_{\mathrm{tan}}}, \qquad r_2|_{\partial\Omega_- \cup \partial\Omega_{\mathrm{tan}}} = -m_2|_{\partial\Omega_- \cup \partial\Omega_{\mathrm{tan}}}.$$

Since the Carleman estimate with boundary terms will give relatively good control of the boundary values of r_j on $\partial\Omega_+$ or $\partial\Omega_-$ and weak control on $\partial\Omega_{\mathrm{tan}}$, it is enough that m_j are bounded on $\partial\Omega_+$ or $\partial\Omega_-$ but they need to be small as $\tau \to \infty$ (or vanish completely) on $\partial\Omega_{\mathrm{tan}}$. The third condition follows since

$$\lim_{\tau\to\infty} u_1 u_2 = \lim_{\tau\to\infty} m_1 m_2.$$

Thus, to show density of the products $u_1 u_2$, it is enough to show density of the set $\{\lim_{\tau\to\infty} m_1 m_2\}$ for all admissible m_1 and m_2. This leads to the transforms considered in the final step.

3.3. **Injectivity of transforms.** In the density of products approach, one needs to show that any function f in, say, $C(\overline{\Omega})$, for which

$$\int_\Omega f u_1 u_2 \, dx = 0$$

for all solutions $u_j \in H_\Delta(\Omega)$ of $(-\Delta + q_j)u_j = 0$ in Ω, must satisfy $f = 0$. By the discussion in this section, one can try to do this by constructing complex geometrical optics solutions with amplitudes $m_j = m_j(\,\cdot\,;\tau,\alpha)$ depending on the large parameter τ and some additional parameters described by α which ranges over some parameter set A. Assume that for each α, there is a bounded measure μ_α on Ω such that

$$\lim_{\tau\to\infty} m_1(\,\cdot\,;\tau,\alpha) m_2(\,\cdot\,;\tau,\alpha) = \mu_\alpha$$

in the weak topology of measures in Ω. The orthogonality condition then implies that

$$\int_\Omega f \, d\mu_\alpha = 0, \quad \alpha \in A.$$

That is, the transform Tf of f vanishes, where

$$Tf(\alpha) = \int_\Omega f \, d\mu_\alpha, \quad \alpha \in A.$$

If this transform is injective, then $f = 0$ and the density of products follows.

Many different transforms have appeared in this connection. In the full data result of [**SU87**], the transform T was just the usual Fourier transform. In the partial data result [**BU01**] one obtained the Fourier transform in a cone, and also Isakov's approach [**Is07**] results in the Fourier transform. The partial data result [**KSU07**] involves a more complicated transform whose injectivity was proved by analytic microlocal analysis, and this argument was simplified in [**DKSjU07**] where matters reduce to inverting the two-plane Radon transform. In the two-dimensional case, the transform appearing [**Bu08**] and [**IUY10**] is related to stationary phase. Finally, in [**DKSaU09**], [**KS12**] a mixed transform involving the Fourier transform in one direction and X-ray transforms (or more generally attenuated geodesic ray transforms) in other directions was employed.

4. Carleman estimates

As discussed in Section 3, the first step in the construction of complex geometrical optics solutions with controlled boundary behaviour is a Carleman estimate with boundary terms. Here we give the proof of such an estimate for the conjugated operator $e^{\varphi/h}(-\Delta+q)e^{-\varphi/h}$ in Ω, where the limiting Carleman weight φ is a linear function and $h > 0$ is a small parameter (in the previous notation, $h = 1/\tau$). Following [**KSU07**], it is useful to consider a slightly modified weight

$$\varphi_\varepsilon = \varphi + h f_\varepsilon$$

where f_ε is a smooth real valued function in $\overline{\Omega}$ depending on a small parameter ε but independent of h. The convexity of f_ε will lead to improved lower bounds in terms of the $L^2(\Omega)$ norms of u and ∇u. On the other hand, the sign of $\partial_\nu \varphi_\varepsilon$ in the boundary term of the Carleman estimate will allow to control functions on different parts of the boundary. Of special interest is the set $\partial\Omega_{\mathrm{tan}}$ where $\partial_\nu \varphi = 0$, so one has

$$\partial_\nu \varphi_\varepsilon = h \partial_\nu f_\varepsilon \quad \text{on } \partial\Omega_{\mathrm{tan}}.$$

We would like to have $\partial_\nu f_\varepsilon < 0$ on $\partial\Omega_{\mathrm{tan}}$. It is not easy to find a global convex function f_ε satisfying the last condition for a general set $\partial\Omega_{\mathrm{tan}}$. However, splitting f_ε to a convex part whose normal derivative vanishes on $\partial\Omega_{\mathrm{tan}}$ and another part which ensures the correct sign on $\partial\Omega_{\mathrm{tan}}$ will give the required result. This second part can be inserted in the weight if one arranges enough convexity (ε sufficiently small).

We will use the notation $D = -i\nabla$, $(u,v) = (u,v)_{L^2(\Omega)}$, $\|u\| = \|u\|_{L^2(\Omega)}$, and also L^2 inner products and norms on parts of $\partial\Omega$ are indicated by a subscript.

PROPOSITION 4.1. *Let $\Omega \subset \mathbb{R}^n$, $n \geq 2$, be a bounded open set with C^∞ boundary, let $\varphi(x) = \alpha \cdot x$ where $\alpha \in \mathbb{R}^n$ is a unit vector, and let κ be a smooth real valued function in $\overline{\Omega}$ so that $\partial_\nu \kappa = -1$ on $\partial\Omega$. Let also $q \in L^\infty(\Omega)$. There are constants $\varepsilon, C_0, h_0 > 0$ with $h_0 \leq \varepsilon \leq 1$ such that for the weight*

$$\varphi_\varepsilon = \varphi + \frac{h}{\varepsilon}\frac{\varphi^2}{2} + h\kappa$$

where $0 < h \leq h_0$, one has

$$\frac{h^3}{C_0}(|\partial_\nu \varphi_\varepsilon| \partial_\nu u, \partial_\nu u)_{\partial\Omega_-(\varphi_\varepsilon)} + \frac{h^2}{C_0}(\|u\|^2 + \|hDu\|^2) \leq \|e^{\varphi/h}((hD)^2 + h^2 q)(e^{-\varphi/h}u)\|^2 + h^3(|\partial_\nu \varphi_\varepsilon| \partial_\nu u, \partial_\nu u)_{\partial\Omega_+(\varphi_\varepsilon)}$$

for any $u \in C^\infty(\overline{\Omega})$ with $u|_{\partial\Omega} = 0$.

PROOF. It is sufficient to consider the case $\alpha = e_1$, so $\varphi(x) = x_1$ and

$$\varphi_\varepsilon = x_1 + \frac{h}{\varepsilon}\frac{x_1^2}{2} + h\kappa.$$

Let $P_{\varphi_\varepsilon} = e^{\varphi_\varepsilon/h}(hD)^2 e^{-\varphi_\varepsilon/h}$, and write

$$P_{\varphi_\varepsilon} = A + iB$$

where A and B are the formally self-adjoint operators

$$A = (hD)^2 - |\nabla\varphi_\varepsilon|^2,$$

$$B = \nabla\varphi_\varepsilon \circ hD + hD \circ \nabla\varphi_\varepsilon = 2\nabla\varphi_\varepsilon \cdot hD + \frac{h}{i}\Delta\varphi_\varepsilon.$$

Then, using the condition $u|_{\partial\Omega} = 0$,

$$\begin{aligned}\|P_{\varphi_\varepsilon} u\|^2 &= ((A+iB)u, (A+iB)u) \\ &= \|Au\|^2 + \|Bu\|^2 + i(Bu, Au) - i(Au, Bu) \\ &= \|Au\|^2 + \|Bu\|^2 + (i[A,B]u, u) - ih^2(Bu, \partial_\nu u)_{\partial\Omega} \\ &= \|Au\|^2 + \|Bu\|^2 + (i[A,B]u, u) - 2h^3((\partial_\nu\varphi_\varepsilon)\partial_\nu u, \partial_\nu u)_{\partial\Omega}.\end{aligned}$$

We compute the commutator

$$\begin{aligned} i[A,B]u &= h\big[((hD)^2-|\nabla\varphi_\varepsilon|^2)(2\nabla\varphi_\varepsilon\cdot\nabla u+(\Delta\varphi_\varepsilon)u) \\ &\quad -(2\nabla\varphi_\varepsilon\cdot\nabla+\Delta\varphi_\varepsilon)((hD)^2u-|\nabla\varphi_\varepsilon|^2u)\big] \\ &= h\big[2\nabla(hD)^2\varphi_\varepsilon\cdot\nabla u+4hD\sum_{k=1}^n\partial_k\varphi_\varepsilon\cdot hD\partial_k u+((hD)^2\Delta\varphi_\varepsilon)u \\ &\quad +2hD\Delta\varphi_\varepsilon\cdot hDu+2\nabla\varphi_\varepsilon\cdot\nabla(|\nabla\varphi_\varepsilon|^2)u\big] \\ &= h\big[4(\varphi_\varepsilon''\nabla\varphi_\varepsilon\cdot\nabla\varphi_\varepsilon)u-4h^2\sum_{j,k=1}^n\partial_{jk}\varphi_\varepsilon\partial_{jk}u-4h^2\nabla\Delta\varphi_\varepsilon\cdot\nabla u \\ &\quad -h^2(\Delta^2\varphi_\varepsilon)u\big]. \end{aligned}$$

Here φ_ε'' is the Hessian matrix of φ_ε. Integrating by parts once, using again that $u|_{\partial\Omega}=0$, yields

$$\begin{aligned} (i[A,B]u,u) = 4h^3(\varphi_\varepsilon''\nabla u,\nabla u)+4h((\varphi_\varepsilon''\nabla\varphi_\varepsilon\cdot\nabla\varphi_\varepsilon)u,u) \\ -h^3((\Delta^2\varphi_\varepsilon)u,u). \end{aligned}$$

Now we use that φ is a linear function, so that $\varphi_\varepsilon''=\frac{h}{\varepsilon}e_1\otimes e_1+h\kappa''$. Combining this with the previous computation for $\|P_{\varphi_\varepsilon}u\|^2$, we obtain

$$\begin{aligned} \|P_{\varphi_\varepsilon}u\|^2 &= \|Au\|^2+\|Bu\|^2+\frac{4h^4}{\varepsilon}\|\partial_1 u\|^2+4h^4(\kappa''\nabla u,\nabla u) \\ &\quad +\frac{4h^2}{\varepsilon}\|(\partial_1\varphi_\varepsilon)u\|^2+4h^2((\kappa''\nabla\varphi_\varepsilon\cdot\nabla\varphi_\varepsilon)u,u)-h^4((\Delta^2\kappa)u,u) \\ &\quad -2h^3((\partial_\nu\varphi_\varepsilon)\partial_\nu u,\partial_\nu u)_{\partial\Omega}. \end{aligned}$$

We have $\nabla\varphi_\varepsilon=\nabla\varphi+h(\frac{\varphi\nabla\varphi}{\varepsilon}+\nabla\kappa)$. Assume that h_0 is so small that

$$\left|h\left(\frac{\varphi\nabla\varphi}{\varepsilon}+\nabla\kappa\right)\right|\le 1/2 \text{ on } \overline{\Omega} \text{ when } 0<h<h_0.$$

Then $1/2\le|\nabla\varphi_\varepsilon|\le 3/2$ on $\overline{\Omega}$, and for some constant $C_\kappa\ge 1$

$$\begin{aligned} \|P_{\varphi_\varepsilon}u\|^2 &\ge \|Au\|^2+\|Bu\|^2+\frac{4h^2}{\varepsilon}\|hD_1u\|^2-C_\kappa h^2\|hDu\|^2 \\ &\quad +\frac{h^2}{\varepsilon}\|u\|^2-C_\kappa h^2\|u\|^2-C_\kappa h^4\|u\|^2-2h^3((\partial_\nu\varphi_\varepsilon)\partial_\nu u,\partial_\nu u)_{\partial\Omega}. \end{aligned}$$

We obtain a further lower bound for $\|hDu\|^2$ from the term $\|Au\|^2$: given any $M>0$, and using that $u|_{\partial\Omega}=0$, one has

$$\begin{aligned} h^2\|hDu\|^2 &= h^2((hD)^2u,u)=h^2(Au,u)+h^2(|\nabla\varphi_\varepsilon|^2u,u) \\ &\le \frac{1}{2M}\|Au\|^2+\frac{Mh^4}{2}\|u\|^2+4h^2\|u\|^2. \end{aligned}$$

From this lower bound for $\|Au\|^2$ and the trivial estimate $\|Bu\|^2 \geq 0$, it follows that

$$\begin{aligned}\|P_{\varphi_\varepsilon} u\|^2 \geq 2Mh^2\|hDu\|^2 - M^2h^4\|u\|^2 - 8Mh^2\|u\|^2 + \frac{4h^2}{\varepsilon}\|hD_1 u\|^2 \\ - C_\kappa h^2\|hDu\|^2 + \frac{h^2}{\varepsilon}\|u\|^2 - C_\kappa h^2\|u\|^2 - C_\kappa h^4\|u\|^2 \\ - 2h^3((\partial_\nu \varphi_\varepsilon)\partial_\nu u, \partial_\nu u)_{\partial\Omega}.\end{aligned}$$

Choosing $M = (16\varepsilon)^{-1}$, choosing ε sufficiently small depending on C_κ and then choosing h_0 sufficiently small depending on ε implies that

$$\|P_{\varphi_\varepsilon} u\|^2 \geq \frac{h^2}{16\varepsilon}(\|u\|^2 + \|hDu\|^2) - 2h^3((\partial_\nu \varphi_\varepsilon)\partial_\nu u, \partial_\nu u)_{\partial\Omega}$$

when $0 < h < h_0$. By further decreasing h_0 we may add a potential $q \in L^\infty(\Omega)$ to get the estimate

$$2\|(P_{\varphi_\varepsilon} + h^2 q)u\|^2 \geq \frac{h^2}{100\varepsilon}(\|u\|^2 + \|hDu\|^2) - 2h^3((\partial_\nu \varphi_\varepsilon)\partial_\nu u, \partial_\nu u)_{\partial\Omega}$$

Finally, we replace u by $e^{x_1^2/2\varepsilon+\kappa}u$, where $u \in C^\infty(\overline{\Omega})$ and $u|_{\partial\Omega} = 0$, and use the fact that

$$1/C \leq e^{x_1^2/2\varepsilon+\kappa} \leq C \quad \text{on } \overline{\Omega}.$$

The required estimate follows. □

We now pass from φ_ε to φ in the boundary terms of the previous result, making use of the special properties of φ_ε on $\partial\Omega$. Note that the factor h^4 in the boundary term on $\{x \in \partial\Omega\,;\, -\delta < \partial_\nu\varphi(x) < h/3\}$ below is weaker than the factor h^3 in the other boundary terms. This follows from the fact that $\partial_\nu\varphi_\varepsilon = h\partial_\nu\kappa = -h$ in the set where $\partial_\nu\varphi$ vanishes, so one only has weak control near $\partial\Omega_{\tan}(\varphi)$.

Proposition 4.2. *Let $q \in L^\infty(\Omega)$, and let $\varphi(x) = \pm x_1$. There exist constants $C_0, h_0 > 0$ such that whenever $0 < h \leq h_0$ and $\delta > 0$, one has*

$$\begin{aligned}\frac{\delta h^3}{C_0}\|\partial_\nu u\|^2_{L^2(\{\partial_\nu\varphi \leq -\delta\})} + \frac{h^4}{C_0}\|\partial_\nu u\|^2_{L^2(\{-\delta<\partial_\nu\varphi<h/3\})} + \frac{h^2}{C_0}(\|u\|^2 + \|h\nabla u\|^2) \\ \leq \|e^{\varphi/h}(-h^2\Delta + h^2 q)(e^{-\varphi/h}u)\|^2 + h^3\|\partial_\nu u\|^2_{L^2(\{\partial_\nu\varphi \geq h/3\})}\end{aligned}$$

for any $u \in C^\infty(\overline{\Omega})$ with $u|_{\partial\Omega} = 0$.

Proof. Note that

$$\partial_\nu\varphi_\varepsilon = \left(1 + \frac{h}{\varepsilon}\varphi\right)\partial_\nu\varphi + h\partial_\nu\kappa = \left(1 + \frac{h}{\varepsilon}\varphi\right)\partial_\nu\varphi - h.$$

We choose h_0 so small that whenever $h \leq h_0$, one has for $x \in \overline{\Omega}$

$$\frac{1}{2} \leq 1 + \frac{h}{\varepsilon}\varphi(x) \leq \frac{3}{2}.$$

On the set where $\partial_\nu\varphi(x) \leq -\delta$, we have

$$|\partial_\nu\varphi_\varepsilon| \geq \delta/2.$$

If $-\delta < \partial_\nu\varphi < h/3$, we use the estimate

$$|\partial_\nu\varphi_\varepsilon| \geq h/2.$$

Moreover, $|\partial_\nu \varphi_\varepsilon| \leq C_0$ on $\partial\Omega$. Since $\{\partial_\nu \varphi < h/3\} \subset \{\partial_\nu \varphi_\varepsilon < 0\}$ and $\{\partial_\nu \varphi_\varepsilon \geq 0\} \subset \{\partial_\nu \varphi \geq h/3\}$, the result follows from Proposition 4.1 after replacing C_0 by some larger constant. □

We can now obtain a solvability result from the previous Carleman estimate in a standard way by duality (see [**BU01**, **KSU07**, **NS10**]). There is a slight technical complication since the solution will be in L^2 but not in H^1. To remedy this we will work with the space

$$H_\Delta(\Omega) = \{u \in L^2(\Omega)\,;\, \Delta u \in L^2(\Omega)\}$$

with norm $\|u\|_{H_\Delta} = \|u\|_{L^2} + \|\Delta u\|_{L^2}$. It is known (see [**BU01**]) that $H_\Delta(\Omega)$ is a Hilbert space having $C^\infty(\overline{\Omega})$ as a dense subset, and there is a well defined bounded trace operator from $H_\Delta(\Omega)$ to $H^{-1/2}(\partial\Omega)$ and a normal derivative operator from $H_\Delta(\Omega)$ to $H^{-3/2}(\partial\Omega)$. We also recall that if $u \in H_\Delta(\Omega)$ and $u|_{\partial\Omega} \in H^{3/2}(\partial\Omega)$, then $u \in H^2(\Omega)$.

PROPOSITION 4.3. *Let $\Omega \subset \mathbb{R}^n$ be a bounded open set with C^∞ boundary, let $q \in L^\infty(\Omega)$, and let $\varphi(x) = \pm x_1$. There exist constants $C_0, \tau_0 > 0$ such that when $\tau \geq \tau_0$ and $\delta > 0$, then for any $f \in L^2(\Omega)$ and $f_- \in L^2(S_- \cup S_0)$ there exists $u \in L^2(\Omega)$ satisfying $e^{\tau\varphi}u \in H_\Delta(\Omega)$ and $e^{\tau\varphi}u|_{\partial\Omega} \in L^2(\partial\Omega)$ such that*

$$e^{-\tau\varphi}(-\Delta + q)(e^{\tau\varphi}u) = f \quad \text{in } \Omega, \qquad e^{\tau\varphi}u|_{S_-\cup S_0} = e^{\tau\varphi}f_-,$$

and

$$\|u\|_{L^2(M)} \leq C_0(\tau^{-1}\|f\|_{L^2(\Omega)} + (\delta\tau)^{-1/2}\|f_-|_{S_-}\|_{L^2(S_-)} + \|f_-|_{S_0}\|_{L^2(S_0)}).$$

Here $S_\pm$ and S_0 are the following subsets of $\partial\Omega$:

$$S_- = \{\partial_\nu\varphi \leq -\delta\}, S_0 = \{-\delta < \partial_\nu\varphi < 1/(3\tau)\}, S_+ = \{\partial_\nu\varphi \geq 1/(3\tau)\}.$$

PROOF. Write $Lv = e^{\tau\varphi}(-\Delta + \bar{q})(e^{-\tau\varphi}v)$ and $\tau = 1/h$, $\tau_0 = 1/h_0$. We rewrite the Carleman estimate of Proposition 4.2 as

$$\begin{aligned}(\delta\tau)^{1/2}\|\partial_\nu v\|_{L^2(S_-)} + \|\partial_\nu v\|_{L^2(S_0)} + \tau\|v\| + \|\nabla v\| \\ \leq C_0\|Lv\| + C_0\tau^{1/2}\|\partial_\nu v\|_{L^2(S_+)}.\end{aligned}$$

This is valid for any $\delta > 0$, provided that $\tau \geq \tau_0$ and $v \in C^\infty(\overline{\Omega})$ with $v|_{\partial\Omega} = 0$.

Consider the following subspace of $L^2(\Omega) \times L^2(S_+)$:

$$X = \{(Lv, \partial_\nu v|_{S_+})\,;\, v \in C^\infty(\overline{\Omega}),\ v|_{\partial\Omega} = 0\}.$$

Any element of X is uniquely represented as $(Lv, \partial_\nu v|_{S_+})$ where $v|_{\partial\Omega} = 0$ by the Carleman estimate. Define a linear functional $l : X \to \mathbb{C}$ by

$$l(Lv, \partial_\nu v|_{S_+}) = (v, f)_{L^2(\Omega)} - (\partial_\nu v, f_-)_{L^2(S_-\cup S_0)}.$$

By the Carleman estimate, we have

$$\begin{aligned}|l(Lv, \partial_\nu v|_{S_+})| &\leq \|v\|\|f\| + \|\partial_\nu v\|_{L^2(S_-)}\|f_-\|_{L^2(S_-)} \\ &\quad + \|\partial_\nu v\|_{L^2(S_0)}\|f_-\|_{L^2(S_0)} \\ &\leq C_0(\tau^{-1}\|f\| + (\delta\tau)^{-1/2}\|f_-\|_{L^2(S_-)} + \|f_-\|_{L^2(S_0)}) \\ &\quad \times (\|Lv\| + \tau^{1/2}\|\partial_\nu v\|_{L^2(S_+)}).\end{aligned}$$

The Hahn-Banach theorem implies that l extends to a continuous linear functional $\bar{l} : L^2(\Omega) \times \tau^{-1/2}L^2(S_+) \to \mathbb{C}$ such that

$$\|\bar{l}\| \leq C_0(\tau^{-1}\|f\| + (\delta\tau)^{-1/2}\|f_-\|_{L^2(S_-)} + \|f_-\|_{L^2(S_0)}).$$

By the Riesz representation theorem, there exist functions $u \in L^2(\Omega)$ and $u_+ \in L^2(S_+)$ satisfying $\bar{l}(w, w_+) = (w, u)_{L^2(\Omega)} + (w_+, u_+)_{L^2(S_+)}$. Moreover,

$$\|u\|_{L^2(\Omega)} + \tau^{-1/2}\|u_+\|_{L^2(S_+)} \leq C_0(\tau^{-1}\|f\| + (\delta\tau)^{-1/2}\|f_-\|_{L^2(S_-)} + \|f_-\|_{L^2(S_0)}).$$

If $v \in C^\infty(\overline{\Omega})$ and $v|_{\partial\Omega} = 0$, we have

$$(Lv, u)_{L^2(\Omega)} + (\partial_\nu v, u_+)_{L^2(S_+)} = (v, f)_{L^2(\Omega)} - (\partial_\nu v, f_-)_{L^2(S_- \cup S_0)}. \tag{4.1}$$

Choosing v compactly supported in Ω, it follows that $L^* u = f$, or

$$e^{-\tau\varphi}(-\Delta + q)(e^{\tau\varphi}u) = f \quad \text{in } \Omega.$$

Furthermore, $e^{\tau\varphi}u \in H_\Delta(\Omega)$.

If $w, v \in C^\infty(\overline{\Omega})$ with $v|_{\partial\Omega} = 0$, an integration by parts gives

$$(Lv, w) = -(e^{-\tau\varphi}\partial_\nu v, e^{\tau\varphi}w)_{L^2(\partial\Omega)} + (v, L^* w).$$

Given our solution u, we choose $u_j \in C^\infty(\overline{\Omega})$ so that $e^{\tau\varphi}u_j \to e^{\tau\varphi}u$ in $H_\Delta(\Omega)$. Applying the above formula with $w = u_j$ and taking the limit, we see that

$$(Lv, u) = -(e^{-\tau\varphi}\partial_\nu v, e^{\tau\varphi}u)_{L^2(\partial\Omega)} + (v, L^* u)$$

for $v \in C^\infty(\overline{\Omega})$ with $v|_{\partial\Omega} = 0$. Combining this with (4.1), using that $L^* u = f$, gives

$$(\partial_\nu v, f_-)_{L^2(S_- \cup S_0)} + (\partial_\nu v, u_+)_{L^2(S_+)} = (e^{-\tau\varphi}\partial_\nu v, e^{\tau\varphi}u)_{L^2(\partial\Omega)}.$$

Since $\partial_\nu v$ can be chosen arbitrarily, it follows that

$$e^{\tau\varphi}u|_{S_- \cup S_0} = e^{\tau\varphi}f_-, \quad e^{\tau\varphi}u|_{S_+} = e^{\tau\varphi}u_+.$$

We also see that $e^{\tau\varphi}u|_{\partial\Omega} \in L^2(\partial\Omega)$. □

5. Complex geometrical optics

We will now describe a construction of CGO solutions that is relevant for the proof of Theorem 2.7. Suppose that $\Omega \subset \mathbb{R} \times \Omega_0$ where Ω_0 is a bounded open set with C^∞ boundary in $\mathbb{R}^2$. Let $\varphi(x) = x_1$, and let E be an open subset of $\partial\Omega_0$ such that Γ_i satisfies

$$\Gamma_i \subset \mathbb{R} \times (\partial\Omega_0 \setminus E).$$

Let also $q \in L^\infty(\Omega)$. We wish to construction a solution $u \in H_\Delta(\Omega)$ of the equation

$$(-\Delta + q)u = 0 \text{ in } \Omega$$

where

$$u = e^{-sx_1}(m + r_0)$$

and s is a *slightly complex large frequency,*

$$s = \tau + i\lambda$$

where $\tau > 0$ (eventually $\tau \to \infty$) and $\lambda \in \mathbb{R}$ is fixed. The use of a slightly complex frequency instead of a real frequency allows to introduce another real parameter λ

in the CGO solutions, which makes it possible to employ the Fourier transform in the x_1 variable.

Inserting the ansatz for u in the equation, we need to solve

$$e^{sx_1}(-\Delta+q)(e^{-sx_1}r_0) = f \text{ in } \Omega$$

where

$$f = -e^{sx_1}(-\Delta+q)(e^{-sx_1}m) = -(-\Delta+2s\partial_1 - s^2+q)m.$$

It will be useful to look for an amplitude m independent of x_1, so $m = m(x')$ where $x = (x_1, x')$ and $x' = (x_2, x_3) \in \mathbb{R}^2$. Then f has the simpler form

$$f = (-\Delta_{x'} - s^2 + q)m.$$

We seek for an amplitude $m \in C^2(\overline{\Omega}_0)$ satisfying

$$\|(-\Delta_{x'} - s^2)m\|_{L^2(\Omega_0)} = O(1), \quad \|m\|_{L^2(\Omega_0)} = O(1),$$

and the boundary values should satisfy

$$\|m\|_{L^2(E)} = O(1), \quad \|m\|_{L^2(\partial\Omega_0\setminus E)} = o(1)$$

as $\tau \to \infty$. These conditions have been chosen to be compatible with Proposition 4.3, and they can be interpreted so that m should be an *approximate eigenfunction*, or *quasimode*, of the Laplacian in Ω_0 with frequency s. In fact, we will arrange so that

$$m|_{\partial\Omega_0\setminus E} = 0.$$

If we can find such an m then Proposition 4.3 together with the fact that the inaccessible part Γ_i satisfies $\Gamma_i \subset \mathbb{R} \times (\partial\Omega_0 \setminus E)$ will allow to find a correction term r with $\|r\|_{L^2(\Omega)} = o(1)$ as $\tau \to \infty$.

Let us now describe one construction of such an m. We choose a straight line γ in $\mathbb{R}^2$ that intersects $\overline{\Omega}_0$ but does not meet $\partial\Omega_0 \setminus E$, and will construct an amplitude m that concentrates on γ. We look for m in the form

$$m(x') = e^{is\psi(x')}a(x').$$

A computation gives that

$$(-\Delta_{x'} - s^2)m = e^{is\psi}\left[s^2(|\nabla\psi|^2 - 1)a - is(2\nabla\psi\cdot\nabla a + (\Delta_{x'}\psi)a) - \Delta_{x'}a\right].$$

It is enough to choose ψ and a so that the following equations are valid in Ω_0:

$$|\nabla\psi|^2 = 1, \qquad 2\nabla\psi\cdot\nabla a + (\Delta_{x'}\psi)a = 0.$$

The first equation is an eikonal equation, and distance functions are solutions. Choose a ball $\hat{\Omega}_0$ with $\Omega_0 \subset\subset \hat{\Omega}_0$, and choose some point $x_0' \in \hat{\Omega}_0 \setminus \overline{\Omega}_0$ that lies the line γ. Define

$$\psi(x') = |x' - x_0'|, \quad x' \in \overline{\Omega}_0.$$

Then $\psi \in C^\infty(\overline{\Omega}_0)$ satisfies $|\nabla\psi| = 1$. The second equation above is a transport equation for a, and has the solution

$$a(r,\theta) = r^{-1/2}b(\theta)$$

where (r,θ) are polar coordinates in $\mathbb{R}^2$ with center at x_0', and b is any function in $C^\infty(S^1)$. (Note that $\psi(r,\theta) = r$ in these coordinates.) The point is that if the line γ is given in the (r,θ) coordinates by $r \mapsto (r,\theta_0)$, then choosing b supported very close to θ_0 and independent of τ will result in the boundary conditions

$$\|m\|_{L^2(E)} = O(1), \quad m|_{\partial\Omega_0\setminus E} = 0$$

as $\tau \to \infty$.

Combining the above amplitude construction with Proposition 4.3 results in the following existence result for CGO solutions. (See [**KS12**, Section 6] for the full details of the proof.)

PROPOSITION 5.1. *Suppose that $\Omega \subset \mathbb{R} \times \Omega_0$ where Ω_0 is a bounded open set with C^∞ boundary in $\mathbb{R}^2$. Let $\varphi(x) = \pm x_1$ and decompose $\partial\Omega_{\tan}(\varphi)$ as $\Gamma_a \cup \Gamma_i$ where the closed set Γ_i satisfies, for some open subset E of $\partial\Omega_0$,*

$$\Gamma_i \subset \mathbb{R} \times (\partial\Omega_0 \setminus E).$$

Let also $q \in L^\infty(\Omega)$. Choose a ball $\hat{\Omega}_0$ with $\Omega_0 \subset\subset \hat{\Omega}_0$, let $x_0' \in \hat{\Omega}_0 \setminus \overline{\Omega}_0$, and let (r,θ) be polar coordinates in $\mathbb{R}^2$ with center at x_0'. Let $\theta_0 \in S^1$ be such that the line $r \mapsto (r,\theta_0)$ meets $\overline{\Omega}_0$ but not $\partial\Omega_0 \setminus E$. There exists a solution $u \in H_\Delta(\Omega)$ of the equation

$$(-\Delta + q)u = 0 \quad \text{in } \Omega$$

having the form

$$u = e^{-s\varphi}(e^{isr}r^{-1/2}b(\theta) + r_0)$$

where $b \in C^\infty(S^1)$ is supported very close to θ_0,

$$\|r_0\|_{L^2(\Omega)} = o(1)$$

as $\tau \to \infty$, and

$$\mathrm{supp}(u|_{\partial\Omega}) \subset \Gamma$$

if Γ is any open set in $\partial\Omega$ with $\partial\Omega_\mp(\varphi) \cup \Gamma_a \subset \Gamma$.

6. Uniqueness results

We now describe how to complete the outline given in Section 3 and prove Theorem 2.7 using an injectivity result for a certain transform (in the end of the section we discuss briefly the proofs of Theorems 2.8–2.10). If f is a piecewise continuous compactly supported function in $\mathbb{R}^3$, the relevant transform of f is given by

$$Tf(\lambda,\gamma) = \int_{-\infty}^{\infty} e^{-2\lambda t}\left[\int_{-\infty}^{\infty} e^{-2i\lambda x_1} f(x_1,\gamma(t))\,dx_1\right]dt,$$

where $\lambda \in \mathbb{R}$ and γ is a line in $\mathbb{R}^2$. That is, we are taking the Fourier transform of f in the x_1 variable with frequency 2λ and then taking the attenuated X-ray transform, with constant attenuation -2λ, along lines in $\mathbb{R}^2_{x'}$. We choose the parametrization $\gamma(t) = \sigma\omega_\perp + t\omega$, where $\omega \in S^1$ is the direction vector of the line γ, $\omega_\perp$ is the counterclockwise rotation of ω by 90 degrees, and σ is the signed distance to the origin. (The choice of parametrization will not be too important below.)

PROPOSITION 6.1. *Suppose that $\Omega \subset \mathbb{R} \times \Omega_0$ where Ω_0 is a bounded open set with C^∞ boundary in $\mathbb{R}^2$. Let $\varphi(x) = x_1$, and let E be an open subset of $\partial\Omega_0$ such that Γ_i satisfies*

$$\Gamma_i \subset \mathbb{R} \times (\partial\Omega_0 \setminus E).$$

If $q_1, q_2 \in C(\overline{\Omega})$, if $\partial\Omega_- \cup \Gamma_a \subset \Gamma_D$ and $\partial\Omega_+ \cup \Gamma_a \subset \Gamma_N$, and if

$$C_{q_1}^{\Gamma_D,\Gamma_N} = C_{q_2}^{\Gamma_D,\Gamma_N},$$

then

$$T(q_1 - q_2)(\lambda,\gamma) = 0$$

for all $\lambda \in \mathbb{R}$ and for any line γ in $\mathbb{R}^2$ that does not meet $\partial\Omega_0 \setminus E$. (Here, we extend $q_1 - q_2$ by zero to $\mathbb{R}^2$.)

PROOF. By Lemma 3.1, our assumptions imply that

$$\int_\Omega (q_1 - q_2) u_1 \bar{u}_2 \, dx = 0$$

for any $u_j \in H_\Delta(\Omega)$ satisfying $(-\Delta + q_1)u_1 = (-\Delta + \bar{q}_2)u_2 = 0$ in Ω and

$$\mathrm{supp}(u_1|_{\partial\Omega}) \subset \Gamma_D, \quad \mathrm{supp}(u_2|_{\partial\Omega}) \subset \Gamma_N.$$

Fix $\lambda \in \mathbb{R}$, let $s = \tau + i\lambda$ where $\tau > 0$ is sufficiently large, and fix a line γ in $\mathbb{R}^2$ that does not meet $\partial\Omega_0 \setminus E$. We use Proposition 5.1 to find solutions u_1 and u_2 satisfying the above conditions and having the form

$$u_1 = e^{-sx_1}(e^{isr} r^{-1/2} b(\theta) + r_1),$$
$$u_2 = e^{sx_1}(e^{isr} r^{-1/2} b(\theta) + r_2)$$

where (r, θ) are polar normal coordinates in $\mathbb{R}^2$ whose center is outside of Ω_0 and such that the line γ is given by $r \mapsto (r, \theta_0)$, $b \in C^\infty(S^1)$ is independent of τ and supported near θ_0, and

$$\|r_j\| = o(1) \text{ as } \tau \to \infty.$$

We then have

$$\begin{aligned} 0 &= \lim_{\tau\to\infty} \int_\Omega (q_1 - q_2) u_1 \bar{u}_2 \, dx \\ &= \int_{-\infty}^{\infty} \int_0^\infty \int_{S^1} (q_1 - q_2)(x_1, r, \theta) e^{-2i\lambda x_1} e^{-2\lambda r} |b(\theta)|^2 \, d\theta \, dr \, dx_1 \end{aligned}$$

Here we have extended $q_1 - q_2$ by zero outside Ω. The result follows by choosing b to approximate a delta function at θ_0. □

The uniqueness result now follows from the *Helgason support theorem* for the X-ray transform [**He99**], which states that if f is a piecewise continuous compactly supported function in $\mathbb{R}^2$ that integrates to zero over any line that does not meet a compact convex set $K \subset \mathbb{R}^2$, then $f|_{\mathbb{R}^2 \setminus K} = 0$. The idea of reducing the attenuated X-ray transform to the unattenuated one comes from [**DKLS13**].

PROOF OF THEOREM 2.7. Denote by f the extension of $q_1 - q_2$ by zero to $\mathbb{R}^2$, and let

$$\hat{f}(\xi_1, x') = \int_{-\infty}^{\infty} e^{-ix_1\xi_1} f(x_1, x') \, dx_1.$$

Let also K be the convex hull of $\partial\Omega_0 \setminus E$ in $\mathbb{R}^2$. By Proposition 6.1, it is enough to show that if

$$Tf(\lambda, \gamma) = 0$$

for all $\lambda \in \mathbb{R}$ and for any line γ that does not meet K, then $f(x_1, x') = 0$ whenever $x_1 \in \mathbb{R}$ and $x' \in \mathbb{R}^2 \setminus K$.

The condition $Tf(\lambda, \gamma) = 0$ implies that

$$\int_{-\infty}^{\infty} e^{-2\lambda t} \hat{f}(2\lambda, \gamma(t)) \, dt = 0$$

for any line γ that does not meet K. Setting $\lambda = 0$, the Helgason support theorem implies that $\hat{f}(0, x') = 0$ for $x' \in \mathbb{R}^2 \setminus K$. We now differentiate the above identity with respect to λ and set $\lambda = 0$ to obtain

$$\int_{-\infty}^{\infty} \left[(-2t)\hat{f}(0, \gamma(t)) + 2\frac{\partial \hat{f}}{\partial \xi_1}(0, \gamma(t)) \right] dt = 0.$$

But we already saw that $\hat{f}(0, x') = 0$ for $x' \in \mathbb{R}^2 \setminus K$. The Helgason support theorem then gives that

$$\frac{\partial \hat{f}}{\partial \xi_1}(0, x') = 0 \text{ for } x' \in \mathbb{R}^2 \setminus K.$$

Repeating this argument shows that

$$\left(\frac{\partial}{\partial \xi_1} \right)^k \hat{f}(0, x') = 0 \text{ for } x' \in \mathbb{R}^2 \setminus K$$

for all $k \geq 0$. But $x_1 \mapsto f(x_1, x')$ is compactly supported, hence $\xi_1 \mapsto \hat{f}(\xi_1, x')$ is analytic, and we obtain that $\hat{f}(\xi_1, x') = 0$ for all $\xi_1 \in \mathbb{R}$ and $x' \in \mathbb{R}^2 \setminus K$. The result follows upon inverting the Fourier transform in x_1. □

Theorems 2.8 and 2.9 can be proved by the same general argument as above. However, the fact that one has nonlinear limiting Carleman weights (the log weight and arg weight) leads, after a suitable conformal scaling, to a situation where the original domain $\overline{\Omega}$ with Euclidean metric is replaced by a compact Riemannian manifold (M, g) with smooth boundary, and the condition

$$\Omega \subset \mathbb{R} \times \Omega_0$$

is replaced by the condition

$$(M, g) \subset (\mathbb{R} \times M_0, g), \quad g = c(e \oplus g_0)$$

where c is a positive function (conformal factor), and the transversal domain $\overline{\Omega}_0$ with Euclidean metric is replaced by a transversal manifold (M_0, g_0) that is compact with smooth boundary. Also the integrals over lines in $\overline{\Omega}_0$ are replaced by integrals over geodesics in (M_0, g_0). This setup is similar to the results for the anisotropic Calderón problem in [**DKSaU09**], [**DKLS13**]. For the log weight, (M_0, g_0) will be a domain in the hemisphere, and for the arg weight (M_0, g_0) will be a domain in hyperbolic space. Correspondingly, one uses geodesics in the sphere and in hyperbolic space. In both cases, if (M_0, g_0) is chosen to have strictly convex boundary, then it is a simple manifold with real-analytic metric and analogues of the Helgason support theorem are available. The full proofs are given in [**KS12**].

The proof of Theorem 2.10 requires a more general complex geometrical optics construction than the one described above. The reason is that one wants to get information on integrals along broken rays that may touch the inaccessible part Γ_i, and one then needs amplitudes in the complex geometrical optics solutions that concentrate along these broken rays and are very small on the inaccessible part. Such amplitudes may obtained from a Gaussian beam quasimode construction. This construction in connection with complex geometrical optics solutions is employed in [**DKLS13**] for non-reflected geodesics (full data case) and in [**KS12**] for reflected geodesics (partial data case), and we refer to these papers for the details.

7. The linearized case

Here we sketch the proof of Theorem 2.6. A first reduction is to show that it suffices to prove a local version of the theorem (see [**DKSjU09**, Section 2]). The global version follows from the local one by using ideas in the spirit of the Runge approximation theorem, developed in an unpublished work of Alessandrini, Isozaki, and Uhlmann. Thus, matters reduce to proving the following "local" version.

THEOREM 7.1. *Let $\Omega \subset \mathbb{R}^n$, $n \geq 2$, be a bounded connected open set with C^∞ boundary. Let $x_0 \in \partial\Omega$ and let F be the complement of some open boundary neighborhood of x_0. There exists $\delta > 0$ such that given any $f \in L^\infty(\Omega)$, if we have the cancellation property*

$$\int_\Omega f u_1 u_2 \, dx = 0$$

for all $u_j \in C^\infty(\overline{\Omega})$ with $\Delta u_j = 0$ in Ω and $u_j|_F = 0$, then $f = 0$ in $B(x_0, \delta) \cap \Omega$.

The next step is to use conformal transformations (in particular Kelvin transforms) of harmonic functions to reduce to the following situation: $x_0 = 0$, $\Omega \subset \{x \in \mathbb{R}^n \,;\, |x + e_1| < 1\}$ where $e_1 = (1, 0, \ldots, 0)$ is the first coordinate vector, the tangent hyperplane to 0 is given by $\{x_1 = 0\}$, and $F \subset \{x \in \partial\Omega \,;\, x_1 \leq -2c\}$ for some $c > 0$. See [**DKSjU09**, Section 3] for this reduction.

From this point on, the proof is inspired by the proof of Kashiwara's "watermelon theorem" in analytic microlocal analysis. We introduce the Segal-Bargmann transform of a function $f \in L^\infty(\mathbb{R}^n)$ with compact support by

$$Tf(z) = \int_{\mathbb{R}^n} e^{-\frac{1}{2h}(z-y)^2} f(y) \, dy, \quad z = x + i\xi \in \mathbb{C}^n.$$

The Segal-Bargmann transform is related to the microlocal analysis of analytic singularities of a distribution. We mention the a priori exponential bound

$$|Tf(z)| \leq (2\pi h)^{n/2} e^{\frac{1}{2h}|\mathrm{Im}\, z|^2} \|f\|_{L^\infty} \tag{7.1}$$

and the fact that if f is supported in $\{x_1 \leq 0\}$, we can improve this to

$$|Tf(z)| \leq (2\pi h)^{n/2} e^{\frac{1}{2h}|\mathrm{Im}\, z|^2 - |\mathrm{Re}\, z_1|^2} \|f\|_{L^\infty} \tag{7.2}$$

when $\mathrm{Re}\, z_1 \geq 0$. Both of these bounds are straightforward. Note also that when we multiply by $(2\pi)^{-n/2}$, when $z = x \in \mathbb{R}^n$, we obtain the Gaussian heat kernel, and hence if f has compact support we have

$$\lim_{h \to 0} (2\pi h)^{-n/2} Tf(x) = f(x) \quad \text{for a.e. } x \in \mathbb{R}^n. \tag{7.3}$$

The strategy is to use the cancellation property (we extend f to be 0 outside of Ω) to obtain an exponential decay of Tf when $x \in \mathbb{R}^n$ is close to 0, thus yielding the vanishing of f near 0, as desired. In order to accomplish this, let us temporarily consider the case $n = 2$ and define $\gamma = ie_1 + e_2 = (i, 1) \in \mathbb{C}^2$. Note that $\{\gamma, \bar{\gamma}\}$ is a basis of $\mathbb{C}^2$, and that $\{\zeta \in \mathbb{C}^2 \,;\, \zeta \cdot \zeta = 0\}$ is the union of two complex lines $\mathbb{C}\gamma \cup \mathbb{C}\overline{\gamma}$. Hence, it is easy to see if $\varepsilon > 0$ is small enough, then any $z \in \mathbb{C}^2$ with $|z - 2ie_1| < 2\varepsilon$ may be decomposed as a sum of the form

$$z = \zeta + \eta, \quad \zeta \cdot \zeta = \eta \cdot \eta = 0, \tag{7.4}$$

where $|\zeta - \gamma| \leq C\varepsilon$, $|\eta + \bar{\gamma}| \leq C\varepsilon$. This last fact extends to $\mathbb{C}^n$, $n \geq 2$, see [**DKSjU09**, Section 3].

Recall that the exponentials $e^{-i\frac{\zeta\cdot x}{h}}$, $\zeta \in \mathbb{C}^n$ $\zeta\cdot\zeta = 0$, are harmonic functions. We need to modify these by adding correction terms to obtain harmonic functions u satisfying the boundary requirement $u|_F = 0$. Let $\chi \in C_c^\infty(\mathbb{R}^n)$ be a cutoff function which is supported in $\{x_1 \leq -c\}$ and equals 1 on $\{x_1 \leq -2c\}$. Consider the solution $w = w(\,\cdot\,;\zeta)$ of the Dirichlet problem

$$\Delta w = 0 \text{ in } \Omega, \quad w|_{\partial\Omega} = -e^{-i\frac{\zeta\cdot x}{h}}\chi|_{\partial\Omega}.$$

We have the following bound on w:

$$\begin{aligned}\|w\|_{H^1(\Omega)} &\leq C_1\|e^{-i\frac{\zeta\cdot x}{h}}\chi\|_{H^{1/2}(\partial\Omega)} \\ &\leq C_2(1+h^{-1}|\zeta|)^{1/2}e^{-c\frac{\operatorname{Im}\zeta_1}{h}}e^{\frac{1}{h}|\operatorname{Im}\zeta'|}\end{aligned} \tag{7.5}$$

where $\operatorname{Im}\zeta_1 \geq 0$, from the choice of χ and our normalization of Ω.

We have the cancellation property

$$\int_\Omega f(x)u(x,\zeta)u(x,\eta)\,dx = 0, \quad \zeta\cdot\zeta = \eta\cdot\eta = 0,$$

where $u(x,\zeta) = e^{-i\frac{\zeta\cdot x}{h}} + w(x,\zeta)$, which is a harmonic function in $C^\infty(\overline{\Omega})$ and satisfies $u|_F = 0$. The identity above can be rewritten as

$$\begin{aligned}\int_\Omega f(x)e^{-\frac{i}{h}x\cdot(\zeta+\eta)}\,dx = &-\int_\Omega f(x)e^{-\frac{i}{h}x\cdot\zeta}w(x,\eta)\,dx \\ &-\int_\Omega f(x)e^{-\frac{i}{h}x\cdot\eta}w(x,\zeta)\,dx - \int_\Omega f(x)w(x,\zeta)w(x,\eta)\,dx.\end{aligned}$$

Using the estimate (7.5), we obtain

$$\begin{aligned}\left|\int_\Omega f(x)e^{-\frac{i}{h}x\cdot(\zeta+\eta)}\,dx\right| &\leq C_3\|f\|_{L^\infty}(1+h^{-1}|\eta|)^{1/2}(1+h^{-1}|\zeta|)^{-1/2} \\ &\quad\times e^{-\frac{c}{h}\min\{\operatorname{Im}\zeta_1,\operatorname{Im}\eta_1\}}e^{\frac{1}{h}(|\operatorname{Im}\zeta'|+|\operatorname{Im}\eta'|)},\end{aligned}$$

when $\operatorname{Im}\zeta_1 \geq 0$, $\operatorname{Im}\eta_1 \geq 0$, and $\zeta\cdot\zeta = \eta\cdot\eta = 0$. In particular, if $|\zeta - a\gamma| \leq C\varepsilon a$, $|\eta + a\bar{\gamma}| < C\varepsilon a$, $\varepsilon \leq 1/(2C)$, then

$$\left|\int_\Omega f(x)e^{-\frac{i}{h}x\cdot(\zeta+\eta)}\,dx\right| \leq C_4h^{-1}\|f\|_{L^\infty(\Omega)}e^{-\frac{ca}{2h}}e^{\frac{2C\varepsilon a}{h}}.$$

Hence if $z \in \mathbb{C}^n$ and $|z - 2iae_1| < 2\varepsilon a$, ε small enough, using a rescaled version of the decomposition 7.4 gives

$$\left|\int_\Omega f(x)e^{-\frac{i}{h}x\cdot z}\,dx\right| \leq C_4h^{-1}\|f\|_{L^\infty(\Omega)}e^{-\frac{ca}{h}}e^{\frac{2C\varepsilon a}{h}}. \tag{7.6}$$

To relate the last estimate to the given estimates on the Segal-Bargmann transform of f, we use the well-known formula

$$e^{-\frac{1}{2h}(z-y)^2} = e^{-\frac{z^2}{2h}}(2\pi h)^{-n/2}\int e^{-\frac{t^2}{2h}}e^{-\frac{i}{h}y\cdot(t+iz)}\,dt$$

which gives

$$Tf(z) = (2\pi h)^{-n/2}\int\int e^{-\frac{1}{2h}(z^2+t^2)}e^{-\frac{i}{h}y\cdot(t+iz)}f(y)\,dt\,dy.$$

For our f, supported in Ω and verifying the cancellation of the integral in our hypothesis, the estimate (7.6) and the formula above allow us to improve the estimate (7.2). If $\operatorname{Re} z_1 \geq 0$, then

$$|Tf(z)| \leq (2\pi h)^{-n/2} \int e^{\frac{1}{2h}(|\operatorname{Im} z|^2 - |\operatorname{Re} z|^2 - t^2)} \left| \int e^{-\frac{i}{h} y\cdot(t+iz)} \, dy \right| dt$$
$$\leq \frac{e^{\frac{1}{2h}(|\operatorname{Im} z|^2 - |\operatorname{Re} z|^2)}}{(2\pi h)^{n/2}} \left[\int_{|t|\leq \varepsilon a} e^{-\frac{t^2}{2h}} \left| \int e^{-\frac{i}{h} y\cdot(t+iz)} f(y)\, dy \right| dt + \int_{|t|\geq \varepsilon a} e^{-\frac{t^2}{2h}} \left| \int e^{-\frac{i}{h} y\cdot(t+iz)} f(y)\, dy \right| dt \right]$$

Using (7.6) with z replaced by $t+iz$, when $|z - 2ae_1| < \varepsilon a$ and $|t| \leq \varepsilon a$ we obtain

$$\text{(7.7)} \quad |Tf(z)| \leq e^{\frac{1}{2h}(|\operatorname{Im} z|^2 - |\operatorname{Re} z|^2)} \left[\sup_{|t|\leq \varepsilon a} \left| \int e^{-\frac{i}{h} y\cdot(t+x)} f(y)\, dy \right| + \sqrt{2} e^{\frac{1}{h}|\operatorname{Re} z'|} e^{-\frac{\varepsilon^2 a^2}{4h}} \int |f(y)|\, dy \right]$$
$$\leq C_5 h^{-1} \|f\|_{L^\infty(\Omega)} e^{\frac{1}{2h}(|\operatorname{Im} z|^2 - |\operatorname{Re} z|^2)} \left[e^{-\frac{ac}{2h}} e^{\frac{2C\varepsilon a}{h}} + e^{-\frac{\varepsilon^2 a^2}{4h}} e^{\frac{\varepsilon a}{h}} \right],$$

provided that $|z - 2ae_1| < \varepsilon a$. Choosing $\varepsilon < c/(8C)$ and $a > (c + 4\varepsilon)/(\varepsilon^2)$, we obtain the bound

$$|Tf(z)| \leq 2C_5 h^{-1} \|f\|_{L^\infty(\Omega)} e^{\frac{1}{2h}(|\operatorname{Im} z|^2 - |\operatorname{Re} z|^2 - \frac{ca}{2})}.$$

Combining (7.1), (7.2) and (7.6) we have

$$e^{-\frac{\Phi(z_1)}{2h}} |Tf(z_1, x')| \leq \frac{C}{h} \|f\|_{L^\infty(\Omega)} \begin{cases} 1, & z_1 \in \mathbb{C}, \\ e^{-\frac{ca}{4h}}, & |z_1 - 2a| \leq \frac{\varepsilon a}{2}, |x'| < \frac{\varepsilon a}{2} \end{cases}$$

where $x' \in \mathbb{R}^{n-1}$ and

$$\Phi(z_1) = \begin{cases} (\operatorname{Im} z_1)^2, & \text{when } \operatorname{Re} z_1 \leq 0, \\ (\operatorname{Im} z_1)^2 - (\operatorname{Re} z_1)^2, & \text{when } \operatorname{Re} z_1 \geq 0. \end{cases}$$

Now, inspired by the proof of the "watermelon theorem", we use the following:

LEMMA 7.2. *Let $b, L > 0$. Let F be an entire function in $\mathbb{C}$ such that*

$$e^{-\frac{\Phi(s)}{h}} |F(s)| \leq \begin{cases} 1, & s \in \mathbb{C}, \\ e^{-\frac{c}{2h}}, & \text{when } |s - L| \leq b. \end{cases}$$

Then for all $r \geq 0$, there exist $c', \delta > 0$ (independent of F) such that

$$|F(s)| \leq e^{-\frac{c'}{2h} + \frac{(\operatorname{Im} s)^2}{2h}}$$

when $|\operatorname{Re} s| \leq \delta$ and $|\operatorname{Im} s| \leq r$.

The proof of the lemma rests on a harmonic majorization argument, which exploits the subharmonicity of $-(\operatorname{Im} s)^2 + (\operatorname{Re} s)^2$ (see [**DKSjU09**, Lemma 4.1 and Remark 4.2] for the details of the proof). Next we apply Lemma 7.2 to

$$F(s) = \frac{hTf(s, x')}{C\|f\|_{L^\infty(\Omega)}}.$$

We obtain in particular that

$$|Tf(x)| \leq Ch^{-1} \|f\|_{L^\infty(\Omega)} e^{-\frac{c'}{2h}}, \quad x \in \Omega, |x_1| < \delta$$

for δ small. Multiplying by $(2\pi h)^{-n/2}$ and letting $h \to 0$, using (7.3), we deduce that $f(x) = 0$ for $x \in \Omega$, $-\delta \leq x_1 \leq 0$. The completes the proof of Theorem 7.1 and hence of Theorem 2.6.

8. Open problems

The following is a list of some questions related to the partial data problem that are open, as far as we know, at the time of writing this. The first question concerns the local data problem for $n \geq 3$. (As discussed above, this result for $n = 2$ is known at least for sufficiently regular coefficients.)

Question 1. (Local data in dimensions $n \geq 3$) If Ω is a bounded domain in $\mathbb{R}^n$, $n \geq 3$, if Γ is an arbitrary nonempty open subset of $\partial\Omega$, and if $q_1, q_2 \in L^\infty(\Omega)$, show that $C_{q_1}^{\Gamma,\Gamma} = C_{q_2}^{\Gamma,\Gamma}$ implies $q_1 = q_2$.

The next question concerns data on disjoint sets. The general case is open even for the linearized problem in any dimension. Partial results for $n = 2$ are given in [**IUY11b**].

Question 2. (Data on disjoint sets in dimensions $n \geq 2$) If Ω is a bounded domain in $\mathbb{R}^n$, $n \geq 2$, if Γ_D and Γ_N are arbitrary disjoint open subsets of $\partial\Omega$, and if $q_1, q_2 \in L^\infty(\Omega)$, show that $C_{q_1}^{\Gamma_D,\Gamma_N} = C_{q_2}^{\Gamma_D,\Gamma_N}$ implies $q_1 = q_2$.

One can also ask for optimal regularity conditions for the coefficients in partial data results. If $n = 2$ the full data result is known for L^∞ conductivities [**AP06**].

Question 3. (Local data for nonsmooth conductivities) If Ω is a bounded domain in $\mathbb{R}^2$, if Γ is an arbitrary nonempty open subset of $\partial\Omega$, and if $\gamma_1, \gamma_2 \in L^\infty(\Omega)$ satisfy $\gamma_1, \gamma_2 \geq c > 0$, show that $C_{\gamma_1}^{\Gamma,\Gamma} = C_{\gamma_2}^{\Gamma,\Gamma}$ implies $\gamma_1 = \gamma_2$.

The next question concerns optimal stability for partial data results. In the full data case it is known that in general one has a logarithmic modulus of continuity for determining an unknown coefficient from boundary measurements, and this result is optimal. We refer to the survey [**Al07**]. Stability results for partial boundary measurements based on the reflection approach [**HW07**], [**Ca11**] and in the case where the coefficient is known near the boundary [**AK12**] also have logarithmic stability. However, available results for the Carleman estimate approach [**HW06**], [**Tz08**], [**CDR12**] seem to involve weaker moduli of continuity (log log or worse), and one can ask if logarithmic stability still holds.

Question 4. (Optimal stability for the Carleman estimate approach) What is the optimal stability for the partial data uniqueness result of [**KSU07**]?

Finally, there are many open questions related to partial data for elliptic systems. In the introduction we mentioned results for systems when $n = 2$ and for the reflection approach when $n \geq 3$, but it seems that there are no partial data results for systems using the Carleman estimate approach in $n \geq 3$ except for [**ST10**]. As an example, one can consider the time-harmonic Maxwell equations for which the full data result is known [**OPS93**].

Question 5. (Carleman estimate approach for systems) Prove an analogue of the partial data result of [**KSU07**] for the time-harmonic Maxwell equations with scalar coefficients as in [**OPS93**].

References

[AGTU11] P. Albin, C. Guillarmou, L. Tzou, G. Uhlmann, *Inverse boundary problems for systems in two dimensions*, Ann. Inst. H. Poincaré (to appear), arXiv:1105.4565.

[Al07] G. Alessandrini, *Open issues of stability for the inverse conductivity problem*, J. Inverse Ill-Posed Probl. **15** (2007), no. 5, 451–460, DOI 10.1515/jiip.2007.025. MR2367859 (2008k:35483)

[AK12] Giovanni Alessandrini and Kyoungsun Kim, *Single-logarithmic stability for the Calderón problem with local data*, J. Inverse Ill-Posed Probl. **20** (2012), no. 4, 389–400, DOI 10.1515/jip-2012-0014. MR2984474

[AU04] Habib Ammari and Gunther Uhlmann, *Reconstruction of the potential from partial Cauchy data for the Schrödinger equation*, Indiana Univ. Math. J. **53** (2004), no. 1, 169–183, DOI 10.1512/iumj.2004.53.2299. MR2048188 (2005h:35351)

[ALP05] Kari Astala, Lassi Päivärinta, and Matti Lassas, *Calderón's inverse problem for anisotropic conductivity in the plane*, Comm. Partial Differential Equations **30** (2005), no. 1-3, 207–224, DOI 10.1081/PDE-200044485. MR2131051 (2005k:35421)

[AP06] Kari Astala and Lassi Päivärinta, *Calderón's inverse conductivity problem in the plane*, Ann. of Math. (2) **163** (2006), no. 1, 265–299, DOI 10.4007/annals.2006.163.265. MR2195135 (2007b:30019)

[Be09] Hajer Ben Joud, *A stability estimate for an inverse problem for the Schrödinger equation in a magnetic field from partial boundary measurements*, Inverse Problems **25** (2009), no. 4, 045012, 23, DOI 10.1088/0266-5611/25/4/045012. MR2482163 (2010h:35411)

[Bu08] A. L. Bukhgeim, *Recovering a potential from Cauchy data in the two-dimensional case*, J. Inverse Ill-Posed Probl. **16** (2008), no. 1, 19–33, DOI 10.1515/jiip.2008.002. MR2387648 (2008m:30049)

[BU01] Alexander L. Bukhgeim and Gunther Uhlmann, *Recovering a potential from partial Cauchy data*, Comm. Partial Differential Equations **27** (2002), no. 3-4, 653–668, DOI 10.1081/PDE-120002868. MR1900557 (2003d:35262)

[Ca80] Alberto-P. Calderón, *On an inverse boundary value problem*, (Rio de Janeiro, 1980), Soc. Brasil. Mat., Rio de Janeiro, 1980, pp. 65–73. MR590275 (81k:35160)

[Ca11] Pedro Caro, *On an inverse problem in electromagnetism with local data: stability and uniqueness*, Inverse Probl. Imaging **5** (2011), no. 2, 297–322, DOI 10.3934/ipi.2011.5.297. MR2805355 (2012k:35595)

[CDR12] P. Caro, D. Dos Santos Ferreira, A. Ruiz, *Stability estimates for the Radon transform with restricted data and applications*, preprint (2012), arXiv:1211.1887.

[COS09] Pedro Caro, Petri Ola, and Mikko Salo, *Inverse boundary value problem for Maxwell equations with local data*, Comm. Partial Differential Equations **34** (2009), no. 10-12, 1425–1464, DOI 10.1080/03605300903296272. MR2581979 (2010m:35558)

[Ch11] Francis Jee-Hyun Chung, *A partial data result for the magnetic Schrodinger inverse problem*, ProQuest LLC, Ann Arbor, MI, 2012. Thesis (Ph.D.)–The University of Chicago. MR3054984

[Ch12] Francis Jee-Hyun Chung, *A partial data result for the magnetic Schrodinger inverse problem*, ProQuest LLC, Ann Arbor, MI, 2012. Thesis (Ph.D.)–The University of Chicago. MR3054984

[DKSaU09] David Dos Santos Ferreira, Carlos E. Kenig, Mikko Salo, and Gunther Uhlmann, *Limiting Carleman weights and anisotropic inverse problems*, Invent. Math. **178** (2009), no. 1, 119–171, DOI 10.1007/s00222-009-0196-4. MR2534094 (2010h:58033)

[DKSjU07] David Dos Santos Ferreira, Carlos E. Kenig, Johannes Sjöstrand, and Gunther Uhlmann, *Determining a magnetic Schrödinger operator from partial Cauchy data*, Comm. Math. Phys. **271** (2007), no. 2, 467–488, DOI 10.1007/s00220-006-0151-9. MR2287913 (2008a:35044)

[DKSjU09] David Dos Santos Ferreira, Carlos E. Kenig, Johannes Sjöstrand, and Gunther Uhlmann, *On the linearized local Calderón problem*, Math. Res. Lett. **16** (2009), no. 6, 955–970, DOI 10.4310/MRL.2009.v16.n6.a4. MR2576684 (2011d:35515)

[DKLS13] D. Dos Santos Ferreira, Y. Kurylev, M. Lassas, M. Salo, *The Calderón problem in transversally anisotropic geometries*, preprint (2013), arXiv:1305.1273.

[Es04] G. Eskin, *Inverse boundary value problems in domains with several obstacles*, Inverse Problems **20** (2004), no. 5, 1497–1516, DOI 10.1088/0266-5611/20/5/011. MR2109132 (2005h:35358)

[Fa66] L.D. Faddeev, *Increasing solutions of the Schrödinger equation*, Sov. Phys. Dokl. **10** (1966), 1033–1035.

[Fa07] Ines Kamoun Fathallah, *Stability for the inverse potential problem by the local Dirichlet-to-Neumann map for the Schrödinger equation*, Appl. Anal. **86** (2007), no. 7, 899–914, DOI 10.1080/00036810701497067. MR2355545 (2008i:35246)

[Ge08] Bastian Gebauer, *Localized potentials in electrical impedance tomography*, Inverse Probl. Imaging **2** (2008), no. 2, 251–269, DOI 10.3934/ipi.2008.2.251. MR2395143 (2009d:35338)

[GT11a] Colin Guillarmou and Leo Tzou, *Calderón inverse problem with partial data on Riemann surfaces*, Duke Math. J. **158** (2011), no. 1, 83–120, DOI 10.1215/00127094-1276310. MR2794369 (2012f:35574)

[GT11b] Colin Guillarmou and Leo Tzou, *Identification of a connection from Cauchy data on a Riemann surface with boundary*, Geom. Funct. Anal. **21** (2011), no. 2, 393–418, DOI 10.1007/s00039-011-0110-2. MR2795512 (2012k:58027)

[GT13] C. Guillarmou, L. Tzou, *Survey on Calderón inverse problem in 2 dimensions*, in Inverse Problems and Applications: Inside Out II (edited by G. Uhlmann), MSRI Publications **60**, p. 119–166, Cambridge University Press, 2013.

[HT13] Boaz Haberman and Daniel Tataru, *Uniqueness in Calderón's problem with Lipschitz conductivities*, Duke Math. J. **162** (2013), no. 3, 496–516, DOI 10.1215/00127094-2019591. MR3024091

[Hä98] Peter Hähner, *A uniqueness theorem for an inverse scattering problem in an exterior domain*, SIAM J. Math. Anal. **29** (1998), no. 5, 1118–1128 (electronic), DOI 10.1137/S0036141097318614. MR1613735 (98m:35223)

[HW06] Horst Heck and Jenn-Nan Wang, *Stability estimates for the inverse boundary value problem by partial Cauchy data*, Inverse Problems **22** (2006), no. 5, 1787–1796, DOI 10.1088/0266-5611/22/5/015. MR2261266 (2007g:35270)

[HW07] H. Heck, J.-N. Wang, *Optimal stability estimate of the inverse boundary value problem by partial measurements*, preprint (2007), arXiv:0708.3289.

[He99] Sigurdur Helgason, *The Radon transform*, 2nd ed., Progress in Mathematics, vol. 5, Birkhäuser Boston Inc., Boston, MA, 1999. MR1723736 (2000m:44003)

[Hu13] M. Hubenthal, *The broken ray transform on the square*, preprint (2013), arXiv:1302.6193.

[HPS12] Nuutti Hyvönen, Petteri Piiroinen, and Otto Seiskari, *Point measurements for a Neumann-to-Dirichlet map and the Calderón problem in the plane*, SIAM J. Math. Anal. **44** (2012), no. 5, 3526–3536, DOI 10.1137/120872164. MR3023421

[Il12] Joonas Ilmavirta, *Broken ray tomography in the disc*, Inverse Problems **29** (2013), no. 3, 035008, 17, DOI 10.1088/0266-5611/29/3/035008. MR3040563

[Il13] J. Ilmavirta, *A reflection approach to the broken ray transform*, preprint (2013), arXiv:1306.0341.

[IUY10] Oleg Yu. Imanuvilov, Gunther Uhlmann, and Masahiro Yamamoto, *The Calderón problem with partial data in two dimensions*, J. Amer. Math. Soc. **23** (2010), no. 3, 655–691, DOI 10.1090/S0894-0347-10-00656-9. MR2629983 (2012c:35472)

[IUY11a] Oleg Y. Imanuvilov, Gunther Uhlmann, and Masahiro Yamamoto, *Determination of second-order elliptic operators in two dimensions from partial Cauchy data*, Proc. Natl. Acad. Sci. USA **108** (2011), no. 2, 467–472, DOI 10.1073/pnas.1011681107. MR2770947 (2012a:35364)

[IUY11b] Oleg Yu. Imanuvilov, Gunther Uhlmann, and Masahiro Yamamoto, *Inverse boundary value problem by measuring Dirichlet data and Neumann data on disjoint sets*, Inverse Problems **27** (2011), no. 8, 085007, 26, DOI 10.1088/0266-5611/27/8/085007. MR2819949 (2012c:78002)

[IY12a] O. Yu. Imanuvilov and M. Yamamoto, *Inverse boundary value problem for Schrödinger equation in two dimensions*, SIAM J. Math. Anal. **44** (2012), no. 3, 1333–1339, DOI 10.1137/11083736X. MR2982714

[IY12b] O. Yu. Imanuvilov and M. Yamamoto, *Inverse problem by Cauchy data on an arbitrary sub-boundary for systems of elliptic equations*, Inverse Problems **28** (2012), no. 9, 095015, 30, DOI 10.1088/0266-5611/28/9/095015. MR2972464

[IY13a] O. Yu. Imanuvilov and M. Yamamoto, *Inverse boundary value problem for the Schrödinger equation in a cylindrical domain by partial boundary data*, Inverse Problems **29** (2013), no. 4, 045002, 8, DOI 10.1088/0266-5611/29/4/045002. MR3041540

[IY13b] O. Yu. Imanuvilov and M. Yamamoto, *Inverse boundary value problem for the Schrödinger equation in a cylindrical domain by partial boundary data*, Inverse Problems **29** (2013), no. 4, 045002, 8, DOI 10.1088/0266-5611/29/4/045002. MR3041540

[Is88] Victor Isakov, *On uniqueness of recovery of a discontinuous conductivity coefficient*, Comm. Pure Appl. Math. **41** (1988), no. 7, 865–877, DOI 10.1002/cpa.3160410702. MR951742 (90f:35205)

[Is07] Victor Isakov, *On uniqueness in the inverse conductivity problem with local data*, Inverse Probl. Imaging **1** (2007), no. 1, 95–105, DOI 10.3934/ipi.2007.1.95. MR2262748 (2007m:35273)

[KKL01] Alexander Katchalov, Yaroslav Kurylev, and Matti Lassas, *Inverse boundary spectral problems*, Chapman & Hall/CRC Monographs and Surveys in Pure and Applied Mathematics, vol. 123, Chapman & Hall/CRC, Boca Raton, FL, 2001. MR1889089 (2003e:58045)

[KS12] C.E. Kenig, M. Salo, *The Calderón problem with partial data on manifolds and applications*, preprint (2012), arXiv:1211.1054.

[KSU11a] Carlos E. Kenig, Mikko Salo, and Gunther Uhlmann, *Inverse problems for the anisotropic Maxwell equations*, Duke Math. J. **157** (2011), no. 2, 369–419, DOI 10.1215/00127094-1272903. MR2783934 (2012d:35408)

[KSU11b] Carlos E. Kenig, Mikko Salo, and Gunther Uhlmann, *Reconstructions from boundary measurements on admissible manifolds*, Inverse Probl. Imaging **5** (2011), no. 4, 859–877, DOI 10.3934/ipi.2011.5.859. MR2852376 (2012k:58038)

[KSU07] Carlos E. Kenig, Johannes Sjöstrand, and Gunther Uhlmann, *The Calderón problem with partial data*, Ann. of Math. (2) **165** (2007), no. 2, 567–591, DOI 10.4007/annals.2007.165.567. MR2299741 (2008k:35498)

[Kn06] Kim Knudsen, *The Calderón problem with partial data for less smooth conductivities*, Comm. Partial Differential Equations **31** (2006), no. 1-3, 57–71, DOI 10.1080/03605300500361610. MR2209749 (2006k:35303)

[KS07] Kim Knudsen and Mikko Salo, *Determining nonsmooth first order terms from partial boundary measurements*, Inverse Probl. Imaging **1** (2007), no. 2, 349–369, DOI 10.3934/ipi.2007.1.349. MR2282273 (2008k:35500)

[KV84] Robert Kohn and Michael Vogelius, *Determining conductivity by boundary measurements*, Comm. Pure Appl. Math. **37** (1984), no. 3, 289–298, DOI 10.1002/cpa.3160370302. MR739921 (85f:80008)

[KV85] R. V. Kohn and M. Vogelius, *Determining conductivity by boundary measurements. II. Interior results*, Comm. Pure Appl. Math. **38** (1985), no. 5, 643–667, DOI 10.1002/cpa.3160380513. MR803253 (86k:35155)

[KLU12] Katsiaryna Krupchyk, Matti Lassas, and Gunther Uhlmann, *Inverse problems with partial data for a magnetic Schrödinger operator in an infinite slab and on a bounded domain*, Comm. Math. Phys. **312** (2012), no. 1, 87–126, DOI 10.1007/s00220-012-1431-1. MR2914058

[LO10] Matti Lassas and Lauri Oksanen, *An inverse problem for a wave equation with sources and observations on disjoint sets*, Inverse Problems **26** (2010), no. 8, 085012, 19, DOI 10.1088/0266-5611/26/8/085012. MR2661691 (2011m:35422)

[LO12] M. Lassas, L. Oksanen, *Inverse problem for the Riemannian wave equation with Dirichlet data and Neumann data on disjoint sets*, preprint (2012), arXiv:1208.2105.

[LiU10] Xiaosheng Li and Gunther Uhlmann, *Inverse problems with partial data in a slab*, Inverse Probl. Imaging **4** (2010), no. 3, 449–462, DOI 10.3934/ipi.2010.4.449. MR2671106 (2011f:35361)

[Na96] Adrian I. Nachman, *Global uniqueness for a two-dimensional inverse boundary value problem*, Ann. of Math. (2) **143** (1996), no. 1, 71–96, DOI 10.2307/2118653. MR1370758 (96k:35189)

[NS10] Adrian Nachman and Brian Street, *Reconstruction in the Calderón problem with partial data*, Comm. Partial Differential Equations **35** (2010), no. 2, 375–390, DOI 10.1080/03605300903296322. MR2748629 (2012b:35368)

[OPS93] Petri Ola, Lassi Päivärinta, and Erkki Somersalo, *An inverse boundary value problem in electrodynamics*, Duke Math. J. **70** (1993), no. 3, 617–653, DOI 10.1215/S0012-7094-93-07014-7. MR1224101 (94i:35196)

[ST10] Mikko Salo and Leo Tzou, *Inverse problems with partial data for a Dirac system: a Carleman estimate approach*, Adv. Math. **225** (2010), no. 1, 487–513, DOI 10.1016/j.aim.2010.03.003. MR2669360 (2011g:35432)

[SU87] John Sylvester and Gunther Uhlmann, *A global uniqueness theorem for an inverse boundary value problem*, Ann. of Math. (2) **125** (1987), no. 1, 153–169, DOI 10.2307/1971291. MR873380 (88b:35205)

[Tz08] Leo Tzou, *Stability estimates for coefficients of magnetic Schrödinger equation from full and partial boundary measurements*, Comm. Partial Differential Equations **33** (2008), no. 10-12, 1911–1952, DOI 10.1080/03605300802402674. MR2475324 (2010b:35492)

[Uh98] Gunther Uhlmann, *Inverse boundary value problems for partial differential equations*, Proceedings of the International Congress of Mathematicians, Vol. III (Berlin, 1998), 1998, pp. 77–86 (electronic). MR1648142 (99i:35178)

[Uh09] G. Uhlmann, *Electrical impedance tomography and Calderón's problem*, Inverse Problems **25** (2009), 123011.

[Zh12] Guo Zhang, *Uniqueness in the Calderón problem with partial data for less smooth conductivities*, Inverse Problems **28** (2012), no. 10, 105008, 18, DOI 10.1088/0266-5611/28/10/105008. MR2987911

Department of Mathematics, University of Chicago
E-mail address: cek@math.uchicago.edu

Department of Mathematics and Statistics, University of Jyväskylä
E-mail address: mikko.j.salo@jyu.fi

Contemporary Mathematics
Volume **615**, 2014
http://dx.doi.org/10.1090/conm/615/12278

Local reconstruction of a Riemannian manifold from a restriction of the hyperbolic Dirichlet-to-Neumann operator

Matti Lassas and Lauri Oksanen

Dedicated to Gunther Uhlmann on the occasion of his 60th birthday

ABSTRACT. We consider a wave equation with a non-symmetric first order perturbation on a Riemannian manifold (M, g) and a restriction $\Lambda^T_{\mathcal{S},\mathcal{R}}$ of the associated Dirichlet-to-Neumann operator. Here $\mathcal{S}$ and $\mathcal{R}$ are open sets in ∂M and the restriction $\Lambda^T_{\mathcal{S},\mathcal{R}}$ corresponds to the case where the Dirichlet data is supported on $(0,T) \times \mathcal{S}$ and the Neumann data is measured on $(0,T) \times \mathcal{R}$. In the case $\overline{\mathcal{R}} \cap \overline{\mathcal{S}} = \emptyset$, we show that $\Lambda^T_{\mathcal{S},\mathcal{R}}$ determines the manifold (M, g) in a neighborhood of the set $\mathcal{R}$ assuming that the wave equation is exactly controllable from the set of sources $\mathcal{S}$ in time $T/2 > 0$.

1. Introduction

Let (M, g) be a smooth, connected and compact Riemannian manifold of dimension n and boundary ∂M, let $A = \sum_{j=1}^n A^j(x)\partial_j$ be a smooth complex valued vector field on M and let q be a smooth complex valued function on M. We consider the wave equation with Dirichlet data $f \in C_0^\infty((0,\infty) \times \partial M)$,

$$\begin{aligned}
(1.1)\qquad & (\partial_t^2 - \Delta_g + A + q)u(t,x) = 0, && \text{in } (0,\infty) \times M,\\
& u|_{(0,\infty)\times\partial M} = f, && \text{in } (0,\infty) \times \partial M,\\
& u|_{t=0} = \partial_t u|_{t=0} = 0, && \text{in } M,
\end{aligned}$$

and denote by $u^f = u(t,x)$ the solution of (1.1). For open and nonempty sets $\mathcal{S}, \mathcal{R} \subset \partial M$ and $T \in (0,\infty]$ we define the restricted Dirichlet-to-Neumann operator,

$$\Lambda^T_{\mathcal{S},\mathcal{R}} : f \mapsto \partial_\nu u^f|_{(0,T)\times\mathcal{R}}, \quad f \in C_0^\infty((0,T) \times \mathcal{S}).$$

When f is regarded as a boundary source, the operator $\Lambda^T_{\mathcal{S},\mathcal{R}}$ models boundary measurements for the wave equation with sources producing the waves on $(0,T)\times\mathcal{S}$ and the waves being observed on $(0,T)\times\mathcal{R}$. We consider the inverse boundary value problem to determine (M, g) from $\Lambda^T_{\mathcal{S},\mathcal{R}}$.

In this paper we focus on the case $\overline{\mathcal{S}} \cap \overline{\mathcal{R}} = \emptyset$. This kind of problems with disjoint partial data have remained open to large extent. We are aware of only few previous results: in [**29**] Rakesh proved that the coefficients of a wave equation on

---

1991 *Mathematics Subject Classification.* Primary: 35R30.

Key words and phrases. Inverse problems, wave equation, partial data.

The first author was supported in part by Finnish Centre of Excellence in Inverse Problems Research, Academy of Finland project COE 250215.

a one-dimensional interval are determined by boundary measurements with sources supported on one end of the interval and the waves observed on the other end, and in [**13**] Imanuvilov, Uhlmann, and Yamamoto proved that the potential of a Schrödinger equation on a two-dimensional domain homeomorphic to a disc, where the boundary is partitioned into eight clockwise-ordered parts $\Gamma_1, \Gamma_2, \dots, \Gamma_8$, is determined by boundary measurements with sources supported on $\mathcal{S} = \Gamma_2 \cup \Gamma_6$ and fields observed on $\mathcal{R} = \Gamma_4 \cup \Gamma_8$. In our earlier paper [**24**] we consider the inverse problem for wave equation (1.1) in the case when the lower order terms vanish, that is, $A = 0$ and $q = 0$, and one is given the map $\Lambda^T_{\mathcal{S},\mathcal{R}}$ for large enough T and $\overline{\mathcal{S}} \cap \overline{\mathcal{R}} = \emptyset$.

The study of problems with disjoint partial data is motivated by applications as it is not always possible to observe fields on the same area where sources are controlled. For example in oil exploration, explosives are used as sources and hence it is difficult to measure waves near the sources.

The two main approaches to solve various inverse problems for the wave equation are the Boundary Control method [**2**] and the method based on complex geometric optics solutions [**26, 32**]. The latter method was developed to solve Calderón's inverse problem [**6**] that is an inverse problem for an isotropic elliptic equation. It is not known how to generalize complex geometric optics solutions for general anisotropic problems, see the discussion in [**7**]. On the other hand, the Boundary Control method can be used to solve inverse problems for anisotropic hyperbolic equations [**3**] but it does not generalize for problems related to elliptic equations. The result in the present paper is based on the Boundary Control method. For problems involving restricted Dirichlet-to-Neumann maps, we refer to the results [**14, 23**] based on the Boundary Control method the results [**10–12, 16, 25**] based on complex geometric optics solutions. For inverse problems involving first order terms, see the results [**4, 8, 19–21**] concerning hyperbolic equations and the results [**9, 17, 18, 27, 28, 30, 31**] concerning elliptic equations.

1.1. Statement of the results. We will work on under the assumption that (M, g) is exactly controllable from $\mathcal{S}$ in time T, that is, the map

$$\mathcal{U} : L^2((0,T) \times \mathcal{S}) \to L^2(M) \times H^{-1}(M), \qquad \mathcal{U}(f) = (u^f(T), \partial_t u^f(T)) \tag{1.2}$$

is surjective. The condition by Bardos, Lebeau and Rauch gives a geometric characterization of exact controllability [**1, 5**]. In particular, if M has a strictly convex boundary, then exact controllability is valid when every generalized geodesic of length T intersects $\mathcal{S}$. The generalized geodesics are continued by normal reflection on the boundary and also include geodesics of $(\partial M, g)$. We refer to [**1**] for the precise formulation of the geometric condition.

In order to formulate our main result, let us introduce some notation. We write $SM := \{\xi \in TM;\ |\xi|_g = 1\}$, that is, SM is the unit sphere bundle, and define $\partial_\pm SM := \{\xi \in \partial SM;\ (\xi, \mp\nu)_g > 0\}$, where ν is the interior unit normal vector on ∂M. We define the *exit time* for $(x, \xi) \in SM \setminus \overline{\partial_+ SM}$,

$$\tau_M(x, \xi) := \inf\{s \in (0, \infty);\ \gamma(s; x, \xi) \in \partial M\},$$

where $\gamma(\cdot; x, \xi)$ is the geodesic with the initial data $\gamma(0) = x$, $\dot\gamma(0) = \xi$. We denote by $d(x, y)$, $x, y \in M$, the Riemannian distance function of (M, g). Moreover, we

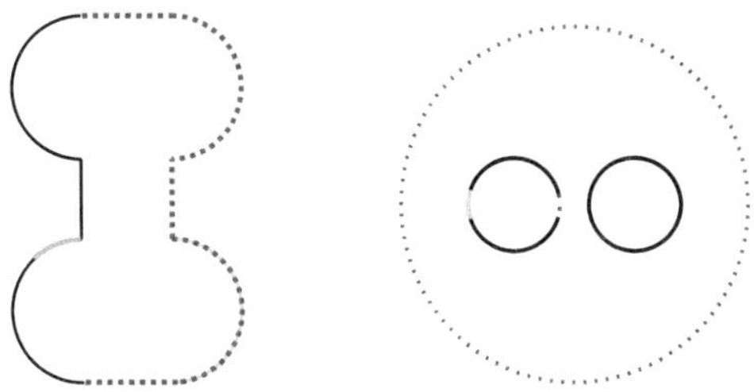

FIGURE 1. The dogbone region (left) and Ikawa's bowling ball (right) both with the Euclidiean metric satisfy the exact controllability assumption of Theorem 1.1 when $\mathcal{S}$ is the dashed blue part of the boundary and $\mathcal{R}$ is the light red or green part of the boundary, see [**1**]. Notice that in the case of Ikawa's bowling ball, Theorem 1.1 allows us to reconstruct the hole on right from data on the outer boundary and on the boundary of the left hole only. This means that we are able to recover non-trivial topology that we are not aware of a priori.

define the strip

$$\begin{aligned} \mathcal{N}_\mathcal{R} &:= \{(s,y) \in (0,\infty)\times\mathcal{R};\ s<\sigma_\mathcal{R}(y)\}, \\ \sigma_\mathcal{R}(y) &:= \max\{s \in (0,\tau_M(y,\nu)];\ d(\gamma(s;y,\nu),\mathcal{R}) = s\}, \end{aligned} \tag{1.3}$$

and denote by $M_\mathcal{R} \subset M$ the image of $\mathcal{N}_\mathcal{R}$ under the map

$$(s,y) \mapsto \gamma(s;y,\nu). \tag{1.4}$$

Notice that $\sigma_\mathcal{R}(y) > 0$, see e.g. [**15**, p. 50]. Our main result is the following:

THEOREM 1.1. *Suppose that (M,g) is exactly controllable from $\mathcal{S}$ in time $T>0$. Then $\Lambda^{2T}_{\mathcal{S},\mathcal{R}}$ determines the Riemannian manifold $(M_\mathcal{R}, g)$.*

In the paper [**24**] we show that $\Lambda^{T}_{\mathcal{S},\mathcal{R}}$, for large enough T, determines the whole Riemannian manifold (M,g) under the further assumption that there are no lower order terms, that is, $A=0$ and $q=0$ identically. The global reconstruction in the presence of lower order terms remains an open question.

2. The proof

Let us consider the adjoint equation

$$\begin{aligned} &(\partial_t^2 - \Delta_g + A^* + \overline{q})v(t,x) = 0, && \text{in } (0,\infty)\times M, \\ &v|_{(0,\infty)\times\partial M} = f, && \text{in } (0,\infty)\times \partial M, \\ &v|_{t=0} = \partial_t v|_{t=0} = 0, && \text{in } M, \end{aligned} \tag{2.1}$$

where $A^* := -\overline{A} - \mathrm{div}\overline{A}$. Here, in the local coordinates,

$$\mathrm{div} A = \sum_{j=1}^n (\det g)^{-1/2}\partial_j((\det g)^{1/2}A^j).$$

We write $v^h := v(t,x)$ and define

$$B^T_{\mathcal{R},\mathcal{S}} f := (\partial_\nu v^f + (\overline{A},\nu)_g v^f)|_{(0,T)\times\mathcal{S}}, \quad f \in C_0^\infty((0,T)\times\mathcal{R}).$$

Moreover, let us denote by dV and dS the Riemannian volume measures of (M, g) and $(\partial M, g)$ respectively. All the integrations below are with respect to these measures unless explicitly stated otherwise. Notice that we do not assume dS to be known a priori.

LEMMA 2.1. *Let $d\widetilde{S}$ be a smooth positive measure on $\overline{\mathcal{S} \cup \mathcal{R}}$. Then there is a strictly positive function $\mu \in C^\infty(\overline{\mathcal{S} \cup \mathcal{R}})$, unknown to us, such that $d\widetilde{S} = \mu dS$ and the operator $\Lambda^{2T}_{\mathcal{S},\mathcal{R}}$ determines for any given functions $f \in C_0^\infty((0,T) \times \mathcal{R})$ and $\phi \in C_0^\infty((0,T) \times \mathcal{S})$ the inner product*

$$I_T(f, g) = (v^{\mu f}(T), u^\phi(T))_{L^2(M)}.$$

PROOF. Let us show first that $\Lambda^T_{\mathcal{S},\mathcal{R}}$ determines $\mu^{-1} B^T_{\mathcal{R},\mathcal{S}} \mu$. To this purpose let us consider the backwards wave equation

$$\begin{aligned} (\partial_t^2 - \Delta_g + A + q) w(t, x) &= 0, && \text{in } (0, T) \times M, \\ w|_{(0,\infty) \times \partial M} &= \psi, && \text{in } (0, T) \times \partial M, \\ w|_{t=T} = \partial_t w|_{t=T} &= 0, && \text{in } M. \end{aligned} \tag{2.2}$$

Notice that if $\phi(t, x) = \psi(T - t, x)$ then $w(t, x) = u^\phi(T - t, x)$ gives the solution. Suppose that $\psi \in C_0^\infty((0, T) \times \mathcal{S})$. Then

$$\begin{aligned} 0 &= ((\partial_t^2 - \Delta_g + A^* + \overline{q}) v^{\mu f}, w)_{L^2((0,T) \times M)} \\ &\quad - (v^{\mu f}, (\partial_t^2 - \Delta_g + A + q) w)_{L^2((0,T) \times M)} \\ &= (B^T_{\mathcal{R},\mathcal{S}} \mu f, \psi)_{L^2((0,T) \times \mathcal{S})} - (\mu f, \partial_\nu w)_{L^2((0,T) \times \mathcal{R})} \\ &= (\mu^{-1} B^T_{\mathcal{R},\mathcal{S}} \mu f, \psi)_{L^2((0,T) \times \mathcal{S}; dt \otimes d\widetilde{S})} - (f, \partial_\nu w)_{L^2((0,T) \times \mathcal{R}; dt \otimes d\widetilde{S})}. \end{aligned}$$

Moreover, $\partial_\nu w(t, x) = \Lambda^T_{\mathcal{S},\mathcal{R}} \phi(T - t, x)$ on $(0, T) \times \mathcal{R}$. Hence $\Lambda^T_{\mathcal{S},\mathcal{R}}$ determines $\mu^{-1} B^T_{\mathcal{R},\mathcal{S}} \mu$.

Let $t \in (0, T)$ and $s \in (0, 2T)$. We have

$$\begin{aligned} &(\partial_t^2 - \partial_s^2)(v^{\mu f}(t), u^\phi(s))_{L^2(M)} \\ &= ((\Delta_g - A^* - \overline{q}) v^{\mu f}(t), u^\phi(s))_{L^2(M)} - (v^{\mu f}(t), (\Delta_g - A - q) u^\phi(s))_{L^2(M)} \\ &= -(\mu^{-1} B^T_{\mathcal{R},\mathcal{S}} \mu f(t), \phi(s))_{L^2(\mathcal{S}; d\widetilde{S})} + (f(t), \Lambda^{2T}_{\mathcal{S},\mathcal{R}} \phi(s))_{L^2(\mathcal{R}; d\widetilde{S})}. \end{aligned} \tag{2.3}$$

We see that $I(t, s) := (v^{\mu f}(t), u^\phi(s))$ satisfies a one-dimensional wave equation (2.3) in the domain $(t, s) \in \mathbb{R}_+ \times \mathbb{R}_+$ with a known right-hand side. By solving this equation with vanishing initial conditions at $t = 0$, that is, $I(t, s)|_{t=0} = 0$ and $\partial_t I(t, s)|_{t=0} = 0$, and the boundary condition $I(t, s)|_{s=0} = 0$ we obtain the claim. $\square$

For $\Gamma \subset \partial M$ and $h : \Gamma \to \mathbb{R}$ we define the modified domain of influence

$$M(\Gamma, h) := \{x \in M;\ \inf_{y \in \Gamma} (d(x, y) - h(y)) \le 0\},$$

and write for $T > 0$

$$\mathcal{B}(\Gamma, h; T) := \{(t, y) \in (0, T) \times \Gamma;\ T - h(y) < t\}.$$

We extend the notations $M(\Gamma, h)$ and $\mathcal{B}(\Gamma, h; T)$ for constants $h \in \mathbb{R}$ by interpreting h as a constant function. Moreover, we define $M(x, h)$ by $M(\{x\}, h)$ for points $x \in \partial M$.

LEMMA 2.2. *Suppose that (M,g) is exactly controllable from $\mathcal{S}$ in time $T>0$. Let $\mu \in C^\infty(\overline{\mathcal{R}})$ be strictly positive and let $h : \overline{\mathcal{R}} \to [0,T]$ be piecewise continuous. Moreover, let $\Gamma_0 \subset \mathcal{R}$ be open and let $s_0 \in (0,T]$. Then the following properties are equivalent:*

(i) $M(\Gamma_0, s_0) \subset M(\mathcal{R}, h)$.

(ii) *For all $f_0 \in C_0^\infty(\mathcal{B}(\Gamma_0, s_0; T))$ there is $(f_j)_{j=1}^\infty \subset C_0^\infty(\mathcal{B}(\mathcal{R}, h; T))$ such that*

$$\lim_{j\to\infty} (v^{\mu f_j}(T) - v^{\mu f_0}(T), u^\phi(T))_{L^2(M)} = 0, \quad \phi \in L^2((0,T)\times \mathcal{S}).$$

PROOF. As $f \mapsto \mu f$ is a bijection on $C_0^\infty(\mathcal{B}(\Gamma_0, s_0; T))$ and on $C_0^\infty(\mathcal{B}(\mathcal{R}, h; T))$, we can assume, without loss of generality, that $\mu = 1$ identically. By exact controllablity we have that the convergence in (ii) is equivalent with the weak convergence $v^{f_j}(T) \to v^{f_0}(T)$ in $L^2(M)$.

By transposing Tataru's unique continuation [**33**], we see that

$$\{v^f(T);\ f \in C_0^\infty(\mathcal{B}(\mathcal{R}, h; T))\} \tag{2.4}$$

is dense in $\{w \in L^2(M);\ \mathrm{supp}(w) \subset M(\mathcal{R}, h)\}$. Together with finite speed of propagation, this yields the implication from (i) to (ii).

We will now show that (ii) implies (i). To this end, let us assume that (i) does not hold and let $x \in M(\Gamma_0, s_0) \setminus M(\mathcal{R}, h)$. As $M(\Gamma, h)$ is closed, there is a neighborhood U of x such that $U \cap M(\mathcal{R}, h) = \emptyset$. By moving from x towards Γ_0 we see easily that $U \cap M(\Gamma_0, s_0)^{\mathrm{int}}$ is nonempty. Hence there is a nonempty open $V \subset M(\Gamma_0, s_0)$ such that $V \cap M(\mathcal{R}, h) = \emptyset$. Then analogously to the density (2.4), there is a smooth function f_0 that is supported in $\mathcal{B}(\Gamma_0, s_0; T)$ and satisfies $\int_V v^{f_0}(T) dV(x) \neq 0$. However, by finite speed of propagation $v^f(T)|_V = 0$ for any f supported in $\mathcal{B}(\mathcal{R}, h; T)$. Thus

$$(v^{f_0}(T) - v^f(T), 1_V)_{L^2(M)} = (v^{f_0}(T), 1_V)_{L^2(M)} \neq 0,$$

for all f supported in $\mathcal{B}(\mathcal{R}, h; T)$ and (ii) does not hold. □

We need the modified distance functions in order to reconstruct $\sigma_\mathcal{R}$ with the help of the next lemma. Let us recall that $\sigma_\mathcal{R}(y)$ indicates the distance when the normal geodesic starting from $y \in \mathcal{R}$ hits to the boundary or to a point on the cut locus, see (1.3). The main difficulty when reconstructing $\sigma_\mathcal{R}(y)$ is that the normal geodesic may intersect ∂M tangentially and this is delicate to dectect from the boundary data. We note that in [**22**] the tangential boundary intersection of normal geodesics and the boundary ∂M was not correctly considered. The appropriate result is given by Lemma 2.3 below.

LEMMA 2.3. *Let $y \in \mathcal{R}$, $h \in C^1(\overline{\mathcal{R}})$ and define*

$$\widetilde{\sigma}_\mathcal{R}(y; h) := \sup\{t \in (0,\infty);\ M(y,t) \not\subset M(\Gamma, s+h) \text{ for all } s < t\}.$$

Let $y_0 \in \mathcal{R}$. Then

$$\liminf_{(y,h)\to(y_0,0)} \widetilde{\sigma}_\mathcal{R}(y; h) = \sigma_\mathcal{R}(y_0; 0),$$

where the lim inf *is taken over all $(y,h) \in \mathcal{R} \times C^1(\overline{\mathcal{R}})$ such that $h(y) = 0$.*

For the proof of this lemma we refer to [**24**, Lemma 14]. The critical case $d(\gamma(s; y_0, \nu), y_1) = t$ of the following lemma is illustrated in the Figure 2.

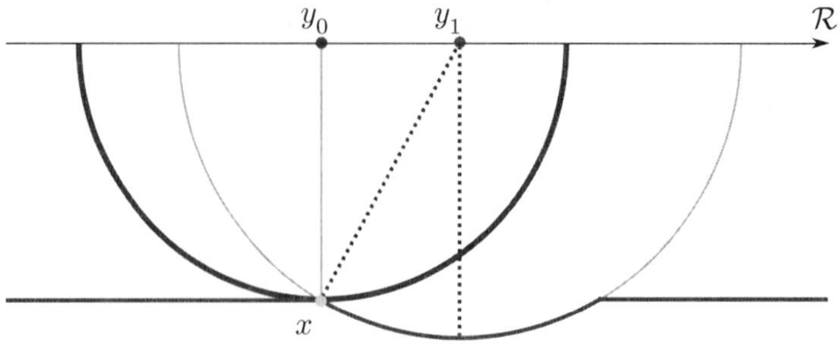

FIGURE 2. Lemma 2.4 allows us to determine the distance between $x = \gamma(s; y_0, \nu)$ and y_1 by looking at relations between domains of influences. Here $\partial M(y_0, s)$ is the red half circle and $\partial(M(\mathcal{R}, s - \delta) \cup M(y_1, t + \epsilon))$ is the blue piecewise smooth curve.

LEMMA 2.4. *Suppose that $y_0, y_1 \in \mathcal{R}$, $t > 0$ and let $0 < s < \sigma_{\mathcal{R}}(y_0)$. Then the following properties are equivalent:*

(i) *$d(\gamma(s; y_0, \nu), y_1) \le t$.*
(ii) *For all $\epsilon > 0$ there is $\delta > 0$ such that*

$$M(y_0, s) \subset M(\mathcal{R}, s - \delta) \cup M(y_1, t + \epsilon).$$

PROOF. The assumption $s < \sigma_{\mathcal{R}}(y_0)$ implies that the boundary normal coordinates (1.4) are valid in a neighborhood of $\gamma(s; y_0, \nu)$. Using these coordinates we see that

$$M(y_0, s) \setminus \bigcup_{\delta > 0} M(\mathcal{R}, s - \delta) = \{\gamma(s; y_0, \nu)\}, \tag{2.5}$$

and the lemma follows using arguments similar to [**24**, Lemma 7]. □

PROOF OF THEOREM 1.1. Notice that the geometric characterization of the exact controllability by Bardos, Lebeau and Rauch implies that

$$T \ge \sup_{y \in \mathcal{R}} \sigma_{\mathcal{R}}(y).$$

Indeed, if this is not the case then there is $y \in \mathcal{R}$ such that the geodesic from (y, ν) does not intersect $\mathcal{S} \subset \partial M$ in the interval $[0, T]$.

By Lemma 2.1 the operator $\Lambda_{\mathcal{S},\mathcal{R}}^{2T}$ determines whether (ii) of Lemma 2.2 holds or not. By letting Γ_0 tend to $\{y_0\}$ we see that relations of form (ii) of Lemma 2.2 determine the function $\widetilde{\sigma}_{\mathcal{R}}$ of Lemma 2.3. Thus $\Lambda_{\mathcal{S},\mathcal{R}}^{2T}$ determines the function $\sigma_{\mathcal{R}}$. In particular, we can determine the set $\mathcal{N}_{\mathcal{R}}$ where the boundary normal coordinates (1.4) are valid.

We may choose a piecewise constant function h so that (i) in Lemma 2.2 becomes

$$M(\Gamma_0, s_0) \subset M(\mathcal{R}, s_1) \cup M(\Gamma_1, t).$$

Furthermore, we may let Γ_j tend to $\{y_j\}$ for $j = 0$ and $j = 1$. Hence we see that $\Lambda_{\mathcal{S},\mathcal{R}}^{2T}$ determines whether (ii) of Lemma 2.4 holds or not. Thus Lemma 2.4 yields that $\Lambda_{\mathcal{S},\mathcal{R}}^{2T}$ determines the distances

$$dist((s, y_0), (0, y_1)), \quad y_0, y_1 \in \mathcal{R}, \ s < \sigma_{\mathcal{R}}(y_0),$$

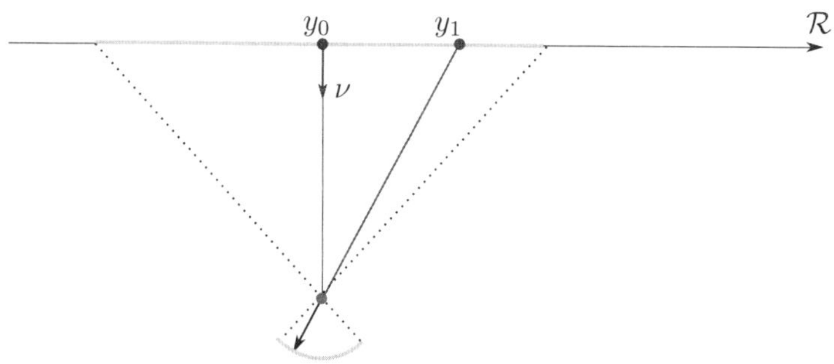

FIGURE 3. Let us denote $x = \gamma(s; y_0, \nu)$ where $s > 0$ is small and let y_1 be in a small neighborhood of y_0. Differentiating the distance $d(x, y_1)$ with respect to x gives a unit covector at x. By moving the point y_1, we get an open set on the unit cosphere $S^*_x M_{\mathcal{R}}$.

where *dist* denotes the distance function d in the boundary normal coordinates.

By [**15**, Lemma 2.15] the image of a sufficiently small neighborhood of y_0 in $\mathcal{R}$ under the map

$$y_1 \mapsto d_{(s,y_0)}\,(dist((s, y_0), (0, y_1)))$$

is an open set on the unit cosphere $S^*_{(s,y_0)} M_{\mathcal{R}}$, see Figure 3 for an illustration. Above, $d_{(s,y_0)}$ denotes the exterior differential on $\mathbb{R} \times \partial M$ evaluated at the point (s, y_0). Moreover, it is clear that a nonempty open part of the cosphere at (s, y_0) determines the metric tensor g at (s, y_0). Thus $\Lambda^{2T}_{\mathcal{S},\mathcal{R}}$ determines the metric tensor in the boundary normal coordinates on $M_{\mathcal{R}}$. $\square$

If $\mathcal{S} \cap \mathcal{R} \neq \emptyset$, then the lower order terms A and q can be determined on $M_{\mathcal{R}}$ modulo a gauge transformation [**21**]. The key point is to construct such boundary sources $(f_j)_{j=1}^{\infty}$ that the corresponding solutions $v^{\mu f_j}(T)$ converge weakly to a function that is localized around a point. This is done as follows. Let $y_0 \in \mathcal{S} \cap \mathcal{R}$ and let Γ_0 be a small neighborhood of y_0 on ∂M. Let $s, \delta > 0$ be small and consider such sources $(f_j)_{j=1}^{\infty} \subset C_0^{\infty}(\mathcal{B}(\Gamma_0, s; T))$ that

$$\lim_{j\to\infty} (v^{\mu f_j}, u^{\phi}(T))_{L^2(M)} = 0, \quad \text{for all } \phi \in C_0^{\infty}(\mathcal{B}(\mathcal{S} \cap \mathcal{R}, s - \delta; T)),$$

and that $(v^{\mu f_j})_{j=1}^{\infty}$ converges weakly in $L^2(M)$ to a function, say to w. Then w is supported in $M(\Gamma_0, s) \setminus M(\mathcal{S} \cap \mathcal{R}, s - \delta)$ and we may localize to $\gamma(s; y_0, \nu)$ as in (2.5) by letting Γ_0 tend to $\{y_0\}$ and δ to zero.

However, if $\mathcal{S}$ and $\mathcal{R}$ are far away from each other, then orthogonality conditions as above are not straightforward to exploit and we do not know how to reconstruct the lower order terms on $M_{\mathcal{R}}$. This prevents us also from proceeding to the global reconstruction of (M, g).

References

[1] Claude Bardos, Gilles Lebeau, and Jeffrey Rauch, *Sharp sufficient conditions for the observation, control, and stabilization of waves from the boundary*, SIAM J. Control Optim. **30** (1992), no. 5, 1024–1065, DOI 10.1137/0330055. MR1178650 (94b:93067)

[2] M. I. Belishev, *An approach to multidimensional inverse problems for the wave equation* (Russian), Dokl. Akad. Nauk SSSR **297** (1987), no. 3, 524–527; English transl., Soviet Math. Dokl. **36** (1988), no. 3, 481–484. MR924687 (89c:35152)

[3] Michael I. Belishev and Yaroslav V. Kurylev, *To the reconstruction of a Riemannian manifold via its spectral data (BC-method)*, Comm. Partial Differential Equations **17** (1992), no. 5-6, 767–804, DOI 10.1080/03605309208820863. MR1177292 (94a:58199)

[4] Mourad Bellassoued and Mourad Choulli, *Stability estimate for an inverse problem for the magnetic Schrödinger equation from the Dirichlet-to-Neumann map*, J. Funct. Anal. **258** (2010), no. 1, 161–195, DOI 10.1016/j.jfa.2009.06.010. MR2557958 (2011d:35511)

[5] Nicolas Burq and Patrick Gérard, *Condition nécessaire et suffisante pour la contrôlabilité exacte des ondes* (French, with English and French summaries), C. R. Acad. Sci. Paris Sér. I Math. **325** (1997), no. 7, 749–752, DOI 10.1016/S0764-4442(97)80053-5. MR1483711 (98j:93052)

[6] Alberto-P. Calderón, *On an inverse boundary value problem*, (Rio de Janeiro, 1980), Soc. Brasil. Mat., Rio de Janeiro, 1980, pp. 65–73. MR590275 (81k:35160)

[7] David Dos Santos Ferreira, Carlos E. Kenig, Mikko Salo, and Gunther Uhlmann, *Limiting Carleman weights and anisotropic inverse problems*, Invent. Math. **178** (2009), no. 1, 119–171, DOI 10.1007/s00222-009-0196-4. MR2534094 (2010h:58033)

[8] G. Eskin, *Inverse hyperbolic problems with time-dependent coefficients*, Comm. Partial Differential Equations **32** (2007), no. 10-12, 1737–1758, DOI 10.1080/03605300701382340. MR2372486 (2008k:35491)

[9] G. Eskin and J. Ralston, *Inverse scattering problem for the Schrödinger equation with magnetic potential at a fixed energy*, Comm. Math. Phys. **173** (1995), no. 1, 199–224. MR1355624 (96i:81295)

[10] Colin Guillarmou and Leo Tzou, *Calderón inverse problem with partial data on Riemann surfaces*, Duke Math. J. **158** (2011), no. 1, 83–120, DOI 10.1215/00127094-1276310. MR2794369 (2012f:35574)

[11] Oleg Y. Imanuvilov, Gunther Uhlmann, and Masahiro Yamamoto, *Determination of second-order elliptic operators in two dimensions from partial Cauchy data*, Proc. Natl. Acad. Sci. USA **108** (2011), no. 2, 467–472, DOI 10.1073/pnas.1011681107. MR2770947 (2012a:35364)

[12] Oleg Yu. Imanuvilov, Gunther Uhlmann, and Masahiro Yamamoto, *The Calderón problem with partial data in two dimensions*, J. Amer. Math. Soc. **23** (2010), no. 3, 655–691, DOI 10.1090/S0894-0347-10-00656-9. MR2629983 (2012c:35472)

[13] Oleg Yu. Imanuvilov, Gunther Uhlmann, and Masahiro Yamamoto, *Inverse boundary value problem by measuring Dirichlet data and Neumann data on disjoint sets*, Inverse Problems **27** (2011), no. 8, 085007, 26, DOI 10.1088/0266-5611/27/8/085007. MR2819949 (2012c:78002)

[14] A. Katchalov and Ya. Kurylev, *Multidimensional inverse problem with incomplete boundary spectral data*, Comm. Partial Differential Equations **23** (1998), no. 1-2, 55–95, DOI 10.1080/03605309808821338. MR1608496 (99b:35213)

[15] Alexander Katchalov, Yaroslav Kurylev, and Matti Lassas, *Inverse boundary spectral problems*, Chapman & Hall/CRC Monographs and Surveys in Pure and Applied Mathematics, vol. 123, Chapman & Hall/CRC, Boca Raton, FL, 2001. MR1889089 (2003e:58045)

[16] Carlos E. Kenig, Johannes Sjöstrand, and Gunther Uhlmann, *The Calderón problem with partial data*, Ann. of Math. (2) **165** (2007), no. 2, 567–591, DOI 10.4007/annals.2007.165.567. MR2299741 (2008k:35498)

[17] Kim Knudsen and Mikko Salo, *Determining nonsmooth first order terms from partial boundary measurements*, Inverse Probl. Imaging **1** (2007), no. 2, 349–369, DOI 10.3934/ipi.2007.1.349. MR2282273 (2008k:35500)

[18] Katsiaryna Krupchyk, Matti Lassas, and Gunther Uhlmann, *Inverse problems with partial data for a magnetic Schrödinger operator in an infinite slab and on a bounded domain*, Comm. Math. Phys. **312** (2012), no. 1, 87–126, DOI 10.1007/s00220-012-1431-1. MR2914058

[19] Yaroslav Kurylev and Matti Lassas, *Gelf'and inverse problem for a quadratic operator pencil*, J. Funct. Anal. **176** (2000), no. 2, 247–263, DOI 10.1006/jfan.2000.3615. MR1784415 (2001g:47089)

[20] Yaroslav Kurylev and Matti Lassas, *Hyperbolic inverse boundary-value problem and time-continuation of the non-stationary Dirichlet-to-Neumann map*, Proc. Roy. Soc. Edinburgh Sect. A **132** (2002), no. 4, 931–949, DOI 10.1017/S0308210500001943. MR1926923 (2003h:35279)

[21] Yaroslav V. Kurylev and Matti Lassas, *The multidimensional Gel'fand inverse problem for non-self-adjoint operators*, Inverse Problems **13** (1997), no. 6, 1495–1501, DOI 10.1088/0266-5611/13/6/006. MR1484000 (98m:35224)

[22] Yaroslav V. Kurylev and Matti Lassas, *Hyperbolic inverse problem with data on a part of the boundary*, Differential equations and mathematical physics (Birmingham, AL, 1999), AMS/IP Stud. Adv. Math., vol. 16, Amer. Math. Soc., Providence, RI, 2000, pp. 259–272. MR1764756 (2001f:58061)

[23] Matti Lassas and Lauri Oksanen, *An inverse problem for a wave equation with sources and observations on disjoint sets*, Inverse Problems **26** (2010), no. 8, 085012, 19, DOI 10.1088/0266-5611/26/8/085012. MR2661691 (2011m:35422)

[24] Matti Lassas and Lauri Oksanen, *Inverse problem for the Riemannian wave equation with Dirichlet data and Neumann data on disjoint sets*, Submitted. Preprint arXiv:1208.2105 (2012).

[25] Matti Lassas and Gunther Uhlmann, *On determining a Riemannian manifold from the Dirichlet-to-Neumann map* (English, with English and French summaries), Ann. Sci. École Norm. Sup. (4) **34** (2001), no. 5, 771–787, DOI 10.1016/S0012-9593(01)01076-X. MR1862026 (2003e:58037)

[26] Adrian I. Nachman, *Global uniqueness for a two-dimensional inverse boundary value problem*, Ann. of Math. (2) **143** (1996), no. 1, 71–96, DOI 10.2307/2118653. MR1370758 (96k:35189)

[27] Gen Nakamura, Zi Qi Sun, and Gunther Uhlmann, *Global identifiability for an inverse problem for the Schrödinger equation in a magnetic field*, Math. Ann. **303** (1995), no. 3, 377–388, DOI 10.1007/BF01460996. MR1354996 (96m:35336)

[28] Lassi Päivärinta, Mikko Salo, and Gunther Uhlmann, *Inverse scattering for the magnetic Schrödinger operator*, J. Funct. Anal. **259** (2010), no. 7, 1771–1798, DOI 10.1016/j.jfa.2010.06.002. MR2665410 (2011f:35363)

[29] Rakesh, *Characterization of transmission data for Webster's horn equation*, Inverse Problems **16** (2000), no. 2, L9–L24, DOI 10.1088/0266-5611/16/2/102. MR1766761 (2001e:35109)

[30] Mikko Salo, *Semiclassical pseudodifferential calculus and the reconstruction of a magnetic field*, Comm. Partial Differential Equations **31** (2006), no. 10-12, 1639–1666, DOI 10.1080/03605300500530420. MR2273968 (2007j:35236)

[31] Zi Qi Sun, *An inverse boundary value problem for Schrödinger operators with vector potentials*, Trans. Amer. Math. Soc. **338** (1993), no. 2, 953–969, DOI 10.2307/2154438. MR1179400 (93j:35066)

[32] John Sylvester and Gunther Uhlmann, *A global uniqueness theorem for an inverse boundary value problem*, Ann. of Math. (2) **125** (1987), no. 1, 153–169, DOI 10.2307/1971291. MR873380 (88b:35205)

[33] Daniel Tataru, *Unique continuation for solutions to PDE's; between Hörmander's theorem and Holmgren's theorem*, Comm. Partial Differential Equations **20** (1995), no. 5-6, 855–884, DOI 10.1080/03605309508821117. MR1326909 (96e:35019)

UNIVERSITY OF HELSINKI, DEPARTMENT OF MATHEMATICS AND STATISTICS, P.O. BOX 68 FI-00014, HELSINKI, FINLAND

E-mail address: matti.lassas@helsinki.fi

UNIVERSITY OF WASHINGTON, DEPARTMENT OF MATHEMATICS, BOX 354350, SEATTLE, WA 98195-4350

Current address: University College London, Department of Mathematics, Gower Street, London, WC1E 6BT

E-mail address: l.oksanen@ucl.ac.uk

Contemporary Mathematics
Volume **615**, 2014
http://dx.doi.org/10.1090/conm/615/12279

Damping Mechanisms for Regularized Transformation-acoustics Cloaking

Jingzhi Li, Hongyu Liu, and Hongpeng Sun

Dedicated to Professor G. Uhlmann on the occasion of his 60th birthday

ABSTRACT. The regularized transformation-acoustics invisibility cloaking via the 'blow-up a-small-ball' construction is considered. In order to defeat the cloak-busting inclusions, a lossy layer is incorporated into the construction. Two mechanisms have been introduced and investigated in the literature. We generalize the two mechanisms and evaluate the corresponding cloaking performances. The exterior boundary conditions on the lossy layers are assessed and they are compared to those for the sound-soft and sound-hard layers. Moreover, the cloaking of active contents is assessed and the results show that in order to cloak a source/sink, the lossy layer must be properly chosen.

1. Introduction

This paper is concerned with the invisibility cloaking for acoustic waves via the approach of transformation optics (cf. [**GL**, **GLU**, **LE**, **P**]). There are many theoretical and experimental developments in this field and we refer to [**C**, **GK**, **GKL**, **NO**, **U**, **Y**] for the state-of-the-art surveys.

The method of transformation acoustics makes use of the transformation properties of the acoustical parameters, namely the density tensor and the acoustic modulus. The invisibility construction is based on a blow-up transformation between the virtual space and the physical space. For the ideal/perfect invisibility cloaking considered in [**LE**, **P**], it is a singular 'blow-up-a-point' transformation. The cloaking media achieved in this way possess singular structures. The singularity poses much challenge to both theoretical analysis and practical construction. In order to avoid the singular structure, it is natural to introduce regularization into the construction. Instead of the perfect cloak, one considers the approximate cloak or near cloak. Various regularized near-cloaking schemes have been proposed for the acoustic cloaking. In [**G**, **G1**, **R**], the regularization relies on truncation of singularities,

2010 *Mathematics Subject Classification.* Primary 78A46, 35J25; Secondary 35P25, 65N21,76Q05.

Key words and phrases. Acoustic scattering, approximate invisibility cloaking, transformation acoustics, damping mechanisms.

The work of Jingzhi Li was supported by the NSF of China (No. 11201453 and 91130022).

The work of Hongyu Liu was supported by NSF grant, DMS 1207784.

The work of Hongpeng Sun was partially supported by the NSF of China (No. 10990012) and the SFB Research Center "Mathematical Optimization and Application in Biomedical Sciences".

whereas in [**KO**, **KS**, **L1**] the 'blow-up-a-point' transformation in [**GLU**, **LE**, **P**] is regularized to be the 'blow-up-a-small-region' transformation. However, it is shown in [**KL**] that the two schemes are equivalent to each other. Hence, in the present study, we shall only consider the 'blow-up-a-small-region' construction. Particularly, it is shown in [**KO**] that if one intends to nearly cloak an arbitrary content, it is necessary to incorporate a certain mechanism to defeat the cloak-busting inclusions. It is proposed in [**KO**] to incorporate a lossy layer right between the cloaked region and the cloaking region, and the lossy layer possesses a large loss parameter in the virtual space, which shall be referred to as a *high-loss scheme* in the present paper. In [**LLS**, **LS**], the authors proposed to employ a lossy layer with a high density tensor instead, and this shall be referred to as a *high-density scheme* in the present paper. The high-density construction is shown to produce significantly enhanced accuracy of approximation compared to that of the high-loss construction. Special mechanisms of different nature are also developed in [**A1**, **A2**, **A3**, **L1**], where a sound-hard or sound-soft layer is employed to isolate the cloaked region from the exterior space and they produce a certain homogeneous boundary condition on the interior interface of the cloaking layer. It is heuristically pointed out in [**KO**, **LS**] that the sound-soft (res. sound-hard) layer is an ideal state corresponding to the limiting case of a high-lossy (res. high-density) layer. In the present article, we shall give a more rigorous and quantitative study in this aspect. Before we proceed to discuss more about our results, we briefly introduce the mathematical formulation of the acoustic cloaking problem under our consideration.

Let Σ be a bounded Lipschitz domain in $\mathbb{R}^N$, $N = 2, 3$. For a *regular* acoustic medium supported in Σ, we denote it by $(\Sigma; g, q)$, where $g = (g^{ij})$ is a symmetric-matrix valued function and q is a scalar function, and they satisfy

$$\lambda|\xi|^2 \leq g(x)\xi \cdot \xi \leq \lambda^{-1}|\xi|^2, \quad \Re q(x) \geq \lambda, \quad \Im q(x) \geq 0 \quad \text{for a.e. } x \in \Sigma, \tag{1.1}$$

where $0 < \lambda < 1$ is a constant, $\Re$ and $\Im$ denote taking the real and imaginary parts, respectively. The time-harmonic acoustic wave propagation in $(\Sigma; g, q)$ is governed by the following Helmholtz equation

$$\begin{cases} \nabla \cdot (g\nabla u) + \omega^2 qu = f & \text{in } \Sigma, \\ \displaystyle\sum_{i,j=1}^{N} \nu_i g^{ij} \partial_j u = \psi \in H^{-1/2}(\partial\Sigma) & \text{on } \partial\Sigma, \end{cases} \tag{1.2}$$

where $\nu = (\nu_i)_{i=1}^N$ is the outward unit normal to Σ, $\omega \in \mathbb{R}_+$ is called the wave number, $f \in H^{-1}(\Sigma)$ denotes a source/sink term inside Σ, and $u \in H^1(\Sigma)$ denotes the wave pressure (cf. [**CK**, **NE**]). Associated with (1.2), we introduce the Neumann-to-Dirichlet (NtD) operator Λ as follows

$$\Lambda(\psi) = u|_{\partial\Sigma} : H^{-1/2}(\Sigma) \to H^{1/2}(\Sigma), \tag{1.3}$$

where u is the solution to (1.2). It is always assumed that ω^2 is not an eigenvalue so that we have a well-defined boundary operator Λ. Here, we remark that Λ is a linear operator if $f = 0$ in (1.2), and generally nonlinear if $f \neq 0$. Next, we let $\widetilde{\Sigma}$ be a bounded Lipschitz domain as well and it is assumed that there exists a bi-Lipschitz transformation F which maps Σ to $\widetilde{\Sigma}$. Let $(\widetilde{\Sigma}; \widetilde{g}, \widetilde{q})$ be a regular medium defined to be the *push-forward* of $(\Sigma; g, q)$ according to the following formulae

$$
\begin{aligned}
\widetilde{g} = F_* g &:= [(|\det(DF)|)^{-1}\left(DF \cdot g \cdot DF^T\right)] \circ F^{-1}, \\
\widetilde{q} = F_* q &:= [(|\det(DF)|)^{-1}\, q] \circ F^{-1}, \\
\widetilde{f} = F_* f &:= [(|\det(DF)|)^{-1}\, f] \circ F^{-1},
\end{aligned} \tag{1.4}
$$

where DF denotes the Jacobian matrix of F. In the following, we shall write $(\widetilde{\Sigma}; \widetilde{g}, \widetilde{q}, \widetilde{f}) := F_*(\Sigma; g, q, f)$ to denote the push-forward introduced above. Then, the following result is in order (see [**KO**, **L1**, **G2**, **GK**]). Let $F: \Sigma \to \Sigma$ be a bi-Lipschitz mapping such that $F|_{\partial\Sigma} = \text{Identity}$, and set $(\Sigma; \widetilde{g}, \widetilde{q}, \widetilde{f}) = F_*(\Sigma; g, q, f)$. Then $u \in H^1(\Sigma)$ satisfies (1.2) iff $\widetilde{u} \in H^1(\Sigma)$ also satisfies (1.2) but with (g, q, f) replaced by $(\widetilde{g}, \widetilde{q}, \widetilde{f})$. Moreover, one has that $\Lambda = \widetilde{\Lambda}$, where $\widetilde{\Lambda}$ is the NtD operator associated with $(\Sigma; \widetilde{g}, \widetilde{q}, \widetilde{f})$

Now, let Ω and D be two simply connected Lipschitz domains in $\mathbb{R}^N$, $N = 2, 3$. It is assumed that D is convex containing the origin and $D \Subset \Omega$. Let $D_\rho := \{\rho x; x \in D\}$ for any $\rho \in \mathbb{R}_+$. We assume that there exists a bi-Lipschitz mapping $F_\rho^{(1)}$ such that

$$
F_\rho^{(1)} : \Omega \backslash D_\rho \to \Omega \backslash D, \quad F_\rho^{(1)}|_{\partial\Omega} = \text{Identity}. \tag{1.5}
$$

Let $F_\rho^{(2)} = x/\rho : D_\rho \to D$ and

$$
F_\rho(x) := \begin{cases} F_\rho^{(1)}(x), & x \in \Omega \backslash D_\rho, \\ F_\rho^{(2)}(x), & x \in D_\rho. \end{cases} \tag{1.6}
$$

Set

$$
(\Omega \backslash \overline{D}; g, q) = (\Omega \backslash \overline{D}; g_c, q_c) := (F_\rho)_*(\Omega \backslash \overline{D}_\rho; I, 1), \tag{1.7}
$$

and

$$
(D \backslash \overline{D}_{1/2}; g, q) = (D \backslash \overline{D}_{1/2}; g_l, q_l) := (F_\rho)_*(D_\rho \backslash D_{\rho/2}; g_l', q_l'), \tag{1.8}
$$

with

$$
g_l' = \gamma \rho^s I, \quad q_l' = \alpha + i\beta\rho^t, \quad s, t \in \mathbb{R}, \tag{1.9}
$$

where α, β and γ are positive constants independent of ρ. Finally, we also let $(D_{1/2}; g, q) = (D_{1/2}; g_a, q_a)$ be an arbitrary but regular medium supported inside $D_{1/2}$, and $f_a \in H^{-1}(D_{1/2})$ be a source/sink term located inside $D_{1/2}$. $(\Omega; g, q, f_a)$ describes a cloaking device with the cloaking region $(\Omega \backslash \overline{D}; g_c, q_c)$, the cloaked region $(D_{1/2}; g_a, q_a, f_a)$, and the lossy layer $(D \backslash \overline{D}_{1/2}; g_l, q_l)$ right between the cloaking and cloaked regions. For the source-free case, namely $f_a \equiv 0$, the above cloaking construction has been investigated in [**KO**] with $(s, t) = (0, -2)$ in (1.9), and [**LLS**, **LS**] with $(s, t) = (2, 0)$ in (1.9), respectively. In order to present a quick discussion on the results obtained therein, we introduce the 'free-space' Helmholtz equation

$$
\begin{cases} \Delta u_0 + \omega^2 u_0 = 0 & \text{in } \Omega, \\ \dfrac{\partial u_0}{\partial \nu} = \psi \in H^{-1/2}(\partial\Omega) & \text{on } \partial\Omega. \end{cases} \tag{1.10}
$$

We let Λ_0 be the NtD map associated with (1.10) and $\Lambda_\rho := \Lambda_{g,q,f_a}$ be the NtD map associated with the cloaking device $(\Omega; g, q, f_a)$. For $f_a \equiv 0$, it is shown in

[**KO**] with $(s,t)=(0,-2)$ in (1.9) that

$$\|\Lambda_\rho-\Lambda_0\|_{\mathcal{L}(H^{-1/2}(\partial\Omega),H^{1/2}(\partial\Omega))}\le C\begin{cases}|\ln\rho|^{-1}, & N=2\\ \rho, & N=3\end{cases}; \tag{1.11}$$

and in [**LLS, LS**] with $(s,t)=(2,0)$ in (1.9) that

$$\|\Lambda_\rho-\Lambda_0\|_{\mathcal{L}(H^{-1/2}(\partial\Omega),H^{1/2}(\partial\Omega))}\le C\rho^N. \tag{1.12}$$

In both (1.11) and (1.12), the estimates are shown to be optimal and the constant C is shown to be independent of (g_a,q_a).

In this work, we shall consider the above two constructions with general $(s,0)$ and $(0,t)$ in (1.9). As mentioned earlier, they shall be referred to as the *high-density scheme* and the *high-loss scheme*, respectively. Also, as discussed earlier, we are particularly interested in assessing the wave field u on the exterior boundary of the lossy layer, namely ∂D, and compare with those for the *sound-soft* and *sound-hard* layers. For a sound-soft layer, the wave pressure vanishes on ∂D and one would have a homogeneous Dirichlet boundary condition, and for a sound-hard layer, the wave velocity vanishes on ∂D and one would have a homogeneous Neumann boundary condition. The results obtained would clarify some heuristic beliefs in the literature that a high-lossy layer is a finite-realization of a sound-soft layer and a high-density layer is a finite-realization of a sound-hard layer. Moreover we shall consider a delta-point source, $f_a=\eta_0\delta_0$, located inside the cloak region, and the obtained results show that in order to cloak active contents, an appropriate lossy layer must be chosen. Furthermore, the results shall clarify some points in [**KL**] on the interior active cloaking and in [**L2**] on the general study of cloaking of active contents.

In the rest of the paper, we shall consider our study by taking $\Omega=B_R$ and $D=B_1$, where $R>1$ and B_s denotes a central ball of radius $s\in\mathbb{R}_+$. Moreover, we shall let g_a be a positive constant multiple of the identity matrix and q_a be a positive constant. With no confusion in the context, we shall just take g_a as a positive constant. We set

$$(\Omega;g',q',f_a')=(F_\rho^{-1})_*(\Omega;g,q,f_a), \tag{1.13}$$

and $u_\rho=(F_\rho)^*u$. Clearly, we have $(B_R\backslash\overline{B}_\rho;g',q')=(B_R\backslash\overline{B}_\rho;I,1)$, $(B_\rho\backslash\overline{B}_{\rho/2};g',q')=(B_\rho\backslash\overline{B}_{\rho/2};g_l',q_l')$ and, $(B_{\rho/2};g',q')=(B_{\rho/2};g_a',q_a')$ with $g_a'=\rho^{2-N}g_a$ and $q_a'=\rho^{-N}q_a$. Moreover,

$$\begin{cases}\nabla\cdot(g'\nabla u_\rho)+\omega^2q'u_\rho=f_a' & \text{in } B_R,\\ \displaystyle\sum_{i,j=1}^N \nu_i(g')^{ij}\partial_j u_\rho=\psi & \text{on } \partial B_R,\end{cases} \tag{1.14}$$

Clearly, in order to assess the cloaking performance, it suffices to evaluate $\|u_\rho-u_0\|_{H^{1/2}(\partial B_R)}$. For the convenience of the following discussion, we recall that it holds in $\mathbb{R}^2$ (cf. [**LS**])

$$\psi(x)=\sum_{n=-\infty}^{\infty}\psi_n(R)e^{in\theta}\in H^{-1/2}(\partial B_R),$$

we have (see, e.g. [**LS**])

$$\|\psi\|^2_{H^{-1/2}(\partial B_R)} = \sum_{n=-\infty}^{\infty} (1+n^2/R^2)^{-1/2}|\psi_n\sqrt{2\pi R}|^2. \tag{1.15}$$

Then the 'free-space' solution $u_0(x) \in H^1(B_R)$ of (1.10) is given by

$$u_0(x) = \sum_{n=-\infty}^{\infty} \frac{\psi_n J_n(\omega|x|)}{\omega J_n'(\omega R)} e^{in\theta}, \quad x \in B_R. \tag{1.16}$$

2. The high-density scheme

We shall first consider consider the high-density construction, namely $(B_R; g, q, f_a)$ with $(s, 0)$ in (1.9), $s \in \mathbb{R}$. In the sequel, $(B_R; g', q', f_a')$ shall be referred to as the virtual image of $(B_R; g, q, f_a)$ and (1.14) shall be referred to as the Helmholtz equation in the virtual space.

2.1. The case with $g_l' = \gamma\rho^{2-\delta}$, $0 \le \delta < 2$, $q_l' = (\alpha + i\beta)$ and $f_a = 0$. We first consider the 2D case with $\delta = 0$. Let $\omega_a = \omega\sqrt{q_a'/g_a'} = \omega\sqrt{q_a/g_a}\rho^{-1}$ and $\omega_l = \omega\sqrt{q_l'/g_l'} = \omega\sqrt{(\alpha+i\beta)/\gamma}\rho^{-1}$, where we shall choose the branch of $\sqrt{(\alpha+i\beta)/\gamma}$ with $\Im\sqrt{(\alpha+i\beta)/\gamma} > 0$ throughout the rest of the paper. We see that $\omega_l\rho$, $\omega_a\rho$ are constants independent of ρ.

PROPOSITION 2.1. *In $\mathbb{R}^2$, when $g_l' = \gamma\rho^2$, $q_l' = (\alpha + i\beta)$ and $f_a = 0$, we have*

$$\begin{cases} \|u_\rho - u_0\|_{H^{1/2}(\partial B_R)} \le C\rho^2 \|\psi\|_{H^{-1/2}(\partial B_R)}, \\ \left\|\dfrac{\partial u_R^+(\rho\cdot)}{\partial\nu}\right\|_{H^{-1/2}(\partial B_1)} \le C\rho \|\psi\|_{H^{-1/2}(\partial B_R)}. \end{cases} \tag{2.1}$$

REMARK 2.2. This means the near-cloak in this case is within ρ^2-approximation of the ideal cloak. Moreover, we see that the lossy layer with $(s,t) = (2,0)$ is an approximation to a sound-hard layer.

PROOF. The proposition can be proved in a similar manner to that given in [**LS**]. However, for our subsequent study, we shall present the proof in some detail. Assume the solution to the virtual problem (1.14) is given by

(2.2)

$$u_\rho(x) = \begin{cases} \sum_{n=-\infty}^{\infty} e_n J_n(\omega_a|x|)e^{in\theta}, & x \in B_{\rho/2}, \\ \sum_{n=-\infty}^{\infty} c_n J_n(\omega_l|x|)e^{in\theta} + \sum_{n=-\infty}^{\infty} d_n H_n^{(1)}(\omega_l|x|)e^{in\theta}, & x \in B_\rho\backslash\overline{B}_{\rho/2}, \\ \sum_{n=-\infty}^{\infty} a_n J_n(\omega|x|)e^{in\theta} + \sum_{n=-\infty}^{\infty} b_n H_n^{(1)}(\omega|x|)e^{in\theta}, & x \in B_R\backslash\overline{B}_\rho. \end{cases}$$

Set $u_a = u_\rho|_{B_{\rho/2}}$, $u_l = u_\rho|_{B_\rho\setminus\overline{B}_{\rho/2}}$ and $u_R = u|_{B_R\setminus\overline{B}_\rho}$. By the standard transmission conditions on $\partial B_{\rho/2}, \partial B_\rho$ and the boundary condition on ∂B_R, we have

$$
\begin{cases}
u_a(x) = u_l(x),\ g_a' \dfrac{\partial u_a(x)}{\partial\nu(x)} = g_l' \dfrac{\partial u_l(x)}{\partial\nu(x)}, & x \in \partial B_{\rho/2}, \\
u_l(x) = u_R(x),\ g_l' \dfrac{\partial u_l(x)}{\partial\nu(x)} = \dfrac{\partial u_R(x)}{\partial\nu(x)}, & x \in \partial B_\rho, \\
\dfrac{\partial u_R(x)}{\partial\nu(x)} = \psi(x),\ x \in \partial B_R.
\end{cases}
\tag{2.3}
$$

Plugging the series representations (2.2) into (2.3), we have the following linear system of equations for the coefficients,

$$
\begin{cases}
e_n J_n(\omega_a\rho/2) = c_n J_n(\omega_l\rho/2) + d_n H_n^{(1)}(\omega_l\rho/2), \\
\sqrt{g_a'q_a'} e_n J_n'(\omega_a\rho/2) = \sqrt{g_l'q_l'}[c_n J_n'(\omega_l\rho/2) + d_n H_n^{(1)'}(\omega_l\rho/2)], \\
c_n J_n(\omega_l\rho) + d_n H_n^{(1)}(\omega_l\rho) = a_n J_n(\omega\rho) + b_n H_n^{(1)}(\omega\rho), \\
\sqrt{g_l'q_l'}[c_n J_n'(\omega_l\rho) + d_n H_n^{(1)'}(\omega_l\rho)] = a_n J_n'(\omega\rho) + b_n H_n^{(1)'}(\omega\rho), \\
a_n\omega J_n'(\omega R) + b_n\omega H_n^{(1)'}(\omega R) = \psi_n.
\end{cases}
\tag{2.4}
$$

Let $A = \sqrt{\frac{g_a'q_a'}{g_l'q_l'}} = \sqrt{\frac{g_aq_a}{\gamma(\alpha+i\beta)}}\rho^{-2}$ and notice that $\omega_l\rho$ and $\omega_a\rho$ are all constants. From the first two equations of (2.4) we have

$$
\begin{cases}
d_n = -\frac{J_n(\omega_l\rho/2)}{H_n^{(1)}(\omega_l\rho/2)} c_n \quad \text{if } J_n(\omega_a\rho/2) = 0, \\
d_n = -\frac{J_n'(\omega_l\rho/2) - AJ_n(\omega_l\rho/2)\frac{J_n'(\omega_a\rho/2)}{J_n(\omega_a\rho/2)}}{H_n^{(1)'}(\omega_l\rho/2) - AH_n^{(1)}(\omega_l\rho/2)\frac{J_n'(\omega_a\rho/2)}{J_n(\omega_a\rho/2)}} c_n \quad \text{if } J_n(\omega_a\rho/2) \neq 0.
\end{cases}
\tag{2.5}
$$

If $\Im q_a' \neq 0$, we have $J_n(\omega_a\rho/2) \neq 0$. Denoting the expressions before c_n in (2.5) by Υ_n, namely $d_n := \Upsilon_n c_n$. So while $\rho \to +0$, we have

$$
\begin{cases}
\Upsilon_n = -\frac{J_n(\omega_l\rho/2)}{H_n^{(1)}(\omega_l\rho/2)} \quad \text{if } J_n(\omega_a\rho/2) = 0, \\
\Upsilon_n \sim -\frac{J_n(\omega_l\rho/2)}{H_n^{(1)}(\omega_l\rho/2)} \quad \text{if } J_n(\omega_a\rho/2) \neq 0.
\end{cases}
\tag{2.6}
$$

Substituting d_n into the third and fourth equations of (2.4), we have by straightforward calculations

$$
b_n = -\frac{\sqrt{g_l'q_l'}\frac{J_n'(\omega_l\rho)+\Upsilon_n H_n^{(1)'}(\omega_l\rho)}{J_n(\omega_l\rho)+\Upsilon_n H_n^{(1)}(\omega_l\rho)} J_n(\omega\rho) - J_n'(\omega\rho)}{\sqrt{g_l'q_l'}\frac{J_n'(\omega_l\rho)+\Upsilon_n H_n^{(1)'}(\omega_l\rho)}{J_n(\omega_l\rho)+\Upsilon_n H_n^{(1)}(\omega_l\rho)} H_n^{(1)}(\omega\rho) - H_n^{(1)'}(\omega\rho)} a_n.
\tag{2.7}
$$

Let Γ_n denote the expression before a_n in (2.7), namely $b_n := \Gamma_n a_n$. Then we have

$$
\mathcal{H}_n(\rho) = \frac{J_n'(\omega_l\rho) + \Upsilon_n H_n^{(1)'}(\omega_l\rho)}{J_n(\omega_l\rho) + \Upsilon_n H_n^{(1)}(\omega_l\rho)}.
\tag{2.8}
$$

As $\rho \to +0$,

$$
\mathcal{H}_n(\rho) \to \mathcal{H}_n^0 = \frac{J_n'(\omega_l\rho) + \Upsilon_n^0 H_n^{(1)'}(\omega_l\rho)}{J_n(\omega_l\rho) + \Upsilon_n^0 H_n^{(1)}(\omega_l\rho)},
$$

where $\Upsilon_n^0 = -\frac{J_n(\omega_l\rho/2)}{H_n^{(1)}(\omega_l\rho/2)}$. Plugging (2.7) into the last equation in (2.4), we have

$$u_R(x) = \sum_{n=-\infty}^{\infty} \frac{\psi_n[J_n(\omega R) + \Gamma_n H_n^{(1)}(\omega R)]}{\omega[J_n'(\omega R) + \Gamma_n H_n^{(1)\prime}(\omega R)]} e^{in\theta}, \quad x \in \partial B_R. \tag{2.9}$$

Hence,

(2.10)

$$[u_\rho(x) - u_0(x)]\,|_{\partial B_R} = \sum_{n=-\infty}^{\infty} \frac{\psi_n J_n(\omega R)}{\omega J_n'(\omega R)} \left[\frac{\Gamma_n \left[\frac{H_n^{(1)}(\omega R)}{J_n(\omega R)} - \frac{H_n^{(1)\prime}(\omega R)}{J_n'(\omega R)}\right]}{1 + \frac{\Gamma_n H_n^{(1)\prime}(\omega R)}{J_n'(\omega R)}} \right] e^{in\theta}.$$

Therefore we have

(2.11)

$$\begin{aligned}
\|u_\rho - u_0\|^2_{H^{1/2}(\partial B_R)} &= \sum_{n=-\infty}^{\infty} \left(1 + \frac{n^2}{R^2}\right)^{1/2} \left| \frac{\psi_n J_n(\omega R)}{\omega J_n'(\omega R)} \widetilde{h}_n \sqrt{2\pi R} \right|^2 \\
&\leq \left\{ \sum_{n=-\infty}^{\infty} \left(1 + \frac{n^2}{R^2}\right)^{-1/2} |\psi_n \sqrt{2\pi R}|^2 \right\} \left\{ \sum_{n=-\infty}^{\infty} \left(1 + \frac{n^2}{R^2}\right) \left| \frac{J_n(\omega R)}{\omega J_n'(\omega R)} \widetilde{h}_n \right|^2 \right\} \\
&\leq \left\{ \sum_{n=-\infty}^{\infty} \left(1 + \frac{n^2}{R^2}\right) \left| \frac{J_n(\omega R)}{\omega J_n'(\omega R)} \widetilde{h}_n \right|^2 \right\} \|\psi\|^2_{H^{-1/2}(\partial B_R)},
\end{aligned}$$

where

$$\widetilde{h}_n := \left| \frac{\Gamma_n \left[\frac{H_n^{(1)}(\omega R)}{J_n(\omega R)} - \frac{H_n^{(1)\prime}(\omega R)}{J_n'(\omega R)}\right]}{1 + \frac{\Gamma_n H_n^{(1)\prime}(\omega R)}{J_n'(\omega R)}} \right|.$$

Since $H_{-n}^{(1)}(\omega r) = (-1)^n H_n^{(1)}(\omega r)$, we only need consider $n \geq 0$ in estimating the series in the last inequality of (2.11). By the fact that Υ_n,$\mathcal{H}_n$ are all constants which together with the asymptotic behaviors of $J_n(\omega\rho)$, $H_n^{(1)}(\omega\rho)$, $J_n'(\omega\rho)$, $H_n^{(1)\prime}(\omega\rho)$ as $\rho \to +0$ (cf. [**AS**, **LZ**]), one can further show

$$\begin{cases} \Gamma_0 \sim -i(\sqrt{\gamma(\alpha + i\beta)}\mathcal{H}_n^0 + \omega/2)\frac{\pi\omega\rho^2}{2}, & n = 0, \\ \Gamma_n \sim \pi i \frac{(\omega\rho)^2 n}{(2^n n!)^2}, & n \geq 1. \end{cases} \tag{2.12}$$

Then using the estimates in (2.12), together with the use of the asymptotic developments of the Bessel and Hankel functions for large n (cf. [**AS**]), one can verify that there exists a sufficiently large integer N_1 such that

$$\begin{cases} \widetilde{h}_0 \sim -i(\sqrt{\gamma(\alpha + i\beta)}\mathcal{H}_n^0 + \omega/2)\frac{\pi\omega\rho^2}{2}\left[\frac{H_0^{(1)}(\omega R)}{J_0(\omega R)} - \frac{H_0^{(1)\prime}(\omega R)}{J_0'(\omega R)}\right], & n = 0, \\ \widetilde{h}_n \sim \pi i \frac{(\omega\rho)^2 n}{(2^n n!)^2}\left[\frac{H_n^{(1)}(\omega R)}{J_n(\omega R)} - \frac{H_n^{(1)\prime}(\omega R)}{J_n'(\omega R)}\right], & 1 \leq n \leq N_1, \\ \widetilde{h}_n \sim 2(\frac{\rho}{R})^{2n}, & n > N_1. \end{cases} \tag{2.13}$$

Hence from (2.13), we readily see that there exists a constant C_1 independent of ρ for ρ sufficiently small such that

$$|\widetilde{h}_n| \leq C_1\rho^2, \quad n \leq N_1, \tag{2.14}$$

and for $n > N_1$

$$|\widetilde{h}_n| \leq 2\left(\frac{\rho}{R}\right)^{2n}. \tag{2.15}$$

Here it is emphasized that due to the asymptotic developments of $\widetilde{h}_0$ and $\widetilde{h}_1$, (2.14) is the best estimate one could achieve, namely $C_1\rho^2$ could not be improved. Now, using (2.14), we see that

$$\sum_{n=0}^{N_1}\left(1+\frac{n^2}{R^2}\right)\left|\frac{J_n(\omega R)}{\omega J_n'(\omega R)}\widetilde{h}_n\right|^2 \le C_2\rho^4. \tag{2.16}$$

Let N_1 be sufficiently large such that $|\frac{J_n(\omega R)}{\omega J_n'(\omega R)}| < 1$ for $n > N_1$, then for $\rho < \min\{R/4, 1\}$

$$\begin{aligned}&\sum_{n>N_1}\left(1+\frac{n^2}{R^2}\right)\left|\frac{J_n(\omega R)}{\omega J_n'(\omega R)}\widetilde{h}_n\right|^2\\ \le&\frac{\rho^4}{R^4}\sum_{n>N_1}\left(1+\frac{n^2}{R^2}\right)\left|2\left(\frac{\rho}{R}\right)^{2(n-1)}\right|^2 < C_3\rho^4.\end{aligned} \tag{2.17}$$

Combining (2.11), (2.16) and (2.17), we have

$$\|u_\rho - u_0\|_{H^{1/2}(\partial B_R)} \le C\rho^2\|\psi\|_{H^{-1/2}(\partial B_R)}. \tag{2.18}$$

Moreover, from the optimality of the estimate (2.14), we readily see the sharpness of (2.18). For the normal derivatives on ∂B_ρ, we have

$$\left.\frac{\partial u_R^+}{\partial \nu}\right|_{\partial B_\rho} = \sum_{n=-\infty}^{\infty}\omega l_n e^{in\theta} \tag{2.19}$$

where

$$l_n := a_n J_n'(\omega\rho) + b_n H_n^{(1)'}(\omega\rho).$$

By (2.4), (2.7) and (2.8), we have

$$l_n = \frac{\psi_n\sqrt{g_l'q_l'}\mathcal{H}_n(\rho)}{\omega[J_n'(\omega R)+\Gamma_n H_n^{(1)'}(\omega R)]}\cdot\frac{J_n'(\omega\rho)H_n^{(1)}(\omega\rho) - J_n(\omega\rho)H_n^{(1)'}(\omega\rho)}{\sqrt{g_l'q_l'}\mathcal{H}_n(\rho)H_n^{(1)}(\omega\rho) - H_n^{(1)'}(\omega\rho)}. \tag{2.20}$$

By the Wronskian $J_n(t)Y_n'(t) - J_n'(t)Y_n(t) = \frac{2}{\pi t}$, we have

$$l_n = \frac{\psi_n\sqrt{g_l'q_l'}\mathcal{H}_n(\rho)}{\omega[J_n'(\omega R)+\Gamma_n H_n^{(1)'}(\omega R)]}\frac{1}{\sqrt{g_l'q_l'}\mathcal{H}_n(\rho)H_n^{(1)}(\omega\rho) - H_n^{(1)'}(\omega\rho)}\cdot\frac{-2i}{\pi\omega\rho}.$$

By (2.12) and $\sqrt{g_l'q_l'} = \rho\sqrt{\gamma(\alpha+i\beta)}$, we have

$$\|\frac{\partial u_R^+(\rho\cdot)}{\partial\nu}\|_{H^{-1/2}(\partial B_1)} \le C\rho\|\psi\|_{H^{-1/2}(\partial B_R)}. \tag{2.21}$$

where C is a generic constant that does not depend on ρ. □

For the 3D case, similar to Proposition 2.1, one has

REMARK 2.3. In $\mathbb{R}^3$, for $g_l' = \gamma\rho^2$, $q_l' = \alpha + i\beta$ and $f_a = 0$, there also exists a positive constant C independent of ρ and $\psi \in H^{-1/2}(\partial B_R)$, such that

$$\begin{aligned}&\|u_\rho - u_0\|_{H^{1/2}(\partial B_R)} \le C\rho^3\|\psi\|_{H^{-1/2}(\partial B_R)},\\ &\left\|\frac{\partial u_R^+(\rho\cdot)}{\partial\nu}\right\|_{H^{-1/2}(\partial B_1)} \le C\rho\|\psi\|_{H^{-1/2}(\partial B_R)},\end{aligned} \tag{2.22}$$

where C is independent of ρ and ψ.

Next we consider the case with $0 < \delta < 2$ and we have

PROPOSITION 2.4. *In $\mathbb{R}^2$, when $g_l' = \gamma\rho^{2-\delta}$, $q_l' = \alpha + i\beta$, $0 < \delta < 2$, and $f_a = 0$, we have that for any $\psi \in H^{-1/2}(\partial B_R)$,*

(2.23)

$$\|u_\rho - u_0\|_{H^{1/2}(\partial B_R)} \le C\rho^{2-\delta}\|\psi\|_{H^{-1/2}(\partial B_R)}, \|\frac{\partial u_R^+(\rho\cdot)}{\partial\nu}\|_{H^{-1/2}(\partial B_1)} \sim C\rho^{1-\delta}\|\psi\|_{H^{-1/2}(\partial B_R)},$$

where C is a positive constant independent of ρ and ψ.

REMARK 2.5. This means the near-cloak in this case is within $\rho^{2-\delta}$-approximation of the ideal cloak. Moreover, when $0 < \delta < 1$, the lossy layer tends to be a sound hard layer and when $1 \le \delta < 2$, the layer does not tend to be a sound hard layer.

PROOF. The proof follows from a completely similar argument to that in Proposition 2.1. The only difference one needs special care is that in this case, $\omega_l\rho = \omega\sqrt{\alpha + i\beta/\gamma}\rho^{\frac{\delta}{2}}$, $A = \sqrt{\frac{g_a' q_a'}{g_l' q_l'}} = \sqrt{\frac{g_a q_a}{\gamma(\alpha+i\beta)}}\rho^{-2+\frac{\delta}{2}}$. It can then be verified that

$$\Upsilon_n \to -\frac{J_n(\omega_l\rho/2)}{H_n^{(1)}(\omega_l\rho/2)}, \quad \text{as} \quad \rho \to 0^+.$$

The asymptotic properties of $\mathcal{H}_n(\rho)$ as $\rho \to 0^+$ are as follows

$$\begin{cases} \mathcal{H}_0 \sim \frac{\sqrt{\gamma}\rho^{-\frac{\delta}{2}}}{\omega\sqrt{\alpha+i\beta}\ln 2}, \\ \mathcal{H}_n \sim \frac{n\sqrt{\gamma}\rho^{-\frac{\delta}{2}}}{\omega\sqrt{\alpha+i\beta}}. \end{cases} \tag{2.24}$$

Then one can show that Γ_n has the following asymptotic properties as $\rho \to 0^+$,

$$\begin{cases} \Gamma_0 \sim \dfrac{-i\gamma\pi\rho^{2-\delta}}{2\ln 2}, \\ \Gamma_n \sim \dfrac{-in\pi(\omega\rho)^{2n}}{(2^n n!)^2}. \end{cases} \tag{2.25}$$

Hence the proof is complete. □

REMARK 2.6. Similar to the 2D case, in $\mathbb{R}^3$, when $g_l' = \gamma\rho^{2-\delta}$, $q_l' = \alpha + i\beta$, $0 < \delta < 2$, and $f_a = 0$, one has

$$\|u_\rho - u_0\|_{H^{1/2}(\partial B_R)} \sim \rho^{3-\delta}\|\psi\|_{H^{-1/2}(\partial B_R)}, \|\frac{\partial u_R^+(\rho\cdot)}{\partial\nu}\|_{H^{-1/2}(\partial B_1)} \sim C\rho^{1-\delta}\|\psi\|_{H^{-1/2}(\partial B_R)}.$$

2.2. The case with $g_l' = \gamma\rho^{2-\delta}$, $0 \le \delta < 2$, $q_l' = \alpha + i\beta$ and $f_a = \eta_0\delta_0$.

PROPOSITION 2.7. *In $\mathbb{R}^2$, let $g_l' = \gamma\rho^2$, $q_l' = \alpha + i\beta$, and $f_a = \eta_0\delta_0$, where η_0 is a constant which represents the strength of the source. In this case, we have*

$$\|u_\rho - u_0\|_{H^{1/2}(\partial B_R)} = \mathcal{O}(1). \tag{2.26}$$

This means that in this case the construction produces no near-cloaking effect.

PROOF. We shall make use of the same notations as that in Section 2.1. In the virtual space, one has

(2.27)
$$u_\rho(x)=\begin{cases}\sum_{n=-\infty}^{\infty} e_nJ_n(\omega_a|x|)e^{in\theta}+\frac{\eta_0}{\rho^2}H_0^{(1)}(\omega_a|x|), & x\in B_{\rho/2},\\ \sum_{n=-\infty}^{\infty} c_nJ_n(\omega_l|x|)e^{in\theta}+\sum_{n=-\infty}^{\infty} d_nH_n^{(1)}(\omega_l|x|)e^{in\theta}, & x\in B_\rho\backslash\overline{B}_{\rho/2},\\ \sum_{n=-\infty}^{\infty} a_nJ_n(\omega|x|)e^{in\theta}+\sum_{n=-\infty}^{\infty} b_nH_n^{(1)}(\omega|x|)e^{in\theta}, & x\in B_R\backslash\overline{B}_\rho.\end{cases}$$

By the transmission boundary condition on $\partial B_{\rho/2}$ and ∂B_ρ, we have

(2.28)
$$\begin{cases}e_0J_0(\omega_a\rho/2)+\frac{\eta_0}{\rho^2}H_0^{(1)}(\omega_a\rho/2)=c_0J_0(\omega_l\rho/2)+d_0H_0^{(1)}(\omega_l\rho/2)\,,\\ \sqrt{g_a'q_a'}[e_0J_0'(\omega_a\rho/2)+\frac{\eta_0}{\rho^2}H_0^{(1)'}(\omega_a\rho/2)]=\sqrt{g_l'q_l'}[c_0J_0'(\omega_l\rho/2)+d_0H_0^{(1)'}(\omega_l\rho/2)]\,,\\ c_0J_0(\omega_l\rho)+d_0H_0^{(1)}(\omega_l\rho)=a_0J_0(\omega\rho)+b_0H_0^{(1)}(\omega\rho)\,,\\ \sqrt{g_l'q_l'}[c_0J_0'(\omega_l\rho)+d_0H_0^{(1)'}(\omega_l\rho)]=a_0J_0'(\omega\rho)+b_0H_0^{(1)}(\omega\rho)\,,\\ a_0J_0'(\omega R)+b_0H_0^{(1)'}(\omega R)=\psi_0/\omega\,.\end{cases}$$

For $n\neq 0$, one can show

$$\begin{cases}e_nJ_n(\omega_a\rho/2)=c_nJ_n(\omega_l\rho/2)+d_nH_n^{(1)}(\omega_l\rho/2)\,,\\ \sqrt{g_a'q_a'}e_nJ_n'(\omega_a\rho/2)=\sqrt{g_l'q_l'}[c_nJ_n'(\omega_l\rho/2)+d_nH_n^{(1)'}(\omega_l\rho/2)]\,,\\ c_nJ_n(\omega_l\rho)+d_nH_n^{(1)}(\omega_l\rho)=a_nJ_n(\omega\rho)+b_nH_n^{(1)}(\omega\rho)\,,\\ \sqrt{g_l'q_l'}[c_nJ_n'(\omega_l\rho)+d_nH_n^{(1)'}(\omega_l\rho)]=a_nJ_n'(\omega\rho)+b_nH_n^{(1)}(\omega\rho)\,,\\ a_nJ_n'(\omega R)+b_nH_n^{(1)'}(\omega R)=\psi_n/\omega\,.\end{cases}\tag{2.29}$$

By the third and fourth equations of (2.28), we have

$$\begin{aligned}d_0=&\frac{[J_0(\omega\rho)J_0'(\omega\rho)-\frac{1}{\sqrt{g_l'q_l'}}J_0'(\omega\rho)J_0(\omega_l\rho)]a_0}{H_0^{(1)}(\omega_l\rho)J_0'(\omega_l\rho)-H_0^{(1)'}(\omega_l\rho)J_0(\omega_l\rho)}\\&+\frac{[H_0^{(1)}(\omega\rho)J_0'(\omega_l\rho)-\frac{1}{\sqrt{g_l'q_l'}}H_0^{(1)'}(\omega\rho)J_0(\omega\rho)]b_0}{H_0^{(1)}(\omega_l\rho)J_0'(\omega_l\rho)-H_0^{(1)'}(\omega_l\rho)J_0(\omega_l\rho)},\\ c_0=&\frac{[J_0(\omega\rho)H_0^{(1)'}(\omega_l\rho)-\frac{1}{\sqrt{g_l'q_l'}}J_0'(\omega\rho)H_0^{(1)}(\omega_l\rho)]a_0}{J_0(\omega_l\rho)H_0^{(1)'}(\omega_l\rho)-J_0'(\omega_l\rho)H_0^{(1)}(\omega_l\rho)}\\&+\frac{[H_0^{(1)}(\omega\rho)H_0^{(1)'}(\omega_l\rho)-\frac{1}{\sqrt{g_l'q_l'}}H_0^{(1)'}(\omega\rho)H_0^{(1)}(\omega_l\rho)]b_0}{J_0(\omega_l\rho)H_0^{(1)'}(\omega_l\rho)-J_0'(\omega_l\rho)H_0^{(1)}(\omega_l\rho)}.\end{aligned}\tag{2.30}$$

Let $A=\sqrt{\frac{g_a'q_a'}{g_l'q_l'}}$. Using the Wronskian

$$J_n(t)Y_n'(t)-J_n'(t)Y_n(t)=\frac{2}{\pi t},\tag{2.31}$$

and by virtue of the first and second equation of (2.28), we further have

(2.32)

$$\begin{aligned}&\Big\{[\frac{J_0(\omega_l\frac{\rho}{2})}{J_0(\omega_a\frac{\rho}{2})}J_0'(\omega_a\frac{\rho}{2})-\frac{1}{A}J_0'(\omega_l\frac{\rho}{2})][J_0(\omega\rho)H_0^{(1)'}(\omega_l\rho)-\frac{1}{\sqrt{g_l'q_l'}}J_0'(\omega\rho)H_0^{(1)}(\omega_l\rho)]\\&-[\frac{H_0^{(1)}(\omega_l\frac{\rho}{2})}{J_0(\omega_a\frac{\rho}{2})}J_0'(\omega_a\frac{\rho}{2})-\frac{1}{A}H_0^{(1)'}(\omega_l\frac{\rho}{2})][J_0(\omega\rho)J_0'(\omega_l\rho)-\frac{1}{\sqrt{g_l'q_l'}}J_0'(\omega\rho)J_0(\omega_l\rho)]\Big\}a_0\\&+\Big\{[\frac{J_0(\omega_l\frac{\rho}{2})}{J_0(\omega_a\frac{\rho}{2})}J_0'(\omega_a\frac{\rho}{2})-\frac{1}{A}J_0'(\omega_l\frac{\rho}{2})][H_0^{(1)}(\omega\rho)H_0^{(1)'}(\omega_l\rho)-\frac{1}{\sqrt{g_l'q_l'}}H_0^{(1)'}(\omega\rho)H_0^{(1)}(\omega_l\rho)]\\&-[\frac{H_0^{(1)}(\omega_l\frac{\rho}{2})}{J_0(\omega_a\frac{\rho}{2})}J_0'(\omega_a\frac{\rho}{2})-\frac{1}{A}H_0^{(1)'}(\omega_l\frac{\rho}{2})][H_0^{(1)}(\omega\rho)J_0'(\omega_l\rho)-\frac{1}{\sqrt{g_l'q_l'}}H_0^{(1)'}(\omega\rho)J_0(\omega_l\rho)]\Big\}b_0\\&=\frac{2i}{\pi\omega_l\rho}\frac{\eta_0}{\rho^2}[-H_0^{(1)'}(\omega_a\rho/2)+\frac{J_0'(\omega_a\rho/2)}{J_0(\omega_a\rho/2)}H_0^{(1)}(\omega_a\rho/2)].\end{aligned}$$

In the case under discussion, one has $1/A=\sqrt{\frac{\gamma(\alpha+i\beta)}{g_aq_a}}\rho^2$. We define the following shorthand notations for brevity.

$$\begin{aligned}A_1&=\frac{J_0(\omega_l\frac{\rho}{2})}{J_0(\omega_a\frac{\rho}{2})}J_0'(\omega_a\frac{\rho}{2}),\quad A_2=\frac{H_0^{(1)}(\omega_l\frac{\rho}{2})}{J_0(\omega_a\frac{\rho}{2})}J_0'(\omega_a\frac{\rho}{2}),\\A_3&=-H_0^{(1)'}(\omega_a\rho/2)+\frac{J_0'(\omega_a\rho/2)}{J_0(\omega_a\rho/2)}H_0^{(1)}(\omega_a\rho/2),\\c_0&=\sqrt{\frac{\gamma(\alpha+i\beta)}{g_aq_a}},\quad d_0=1/\sqrt{\gamma(\alpha+i\beta)}.\end{aligned}$$

Since $\omega_l\rho=\omega\sqrt{\frac{\alpha+i\beta}{\gamma}}$ and $\omega_a\rho$ are constants, it can be shown that A_1,A_2,A_3 are all constants as $\rho\to+0$. Then one has that as $\rho\to0^+$,

$$\begin{aligned}&\Big\{[A_1-\rho^2c_0J_0'(\omega_l\frac{\rho}{2})][H_0^{(1)'}(\omega_l\rho)+d_0\frac{\omega}{2}H_0^{(1)}(\omega_l\rho)]\\&-[A_2-\rho^2c_0H_0^{(1)'}(\omega_l\frac{\rho}{2})][J_0'(\omega_l\rho)+d_0\frac{\omega}{2}J_0(\omega_l\rho)]\Big\}a_0\\&+\Big\{[A_1-\rho^2c_0J_0'(\omega_l\frac{\rho}{2})][\frac{2i}{\pi}\ln\frac{\omega\rho}{2}H_0^{(1)'}(\omega_l\rho)-d_02i\frac{1}{\pi\omega\rho^2}H_0^{(1)}(\omega_l\rho)]\\&-[A_2-\rho^2c_0H_0^{(1)'}(\omega_l\frac{\rho}{2})][\frac{2i}{\pi}\ln\frac{\omega\rho}{2}J_0'(\omega_l\rho)-d_02i\frac{1}{\pi\omega\rho^2}J_0(\omega_l\rho)]\Big\}b_0\\&\sim\sqrt{\frac{\gamma}{\alpha+i\beta}}\frac{2i}{\pi\omega}\frac{\eta_0}{\rho^2}A_3.\end{aligned}\tag{2.33}$$

Next, letting $D_0=\sqrt{\frac{\gamma}{\alpha+i\beta}}\frac{2i}{\pi\omega}\eta_0A_3$ and dropping the lower order term, we have

$$\begin{aligned}&\Big\{A_1[H_0^{(1)'}(\omega_l\rho)+d_0\frac{\omega}{2}H_0^{(1)}(\omega_l\rho)]-A_2[J_0'(\omega_l\rho)+d_0\frac{\omega}{2}J_0(\omega_l\rho)]\Big\}a_0\\&+\frac{1}{\rho^2}[-A_1\frac{d_02i}{\pi\omega}H_0^{(1)}(\omega_l\rho)+A_2\frac{d_02i}{\pi\omega}J_0(\omega_l\rho)]b_0\sim\frac{D_0}{\rho^2},\end{aligned}\tag{2.34}$$

In combination with

$$a_0J_0'(\omega R)+b_0H_0^{(1)'}(\omega R)=\psi_0/\omega,\tag{2.35}$$

(2.34) yields that

$$\begin{cases} a_0 \sim \frac{\psi_0/\omega - D_0 H_0^{(1)'}(\omega R)/E_0}{J_0'(\omega R) - C_0\rho^2 H_0^{(1)'}(\omega R)/E_0} \\ b_0 \sim \frac{\rho^2\psi_0/\omega - D_0 J_0'(\omega R)}{\rho^2 H_0^{(1)'}(\omega R) - E_0 J_0'(\omega R)/C_0}, \end{cases} \tag{2.36}$$

by letting

$$C_0 := \{A_1[H_0^{(1)'}(\omega_l\rho) + d_0\frac{\omega}{2}H_0^{(1)}(\omega_l\rho)] - A_2[J_0'(\omega_l\rho) + d_0\frac{\omega}{2}J_0(\omega_l\rho)]\},$$

$$E_0 := -A_1\frac{d_0 2i}{\pi\omega}H_0^{(1)}(\omega_l\rho) + A_2\frac{d_0 2i}{\pi\omega}J_0(\omega_l\rho).$$

Then by the Wronskians (2.31) and (2.35) we have further

$$\begin{aligned} &\left|a_0 J_0(\omega R) + b_0 H_0^{(1)}(\omega R) - \frac{\psi_0 J_0(\omega R)}{\omega J_0'(\omega R)}\right| \\ =&\frac{b_0\omega[H_0^{(1)}(\omega R)J_0'(\omega R) - H_0^{(1)'}(\omega R)J_0(\omega R)]}{\omega J_0'(\omega R)} = \frac{-2ib_0}{\pi\omega R J_0'(\omega R)}. \end{aligned} \tag{2.37}$$

Next, by (2.36), $b_0 \sim \mathcal{O}(1)$ and

$$\left|a_n J_n(\omega R) + b_n H_n^{(1)}(\omega R) - \frac{\psi_n J_n(\omega R)}{\omega J_n'(\omega R)}\right| = \mathcal{O}(\rho^{2n}), \quad n \geq 1.$$

Using the estimates derived above, together with a similar analysis to that for the proof of Propostion 2.1, one can show that

$$\|u_\rho - u_0\|_{H^{1/2}(\partial B_R)} = \mathcal{O}(1). \tag{2.38}$$

□

REMARK 2.8. It can be verified by using (2.36) that in $\mathbb{R}^2$, when $g_l' = \gamma\rho^2$, $q_l' = \alpha + i\beta$, and $f_a = \eta_0\delta_0$, one has $\left\|\frac{\partial u_R^+(\rho\cdot)}{\partial\nu}\right\|_{H^{-1/2}(\partial B_1)} \sim \rho^{-1}$. That is, the normal velocity of the wave field on ∂B_1 would blow up as $\rho \to 0^+$.

PROPOSITION 2.9. *In $\mathbb{R}^2$, when $g_l' = \gamma\rho^{2-\delta}$, $q_l' = \alpha + i\beta$, $0 < \delta < 2$, and $f_a = \eta_0\delta_0$, one has $\|u_\rho - u_0\|_{H^{1/2}(\partial B_R)} = \mathcal{O}(\rho^{-\delta})$. That is, the construction woud not yield a near-cloaking device.*

PROOF. The proof follows from a completely similar argument to that for Proposition 2.7. We only need notice the following significant differences. For this case, equations (2.27)-(2.32) and (2.35) are of the same formulas as the $g_l' = \rho^2$ case, but with different parameters. $1/A = \sqrt{\frac{g_a' q_a'}{g_l' q_l'}} = c_0\rho^{2-\frac{\delta}{2}}$, $\omega_l\rho = \omega\sqrt{\frac{\alpha+i\beta}{\gamma}}\rho^{\frac{\delta}{2}}$, $\sqrt{g_l' q_l'} = \sqrt{(\alpha+i\beta)\gamma}\rho^{1-\frac{\delta}{2}}$. By the asymptotic behaviors of the Bessel and Hankel functions (cf. [**AS**] and [**LZ**]), the equation (2.32) has the following asyptotics,

$$\begin{aligned} &\frac{J_0'(\omega_a\frac{\rho}{2})}{J_0(\omega_a\frac{\rho}{2})}[\frac{2i}{\pi\omega_l\rho}a_0 + \frac{4\ln 2\rho^{\frac{\delta}{2}-2}}{\pi^2\omega\sqrt{(\alpha+i\beta)\gamma}}b_0] \\ \sim&\frac{2i}{\pi\omega_l\rho}\frac{D}{\rho^2}[-H_0^{(1)'}(\omega_a\rho/2) + \frac{J_0'(\omega_a\rho/2)}{J_0(\omega_a\rho/2)}H_0^{(1)}(\omega_a\rho/2)], \end{aligned} \tag{2.39}$$

and this in combination with (2.35) gives

$$b_0 \sim \rho^{-\delta}, \tag{2.40}$$

which readily implies the statement of the proposition. □

REMARK 2.10. In the same setting as Proposition 2.9, one can show that $\left\|\frac{\partial u_R^+(\rho\cdot)}{\partial\nu}\right\|_{H^{-1/2}(\partial B_1)}\sim\rho^{-1-\delta}$. That is, the normal velocity of the wave field would blow up as $\rho\to 0^+$.

For the three dimensional case, we have

PROPOSITION 2.11. *In $\mathbb{R}^3$, when $f_a=\eta_0\delta_0$, $g_l'=\gamma\rho^{2-\delta}$ and $q_l=\alpha+i\beta$, $0\le\delta\le 1$, one has*

$$\|u_\rho-u_0\|_{H^{1/2}(\partial B_R)}\le C_1\rho^{1-\delta}\eta_0+C_2\rho^2\|\psi\|_{H^{-1/2}(\partial B_R)}, \tag{2.41}$$

where C_1 and C_2 are independent of ρ,η_0 and ψ.; whereas when $1<\delta<2$, we have

$$\|u_\rho-u_0\|_{H^{1/2}(\partial B_R)}\sim\mathcal{O}(\rho^{1-\delta}). \tag{2.42}$$

PROOF. The 3D case can be proved in a similar manner to that for the 2D case. Let

$$\psi(x)|_{\partial B_R}=\sum_{n=0}^{\infty}\sum_{m=-n}^{n}\psi_n^mY_n^m(\hat{x})\in H^{-1/2}(\partial B_R),$$

with

$$\|\psi(x)\|^2_{H^{-1/2}(\partial B_R)}=\sum_{n=0}^{\infty}\sum_{m=-n}^{n}(1+n(n+1)/R^2)^{-1/2}|\psi_n^mR|^2<+\infty. \tag{2.43}$$

The free space solution u_0 to (1.14) in $\mathbb{R}^3$ is given by

$$u_0=\sum_{n=0}^{\infty}\sum_{m=-n}^{n}\frac{\psi_n^m}{\omega j_n'(\omega R)}j_n(\omega|x|)Y_n^m(\hat{x}).$$

Noting $g_a'=g_a/\rho$, $q_a'=q_a'/\rho^3$ in $B_{\rho/2}$, similar to (2.2) for the 2D case, the wave fields in the separated domains could be represented as follows

$$\begin{aligned} u_a(x)=&\sum_{n=0}^{\infty}\sum_{m=-n}^{n}e_n^mj_n(\omega_a|x|)Y_n^m(\hat{x})-\frac{i\eta_0}{4\pi\rho^2}h_0^{(1)}(\omega_a|x|),\\ u_l(x)=&\sum_{n=0}^{\infty}\sum_{m=-n}^{n}c_n^mj_n(\omega_l|x|)Y_n^m(\hat{x})+\sum_{n=0}^{\infty}\sum_{m=-n}^{n}d_n^mh_n^{(1)}(\omega_l|x|)Y_n^m(\hat{x}),\\ u_R(x)=&\sum_{n=0}^{\infty}\sum_{m=-n}^{n}a_n^mj_n(\omega|x|)Y_n^m(\hat{x})+\sum_{n=0}^{\infty}\sum_{m=-n}^{n}b_n^mh_n^{(1)}(\omega|x|)Y_n^m(\hat{x}). \end{aligned} \tag{2.44}$$

By using the standard transmission conditions and the boundary condition, one could derive the following linear systems of equations for the coefficients, when $n=0$

$$\begin{cases} e_0^0j_0(\omega_a\rho/2)-\frac{i\eta_0}{4\pi\rho^2}h_0^{(1)}(\omega_a|x|)=c_0^0j_0(\omega_l\rho/2)+d_0^0h_0^{(1)}(\omega_l\rho/2)\,,\\ \sqrt{g_a'q_a'}[e_0^0j_0'(\omega_a\rho/2)-\frac{i\eta_0}{4\pi\rho^2}h_0^{(1)'}(\omega_a|x|)]=\sqrt{g_l'q_l'}[c_0^0j_0'(\omega_l\rho/2)+d_0^0h_0^{(1)'}(\omega_l\rho/2)]\,,\\ c_0^0j_0(\omega_l\rho)+d_0^0h_0^{(1)}(\omega_l\rho)=a_0^0j_0(\omega\rho)+b_0^0h_0^{(1)}(\omega\rho)\,,\\ \sqrt{g_l'q_l'}[c_0^0j_n'(\omega_l\rho)+d_0^0h_0^{(1)'}(\omega_l\rho)]=a_0^0j_0'(\omega\rho)+b_0^0h_0^{(1)'}(\omega\rho)\,,\\ \omega a_0^0j_0'(\omega R)+\omega b_0^0h_0^{(1)'}(\omega R)=\psi_0^0\,, \end{cases} \tag{2.45}$$

and when $n \geq 1$,

$$\begin{cases} e_n^m j_n(\omega_a\rho/2) = c_n^m j_n(\omega_l\rho/2) + d_n^m h_n^{(1)}(\omega_l\rho/2)\,, \\ \sqrt{g_a'q_a'} e_n^m j_n'(\omega_a\rho/2) = \sqrt{g_l'q_l'}[c_n^n j_n'(\omega_l\rho/2) + d_n^m h_n^{(1)'}(\omega_l\rho/2)]\,, \\ c_n^m j_n(\omega_l\rho) + d_n^m h_n^{(1)}(\omega_l\rho) = a_n^m j_n(\omega\rho) + b_n^m h_n^{(1)}(\omega\rho)\,, \\ \sqrt{g_l'q_l'}[c_n^m j_n'(\omega_l\rho) + d_n^m h_n^{(1)'}(\omega_l\rho)] = a_n^m j_n'(\omega\rho) + b_n^m h_n^{(1)'}(\omega\rho)\,, \\ \omega a_n^m j_n'(\omega R) + \omega b_n^m h_n^{(1)'}(\omega R) = \psi_n^m\,. \end{cases} \tag{2.46}$$

Letting $\widetilde{A} = \sqrt{\frac{g_a'q_a'}{g_l'q_l'}}$, and solving (2.46), one has

$$b_n^m = -\frac{\sqrt{g_l'q_l'}\frac{j_n'(\omega_l\rho)+\widetilde{\Upsilon}_n h_n^{(1)'}(\omega_l\rho)}{j_n(\omega_l\rho)+\widetilde{\Upsilon}_n h_n^{(1)}(\omega_l\rho)} j_n(\omega\rho) - j_n'(\omega\rho)}{\sqrt{g_l'q_l'}\frac{j_n'(\omega_l\rho)+\widetilde{\Upsilon}_n h_n^{(1)'}(\omega_l\rho)}{j_n(\omega_l\rho)+\widetilde{\Upsilon}_n h_n^{(1)}(\omega_l\rho)} h_n^{(1)}(\omega\rho) - h_n^{(1)'}(\omega\rho)} a_n^m, \tag{2.47}$$

where

$$\widetilde{\Upsilon}_n := \begin{cases} -\frac{j_n(\omega_l\rho/2)}{h_n^{(1)}(\omega_l\rho/2)} & \text{if } j_n(\omega_a\rho/2) = 0, \\ -\frac{j_n'(\omega_l\rho/2) - \widetilde{A} j_n(\omega_l\rho/2)\frac{j_n'(\omega_a\rho/2)}{j_n(\omega_a\rho/2)}}{h_n^{(1)'}(\omega_l\rho/2) - \widetilde{A} h_n^{(1)}(\omega_l\rho/2)\frac{j_n'(\omega_a\rho/2)}{j_n(\omega_a\rho/2)}} & \text{if } j_n(\omega_a\rho/2) \neq 0. \end{cases} \tag{2.48}$$

Let $\widetilde{\Gamma}_n$ denote the expression before a_n^m in (2.47). Let u_ρ^0, u_0^0 represent the zeroth order Fourier term with Y_n^m on ∂B_R. Then

$$\begin{aligned} &\left([u_\rho(x) - u_0(x)] - [u_\rho^0 - u_0^0]\right)\Big|_{\partial B_R} \\ &= \sum_{n=1}^{\infty}\sum_{m=-n}^{n} \frac{\psi_n^m j_n(\omega R)}{\omega j_n'(\omega R)} \left\{ \frac{\widetilde{\Gamma}_n \left[\frac{h_n^{(1)}(\omega R)}{j_n(\omega R)} - \frac{h_n^{(1)'}(\omega R)}{j_n'(\omega R)}\right]}{1 + \frac{\widetilde{\Gamma}_n h_n^{(1)'}(\omega R)}{j_n'(\omega R)}} \right\} Y_n^m(\hat{x}). \end{aligned} \tag{2.49}$$

Set

$$\widetilde{g}_n = \widetilde{\Gamma}_n \left[\frac{h_n^{(1)}(\omega R)}{j_n(\omega R)} - \frac{h_n^{(1)'}(\omega R)}{j_n'(\omega R)}\right] \Big/ \left[1 + \frac{\widetilde{\Gamma}_n h_n^{(1)'}(\omega R)}{j_n'(\omega R)}\right], \quad n \geq 1,$$

then

$$\begin{aligned} &\left\|[u_\rho(x) - u_0(x)] - [u_\rho^0 - u_0^0]\right\|_{H^{1/2}(\partial B_R)}^2 \\ &= \sum_{n=1}^{\infty}\sum_{m=-n}^{n} \sqrt{1 + \frac{n(n+1)}{R^2}} \left|\frac{\psi_n^m j_n(\omega R)}{\omega j_n'(\omega R)} \widetilde{g}_n R\right|^2 \\ &\leq \left\{\sum_{n=1}^{\infty}\sum_{m=-n}^{n} \frac{1}{\sqrt{1 + \frac{n(n+1)}{R^2}}} |\psi_n^m R|^2\right\} \left\{\sum_{n=1}^{\infty} (1 + \frac{n(n+1)}{R^2}) \left|\frac{j_n(\omega R)\widetilde{g}_n}{\omega j_n'(\omega R)}\right|^2\right\}. \end{aligned} \tag{2.50}$$

By a similar asymptotic argument to the 2D case, one can show

$$\widetilde{g}_n \sim \mathcal{O}(\omega\rho)^{2n+1}, \tag{2.51}$$

which is then applied to the estimate of (2.50), one can further show that

$$\|u_\rho(x) - u_0(x) - [u_\rho^0 - u_0^0]\|_{H^{1/2}(\partial B_R)} \leq C\rho^3 \|\psi - \psi_0^0\|_{H^{-1/2}(\partial B_R)}, \tag{2.52}$$

Next for $n = 0$, by using the Wronskian

$$j_n(t)y_n'(t) - j_n'(t)y_n(t) = \frac{1}{t^2}, \tag{2.53}$$

and the first and second equations of (2.45), we have

(2.54)

$$\begin{aligned}
&\{[\frac{j_0(\omega_l \frac{\rho}{2})}{j_0(\omega_a \frac{\rho}{2})} j_0'(\omega_a \frac{\rho}{2}) - \frac{1}{\tilde{A}} j_0'(\omega_l \frac{\rho}{2})][j_0(\omega\rho)h_0^{(1)'}(\omega_l\rho) - \frac{1}{\sqrt{g_l'q_l'}} j_0'(\omega\rho)h_0^{(1)}(\omega_l\rho)] \\
&- [\frac{h_0^{(1)}(\omega_l \frac{\rho}{2})}{j_0(\omega_a \frac{\rho}{2})} j_0'(\omega_a \frac{\rho}{2}) - \frac{1}{\tilde{A}} h_0^{(1)'}(\omega_l \frac{\rho}{2})][j_0(\omega\rho)j_0'(\omega_l\rho) - \frac{1}{\sqrt{g_l'q_l'}} j_0'(\omega\rho)j_0(\omega_l\rho)]\}a_0^0 \\
&+ \{[\frac{j_0(\omega_l \frac{\rho}{2})}{j_0(\omega_a \frac{\rho}{2})} j_0'(\omega_a \frac{\rho}{2}) - \frac{1}{\tilde{A}} j_0'(\omega_l \frac{\rho}{2})][h_0^{(1)}(\omega\rho)h_0^{(1)'}(\omega_l\rho) - \frac{1}{\sqrt{g_l'q_l'}} h_0^{(1)'}(\omega\rho)h_0^{(1)}(\omega_l\rho)] \\
&- [\frac{h_0^{(1)}(\omega_l \frac{\rho}{2})}{j_0(\omega_a \frac{\rho}{2})} j_0'(\omega_a \frac{\rho}{2}) - \frac{1}{\tilde{A}} h_0^{(1)'}(\omega_l \frac{\rho}{2})][h_0^{(1)}(\omega\rho)j_0'(\omega_l\rho) - \frac{1}{\sqrt{g_l'q_l'}} h_0^{(1)'}(\omega\rho)j_0(\omega_l\rho)]\}b_0^0 \\
&= \frac{1}{4\pi(\omega_l\rho)^2}\frac{\eta_0}{\rho^2}[-h_0^{(1)'}(\omega_a\rho/2) + \frac{j_0'(\omega_a\rho/2)}{j_0(\omega_a\rho/2)} h_0^{(1)}(\omega_a\rho/2)].
\end{aligned}$$

For the first term of $u_\rho - u_0$, by the Wronskian (2.53) and the last equation of (2.45), we have

$$\left| a_0^0 j_0(\omega R)Y_0^0 + b_0^0 h_0^{(1)}(\omega R)Y_0^0 - \frac{\psi_0^0 j_0(\omega R)}{\omega j_0'(\omega R)} Y_0^0 \right| = \left| \frac{b_0^0 Y_0^0}{j_0'(\omega R)(\omega R)^2} \right|. \tag{2.55}$$

Clearly, b_0^0 is the dominant factor for the first term of $u_\rho - u_0$. Since $\omega_l\rho = \omega\sqrt{\frac{\alpha+i\beta}{\gamma}}\rho^{\frac{\delta}{2}}$, $\sqrt{g_l'q_l'} = \sqrt{(\alpha+i\beta)\gamma}\rho^{1-\frac{\delta}{2}}$, and $\widetilde{A} = \sqrt{\frac{g_a q_a}{(\alpha+i\beta)\gamma}}\rho^{\frac{\delta}{2}-3}$, we further have the following from (2.54),

$$\frac{j_0'(\omega_a \frac{\rho}{2})}{j_0'(\omega_a \frac{\rho}{2})}[\frac{ia_0^0}{(\omega_l\rho)^2} - \frac{\rho^{\frac{\delta}{2}-3}}{\omega^2\sqrt{(\alpha+i\beta)\gamma}}\frac{b_0^0}{\omega_l\rho}] \sim \frac{1}{4\pi(\omega_l\rho)^2}\frac{\eta_0}{\rho^2}[-h_0^{(1)'}(\omega_a\rho/2) + \frac{j_0'(\omega_a\rho/2)}{j_0(\omega_a\rho/2)} h_0^{(1)}(\omega_a\rho/2)],$$

which in combination with the last equation in (2.45), implies that there exist constants C_1, C_2 independent of ρ such that

$$b_0^0 \sim C_1\eta_0\rho^{1-\delta} + C_2\psi_0^0\rho^{3-\delta}. \tag{2.56}$$

Thus the proof is completed. □

2.3. The case with $g_l' = \gamma\rho^{2+2\delta}$, $\delta > 0$ and $q_l' = \alpha + i\beta$.

PROPOSITION 2.12. *In $\mathbb{R}^2$, when $g_l' = \gamma\rho^{2+2\delta}$, $\delta > 0$ and $q_l' = \alpha + i\beta$, and $f_a = \eta_0\delta_0$, we have that*

$$\|u_\rho - u_0\|_{H^{1/2}(\partial B_R)} \le C_1\rho^2\|\psi\|_{H^{-1/2}(\partial B_R)} + C_2\rho^\delta e^{-\frac{l_1\rho^{-\delta}}{2}}\eta_0, \tag{2.57}$$

where $l_1 = \Im(\omega\sqrt{(\alpha+i\beta)/\gamma})$ (the positive branch) and C_1 and C_2 are positive constants independent of ψ and η_0. That is, the source/sink term can be exponentially cloaked in this case.

PROOF. We shall make use of a similar argument to that for Propsition 2.7. For the present proof, we have the same equations as those in (2.28), (2.29), but with different g_l',q_l'. Here $\omega_l = \omega\sqrt{q_l'/g_l'} = \omega\rho^{-1-\delta}\sqrt{(\alpha+i\beta)/\gamma}$, $\omega_l\rho = \omega\rho^{-\delta}\sqrt{(\alpha+i\beta)/\gamma}$. Let $\Re\omega_l\rho = \rho^{-\delta}l_0, \Im\omega_l\rho = \rho^{-\delta}l_1$. g_a', q_a', ω_a are the same as the above case. For $n \ne 0$, the equations are the same as those in [**LS**] and a_n, b_n are also the same.

We only need to consider the case $n=0$. Here in equation (2.32), $A=\sqrt{\frac{g_a' q_a'}{g_l' q_l'}}=\sqrt{\frac{g_a q_a}{\gamma(\alpha+i\beta)}}\rho^{-2-\delta}$. Let $B_1=\frac{J_0'(\omega_a\rho/2)}{J_0(\omega_a\rho/2)}$. By the asymptotic behavior of the Bessel functions Hankel functions with large variables and positive imaginary part (cf. [**AS**])

$$\begin{cases} J_n(z)\sim\sqrt{\frac{1}{2\pi z}}e^{|\Im(z)|}e^{i(-\Re(z)+\frac{n\pi}{2}+\frac{\pi}{4})}, & |\arg z|<\pi\,, \\ H_n^{(1)}(z)\sim\sqrt{\frac{2}{\pi z}}e^{-\Im(z)}e^{i(\Re(z)-\frac{n\pi}{2}-\frac{\pi}{4})}, & -\pi<\arg z<2\pi\,, \\ J_n'(z)\sim-\sqrt{\frac{1}{2\pi z}}e^{|\Im(z)|}e^{i(-\Re(z)+\frac{(n+1)\pi}{2}+\frac{\pi}{4})}, & |\arg z|<\pi\,, \\ H_n^{(1)'}(z)\sim-\sqrt{\frac{2}{\pi z}}e^{-\Im(z)}e^{i(\Re(z)-\frac{(n+1)\pi}{2}-\frac{\pi}{4})}, & -\pi<\arg z<2\pi\,. \end{cases} \tag{2.58}$$

With the same notations d_0, c_0 as those in (2.33), we have the following asymptotic expansion as $\rho\to 0^+$,

$$\begin{aligned} & e^{-\frac{l_1\rho^{-\delta}}{2}}\frac{\sqrt{2}}{\pi\omega_l\rho}[B_1+\rho^{2+\delta}c_0e^{i\pi/2}][-e^{i\pi/2}+\rho^{-\delta}d_0\frac{\omega}{2}]e^{i\frac{l_0\rho^{-\delta}}{2}}a_0 \\ & -e^{\frac{l_1\rho^{-\delta}}{2}}\frac{\sqrt{2}}{\pi\omega_l\rho}[B_1+\rho^{2+\delta}c_0e^{-i\pi/2}][-e^{i\pi/2}+\rho^{-\delta}d_0\frac{\omega}{2}]e^{-i\frac{l_0\rho^{-\delta}}{2}}a_0 \\ & -e^{-\frac{l_1\rho^{-\delta}}{2}}\frac{\sqrt{2}}{\pi\omega_l\rho}[B_1+\rho^{2+\delta}c_0e^{i\pi/2}][\frac{2}{\pi}\ln\frac{\omega\rho}{2}+d_0i\frac{2}{\pi\omega\rho^{2+\delta}}]e^{i\frac{l_0\rho^{-\delta}}{2}}b_0 \\ & +e^{\frac{l_1\rho^{-\delta}}{2}}\frac{\sqrt{2}}{\pi\omega_l\rho}[B_1+\rho^{2+\delta}c_0e^{-i\pi/2}][-\frac{2}{\pi}\ln\frac{\omega\rho}{2}+d_0i\frac{2}{\pi\omega\rho^{2+\delta}}]e^{-i\frac{l_0\rho^{-\delta}}{2}}b_0 \\ & \sim\frac{2i}{\pi\omega_l\rho}\frac{\eta_0}{\rho^2}[-H_0^{(1)'}(\omega_a\rho/2)+\frac{J_0'(\omega_a\rho/2)}{J_0(\omega_a\rho/2)}H_0^{(1)}(\omega_a\rho/2)]. \end{aligned} \tag{2.59}$$

Next, by dropping the exponentially decaying terms in (2.59), we further have

$$\begin{aligned} & -e^{\frac{l_1\rho^{-\delta}}{2}}\frac{\sqrt{2}}{\pi\omega_l\rho}[B_1+\rho^{2+\delta}c_0e^{-i\pi/2}][-e^{i\pi/2}+\rho^{-\delta}d_0\frac{\omega}{2}]e^{-i\frac{l_0\rho^{-\delta}}{2}}a_0 \\ & +e^{\frac{l_1\rho^{-\delta}}{2}}\frac{\sqrt{2}}{\pi\omega_l\rho}[B_1+\rho^{2+\delta}c_0e^{-i\pi/2}][-\frac{2}{\pi}\ln\frac{\omega\rho}{2}+c_0i\frac{2}{\pi\omega\rho^{2+\delta}}]e^{-i\frac{l_0\rho^{-\delta}}{2}}b_0 \\ & \sim\frac{2i}{\pi\omega_l\rho}\frac{\eta_0}{\rho^2}[-H_0^{(1)'}(\omega_a\rho/2)+\frac{J_0'(\omega_a\rho/2)}{J_0(\omega_a\rho/2)}H_0^{(1)}(\omega_a\rho/2)]. \end{aligned} \tag{2.60}$$

Let $g_0=\sqrt{2}[B_1+\rho^{2+\delta}c_0e^{-i\pi/2}][-e^{i\pi/2}+\rho^{-\delta}d_0\frac{\omega}{2}]$ and $y=[-H_0^{(1)'}(\omega_a\rho/2)+\frac{J_0'(\omega_a\rho/2)}{J_0(\omega_a\rho/2)}H_0^{(1)}(\omega_a\rho/2)]$. One can show from (2.60) that

$$\begin{aligned} & -a_0+\frac{c_0i4}{d_0\pi\omega^2}\rho^{-2}b_0\sim\frac{1}{g_0}e^{-\frac{l_1\rho^{-\delta}}{2}}e^{i\frac{l_0\rho^{-\delta}}{2}}2i\frac{\eta_0}{\rho^2}y \\ & \sim\frac{iy2\sqrt{2}\eta_0}{B_1d_0\omega}\frac{1}{\rho^{2-\delta}}e^{-\frac{l_1\rho^{-\delta}}{2}}e^{i\frac{l_0\rho^{-\delta}}{2}} \end{aligned} \tag{2.61}$$

Let $k_0 = \frac{c_0 i4}{d_0\pi\omega^2}$, $n_0 = \frac{iy2\sqrt{2}\eta_0}{B_1 d_0\omega}$ and $v(\rho) = \frac{1}{\rho^{2-\delta}} e^{-\frac{l_1\rho^{-\delta}}{2}} e^{i\frac{l_0\rho^{-\delta}}{2}}$. Combining (2.61) and (2.35), we have

$$(2.62)\qquad \begin{cases} a_0 \sim \dfrac{\frac{\psi_0}{\omega} - \frac{n_0 v(\rho)\rho^2 H_0^{(1)'}(\omega R)}{k_0}}{J_0'(\omega R) + \frac{\rho^2 H_0^{(1)'}(\omega R)}{k_0}} \\ b_0 \sim \dfrac{[\frac{\psi_0}{\omega} + n_0 v(\rho) J_0'(\omega R)]\rho^2}{\rho^2 H_0^{(1)'}(\omega R) + k_0 J_0'(\omega R)}. \end{cases}$$

Hence there exists constants C_1, C_2, such that

$$(2.63)\qquad \left| a_0 J_0(\omega R) + b_0 H_0^{(1)}(\omega R) - \frac{\psi_0 J_0(\omega R)}{\omega J_0'(\omega R)} \right| \le C_1 \rho^2 \psi_0 + C_2 \rho^\delta e^{-\frac{l_1\rho^{-\delta}}{2}} \eta_0,$$

and by which the proposition is readily proved. □

REMARK 2.13. It can be easily seen from the proof of Proposition 2.12 that

$$\|\frac{\partial u_R^+(\rho\cdot)}{\partial\nu}\|_{H^{-1/2}(\partial B_1)} \le C\rho,$$

where C depends on ψ and η_0 but does not depend on ρ.

3. The high-loss scheme

In this section, we shall consider the high-loss scheme, namely $g_l' = I$ and $q_l' = 1 + i\rho^t$, $t \in \mathbb{R}$. The cloaking performance in various settings can be assessed by following completely similar arguments to those for the high-density scheme, and we would present those results in the sequel but only sketch the proofs.

PROPOSITION 3.1. *In $\mathbb{R}^2$, when $g_l' = I, q_l' = 1 + i\rho^{-2+\delta}, 0 < \delta < 2$, and $f_a = 0$, we have*

$$(3.1)\qquad \|u_\rho - u_0\|_{H^{1/2}(\partial\Omega)} \le C\frac{1}{\ln\rho}\|\psi\|_{H^{-1/2}(\partial B_R)}, \quad \|u_R^+(\rho\cdot)\|_{H^{1/2}(\partial B_1)} \le C\frac{1}{\ln\rho}\|\psi\|_{H^{-1/2}(\partial B_R)},$$

where C is independent of ρ and ψ.

PROOF. The proof follows from a completely similar argument to that in Proposition 2.1. The following are the major ingredients of the present proof. In this case, $\omega_l\rho = \omega\rho\sqrt{\frac{q_l'}{g_l'}} = \omega\rho^{\frac{\delta}{2}}\sqrt{\rho^{2-\delta}+i}$, $A = \sqrt{\frac{g_a' q_a'}{g_l' q_l'}} = \sqrt{\frac{g_a q_a}{\rho^{2-\delta}+i}}\rho^{-\frac{\delta}{2}}$ and $\sqrt{g_l' q_l'} = \rho^{-1+\frac{\delta}{2}}\sqrt{\rho^{2-\delta}+i}$. As $\rho \to 0$, one has

$$\Upsilon_n \sim -\frac{J_n(\omega_l\rho/2)}{H_n^{(1)}(\omega_l\rho/2)}, \quad \mathcal{H}_n(\rho) = \mathcal{O}(\rho^{-\delta/2}),$$

and

$$(3.2)\qquad \begin{cases} \Gamma_0 \sim \frac{\pi i}{2\ln\frac{\omega\rho}{4}}, \\ \Gamma_n \sim \rho^{2n}. \end{cases}$$

Finally, by supposing $u_R^+(x)|_{\partial B_\rho} = \sum_{n=-\infty}^{\infty} k_n e^{in\theta}$ and using (2.4) and (2.7), one has

$$(3.3)\qquad k_n = \frac{\psi_n}{\omega[J_n'(\omega R) + \Gamma_n H_n^{(1)'}(\omega R)]} \cdot \frac{J_n'(\omega\rho)H_n^{(1)}(\omega\rho) - J_n(\omega\rho)H_n^{(1)'}(\omega\rho)}{\sqrt{g_l' q_l'}\mathcal{H}_n(\rho)H_n^{(1)}(\omega\rho) - H_n^{(1)'}(\omega\rho)}.$$

□

REMARK 3.2. In $\mathbb{R}^3$, when $g_l' = I, q_l' = 1 + i\rho^{-2+\delta}, 0 < \delta < 2$, and $f_a = 0$, we have

$$\|u_\rho - u_0\|_{H^{1/2}(\partial\Omega)} \le C\rho\|\psi\|_{H^{-1/2}(\partial B_R)}, \quad \|u_R^+(\rho\cdot)\|_{H^{1/2}(\partial B_1)} = \mathcal{O}(1). \tag{3.4}$$

Next, we consider the near-cloak by including a delta-point source into the cloaked region.

PROPOSITION 3.3. *In $\mathbb{R}^2$, when $g_l' = I, q_l' = 1 + i\rho^{-2+\delta}, 0 \le \delta < 2$, and $f_a = \eta_0\delta_0$, we have*

$$\|u_\rho - u_0\|_{H^{1/2}(\partial B_R)} = \mathcal{O}\left(\frac{1}{\rho^2 \ln \rho}\right). \tag{3.5}$$

PROOF. We only sketch the proof for the case with $\delta = 0$ and the other case with $0 < \delta < 2$ can be proved in a completely similar manner. In this case, we have $g_a' = g_a$ and $q_a' = \frac{q_a}{\rho^2}$; and $\omega_a = \omega\rho^{-1}\sqrt{\frac{q_a}{g_a}}$ with $\omega_a\rho$ being a constant. $\omega_l = \omega\frac{q_l'}{g_l'} = \omega\sqrt{1+i\rho^{-2}}$ and $\omega_l\rho \to \omega\sqrt{i} = \omega(1+i)/\sqrt{2}$ as $\rho \to +0$. Moreover, $A = \sqrt{\frac{g_a'q_a'}{g_l'q_l'}} = \sqrt{\frac{g_aq_a}{\rho^2+i}} \to \frac{\sqrt{2g_aq_a}}{1+i}$ and $\sqrt{\frac{1}{g_l'q_l'}} = \sqrt{\frac{\rho^2}{\rho^2+i}} \to \frac{\sqrt{2}\rho}{1+i}$ as $\rho \to 0^+$.

It only suffices to consider the case with $n = 0$. Let $A_1' = \frac{J_0(\omega(1+i)/\sqrt{2})}{J_0(\omega_a\frac{\rho}{2})}J_0'(\omega_a\frac{\rho}{2})$, $A_2' = \frac{H_0^{(1)}(\omega(1+i)/\sqrt{2})}{J_0(\omega_a\frac{\rho}{2})}J_0'(\omega_a\frac{\rho}{2})$, $c_0' = \frac{1+i}{\sqrt{2g_aq_a}}$, $z_0 = \omega(1+i)/\sqrt{2}$. The asymptotic counterpart to (2.32) for the present proof is

$$\begin{aligned}
&\Big\{[A_1' - c_0'J_0'(z_0/2)][H_0^{(1)'}(z_0) + \rho^2\frac{\sqrt{2}}{1+i}\frac{\omega}{2}H_0^{(1)}(z_0)] \\
&- [A_2' - c_0'H_0^{(1)'}(z_0/2)][J_0'(z_0) + \rho^2\frac{\sqrt{2}}{1+i}\frac{\omega}{2}J_0(z_0)]\Big\}a_0 \\
&+\Big\{[A_1' - c_0'J_0'(z_0/2)][\frac{2i}{\pi}\ln\frac{\omega\rho}{2}H_0^{(1)'}(z_0) - \frac{\sqrt{2}}{1+i}\frac{2i}{\pi\omega}H_0^{(1)}(z_0)] \\
&- [A_2' - c_0'H_0^{(1)'}(z_0/2)][\frac{2i}{\pi}\ln\frac{\omega\rho}{2}J_0'(z_0) - \frac{\sqrt{2}}{1+i}\frac{2i}{\pi\omega}J_0(z_0)]\Big\}b_0 \\
&\sim\frac{2i}{\pi z_0}\frac{\eta_0}{\rho^2}A_3.
\end{aligned} \tag{3.6}$$

By dropping the lower order terms in (3.6), we have

$$\begin{aligned}
&\Big\{[A_1' - c_0'J_0'(z_0/2)]H_0^{(1)'}(z_0) - [A_2' - c_0'H_0^{(1)'}(z_0/2)]J_0'(z_0)\Big\}a_0 \\
&+\Big\{[A_1' - c_0'J_0'(z_0/2)]\frac{2i}{\pi}H_0^{(1)'}(z_0) - [A_2' - c_0'H_0^{(1)'}(z_0/2)]\frac{2i}{\pi}J_0'(z_0)\Big\}\ln\frac{\omega\rho}{2}b_0 \\
&\sim\frac{2i}{\pi z_0}\frac{\eta_0}{\rho^2}A_3.
\end{aligned} \tag{3.7}$$

Let

$$\begin{aligned}
g_0 =& \frac{[A_1' - c_0'J_0'(z_0/2)]\frac{2i}{\pi}H_0^{(1)'}(z_0) - [A_2' - c_0'H_0^{(1)'}(z_0/2)]\frac{2i}{\pi}J_0'(z_0)}{[A_1' - c_0'J_0'(z_0/2)]H_0^{(1)'}(z_0) - [A_2' - c_0'H_0^{(1)'}(z_0/2)]J_0'(z_0)}, \\
p_0 =& [A_1' - c_0'J_0'(z_0/2)]H_0^{(1)'}(z_0) - [A_2' - c_0'H_0^{(1)'}(z_0/2)]J_0'(z_0).
\end{aligned}$$

We have

$$a_0 + g_0 \ln \frac{\omega\rho}{2} b_0 \sim \frac{2i}{\pi z_0 p_0} \frac{\eta_0}{\rho^2} A_3. \tag{3.8}$$

Let $m_0 = \frac{2i}{\pi z_0 p_0}\eta_0 A_3$. By combining the above estimates with (2.35), we have

$$\begin{cases} a_0 \sim \dfrac{\frac{\psi_0}{\omega} - \frac{m_0 H_0^{(1)\prime}(\omega R)}{\rho^2 \ln \frac{\omega\rho}{2} g_0}}{J_0'(\omega R) - \frac{H_0^{(1)\prime}(\omega R)}{\ln \frac{\omega\rho}{2} g_0}}, \\ b_0 \sim \dfrac{\frac{\psi_0}{\omega} - \frac{m_0}{\rho^2} J_0'(\omega R)}{H_0^{(1)}(\omega R) - g_0 \ln \frac{\omega\rho}{2} J_0'(\omega R)}. \end{cases} \tag{3.9}$$

Finally, using (3.9), we have

$$\left| a_0 J_0(\omega R) + b_0 H_0^{(1)}(\omega R) - \frac{\psi_0 J_0(\omega R)}{\omega J_0'(\omega R)} \right| = \mathcal{O}\left(\frac{1}{\rho^2 \ln \frac{\omega\rho}{2}} \right).$$

□

REMARK 3.4. In $\mathbb{R}^3$, when $g_l' = I, q_l' = 1 + i\rho^{-2+\delta}, 0 \le \delta < 2$, and $f_a = \eta_0\delta_0$, we have

$$\|u_\rho - u_0\|_{H^{1/2}(\partial B_R)} \sim \rho^{-1}. \tag{3.10}$$

3.1. The case with $g_l' = I$, $q_l' = 1 + i\rho^{-2-\delta}$, $\delta > 0$. We first consider the case without the delta-point source presented inside the cloaked region.

PROPOSITION 3.5. *In $\mathbb{R}^2$, when $g_l' = I, q_l' = 1 + i\rho^{-2-\delta}, \delta > 0$, and $f_a = 0$, we have*

$$\|u_\rho - u_0\|_{H^{1/2}(\partial B_R)} \le C\frac{1}{\ln \rho}\|\psi\|_{H^{-1/2}(\partial B_R)}, \|u_R^+(\rho\cdot)\|_{H^{1/2}(\partial B_1)} \le C\frac{\rho^{\delta/2}}{\ln \rho}\|\psi\|_{H^{-1/2}(\partial B_R)}.$$

PROPOSITION 3.6. *In $\mathbb{R}^3$, when $g_l' = I, q_l' = 1 + i\rho^{-2-\delta}, \delta > 0$, and $f_a = 0$, we have*

$$\|u_\rho - u_0\|_{H^{1/2}(\partial B_R)} \le C\rho\|\psi\|_{H^{-1/2}(\partial B_R)}, \ \|u_R^+(\rho\cdot)\|_{H^{1/2}(\partial B_1)} \le C\rho^{\delta/2}\|\psi\|_{H^{-1/2}(\partial B_R)}.$$

Next, we consider the case with a delta-point source presented inside the cloaked region.

PROPOSITION 3.7. *In $\mathbb{R}^2$, when $g_l' = I$, $q_l' = 1 + i\rho^{-2-\delta}$, and $f_a = \eta_0\delta_0$, we have*

$$\|u_\rho - u_0\|_{H^{1/2}(\partial\Omega)} = \mathcal{O}(|\ln \rho|^{-1}). \tag{3.11}$$

PROOF. In this case, g_a', q_a', ω_a are the same to those in Proposition 3.3. $\omega_l\rho = \omega\rho^{-\frac{\delta}{2}}\sqrt{\rho^{2+\delta} + i}$ and $A = \sqrt{\frac{g_a' q_a'}{g_l' q_l'}} = \rho^{-\frac{\delta}{2}}\sqrt{\frac{g_a q_a}{\rho^{2+\delta}+i}}$. $\frac{1}{\sqrt{g_l' q_l'}} = \frac{\rho^{1+\frac{\delta}{2}}}{\sqrt{\rho^{2+\delta}+i}}$. Set $l_0' = l_1' = \frac{\sqrt{2}}{2}$, $c_0^* = \frac{1+i}{\sqrt{2g_a q_a}}$. By using (2.58), one can obtain the asymptotic counterpart to (2.32) for the present proof, and from which one can further show that

$$\left| a_0 J_0(\omega R) + b_0 H_0^{(1)}(\omega R) - \frac{\psi_0 J_0(\omega R)}{\omega J_0'(\omega R)} \right| = \mathcal{O}\left(\frac{1}{\ln \frac{\omega\rho}{2}} \right) + \mathcal{O}\left(\frac{\eta_0}{\ln \frac{\omega\rho}{2}} e^{-\frac{l_1'\rho^{-\delta}}{2}} \rho^\delta \right), \quad n \ne 0.$$

and

$$\left| a_n J_n(\omega R) + b_n H_n^{(1)}(\omega R) - \frac{\psi_n J_n(\omega R)}{\omega J_n'(\omega R)} \right| = \mathcal{O}(\rho^{2|n|}).$$

Using the above results, the proof follows from a completely similar argument to that for Proposition 3.3. □

For the three-dimensional case, we have

PROPOSITION 3.8. *In $\mathbb{R}^3$, when $g_l' = I$, $q_l' = 1 + i\rho^{-2-\delta}$, $\delta > 0$ and $f_a = \eta_0\delta_0$, we have*

$$\|u_\rho - u_0\|_{H^{1/2}(\partial B_R)} = \mathcal{O}(\rho). \tag{3.12}$$

References

[AS] M. Abramowitz and I. A. Stegun, *Handbook of Mathematical Functions*, New York: Dover Publications, 1965.

[A1] Habib Ammari, Hyeonbae Kang, Hyundae Lee, and Mikyoung Lim, *Enhancement of near cloaking using generalized polarization tensors vanishing structures. Part I: The conductivity problem*, Comm. Math. Phys. **317** (2013), no. 1, 253–266, DOI 10.1007/s00220-012-1615-8. MR3010374

[A2] Habib Ammari, Hyeonbae Kang, Hyundae Lee, and Mikyoung Lim, *Enhancement of near-cloaking. Part II: The Helmholtz equation*, Comm. Math. Phys. **317** (2013), no. 2, 485–502, DOI 10.1007/s00220-012-1620-y. MR3010192

[A3] Habib Ammari, Josselin Garnier, Vincent Jugnon, Hyeonbae Kang, Hyundae Lee, and Mikyoung Lim, *Enhancement of near-cloaking. Part III: Numerical simulations, statistical stability, and related questions*, Multi-scale and high-contrast PDE: from modelling, to mathematical analysis, to inversion, Contemp. Math. vol. 577, Amer. Math. Soc. Providence, RI, 2012, pp. 1–24, DOI 10.1090/conm/577/11460. MR2985063

[C] H. Chen and C. T. Chan, *Acoustic cloaking and transformation acoustics*, J. Phys. D: Appl. Phys., **43** (2010), 113001.

[CK] David Colton and Rainer Kress, *Inverse acoustic and electromagnetic scattering theory*, 2nd ed. Applied Mathematical Sciences, vol. 93, Springer-Verlag, Berlin, 1998. MR1635980 (99c:35181)

[G] A. Greenleaf, Y. Kurylev, M. Lassas and G. Uhlmann, *Improvement of cylindrical cloaking with SHS lining*, Optics Express, **15** (2007), 12717–12734.

[G1] A. Greenleaf, Y. Kurylev, M. Lassas and G. Uhlmann, *Isotropic transformation optics: approximate acoustic and quantum cloaking*, New J. Phys., **10** (2008), 115024.

[G2] Allan Greenleaf, Yaroslav Kurylev, Matti Lassas, and Gunther Uhlmann, *Full-wave invisibility of active devices at all frequencies*, Comm. Math. Phys. **275** (2007), no. 3, 749–789, DOI 10.1007/s00220-007-0311-6. MR2336363 (2008g:78016)

[GK] Allan Greenleaf, Yaroslav Kurylev, Matti Lassas, and Gunther Uhlmann, *Invisibility and inverse problems*, Bull. Amer. Math. Soc. (N.S.) **46** (2009), no. 1, 55–97, DOI 10.1090/S0273-0979-08-01232-9. MR2457072 (2010d:35399)

[GKL] Allan Greenleaf, Yaroslav Kurylev, Matti Lassas, and Gunther Uhlmann, *Cloaking devices, electromagnetic wormholes, and transformation optics*, SIAM Rev. **51** (2009), no. 1, 3–33, DOI 10.1137/080716827. MR2481110 (2010b:35484)

[GL] A. Greenleaf, M. Lassas and G. Uhlmann, *Anisotropic conductivities that cannot detected by EIT*, Physiolog. Meas, (special issue on Impedance Tomography), 24, 413 (2003).

[GLU] Allan Greenleaf, Matti Lassas, and Gunther Uhlmann, *On nonuniqueness for Calderón's inverse problem*, Math. Res. Lett. **10** (2003), no. 5-6, 685–693, DOI 10.4310/MRL.2003.v10.n5.a11. MR2024725 (2005f:35316)

[KL] Ilker Kocyigit, Hongyu Liu, and Hongpeng Sun, *Regular scattering patterns from near-cloaking devices and their implications for invisibility cloaking*, Inverse Problems **29** (2013), no. 4, 045005, 25, DOI 10.1088/0266-5611/29/4/045005. MR3042081

[KO] Robert V. Kohn, Daniel Onofrei, Michael S. Vogelius, and Michael I. Weinstein, *Cloaking via change of variables for the Helmholtz equation*, Comm. Pure Appl. Math. **63** (2010), no. 8, 973–1016, DOI 10.1002/cpa.20326. MR2642383 (2011j:78004)

[KS] R. V. Kohn, H. Shen, M. S. Vogelius, and M. I. Weinstein, *Cloaking via change of variables in electric impedance tomography*, Inverse Problems **24** (2008), no. 1, 015016, 21, DOI 10.1088/0266-5611/24/1/015016. MR2384775 (2008m:78014)

[LE] Ulf Leonhardt, *Optical conformal mapping*, Science **312** (2006), no. 5781, 1777–1780, DOI 10.1126/science.1126493. MR2237569

[LLS] Jingzhi Li, Hongyu Liu, and Hongpeng Sun, *Enhanced approximate cloaking by SH and FSH lining*, Inverse Problems **28** (2012), no. 7, 075011, 21, DOI 10.1088/0266-5611/28/7/075011. MR2946799

[L1] Hongyu Liu, *Virtual reshaping and invisibility in obstacle scattering*, Inverse Problems **25** (2009), no. 4, 045006, 16, DOI 10.1088/0266-5611/25/4/045006. MR2482157 (2010d:35044)

[L2] Hongyu Liu, *On near-cloak in acoustic scattering*, J. Differential Equations **254** (2013), no. 3, 1230–1246, DOI 10.1016/j.jde.2012.10.015. MR2997368

[LS] Hongyu Liu and Hongpeng Sun, *Enhanced near-cloak by FSH lining* (English, with English and French summaries), J. Math. Pures Appl. (9) **99** (2013), no. 1, 17–42, DOI 10.1016/j.matpur.2012.06.001. MR3003281

[LZ] Hongyu Liu and Jun Zou, *Zeros of the Bessel and spherical Bessel functions and their applications for uniqueness in inverse acoustic obstacle scattering*, IMA J. Appl. Math. **72** (2007), no. 6, 817–831, DOI 10.1093/imamat/hxm013. MR2372090 (2009e:33013)

[NE] Jean-Claude Nédélec, *Acoustic and electromagnetic equations*, Applied Mathematical Sciences, vol. 144, Springer-Verlag, New York, 2001. Integral representations for harmonic problems. MR1822275 (2002c:35003)

[NO] Andrew N. Norris, *Acoustic cloaking theory*, Proc. R. Soc. Lond. Ser. A Math. Phys. Eng. Sci. **464** (2008), no. 2097, 2411–2434, DOI 10.1098/rspa.2008.0076. MR2429553 (2009k:74044)

[P] J. B. Pendry, D. Schurig, and D. R. Smith, *Controlling electromagnetic fields*, Science **312** (2006), no. 5781, 1780–1782, DOI 10.1126/science.1125907. MR2237570

[R] Z. Ruan, M. Yan, C. W. Neff and M. Qiu, *Ideal cylyindrical cloak: Perfect but sensitive to tiny perturbations*, Phy. Rev. Lett., **99** (2007), no. 11, 113903.

[U] Gunther Uhlmann, *Visibility and invisibility*, ICIAM 07—6th International Congress on Industrial and Applied Mathematics, Eur. Math. Soc., Zürich, 2009, pp. 381–408, DOI 10.4171/056-1/18. MR2588602 (2010m:35571)

[Y] M. Yan, W. Yan and M. Qiu, *Invisibility cloaking by coordinate transformation*, Chapter 4 of *Progress in Optics* Vol. 52, Elsevier, pp. 261–304, 2008.

Faculty of Science, South University of Science and Technology of China, Shenzhen 518055, People's Republic of China.
E-mail address: li.jz@sustc.edu.cn

Department of Mathematics and Statistics, University of North Carolina at Charlotte, Charlotte, North Carolina 28223
E-mail address: hongyu.liuip@gmail.com

Institute for Mathematics and Scientific Computing, University of Graz, Heinrichstr. 36, A-8010 Graz, Austria.
E-mail address: hongpeng.sun@uni-graz.at

Contemporary Mathematics
Volume **615**, 2014
http://dx.doi.org/10.1090/conm/615/12261

Hybrid Inverse Problem for Porous Media

Shari Moskow and John C. Schotland

Abstract. We consider the inverse problem of reconstructing the anisotropic hydrostatic conductivity of a porous medium from internal measurements in the presence of acoustic modulation. We find that the conductivity can be recovered by an algebraic formula with Lipschitz stability.

1. Introduction

We consider the problem of reconstructing the hydrostatic conductivity of a porous medium from measurements of the pressure within the medium. The pressure u obeys the equation

$$-\nabla \cdot \sigma(x)\nabla u = f \quad \text{in} \quad \Omega , \tag{1.1}$$

$$\sigma(x)\nabla u \cdot n = g \quad \text{on} \quad \partial\Omega . \tag{1.2}$$

Here the conductivity σ is taken to be a symmetric, positive definite matrix, f and g correspond to volume and surface sources of fluid, respectively and Ω is a bounded domain in $\mathbb{R}^d$, $d \geq 2$, with a smooth boundary and outward unit normal n. We note that the compatibility condition

$$\int_\Omega f dx + \int_{\partial\Omega} g dx = 0 \tag{1.3}$$

must also be obeyed.

In geophysical applications, (1.1) is used to model the flow of water through an aquifer [**12, 17**]. In this context, it is of interest to determine the conductivity from internal measurements of the pressure within Ω. For the case of an isotropic medium, the usual approach to this problem is to view (1.1) as a transport equation for the scalar field σ of the form

$$\nabla u \cdot \nabla \sigma + \sigma \Delta u = -f . \tag{1.4}$$

Assuming that the vector field ∇u is nonvanishing, (1.4) is readily solved by characteristics. Evidently, if σ is known on all of $\partial\Omega$, this problem is overdetermined; knowledge of σ on the portion of the boundary where $\partial u/\partial n < 0$ is sufficient to

2010 *Mathematics Subject Classification.* Primary 34A55; Secondary 85A25.
Key words and phrases. Inverse problems, hybrid imaging.
The first author was supported in part by the NSF grant DMS-1108858.
The second author was supported in part by the NSF grants DMS-1115574 and DMS-1108969.

determine σ uniquely if the condition

$$\inf_{x\in\Omega} \max\{\Delta u(x), |\nabla u(x)|\} > 0 \tag{1.5}$$

holds [**17**]. We note that (1.5) is not sufficient to assure the stable recovery of σ. This follows from the fact that derivatives are taken of the data u and that the equation for σ is hyperbolic. Thus, numerical methods for reconstructing σ must be suitably regularized [**12**].

In this paper we consider an alternative approach to the above inverse problem which, in some sense, is in the spirt of several recently proposed hybrid imaging modalities [**1–11, 13–16**]. In hybrid imaging (also called coupled-physics or multi-wave imaging), an external field is used to control the material properties of a medium of interest, which is then probed by a second field. In the physical setting we consider, the conductivity of the medium is spatially modulated by an acoustic wave, while internal measurements of the pressure are recorded. We find that it is possible to recover the anisotropic conductivity by an algebraic formula; that is, without solving a partial differential equation such as (1.4). Using this result, it can be seen that the conductivity can be reconstructed with Lipschitz stability.

The remainder of this paper is organized as follows. In section 2 we develop a model for the propagation of an acoustic wave in a porous medium. This result is then utilized to formulate the forward problem. The inverse problem is analyzed in section 3 for both isotropic and anisotropic media.

2. Model

We begin by considering a suspension of particles in which a standing acoustic wave propagates. If the amplitude of the wave is sufficiently small, then each particle will oscillate about its equilibrium position. The acoustic pressure is taken to be the plane wave

$$p = A\cos(k\cdot x + \varphi)\ , \tag{2.1}$$

where A is the amplitude, k is the wavevector and φ is the phase of the wave. If the particles have positions x_i, then their number density is defined as

$$\rho_\epsilon(x) = \sum_i \delta(x - x_i)\ . \tag{2.2}$$

Following [**9**], it can then be seen that

$$\rho_\epsilon(x) = \rho_0(x)\left[1 + \epsilon\cos(k\cdot x + \varphi)\right]\ , \tag{2.3}$$

where ρ_0 is the number density in the absence of the acoustic wave and $\epsilon = A/(\rho c_s^2) \ll 1$ is a small parameter. Thus, the number density of particles is spatially modulated by the acoustic wave.

The conductivity is proportional to the density of conducting particles and is given by

$$\sigma_\epsilon(x) = \sigma_0(x)\left[1 + \epsilon\cos(k\cdot x + \varphi)\right]\ , \tag{2.4}$$

where σ_0 is the conductivity in the absence of the acoustic wave. Eq. (1.1) thus becomes

$$-\nabla\cdot\sigma_\epsilon(x)\nabla u_\epsilon = f \quad \text{in} \quad \Omega\ , \tag{2.5}$$

$$\sigma_\epsilon(x)\nabla u_\epsilon\cdot n = g \quad \text{on} \quad \partial\Omega\ , \tag{2.6}$$

where σ_ϵ is taken to be constant on $\partial\Omega$. The Green's function G obeys the equation

$$\begin{aligned} -\nabla_x \cdot \sigma_0(x)\nabla_x G(x,y) &= \delta(x-y) \quad \text{in} \quad \Omega\ , & (2.7)\\ \sigma_0(x)\nabla_x G(x,y)\cdot n &= 0 \quad \text{on} \quad \partial\Omega\ . & (2.8) \end{aligned}$$

If we multiply (2.5) by $G(x,y)$ and (2.7) by $u_\epsilon(y)$, subtract the resulting equations and integrate over Ω, we find that

$$u_\epsilon(x) = U_\epsilon(x) + \epsilon \int_\Omega \sigma_0(y)\nabla_y G(x,y)\cdot \nabla u_\epsilon(y)\cos(k\cdot y+\varphi)dy\ .$$

Here we have integrated by parts and define

$$U_\epsilon(x) = \int_\Omega G(x,y)f(y)dy + \int_{\partial\Omega} \sigma_\epsilon(y)G(x,y)\nabla u_\epsilon(y)\cdot n dy\ . \tag{2.9}$$

We now introduce asymptotic expansions for u_ϵ and U_ϵ of the form

$$\begin{aligned} u_\epsilon &= u_0 + \epsilon u_1 + \cdots\ , & (2.10)\\ U_\epsilon &= U_0 + \epsilon U_1 + \cdots\ , & (2.11) \end{aligned}$$

which we substitute into (2.9). Upon collecting terms, we obtain

(2.12)

$$u_0(x) = U_0(x)\ ,$$

(2.13)

$$u_j(x) = U_j(x) + \int_\Omega \sigma_0(y)\nabla_y G(x,y)\cdot\nabla u_{j-1}(y)\cos(k\cdot y+\varphi)dy\ , \quad j=1,2,\dots\ .$$

Since f and g are independent of ϵ, it follows that $U_j=0$ for $j>0$. Thus, we have

$$u_\epsilon(x) = u_0(x) + \epsilon\int_\Omega \sigma_0(y)\nabla_y G(x,y)\cdot\nabla u_0(y)\cos(k\cdot y+\varphi)dy + \cdots\ ,$$

where

$$u_0(x) = \int_{\partial\Omega} G(x,y)g(y)dy + \int_\Omega G(x,y)f(y)dy\ . \tag{2.14}$$

We will find it useful to introduce the scattering data $\phi = (u_\epsilon - u_0)/\epsilon$, which can be determined from measurements. By varying the wavevector k and the phase φ, it is possible to recover the functional

$$H_0(x,y) = \sigma_0(y)\nabla_y G(x,y)\cdot\nabla u_0(y)\ , \quad x,y\in\Omega\ , \tag{2.15}$$

by Fourier inversion. That is,

$$H_0(x,y) = \frac{1}{(2\pi)^n}\int e^{-ik\cdot y}\left[\phi(x;k,0) + i\phi(x;k,3\pi/2)\right]dk\ , \tag{2.16}$$

where the dependence of ϕ on k and φ has been made explicit. The recovery of H_0 from measurements is the first step in the solution to the inverse problem. For the remainder of this work, we will assume that H_0 is known.

3. Inverse Problem

We begin by deriving two identities. We define

$$h(x) = \int_{\Omega} H_0(x,y)dy \ . \tag{3.1}$$

Substituting (2.15) into (3.1), integrating by parts and using the boundary condition (2.8), we obtain

$$h(x) = u_0(x) \ , \quad x \in \Omega \ . \tag{3.2}$$

Next, we define

$$H(x) = \int_{\partial\Omega} H_0(y,x)g(y)dy + \int_{\Omega} H_0(y,x)f(y)dy \ . \tag{3.3}$$

Using (2.15), (2.14) and the fact that $G(x,y) = G(y,x)$, we find that

$$H(x) = \sigma_0(x)\nabla u_0(x) \cdot \nabla u_0(x) \ , \quad x \in \Omega \ . \tag{3.4}$$

Since H_0 is known, it follows from the identities (3.2) and (3.4) that u and H are also known everywhere in Ω.

The inverse problem consists of recovering the conductivity σ_0 from measurements of the functional H_0. We will see that that it is possible to obtain an algebraic inversion formula for σ_0. First, we consider the case where σ_0 is isotropic. If ∇u is nonvanishing, (3.4) can then be solved for σ_0. We thus obtain

$$\sigma_0(x) = \frac{H(x)}{|\nabla u_0(x)|^2} \ . \tag{3.5}$$

Making use of (3.2), we find that

$$\sigma_0(x) = \frac{H(x)}{|\nabla h(x)|^2} \ , \tag{3.6}$$

which is the required inversion formula.

It follows immediately from (3.6) that σ_0 can be reconstructed with Lipschitz stability. To see this, suppose that (H,h) and (H',h') are data corresponding to the conductivities σ_0 and σ_0', respectively and

$$\sigma_0(x) = \frac{H(x)}{|\nabla h(x)|^2} \ , \quad \sigma_0'(x) = \frac{H'(x)}{|\nabla h'(x)|^2} \ . \tag{3.7}$$

We also assume that ∇h and $\nabla h'$ are nonvanishing and define

$$C = \sup_{x\in\Omega} \max\left\{ \frac{1}{|\nabla h(x)|^2}, \frac{1}{|\nabla h'(x)|^2} \right\} \ . \tag{3.8}$$

We then have the stability estimate

$$\|\sigma_0 - \sigma_0'\|_{L^\infty(\Omega)} \le C\|H - H'\|_{L^\infty(\Omega)} \ . \tag{3.9}$$

Next, we consider the case where σ_0 is anisotropic. Multiple boundary sources are then required to recover all the independent components of σ_0. Eq. (2.5) thus becomes

$$-\nabla \cdot \sigma_\epsilon(x)\nabla u_\epsilon^{(j)} = f \quad \text{in} \quad \Omega \ , \tag{3.10}$$

$$\sigma_\epsilon(x)\nabla u_\epsilon^{(j)} \cdot n = g_j \quad \text{on} \quad \partial\Omega \ , \quad j = 1, \dots, J \ , \tag{3.11}$$

where the g_j are boundary sources and $u^{(j)}$ are the corresponding fields. Following the previous development, it is possible to recover J functionals of the form

$$H_j(x) = \sigma_0(x)\nabla h_j(x) \cdot \nabla h_j(x) \ , \tag{3.12}$$

where $h_j(x) = u^{(j)}(x)$ is known. We note that (3.12) is a $J \times J$ system of linear equations for the components of σ_0. If J is sufficiently large, these equations can be solved (at least in principle) to reconstruct σ_0. Evidently, a stability estimate analogous to (3.9) also holds.

References

[1] H. Ammari, E. Bonnetier, Y. Capdeboscq, M. Tanter, and M. Fink, *Electrical impedance tomography by elastic deformation*, SIAM J. Appl. Math. **68** (2008), no. 6, 1557–1573, DOI 10.1137/070686408. MR2424952 (2009h:35439)

[2] Habib Ammari, Yves Capdeboscq, Frédéric de Gournay, Anna Rozanova-Pierrat, and Faouzi Triki, *Microwave imaging by elastic deformation*, SIAM J. Appl. Math. **71** (2011), no. 6, 2112–2130, DOI 10.1137/110828241. MR2873260

[3] S. R. Arridge and O. Scherzer, Imaging from coupled physics, Inverse Problems, 28 (2012), p. 080201.

[4] G. Bal, Hybrid Inverse Problems and Systems of Partial Differential Equations, arXiv:1210.0265.

[5] Guillaume Bal, *Inverse transport theory and applications*, Inverse Problems **25** (2009), no. 5, 053001, 48, DOI 10.1088/0266-5611/25/5/053001. MR2501018 (2010d:78021)

[6] G. Bal, Hybrid inverse problems and internal functionals, Inside Out, Cambridge University Press, Cambridge, UK, G. Uhlmann, Editor, 2012.

[7] Guillaume Bal, Eric Bonnetier, François Monard, and Faouzi Triki, *Inverse diffusion from knowledge of power densities*, Inverse Probl. Imaging **7** (2013), no. 2, 353–375, DOI 10.3934/ipi.2013.7.353. MR3063538

[8] Guillaume Bal, Wolf Naetar, Otmar Scherzer, and John Schotland, *The Levenberg-Marquardt iteration for numerical inversion of the power density operator*, J. Inverse Ill-Posed Probl. **21** (2013), no. 2, 265–280, DOI 10.1515/jip-2012-0091. MR3043625

[9] G. Bal and J. C. Schotland, Inverse Scattering and Acousto-Optics Imaging, Phys. Rev. Letters, 104 (2010), p. 043902.

[10] Y. Capdeboscq, J. Fehrenbach, F. de Gournay, and O. Kavian, *Imaging by modification: numerical reconstruction of local conductivities from corresponding power density measurements*, SIAM J. Imaging Sci. **2** (2009), no. 4, 1003–1030, DOI 10.1137/080723521. MR2559157 (2011c:35611)

[11] Bastian Gebauer and Otmar Scherzer, *Impedance-acoustic tomography*, SIAM J. Appl. Math. **69** (2008), no. 2, 565–576, DOI 10.1137/080715123. MR2465856 (2009j:35381)

[12] Robert V. Kohn and Bruce D. Lowe, *A variational method for parameter identification* (English, with French summary), RAIRO Modél. Math. Anal. Numér. **22** (1988), no. 1, 119–158. MR934704 (89d:93036)

[13] Peter Kuchment and Leonid Kunyansky, *2D and 3D reconstructions in acousto-electric tomography*, Inverse Problems **27** (2011), no. 5, 055013, 21, DOI 10.1088/0266-5611/27/5/055013. MR2793832 (2012b:65166)

[14] Peter Kuchment and Dustin Steinhauer, *Stabilizing inverse problems by internal data*, Inverse Problems **28** (2012), no. 8, 084007, 20, DOI 10.1088/0266-5611/28/8/084007. MR2956563

[15] François Monard and Guillaume Bal, *Inverse anisotropic diffusion from power density measurements in two dimensions*, Inverse Problems **28** (2012), no. 8, 084001, 20, DOI 10.1088/0266-5611/28/8/084001. MR2956557

[16] François Monard and Guillaume Bal, *Inverse diffusion problems with redundant internal information*, Inverse Probl. Imaging **6** (2012), no. 2, 289–313, DOI 10.3934/ipi.2012.6.289. MR2942741

[17] Gerard R. Richter, *An inverse problem for the steady state diffusion equation*, SIAM J. Appl. Math. **41** (1981), no. 2, 210–221, DOI 10.1137/0141016. MR628945 (82m:35143)

Department of Mathematics, Drexel University, Philadelphia, Pennsylvania
E-mail address: moskow@math.drexel.edu

Departments of Mathematics and Physics, University of Michigan, Ann Arbor, Michigan
E-mail address: schotland@umich.edu

Contemporary Mathematics
Volume **615**, 2014
http://dx.doi.org/10.1090/conm/615/12259

Efficient Algorithms for Ptychographic Phase Retrieval

Jianliang Qian, Chao Yang, A. Schirotzek, F. Maia, and S. Marchesini

This paper is dedicated to Prof. Gunther Uhlmann on the occasion of his 60th birthday.

ABSTRACT. Ptychography is an emerging non-crystalline diffractive imaging technique that can potentially reach diffraction limited resolution without the need for high resolution lenses. To achieve high resolution one must solve a phase-retrieval inverse problem using the diffraction patterns of many partially overlapping subimage frames. We examine the mathematical formulation of the ptychographic phase retrieval problem, and analyze some of the existing methods for solving the inverse problem. We also discuss a number of practical techniques that can improve the efficiency and robustness of numerical algorithms for solving the ptychographic phase retrieval problem.

1. Introduction

Pytchography is an emerging non-crystalline diffractive imaging technique by which one can deduce the structure of an object from a number of diffraction patterns; see Figure 1. It can be formulated as an inverse problem in which the phase relationship between different parts of a scattered wave disturbance is deduced from the magnitude of the wave that can be physically measured. Its usefulness lies in its ability to recover structure information without using high resolution lenses or defining properties of the scattering medium. The missing phase information is obtained implicitly from the intensity recorded in the diffraction plane through an iterative computational method [**1**–**3**]. In this paper, we examine the mathematical formulation of the ptychographic phase retrieval problem, and analyze some of the existing algorithms for solving this type of inverse problems.

We consider reconstruction of two-dimensional (2D) objects, although the technique we discuss here can in principle be extended to 3D structure elucidation. In a ptychography experiment, one collects a sequence of diffraction images of dimension $m \times m$. Each image frame $y_{\mathbf{x}}(\mathbf{r}')$ represents the magnitude of the Fourier transform of $a(\mathbf{r})\psi_0(\mathbf{r}+\mathbf{x})$, where $a(\mathbf{r})$ is a localized illumination (window) function or a probe, $\psi_0(\mathbf{r})$ is the unknown object of interest, and $\mathbf{x}$ is a translational vector. We can express $y_{\mathbf{x}}$ as

$$y_{\mathbf{x}}(\mathbf{r}') = |\mathcal{F}\{a(\mathbf{r})\psi_0(\mathbf{r}+\mathbf{x})\}|, \tag{1.1}$$

where $\mathcal{F}\{f\}$ denotes the Fourier transform of f with respect to $\mathbf{r}$.

2010 *Mathematics Subject Classification.* Primary 65T50; Secondary 42B99.

Key words and phrases. Diffractive imaging, ptycography, phase retrieval, numerical algorithms.

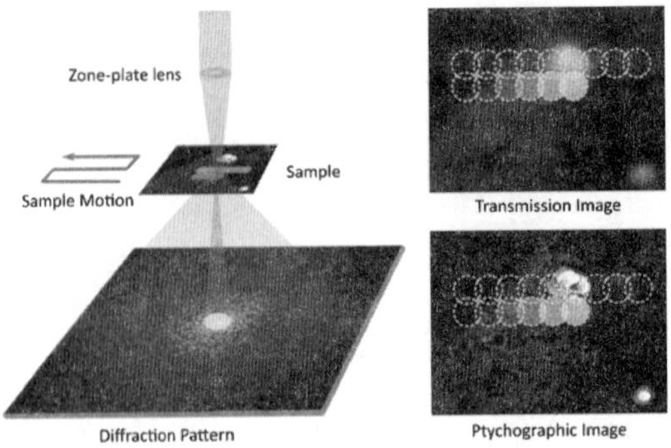

FIGURE 1. A schematic drawing of a ptychography experiment in which a probe scans through a 2D object in an overlapping fashion and produces a sequence of diffraction patterns of the scanned regions.

In order to reconstruct the unknown object, we must retrieve the amplitude and phase of $\psi_0(\mathbf{r})$ from a number of $y_{\mathbf{x}}(\mathbf{r}')$ that are associated with different $\mathbf{x}$'s. A few methods have been proposed to recover $\psi_0(\mathbf{r})$ from ptychographic measurements $y_{\mathbf{x}}(\mathbf{r}')$ [**1**–**5**]. The connection among these methods is not entirely clear from the existing literature. Furthermore, little detail is provided on convergence properties or computational efficiency of these methods.

In this paper, we analyze some of the existing methods for solving ptychographic phase retrieval problem from a numerical optimization point of view. In particular, we examine the local convergence properties of these methods by analyzing the gradient and Hessian of different objective functions, which we present in section 2. We discuss a number of computational details such as weighting and preconditioning that are important for achieving good performance in these methods in section 3. We also describe the connection between optimization based algorithms and projection algorithms that are frequently used in phase retrieval in section 3.2. A number of computational examples are presented in section 4 to illustrate and compare the convergence behavior of several methods for solving the ptychographic phase retrieval problem.

We point out that the ptychographic minimization problem is not globally convex, which means that iterative methods can be trapped at a local minimizer if a poor starting guess is chosen. We show by a numerical example that one way to escape from a local minimizer is to switch to a different objective function in section 4.

We observed that the convergence of the optimization based iterative algorithms used to perform ptychographic phase retrieval become faster when the amount of overlap between two adjacent image frames increases. We provide a preliminary analysis of how the amount of overlap between adjacent frames affects the convergence of iterative optimization algorithms in section 4.

We use standard linear algebra notation whenever possible to describe various quantities evaluated in the iterative algorithms we present. To simplify notation we use a/b to denote an element-wise division between two vectors a and b. Similarly, we use $a \cdot b$ to denote an element-wise multiplication of a and b. We also use a^2 and $a^{1/2}$ occasionally to denote the element-wise square and square root of a

respectively. The conjugate of a complex variable a is denoted by $\bar{a}$. The real part of a is denoted by $\mathrm{Re}(a)$. The conjugate transpose of a matrix (or a vector) A is denoted by A^*. The $|x|$ symbol is reserved for the magnitude (or absolute value) of x. The Euclidean norm of x is denoted by $\|x\| = \sqrt{x^*x}$. We use $\mathrm{Diag}\,(x)$ to represent a diagonal matrix with the vector x on its diagonal.

2. Ptychographic reconstruction as an inverse problem

The phase retrieval problem has a long history in both the optics and inverse problem communities. The uniqueness question of phase retrieval has been investigated under various conditions in different contexts [**6, 8, 9**]. Recently, it has been shown that the problem can be formulated as a low-rank matrix completion problem [**10**] and solved by convex programming techniques if the standard diffraction experiments can be modified to generate additional information. In this paper, we examine efficient algorithms for solving another special class of phase retrieval problems in which the unknown object ψ_0 is recovered from a number of intensity measurements represented by (1.1). For a finite set of translational vectors $\mathbf{x}_i$, we will denote each measurement by

$$b_i = |FQ_i\psi_0|, \quad i = 1, 2, ..., k,$$

where ψ_0 is the sampled unknown object that contains n pixels, b_i is a sampled measurement that contains m pixels, F is the matrix representation of a discrete Fourier transform, and Q_i is an $m \times n$ "illumination matrix" that extracts a frame containing m pixels out of an image containing n pixels. Each row of Q_i contains at most one nonzero element. The nonzero values in Q_i are determined by the illumination function $a(\mathbf{r})$.

Given a set of measurements, b_1, b_2, ..., b_k, we may attempt to recover ψ_0 by solving either the least squares problem

$$\min_{\psi} \rho(\psi) \equiv \min_{\psi} \frac{1}{2} \sum_{i=1}^{k} \||z_i| - b_i\|^2, \quad \text{or} \tag{2.1}$$

$$\min_{\psi} \epsilon(\psi) \equiv \min_{\psi} \frac{1}{2} \sum_{i=1}^{k} \||z_i|^2 - b_i^2\|^2, \tag{2.2}$$

where $z_i \equiv FQ_i\psi$. The advantage of using (2.2) is that it is slightly smoother than (2.1), hence more amenable to analysis. In practice, we found the objective function in (2.1) to be a better choice in terms of computational efficiency in most cases.

To obtain the minimizers of (2.1) or (2.2) using numerical optimization techniques, we often need to evaluate the gradient and possibly the Hessian of these objective functions. Because both (2.1) and (2.2) are real-valued functions of a (potentially) complex vector ψ, one can either take the derivative of (2.1) and (2.2) with respect to the real and imaginary parts of ψ independently or follow the $\mathbb{CR}$-calculus formalism established in [**11, 12**] by treating ψ and $\bar{\psi}$ as two independent variables. The latter approach is what we use throughout this paper.

2.1. Gradient. If we let $r_i \equiv |z_i|^2 - b_i^2$ and define

$$r \equiv \left(r_1^T, r_2^T, \cdots r_k^T\right)^T,$$

we can rewrite (2.2) as $\epsilon(\psi) = r^T r/2$. It is not difficult to show that [**13**]

$$\nabla \epsilon = \sum_{i=1}^{k} Q_i^* F^* \mathrm{Diag}(z_i)[|z_i|^2 - b_i^2]. \tag{2.3}$$

The gradient of the objective function $\rho(\psi)$ in (2.1) is slightly more complicated. By rewriting $|z_i|$ as $(|z_i|^2)^{1/2}$ and using the chain rule, we obtain

$$\nabla \rho(\psi) = \frac{1}{2} \sum_{i=1}^{k} \left[Q_i^* Q_i \psi - Q_i^* F^* \mathrm{Diag}\left(\frac{z_i}{|z_i|}\right) b_i \right]. \tag{2.4}$$

Note that both (2.3) and (2.4) remain real when ψ is real and when b_i is obtained from a discrete Fourier transform of a real image (so that conjugate symmetry is preserved in $\mathrm{Diag}\,(z_i/|z_i|)\, b_i$.)

2.2. Hessian. The Hessians of $\epsilon(\psi)$ and $\rho(\psi)$ provide information on the convexity of these objective functions. Again, because both $\epsilon(\psi)$ and $\rho(\psi)$ are real valued functions of a potentially complex vector ψ, their Hessians are defined as

$$H^f = \begin{pmatrix} H^f_{\psi\psi} & H^f_{\psi\bar\psi} \\ H^f_{\bar\psi\psi} & H^f_{\bar\psi\bar\psi} \end{pmatrix},$$

where

$$H^f_{\psi\psi} \equiv \frac{\partial}{\partial \psi}\left(\frac{\partial f}{\partial \psi}\right)^*, \quad H^f_{\bar\psi\psi} \equiv \frac{\partial}{\partial \bar\psi}\left(\frac{\partial f}{\partial \psi}\right)^*, \quad H^f_{\psi\bar\psi} \equiv \frac{\partial}{\partial \psi}\left(\frac{\partial f}{\partial \bar\psi}\right)^*, \quad H^f_{\bar\psi\bar\psi} \equiv \frac{\partial}{\partial \bar\psi}\left(\frac{\partial f}{\partial \bar\psi}\right)^*,$$

and f is either ϵ or ρ.

It is not difficult to show that [**13**]

$$H^\epsilon = \begin{pmatrix} \sum_i Q_i^* F^* \mathrm{Diag}\left(2|z_i|^2 - b_i^2\right) F Q_i & \sum_i Q_i^* F^* \mathrm{Diag}\left(z_i\right)^2 \overline{F Q_i} \\ \sum_i Q_i^T F^T \mathrm{Diag}\left(\bar z_i\right)^2 F Q_i & \sum_i Q_i^T F^T \mathrm{Diag}\left(2|z_i|^2 - b_i^2\right) \overline{F Q_i} \end{pmatrix}.$$

If we let $t_{ji} \equiv |t_{ji}|e^{i\mu_{ji}}$, $\zeta_{ji} \equiv |\zeta_{ji}|e^{i\theta_{ji}}$ and β_{ji} be the jth component of $t_i = FQ_i\phi$, $z_i = FQ_i\psi$ and b_i respectively, then the curvature $\tau_\epsilon(\psi, \phi)$ at ψ along any direction ϕ can be shown to be

$$\begin{aligned} \tau_\epsilon(\psi,\phi) &= (\phi^* \;\; \phi^T) \begin{pmatrix} H^\epsilon_{\psi\psi} & H^\epsilon_{\psi\bar\psi} \\ H^\epsilon_{\bar\psi\psi} & H^\epsilon_{\bar\psi\bar\psi} \end{pmatrix} \begin{pmatrix} \phi \\ \bar\phi \end{pmatrix} \\ &= 2\sum_{i=1}^{k}\sum_{j=1}^{n} |t_{ji}|^2(|z_{ji}|^2 - \beta_{ji}^2) + 2|t_{ji}|^2|z_{ji}|^2\cos^2(\mu_{ji} - \theta_{ji}). \end{aligned} \tag{2.5}$$

At the minimizer of $\epsilon(\psi)$, $|z_i| = b_i$. So the first term of (2.5) is zero. Because the second term of (2.5) is nonnegative, $\tau \geq 0$, i.e., ϵ is locally convex at the solution. Moreover, the convexity of ϵ is preserved in the area where $|z_{ji}| \geq \beta_{ji}$.

A similar observation can be made from the curvature of ρ. It is not difficult to show that

$$H^\rho = \begin{pmatrix} \sum_{i=1}^{k} W_i^* \left[\frac{1}{2}I - \frac{1}{4}\mathrm{Diag}\left(\frac{b_i}{|z_i|}\right)\right] W_i & \frac{1}{4}\sum_{i=1}^{k} W_i^* \mathrm{Diag}\left(\frac{b_i \cdot z_i^2}{|z_i|^3}\right) \overline{W}_i \\ \frac{1}{4}\sum_{i=1}^{k} W_i^T \mathrm{Diag}\left(\frac{b_i \cdot \bar z_i^2}{|z_i|^3}\right) W_i & \sum_{i=1}^{k} W_i^T \overline{Q}_i \left[\frac{1}{2}I - \frac{1}{4}\mathrm{Diag}\left(\frac{b_i}{|z_i|}\right)\right] \overline{W}_i \end{pmatrix},$$

where $W_i = FQ_i$. It follows that

$$\tau_\rho(\psi, \phi) = \sum_{i=1}^{k}\sum_{j=1}^{n} |t_{ji}|^2 \left(1 - \frac{\beta_{ji}}{|\zeta_{ji}|}\sin^2(\mu_{ji} - \theta_{ji})\right). \tag{2.6}$$

Thus, $\tau_\rho \geq 0$ when $|\zeta_{ji}| \geq \beta_{ji}$ for all $j = 1, 2, ..., n$ and $i = 1, 2, ..., k$. Even if $|\zeta_{ji}|$ is slightly less than β_{ji} for some j and i, τ_ρ may remain positive when the corresponding $\sin^2(\mu_{ji} - \theta_{ji})$ is sufficiently small and other terms in the summation in (2.6) are sufficiently large and positive.

A classical problem encountered in optics is associated with $k = 1$. When only one diffraction image is recorded, experience shows that local minima are common. Regions of negative curvature separate local minima from the global solution [**16**].

3. Iterative algorithms based on nonlinear optimization

Because the gradient and Hessian of (2.1) and (2.2) are relatively easy to evaluate, we may use standard minimization algorithms such as the steepest descent method, the Newton's method and the nonlinear conjugate gradient method to find the solution to the ptychographic reconstruction problem.

3.1. Unconstrained minimization. When the objective function (2.1) or (2.2) is minimized directly, we construct a sequence of approximations to ψ_0 by

$$\psi^{(\ell+1)} = \psi^{(\ell)} + \beta p^{(\ell)}, \tag{3.1}$$

where $p^{(\ell)}$ is a *search direction* along which the objective function (2.1) or (2.2) decreases, and $\beta > 0$ is an appropriate step length chosen through a line search procedure that ensures global convergence [**17**]. Global convergence can also be achieved by introducing an additional inequality constraint that limits the size of the update within a "trust region" [**18**].

To accelerate convergence, we may also introduce an appropriate weighting matrix into least squares objective functions (2.1) and (2.2) by expressing them as

$$\rho(\psi) = \frac{1}{2}\sum_{i=1}^{k}\langle |z_i| - b_i, |z_i| - b_i\rangle_B,$$

and

$$\epsilon(\psi) = \frac{1}{2}\sum_{i=1}^{k}\langle |z_i|^2 - b_i^2, |z_i|^2 - b_i^2\rangle_B$$

respectively, where $\langle x, y\rangle_B = x^* By$ with B being a symmetric positive definite matrix. As we will show in section 4, the choice of $B = \text{Diag}(b_i)^{-1}$ is particularly useful for accelerating the convergence of all iterative methods we have looked at. To maintain numerical stability and reduce noise amplification, it is often necessary to add a small constant to the diagonal of B to prevent it from becoming singular or ill-conditioned. In the presence of noise, the choice of B can be made according to a stochastic characterization of the noise. This leads to a maximum likelihood formulation of the phase retrieval problem [**15**].

Another useful technique for accelerating iterative methods for solving unconstrained minimization problems is preconditioning. Instead of minimizing $\rho(\psi)$ or $\epsilon(\psi)$, we make a change of variable and minimize $\hat{\rho}(\phi)$ and $\hat{\epsilon}(\phi)$, where $\phi = K\psi$, and K is a preconditioner that is usually required to be Hermitian and positive definite. The purpose of introducing the preconditioner K is to reduce the condition number of the Hessian of the objective function. A highly ill-conditioned Hessian often leads to slow convergence of an iterative method. A well-known example is the zig-zag behavior of the steepest descent algorithm when it is applied to the Rosenbrock function.

It follows from the chain rule and (2.4) that the gradient of $\hat{\rho}(\psi)$ is simply

$$\nabla\hat{\rho}(\psi) = \frac{1}{2}K^{-1}\sum_{i=1}^{k}[Q_i^*Q_i\psi - Q_i^*F^*\text{Diag}\left(\frac{z_i}{|z_i|}\right)b_i],$$

where $z_i = FQ_i\psi$.

If we take the preconditioner to be the constant term on the diagonal blocks of $H^{\rho}_{\psi\psi}$, i.e.,

$$K = \sum_{i=1}^{k} Q_i^*Q_i, \tag{3.2}$$

which is a diagonal matrix, the gradient of $\hat{\rho}$ simply becomes

$$\nabla\hat{\rho}(\psi) = \frac{1}{2}\left[\psi - \left(\sum_{i=1}^{k} Q_i^*Q_i\right)^{-1}\left(\sum_{i=1}^{k} Q_i^*F^*\text{Diag}\left(\frac{z_i}{|z_i|}\right)b_i\right)\right],$$

and the corresponding preconditioned steepest descent algorithm with a constant step length of 2 yields the following updating formula:

$$\psi^{(\ell+1)} = \left(\sum_{i=1}^{k} Q_i^*Q_i\right)^{-1}\left(\sum_{i=1}^{k} Q_i^*F^*\text{Diag}\left(\frac{z_i^{(\ell)}}{|z_i^{(\ell)}|}\right)b_i\right),$$

where $z_i^{(\ell)} = FQ_i\psi^{(\ell)}$. This updating formula is identical to that used in the *error reduction* algorithm or *alternate projection* algorithm [**16**], which is guaranteed to converge to at least a local minimizer as shown in section 3.2.

3.2. Fixed-point iteration and projection algorithms. An alternative approach to finding a minimizer of (2.1) is to set its gradient to zero and seek ψ that satisfies the first order necessary condition of the minimization problem. If $\sum_{i=1}^k Q_i^*Q_i$ is nonsingular, by setting $\nabla\rho(\psi)=\frac{1}{2}\sum_{i=1}^k\left[Q_i^*Q_i\psi - Q_i^*F^*\text{Diag}\left(\frac{z_i}{|z_i|}\right)b_i\right]$ to 0, we obtain

$$\psi = f(\psi) \tag{3.3}$$

where

$$f(\psi) = \left(\sum_{i=1}^{k} Q_i^*Q_i\right)^{-1}\left[\sum_{i=1}^{k} Q_i^*F^*\text{Diag}\left(\frac{z_i}{|z_i|}\right)b_i\right]. \tag{3.4}$$

Recall that $z_i \equiv FQ_i\psi$. Clearly, ψ is a fixed point of the function f.

A simple iterative technique one may use to find the solution to (3.4) is the fixed point iteration that has the form

$$\psi^{(\ell+1)} = f(\psi^{(\ell)}).$$

Replacing f with the right hand side of (3.4) yields

$$\psi^{(\ell+1)} = \left(\sum_{i=1}^{k} Q_i^*Q_i\right)^{-1}\left[\sum_{i=1}^{k} Q_i^*F^*\text{Diag}\left(\frac{z_i^{(\ell)}}{|z_i|^{(\ell)}}\right)b_i\right], \tag{3.5}$$

where $z_i^{(\ell)} \equiv FQ_i\psi^{(\ell)}$. This is the same sequence of iterates produced in what is known as the *error reduction* algorithm in standard phase retrieval literature [**16**]. This method is also known as the *alternate projection* algorithm for reasons to be discussed below.

It is easy to verify that the updating formula in (3.5) is identical to that produced by a preconditioned steepest descent algorithm in which the preconditioner K is chosen to be $K = \sum_{i=1}^{k} Q_i^* Q_i$, and a constant step length of 2 is taken at each iteration, i.e.,

$$\psi^{(\ell+1)} = \psi^{(\ell)} - 2\nabla\rho(\psi^{(\ell)}).$$

The sequence of iterates $\{\psi^{(\ell)}\}$ produced by (3.5) is guaranteed to converge to the fixed point of $f(\psi)$ from any starting point $\{\psi^{(0)}\}$, if the spectral radius (i.e., the largest eigenvalue) of the Jacobian of f (with respect to ψ) is strictly less than 1. Because the function f in (3.3) can be viewed as a function of ψ and $\bar{\psi}$, we should examine the Jacobian matrix of the system

$$\psi = \left(\sum_{i=1}^{k} Q_i^* Q_i\right)^{-1} \left[\sum_{i=1}^{k} Q_i^* F^* \mathrm{Diag}\left(\frac{z_i}{|z_i|}\right) b_i\right], \tag{3.6}$$

$$\bar{\psi} = (\sum_{i=1}^{k} Q_i^T \overline{Q}_i)^{-1} \left[\sum_{i=1}^{k} Q_i^T F^T \mathrm{Diag}\left(\frac{\bar{z}_i}{|z_i|}\right) b_i,\right] \tag{3.7}$$

where (3.7) is simply the conjugate of (3.6). It is not difficult to show that this Jacobian matrix has the form

$$J = \begin{pmatrix} K^{-1} & 0 \\ 0 & \overline{K}^{-1} \end{pmatrix} \begin{pmatrix} K - 2H^{\rho}_{\psi\psi} & -2H^{\rho}_{\psi\bar{\psi}} \\ -2H^{\rho}_{\bar{\psi}\psi} & \overline{K} - 2H^{\rho}_{\bar{\psi}\bar{\psi}} \end{pmatrix}, \tag{3.8}$$

where $H^{\rho}_{\psi\psi}$, $H^{\rho}_{\psi\bar{\psi}}$, $H^{\rho}_{\bar{\psi}\psi}$ and $H^{\rho}_{\bar{\psi}\bar{\psi}}$ are as defined in the formula for H^{ρ}.

If (λ, ϕ) is an eigenpair of J, we can easily show that

$$2\begin{pmatrix} H^{\rho}_{\psi\psi} & H^{\rho}_{\psi\bar{\psi}} \\ H^{\rho}_{\bar{\psi}\psi} & H^{\rho}_{\bar{\psi}\bar{\psi}} \end{pmatrix} \begin{pmatrix} \phi \\ \bar{\phi} \end{pmatrix} = (1-\lambda) \begin{pmatrix} K & 0 \\ 0 & \overline{K} \end{pmatrix} \begin{pmatrix} \phi \\ \bar{\phi} \end{pmatrix}.$$

If we again let $t_{ji} \equiv |t_{ji}|e^{i\mu_{ji}}$, $\zeta_{ji} \equiv |\zeta_{ji}|e^{i\theta_{ji}}$ and β_{ji} be the jth component of the vectors $t_i = FQ_i\phi$, $z_i = FQ_i\psi$ and b_i respectively, we can easily show that

$$\lambda = \frac{\sum_{i=1}^{k}\sum_{j=1}^{n} \sin^2(\mu_{ji} - \theta_{ji})|t_{ji}|^2 \beta_{ji}/|\zeta_{ji}|}{\sum_{i=1}^{k}\sum_{j=1}^{n} |t_{ji}|^2}. \tag{3.9}$$

Clearly, when $\beta_{ji} \le |\zeta_{ji}|$ for all $j = 1, 2, ..., m$ and $i = 1, 2, ...n$, $|\lambda| \le 1$, and the fixed point iteration is guaranteed to converge to at least a local minimizer of ρ.

The fixed point of f may also be obtained by applying Newton's method or a quasi-Newton algorithm to seek the root of $r(\psi) = 0$, where $r(\psi) = \psi - f(\psi)$.

This approach is equivalent to applying Newton's method or a quasi-Newton algorithm (with appropriate line search and trust region strategies) to minimize $\rho(\psi)$.

If we multiply (3.6) from the left by Q_i for $i = 1, 2, ..., k$, and let $y^{(\ell)} = Q\psi^{(\ell)}$, where $Q = (Q_1^* \;\; Q_2^* \;\; ... \;\; Q_k^*)^*$, we obtain

$$y^{(\ell+1)} = P_Q P_F(y^{(\ell)}), \tag{3.10}$$

where $P_Q = Q(Q^*Q)^{-1}Q^*$, and

$$P_F(y) = \hat{F}^* \frac{y}{|y|} \cdot b,$$

where $\hat{F} = \text{Diag}\,(F, F, ..., F)$ and $b = (b_1^T \;\; b_2^T \;\; ... \;\; b_k^T)^T$.

Because a fixed point y of $P_Q P_F$ is in the range of Q, which is typically full rank when $mk > n$, we may recover the corresponding fixed point of f from y via the least squares solution $\psi^{(\ell)} = (Q^*Q)^{-1}Q^*y^{(\ell)}$.

This nonlinear map is the composition of a (linear) orthogonal projector P_Q and a (nonlinear) Fourier magnitude projector P_F. A fixed point iteration based on (3.10) is also called *alternating projection* (AP) algorithm in the phase retrieval literature because the approximation to the solution of (3.10) is obtained by applying P_Q and P_F in an alternating fashion.

It is easy to verify that P_F is indeed a projection operator in the sense that

$$\|P_F(y) - y\| \leq \|w - y\| \;\text{ for all }\; w \in \{w | w = P_F(w)\}. \tag{3.11}$$

This property of P_F, together with the fact that P_Q is an orthogonal projection operator, i.e. $\|P_Q y - y\| \leq \|w - y\|$ for all $w \in \text{Range}(Q)$, allows us to show that the residual error $\|P_Q P_F(y^{(\ell)}) - y^{(\ell)}\|$ decreases monotonically in the AP algorithm. The proof of this observation was provided by Fienup in [**19**], which showed that

$$\|y^{(\ell+2)} - y^{(\ell+1)}\| \leq \|y^{(\ell+1)} - y^{(\ell)}\|, \tag{3.12}$$

and the equality in (3.12) holds only when $P_F(y^{(\ell)}) = y^{(\ell)}$, i.e., when convergence is reached.

The inequality (3.12) shows that the AP algorithm converges to a stationary point. However, the convergence can be extremely slow because

$$\|z^{(\ell+1)}\| = \|Fy^{(\ell+1)}\| = \|y^{(\ell+1)}\| = \|P_Q P_F(y^{(\ell)})\| \leq \|P_F(y^{(\ell)})\| = \|b\|,$$

and many of the terms β_{ji}/ζ_{ji}, $i = 1, 2, ..., k$ and $j = 1, 2, ..., m$, in (3.9) may be great than 1. Only when $y^{(\ell)}$ is very close to the fixed point of $P_Q P_F$, the spectral radius of the Jacobian of (3.4) may become much smaller than 1 in (3.9) due to the reduction effect of the $\sin^2(\mu_{ji} - \theta_{ji})$ terms.

The simple alternating projection algorithm has been extended to the hybrid input-output (HIO) algorithm [**19**], the relaxed averaged alternating reflection (RAAR) algorithm [**20**], and many other variants [**16, 21**] in the phase retrieval literature. Just to give a few examples, in the HIO and RAAR algorithms, the approximation to the solutions of (3.7) and (3.10) are updated by

$$\begin{aligned} y^{(\ell+1)} &= [P_Q P_F + (I - P_Q)(I - \beta P_F)]\, y^{(\ell)}, \;\text{ and} \\ y^{(\ell+1)} &= [2\beta P_Q P_F + (1 - 2\beta) P_F + \beta(P_Q - I)]\, y^{(\ell)}, \end{aligned}$$

respectively, where β is a constant often chosen to be between 0 and 1, and the object itself can be recovered from $y^{(\ell)}$ through $\psi^{(\ell+1)} = (Q^*Q)^{-1}Q^*y^{(\ell)}$.

Although these algorithms tend to accelerate the convergence of $y^{(\ell)}$, their convergence behavior is less predictable and not well understood.

4. Numerical examples

In this section, we demonstrate and compare the convergence of iterative algorithms for ptychographic reconstruction using two test images. The first test image is a 256×256 real-valued cameraman image shown in Figure 2. The image is often used in the image processing community to test image reconstruction and restoration algorithms. The second test image is a complex valued image. It also contains 256×256 pixels that correspond to the complex transmission coefficients of a collection of gold balls embedded in some medium. The amplitude and phase angles of these pixels are shown in Figure 3.

FIGURE 2. The cameraman test image.

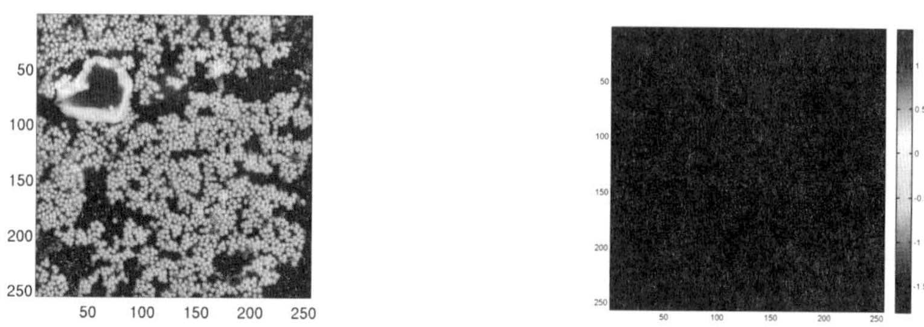

(a) Amplitude

(b) Phase

FIGURE 3. The amplitude and phase of the transmission coefficient of a collection of gold balls.

All numerical examples presented in this paper are performed in MATLAB.

4.1. Numerical comparison of iterative methods. In this section, we show the convergence behavior of a few iterative algorithms for solving the ptychographic reconstruction by numerical experiments. In the cameraman image reconstruction experiment, we choose the illuminating probe $a(\mathbf{r})$ to be a 64×64 binary probe shown in Figure 4(a). The pixels within the 32×32 square at the center of the probe assume the value of 1. All other pixels take the value of 0. The zero padding of the inner 32×32 square ensures that the diffraction pattern of a 64×64 frame associated with this probe is oversampled in the reciprocal space. In the gold ball image reconstruction experiment, the illuminating probe is chosen to be the amplitude of the Fourier transform of an annular ring with inner radius of

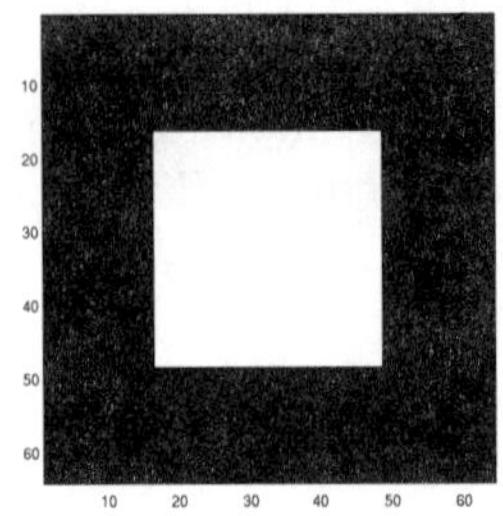

(a) The binary probe used in the reconstruction of the cameraman image.

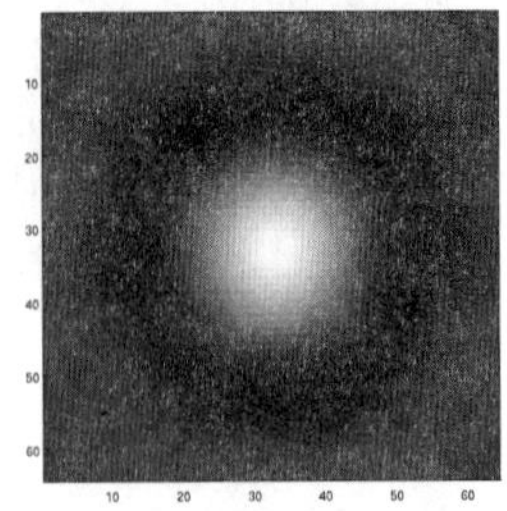

(b) The probe used in the reconstruction of the gold ball image.

FIGURE 4. The illuminating probes $a(\mathbf{r})$ used in ptychographic reconstructions of the cameraman and gold ball images.

$r_1 \approx 5.4$ and outer radius of $r_2 \approx 19.4$. This probe mimics the true illumination used in a physical experiment.

In the cameraman experiment, the probe is translated by 8 pixels at a time in either horizontal or vertical direction. To prepare a stack of k diffraction images b_i, $i = 1, 2, ..., k$, we start from the upper left corner of the true image, extract a 64×64 frame, and multiply it with the probe, and then apply a 2D FFT to the product. The magnitude of transform is recorded and saved before we move either horizontally or vertically to obtain the next frame. If the lower right corner of the frame goes outside of the image (which does not happen in this particular case), we simply "wrap the probe around" the image as if the image is periodically extended. As a result, the total number of diffraction frames we use for each reconstruction is $k = (256/8) \cdot (256/8) = 1024$.

As we will show in section 4.3, the size of translation, which determines the amount of overlap between adjacent frames, has a noticeable effect on the convergence of the iterative reconstruction algorithms.

Figure 5 shows the convergence history of several iterative algorithms discussed in section 3 when they are applied to the diffraction frames extracted from the cameraman image. We plot both the relative residual norm defined by

$$res = \frac{\sqrt{\sum_{i=1}^{k} \||z_i|^{(\ell)} - b_i\|^2}}{\sqrt{\sum_{i=1}^{k} \|b_i\|^2}}, \tag{4.1}$$

where $|z_i|^{(\ell)} = |FQ_i\psi^{(\ell)}|$ and ℓ is the iteration number, and the relative error of the reconstructed image defined by $err = \|\psi^{(\ell)} - \psi_0\|/\|\psi_0\|$. In these runs, an exact line search is used in both the steepest descent (SD) method and the nonlinear conjugate gradient (CG) method. The Steihaug's trust region technique [**23**] is used in the Newton's method (NT). We set the starting guess of the solution ψ_0 to

$$\psi^{(0)} = \left(\sum_{i=1}^{k} Q_i^* Q_i\right)^{-1} \sum_{i=1}^{k} Q_i^* b_i.$$

It is clear from Figure 5 that NT converges much faster than the other algorithms. Its performance is followed by the CG algorithm which is much faster than the

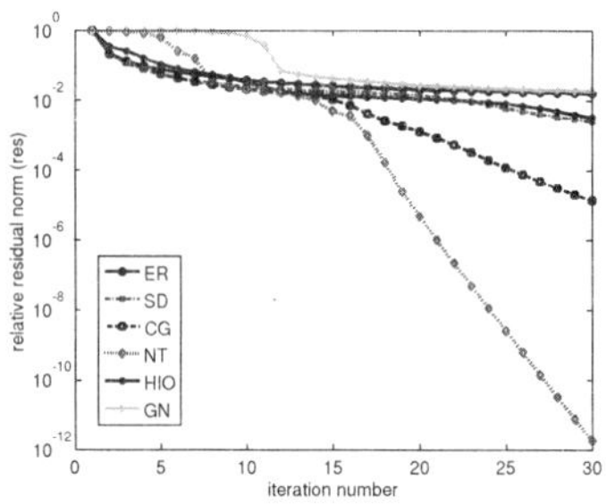

(a) Change of the relative residual norm (res) for the reconstruction of the cameraman image.

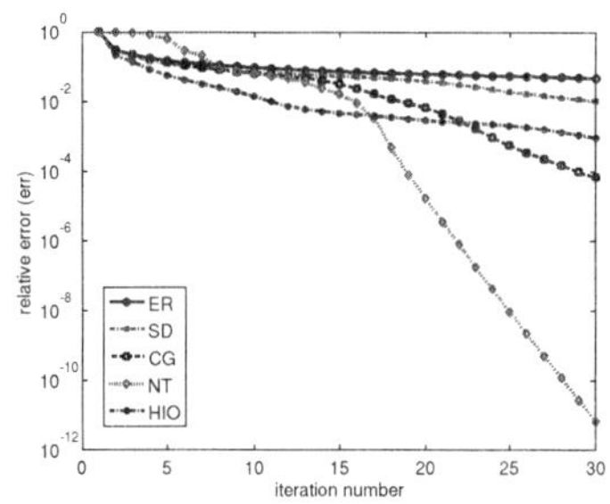

(b) Change of the relative error (err) for the reconstruction of the cameraman image.

FIGURE 5. A comparison of the convergence behavior of different iterative ptychographic reconstruction algorithms for the cameraman image.

error reduction (ER), SD, Gauss-Newton (GN) and the hybrid input-output (HIO) algorithms. Similar convergence behavior is observed when other random starting guesses are used, although occasionally, a random starting guess can lead to stagnation or convergence to a local minimizer. We will discuss this issue in section 4.2. We set the maximum number of iterations allowed in all runs to 30. This is somewhat excessive for both NT and CG algorithms. Typically, when the relative error of the reconstructed image falls below 10^{-3}, it is nearly impossible to visually distinguish the reconstruction from the true image. When the relative error is larger, the reconstructed cameraman images may contain visible artifacts such as those shown in Figures 6(a) and 6(b) which are produced at the end of the 30th ER and SD iterations respectively.

(a) ER reconstruction

(b) SD reconstruction

FIGURE 6. The reconstructed cameraman images by ER and SD algorithms contain visible ringing artifacts.

For the reconstruction of the gold ball image, we choose the starting guess to be

$$\psi^{(0)} = \left(\sum_{i=1}^{k} Q_i^* Q_i \right)^{-1} \sum_{i=1}^{k} Q_i^* \mathrm{Diag}\,(b_i)\, \mathrm{Diag}\,(|u_i|)^{-1} u_i,$$

where u_i is a complex random vector, and the real and imaginary part of each component has a uniform distribution within $[-1, 1]$.

In this experiment, the probe is translated by a larger amount (16 pixels) in either horizontal or vertical direction. Figure 7 shows the convergence history of ER, SD, CG, HIO, and NT. From Figure 7(a), it appears that CG is the best among all the methods we tried. The HIO algorithm performs well in the first 60 iterations, but then stagnates. As we can see from Figure 7 that neither the residual norm nor the relative error associated with HIO changes monotonically. This is not completely surprising because HIO does not try to minimize either objective functions. For this example, the performance of NT lags behind CG by a large margin although both algorithms exhibit monotonic convergence with a more predictable error reduction. We should mention that to measure the relative error associated with a reconstructed gold ball image $\psi^{(\ell)}$, we need to multiply it by a constant phase factor γ first, i.e., the relative error is defined as

$$err = \frac{\|\gamma\psi^{(\ell)} - \psi_0\|}{\|\psi_0\|}.$$

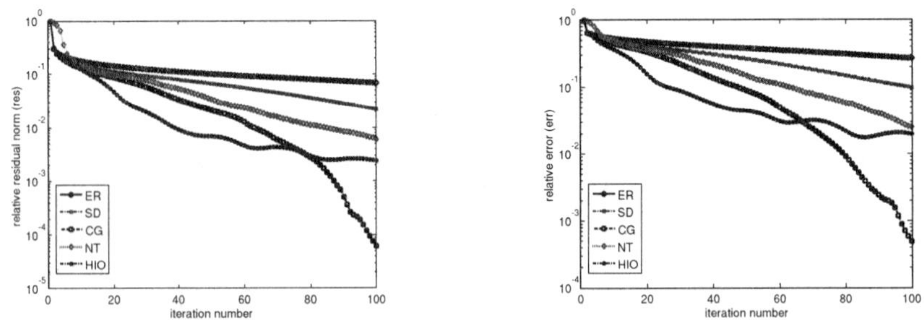

(a) Change of the relative residual norm (res) for the reconstruction of the gold ball image.

(b) Change of the relative error (err) for the reconstruction of the gold ball image.

FIGURE 7. A comparison of the convergence behavior of different iterative ptychographic reconstruction algorithms for the gold ball image.

4.2. Local minimizer and the choice of the objective function. As we indicated in section 2.2, based on the analytic Hessian and curvature expression, neither $\epsilon(\psi)$ nor $\rho(\psi)$ is globally convex. This observation suggests that all iterative optimization algorithm discussed above may converge to a local minimizer. Although we found that in practice, local minimizers are not easy to find, they do exist as the following example shows.

In order to find a local minimizer, we construct many random starting guesses using the MATLAB `rand` function. To save time, we choose to reconstruct a 64×64 subimage of the cameraman image shown in Figure 2. This subimage is shown in Figure 10(a). A 16×16 binary probe that has a value 1 in the 8×8 center of the probe and 0 elsewhere is used. The diffraction stack consisting of 64 diffraction images is obtained by translating the probe 4 pixels a time in either the horizontal or the vertical direction.

Figure 8 shows that one of the random starting guesses leads to the convergence of the CG algorithm to a local minimizer. In particular, the relative residual (4.1) which is proportional to the objective function ρ stagnates around 0.9 after the first 15 iterations (Figure 8(a)), whereas the relative gradient $\|\nabla\rho(\psi^{(\ell)})\|/\|\psi_0\|$ decreases to 10^{-8} after 40 iterations.

Figure 10(b) shows how the reconstructed image compares with the true image for this particular starting guess used. In this case, the local minimizer appears to contain visible artifacts in a small region near top of the tripod. The amplitude of this localized error is also revealed in the relative error plot shown in Figure 9(a). The phase error associated with a particular frame of the reconstruction obtained from

$$\frac{\overline{Q_i\psi}}{|Q_i\psi|}\cdot\frac{Q_i\psi_0}{|Q_i\psi_0|},$$

for some particular Q_i is shown in Figure 9(b).

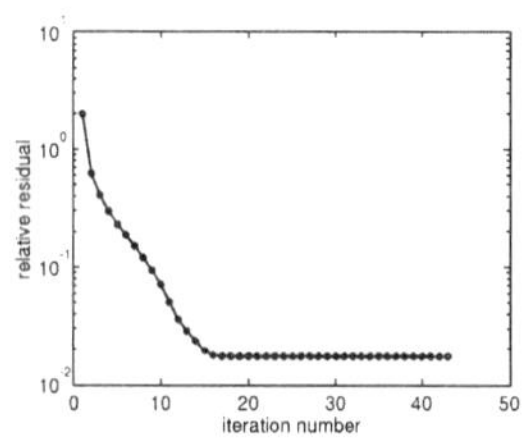

(a) Change of the relative residual norm (res).

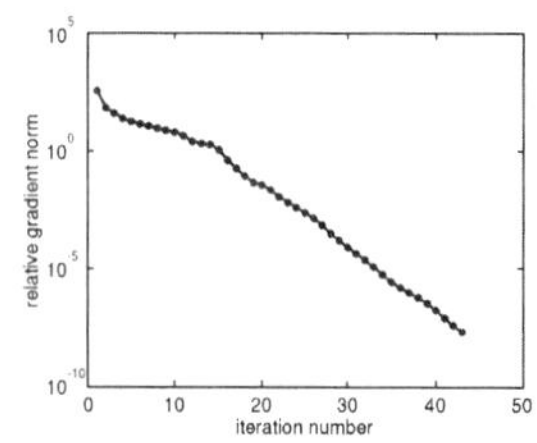

(b) Change of the relative gradient.

FIGURE 8. The convergence of CG to a local minimizer.

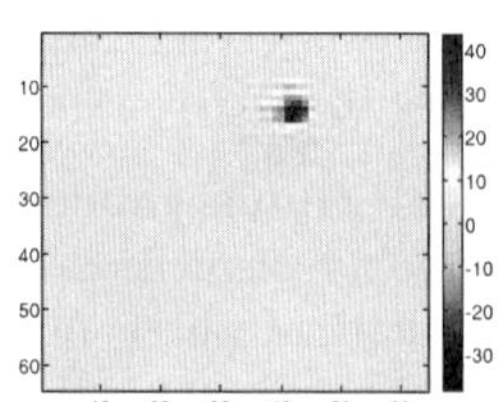

(a) Amplitude error in the reconstruct image

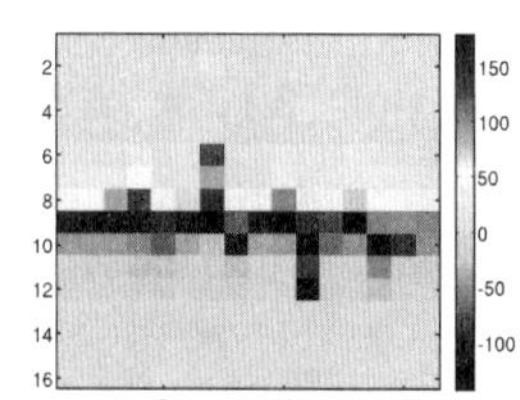

(b) Phase error in degrees associated with a particular frame

FIGURE 9. The error associated with a local minimizer.

We should also note that for this particular starting guess, all methods we tried converged to the same local minimizer. This is not all that surprising. It simply shows (empirically) that a local minimizer of (2.1) exists, and our starting guess is sufficiently close to it.

However, what is interesting is that if we choose to minimize (2.2) by using any one of the iterative methods discussed above from the same starting guess, we are

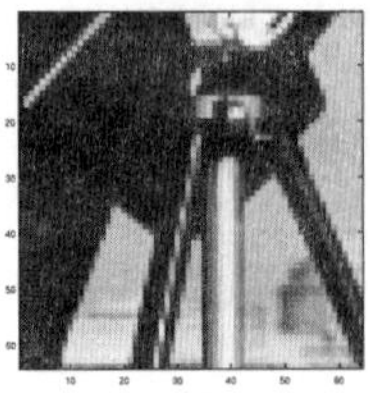

(a) True image.

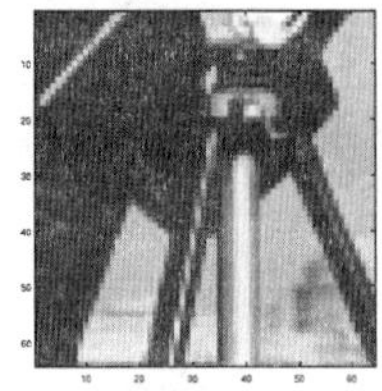

(b) The reconstructed image (a local minimizer).

FIGURE 10. The artifacts produced by a local minimizer of ρ.

able to obtain the correct solution. For example, Figure 11(a) shows that when the NT applied to the weighted (scaled) objective function

$$\tilde{\epsilon}(\psi) = \frac{1}{2}\sum_{i=1}^{k}(|z_i|^2 - b_i^2)^T \mathrm{Diag}\,(b_i)^{-1}\,(|z_i|^2 - b_i^2), \tag{4.2}$$

where $|z_i| = |FQ_i\psi|$ and $b_i = |FQ_i\psi_0|$, an accurate reconstruction can be obtained in roughly 350 iterations. Admittedly, the convergence rate is much slower in this case when compared to the convergence of NT applied to (2.1) from a different starting point. The convergence is even slower if no weighting (or scaling) is used, i.e. when (2.2) is used as the objective function. However, the fact that convergence can be reached for (4.2) but not (2.1) from the same starting point is quite interesting. Furthermore, Figure 11(b) shows that if we take the local minimizer returned from an iterative minimization of (2.1) as the starting guess for minimizing (4.2), convergence can be reached in 12 iterations. This experiment suggests that it may be useful to have a hybrid optimization scheme in which (2.1) is minimized first. If a local minimizer of (2.1) is identified, one can then try to minimize (4.2) starting from the local minimizer of (2.1). A local minimizer can be recognized by examining the norm of gradient, which should be very small, and the objective function (2.1) or (2.2) itself, which is not close to zero at a local minimizer.

4.3. The effect of overlapping on the convergence of iterative algorithm. When there is no overlap between two adjacent frames, the ptychographic phase retrieval problem reduces to that of classical phase retrieval for a number of isolated diffraction images. For this type of problems, optimization based algorithms often converge to a local minimizer. On the other hand, when all frames completely overlap, the phase retrieval problem is equally difficult to solve because overlapping does not provide any new information and phase retrieval is essentially performed on a single diffraction image.

Apart from these two extreme cases, having significant overlap among adjacent frames as we move the probe generally helps improve the convergence of optimization algorithms. The following example show that the amount of overlap has a noticeable effect on the convergence of optimization based iterative algorithms (e.g., CG, NT, SD etc.) A similar observation is also reported in [**24**].

In this example, we try to reconstruct the gold ball image from four different diffraction stacks. Each stack contains a set of 64×64 diffraction frames. These frames are generated by translating the probe shown in Figure 4(b) by different

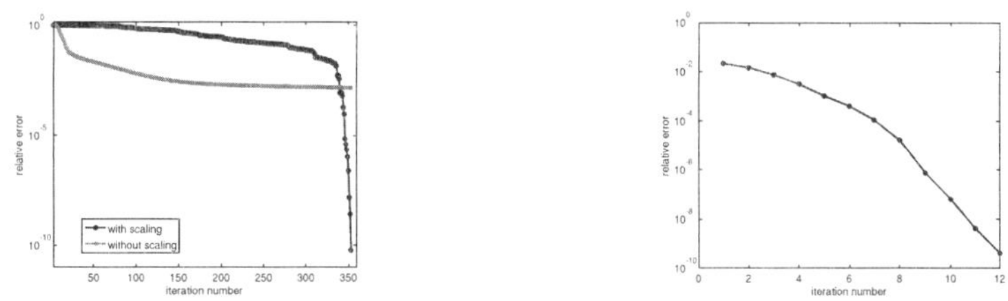

(a) The convergence of the NT algorithm when it is applied to (2.2) (red) and (4.2). The starting guess chosen in these runs is the same one used in the minimization of (2.1).

(b) The convergence of the NT algorithm when the starting guess is chosen to be the local minimizer shown in Figure 10(b)

FIGURE 11. The convergence of the NT algorithm when applied to (2.2) (red) and (4.2) (blue).

amount (Δx) in horizontal and vertical directions. The larger the translation, the smaller the overlap is between two adjacent images. Figure 12(a) shows that CG converges very slowly when the diffraction stack contains diffraction frames obtained by translating the probe 20 pixels at a time (the black curve). Faster convergence is observed when the amount of translation is decreased to $\Delta x = 16, 12, 8$. It is interesting to see from Figure 12(b) that the amount of overlap does not affect the convergence of the HIO algorithm.

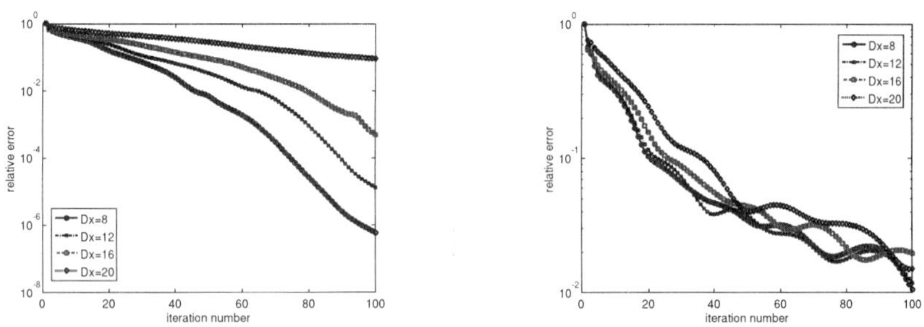

(a) The effect of overlapping on the convergence of CG for the gold ball image reconstruction.

(b) The effect of overlapping on the convergence of HIO for the gold ball image reconstruction.

FIGURE 12. The effect of overlapping on the convergence of CG and HIO algorithms.

To explain the effect of overlapping on the convergence of optimization based iterative algorithms such as the nonlinear CG, we examine the structure of the Hessian of the objective function ρ in (2.1). It is not difficult to show [**13**] that H^ρ can be written as

$$H^\rho = \begin{pmatrix} (\hat{F}Q)^* & \\ & (\hat{F}Q)^T \end{pmatrix} \begin{pmatrix} B_{11} & B_{12} \\ B_{21} & B_{22} \end{pmatrix} \begin{pmatrix} \hat{F}Q & \\ & \overline{\hat{F}Q} \end{pmatrix}, \tag{4.3}$$

where $B_{11} = \overline{B_{22}}$ and $B_{12} = B_{21}^*$ are all diagonal, $\hat{F}$ is a block diagonal matrix of discrete Fourier transforms, i.e. $\hat{F} = \mathrm{Diag}(F, F, ..., F)$, and $Q = (Q_1^* \, Q_2^* \, ... \, Q_k^*)^*$. The diagonal elements of B_{11} and B_{12} are simply $1 - \beta_{ji}/(2\zeta_{ji})$ and $\beta_{ji}\zeta_{ji}^2/(2|\zeta_{ji}|^3)$ respectively for $i = 1, 2, ..., k$ and $j = 1, 2, ..., m$.

We will show that H^ρ is diagonal-dominant when there is a sufficient amount of overlap between adjacent diffraction frames. To simplify our discussion, let us assume for the moment that b_i is a 1D diffraction pattern obtained from a binary probe that illuminates three pixels at a time, and the probe is translated one pixel at a time so that the adjacent image frames overlap by two pixels. In this case, the $\hat{F}Q$ term in (4.3) has the form

$$\begin{pmatrix} f_1 & f_2 & f_3 & \cdots & 0 \\ 0 & f_2 & f_3 & \ddots & \vdots \\ 0 & 0 & f_3 & \ddots & f_k \\ f_1 & 0 & 0 & \ddots & f_k \\ f_1 & f_2 & 0 & \cdots & f_k \end{pmatrix},$$

where f_i is the ith column of F.

As a result, a typical diagonal term of H^ρ has the form

$$H_{i,i}^\rho = f_i^* D_{i-2} f_i + f_i^* D_{i-1} f_i + f_i^* D_i f_i = \mathrm{trace}(D_{i-2} + D_{i-1} + D_i), \tag{4.4}$$

where D_i is a diagonal matrix that contains elements $1 - \beta_{ji}/(2\zeta_{ji})$ for $j = 1, 2, 3$.

When ψ is near the solution, z_i is close to b_i. Hence, D_i is likely to contain positive entries only. Therefore, the diagonal elements of H^ρ are likely to be much larger compared to the nonzero off-diagonal elements which contain terms in the form of either $f_j^* D_i f_\ell$ and its conjugate, where $j \neq \ell$, or $f_j^T E_i f_j$ and its conjugate, where E_i is a diagonal matrix (and part of B_{12}) that contains elements $\beta_{ji}\zeta_{ji}^2/(2|\zeta_{ji}|^3)$ for $j = 1, 2, 3$. Due to the phase difference between f_j and f_ℓ, D_i's do not add up "coherently" on the off-diagonal of H^ρ as they do on the diagonal. Neither do nonzero entries in E_i's add up coherently on the off-diagonal blocks of H^ρ. Hence, the matrix H^ρ becomes diagonal-dominant when there is a large amount of overlap between two adjacent frames. In fact, the diagonal of H^ρ may become so dominant that the spectral property of H^ρ is determined largely by the diagonal part of the matrix, which is typically well conditioned due to the averaging of D_i in (4.4). This observation provides an intuitive explanation on why increasing the amount of overlap between adjacent frames tends to improve the convergence rate of CG and other optimization based iterative ptychographical phase retrieval algorithms. Although this is not a precise analysis of the spectral property of H^ρ, the analysis does match with observations made in our numerical experiments. Moreover, this type of analysis can be extended to the 2D case in which F is represented as a tensor product of two 1D discrete Fourier transforms.

5. Conclusion

We formulated the ptychographic phase retrieval problem as a nonlinear optimization problem and discussed how standard iterative optimization algorithms can be applied to solve this problem. We showed that the optimization problems we solve are not globally convex. Hence standard optimization algorithms can produce local minimizers. However, the Hessians of the objective functions we minimize do have special structures that may be exploited.

We compared the performance of several optimization algorithms and found that both Newton's method with Steihaug's trust region technique and the nonlinear conjugate gradient algorithm are efficient for solving this type of problems. An even more efficient algorithm based on an augmented Lagrangian formulation the problem and the use alternating direction method has recently developed by authors in [**25**].

We demonstrated by a numerical example that the convergence rate of an optimization algorithm depends on the amount of overlapping between two adjacent diffraction frames. We provided a preliminary analysis on why this occurs. More research is needed to provide a more rigorous study on this phenomenon.

In practice, the diffraction measurements often contain some noise. As a result, regularization techniques must be used in the iterative reconstruction algorithms to limit the amplification of noise. For most of the iterative algorithms described in this paper, regularization amounts to terminating the iterations early [**26**]. Other regularization techniques include reformulating the problem as a maximum likelihood estimation problem [**15**] and imposing additional constraints to the optimization problem [**7**], which can be solved by an augmented projection method [**7**]. We will describe and compare these technique in a future publication.

Acknowledgment

This work was supported by the Laboratory Directed Research and Development Program of Lawrence Berkeley National Laboratory (A. S.) and the Applied Mathematical Sciences sub-program of the Office of Energy Research, U.S. Department of Energy, under contract DE-AC02-05CH11231 (C. Y., S. M.), the National Science Foundation Grant (J. Q.) and by the Director, Office of Science, Advanced Scientific Computing Research, of the U.S. Department of Energy under Contract No. DE-AC02-05CH11231 (F.M.). The computational results presented were obtained at the National Energy Research Scientific Computing Center (NERSC), which is supported by the Director, Office of Advanced Scientific Computing Research of the U.S. Department of Energy under contract number DE-AC02-05CH11232.

References

[1] J. M. Rodenburg and R. H. T. Bates. The theory of super-resolution electron microscopy via Wigner-distribution deconvolution. *Phil. Trans. R. Soc. Lond. A*, 339:521–553, 1992.

[2] H. M. L. Faulkner and J. M. Rodenburg. Movable aperture lensless transmission microscopy: a novel phase retrieval algorithm. *Phy. Rev. Lett.*, 93:023903, 2004.

[3] J. M. Rodenburg and H. M. L. Faulkner. A phase retrieval algorithm for shifting illumination. *Appl. Phy. Lett.*, 85:4795–4797, 2004.

[4] M. Guizar-Sicairos and J. R. Fineup. Phase retrieval with transverse translation diversity: a nonlinear optimization approach. *Opt. Express*, 16:7264–7278, 2008.

[5] P. Thibault, M. Dierolf, O. Bunk, A. Menzel, and F. Pfeiffer. Probe retrieval in ptychographic coherent diffractive imaging. *Ultramicroscopy*, 109:338–343, 2009.

[6] Monson H. Hayes, *The reconstruction of a multidimensional sequence from the phase or magnitude of its Fourier transform*, IEEE Trans. Acoust. Speech Signal Process. **30** (1982), no. 2, 140–154, DOI 10.1109/TASSP.1982.1163863. MR654258 (83d:94004)

[7] S. Marchesini, A. Schirotzek, C. Yang, H. Wu and F. Maia. Augmented projections for ptychographic imaging. http://arxiv.org/pdf/1209.4924.pdf, 2013.

[8] Jorge L. C. Sanz, *Mathematical considerations for the problem of Fourier transform phase retrieval from magnitude*, SIAM J. Appl. Math. **45** (1985), no. 4, 651–664, DOI 10.1137/0145038. MR796101 (87h:42024)

[9] Gabriele Bianchi, Richard J. Gardner, and Markus Kiderlen, *Phase retrieval for characteristic functions of convex bodies and reconstruction from covariograms*, J. Amer. Math. Soc. **24** (2011), no. 2, 293–343, DOI 10.1090/S0894-0347-2010-00683-2. MR2748395 (2012d:42023)

[10] Emmanuel J. Candès, Yonina C. Eldar, Thomas Strohmer, and Vladislav Voroninski, *Phase retrieval via matrix completion*, SIAM J. Imaging Sci. **6** (2013), no. 1, 199–225, DOI 10.1137/110848074. MR3032952

[11] Reinhold Remmert, *Theory of complex functions*, Graduate Texts in Mathematics, vol. 122, Springer-Verlag, New York, 1991. Translated from the second German edition by Robert B. Burckel; Readings in Mathematics. MR1084167 (91m:30001)

[12] K. Kreutz-Delgado. *The Complex Gradient Operator and the $\mathbb{CR}$-Calculus*. UCSD, 2003.

[13] C. Yang, J. Qian, A. Schirotzek, F. Maia and S. Marchesini. Iterative Algorithms for Ptychographic Phase Retrieval. arxiv.org/pdf/1105.5628, 2011

[14] Trond Steihaug, *The conjugate gradient method and trust regions in large scale optimization*, SIAM J. Numer. Anal. **20** (1983), no. 3, 626–637, DOI 10.1137/0720042. MR701102 (84g:49047)

[15] P. Thibault and M. Guizar-Sicairos. Maximum-likelihood refinement for coherent diffractive imaging. *New Journal of Physics*, 14(6):063004, 2012.

[16] S. Marchesini. A unified evaluation of iterative projection algorithms for phase retrieval. *Rev. Sci. Instrum.*, 78:011301, Jan 2007.

[17] Jorge J. Moré and David J. Thuente, *Line search algorithms with guaranteed sufficient decrease*, ACM Trans. Math. Software **20** (1994), no. 3, 286–307, DOI 10.1145/192115.192132. MR1367800 (96k:90074)

[18] Jorge Nocedal and Stephen J. Wright, *Numerical optimization*, Springer Series in Operations Research, Springer-Verlag, New York, 1999. MR1713114 (2001b:90002)

[19] J. R. Fienup. Phase retrieval algorithms: a comparison. *Appl. Opt.*, 21:2758–2769, 1982.

[20] D. Russell Luke, *Relaxed averaged alternating reflections for diffraction imaging*, Inverse Problems **21** (2005), no. 1, 37–50, DOI 10.1088/0266-5611/21/1/004. MR2146163 (2006a:78005)

[21] S. Marchesini. Phase-retrieval and saddle-point optimization. *J. Opt. Soc. Am. A*, 24:3289–3296, 2007.

[22] J. M. Rodenburg. Ptychography and related diffractive imaging methods. *Advances in Imaging and Electron Physics*, 150:87184, 2008.

[23] C. T. Kelley, *Iterative methods for optimization*, Frontiers in Applied Mathematics, vol. 18, Society for Industrial and Applied Mathematics (SIAM), Philadelphia, PA, 1999. MR1678201 (2000h:90003)

[24] O. Bunk, M. Dierolf, S. Kynde, I. Johnson, O. Marti, and F. Pfeiffer. Influence of the overlap parameter on the convergence of the ptychographical iterative engine. *Ultramicroscopy*, 108:481–487, 2008.

[25] Zaiwen Wen, Chao Yang, Xin Liu, and Stefano Marchesini, *Alternating direction methods for classical and ptychographic phase retrieval*, Inverse Problems **28** (2012), no. 11, 115010, 18, DOI 10.1088/0266-5611/28/11/115010. MR2992965

[26] Per Christian Hansen, *Discrete inverse problems*, Fundamentals of Algorithms, vol. 7, Society for Industrial and Applied Mathematics (SIAM), Philadelphia, PA, 2010. Insight and algorithms. MR2584074 (2011a:65105)

Department of Mathematics, Michigan State University, East Lansing, Michigan 48824
E-mail address: qian@math.msu.edu

Computational Research Division, Lawrence Berkeley National Laboratory, Berkeley, California 94720
E-mail address: cyang@lbl.gov

Advanced Light Source, Lawrence Berkeley National Laboratory, Berkeley, California 94720
E-mail address: aschirotzek@lbl.gov

NERSC, Lawrence Berkeley National Laboratory, Berkeley, California 94720
E-mail address: frmaia@lbl.gov

Advanced Light Source, Lawrence Berkeley National Laboratory, Berkeley, California 94720
E-mail address: smarchesini@lbl.gov

Contemporary Mathematics
Volume **615**, 2014
http://dx.doi.org/10.1090/conm/615/12288

Matrix elements of Fourier Integral Operator

Steve Zelditch

ABSTRACT. This article is concerned with the semi-classical limits of matrix elements $\langle F\varphi_j, \varphi_j \rangle$ of a Fourier integral operator F with respect to eigenfunctions of the Laplacian Δ_g of a compact Riemannian manifold (M, g). More generally, we consider matrix elements of eigensections of quantum maps. Many results exist for the case where F is a pseudo-differential operator, but matrix elements of Fourier integral operators involve new considerations. The limits reflect the extent to which the canonical relation of F is invariant under the geodesic flow of (M, g). When the canonical relation is almost nowhere invariant, a density one subsequence of the matrix elements tends to zero (related results arose first in the study of quantum ergodic restriction theorems). When the canonical relation is invariant, the limit states are invariant measures on the canonical relation of F and their invariance properties are explained. The invariance properties in the case of Hecke operators answers an old question raised by the author in Zelditch (1988).

One of the main objects of study in quantum ergodicity is the sequence of diagonal matrix elements

$$\rho_j(A) = \langle A\varphi_j, \varphi_j \rangle \tag{1}$$

of zeroth order pseudodifferential operators $A \in \Psi^0(M)$ relative to an orthonormal basis $\{\varphi_j\}$ of eigenfunctions

$$\Delta\varphi_j = \lambda_j^2 \, \varphi_j, \quad \langle \varphi_j, \varphi_k \rangle = 0.$$

of the Laplacian Δ of a compact Riemannian manifold (M, g). The diagonal matrix element (1) define positive linear functionals of mass one,

$$\rho_j : \overline{\Psi}^0 \to \mathbb{R}, \quad \rho(I) = 1, \tag{2}$$

on the norm closure of the space $\Psi^0(M)$ of zeroth order pseudo-differential operators, and are invariant under the wave group in the sense that

$$\rho_j(U^{t*} A U^t) = \rho_j. \tag{3}$$

The well-known consequence of Egorov's theorem is that any weak* limit μ of the sequence $\{\rho_j\}$ lies in the space $\mathcal{M}_I$ of invariant probability measure for the geodesic flow G^t on S^*M, i.e. is a positive linear functional on $C(S^*M)$ with $\mu(1) = 1$ and

Research partially supported by NSF grant # DMS-1206527.

$G^t_* \mu = \mu$. Moreover, one has the local Weyl law

$$\lim_{\lambda \to \infty} \frac{1}{N(\lambda)} \sum_{j: \lambda_j \leq \lambda} \rho_j = \omega_L, \tag{4}$$

where $\omega_L(A) = \int_{S^*M} \sigma_A d\mu_L$ is the Liouville state of integration of the principal symbol of A with respect to normalized Liouville measure. Also, $N(\lambda) = \#\{j : \lambda_j \leq \lambda\}$ is the Weyl counting function and convergence is in the sense of continuous linear functionals on Ψ_0. We refer to [**Z3, Z4, Zw**] for background on these statements. The off-diagonal matrix elements are also important and we refer to [**Z5**] for results on them.

The purpose of the present note is to consider the analogues of these basic results for diagonal (and to a lesser extent, off-diagonal) matrix elements (1) of Fourier integral operators F associated to a closed canonical relation $C \subset T^*M \times T^*M$. That is, we consider the functionals

$$\rho_j : I^r(M \times M, C) \to \mathbb{C}, \quad \rho_j(F) = \langle F\varphi_j, \varphi_j \rangle, \tag{5}$$

where $I^r(M \times M, C)$ is the space of Fourier integral operators of order r with wave front relation along C (see Vol. 4 of [**HoI-IV**] for background and notation). The linear functionals ρ_j (5) are invariant under the wave group in the sense that if $U^t = e^{it\sqrt{\Delta}}$ is the wave group of (M, g), then

$$\rho_j^t(F) := \rho_j(U^{-t} F U^t) = \rho_j(F). \tag{6}$$

In particular, we are interested in the weak* limits of the sequence $\{\rho_j\}$.

Definition: We define a weak* limit ρ_∞ of the functionals $\{\rho_j\}$ to be a functional on $I^0(M \times M, C)$ so that $\rho_j(F) \to \rho_\infty(F)$ for all F in this class.

It is shown in Proposition 1 that $\rho_\infty(F)$ depends only on the principal symbol of F, i.e. on the half-density symbol on the associated canonical relation and defines a measure on the symbols. We therefore use a convenient abuse of notation and identity the state and the measure, i.e. we put

$$\rho_\infty(F) = \rho_\infty(\sigma_F). \tag{7}$$

The motivation for this problem comes from several sources, for instance:

- (i) The question of weak * limits for matrix elements of Fourier integral operators arose in recent work on quantum ergodic restriction theorems [**TZ, TZ2, DZ**], and more recently for ray-splitting in [**JSS**]. For the F in those articles, the underlying canonical relation is a local canonical graph. A key point was that $\rho_j(F) \to 0$ along a subsequence of density one when the local canonical graphs are 'almost nowhere invariant' under the geodesic flow. We say that $F\varphi_j$ is almost orthogonal to φ_j. One aim of this note is to understand such almost orthogonality more systematically.
- (ii) Hecke operators T_p are Fourier integral operators associated to local canonical graphs, namely C is the lift to G/Γ of the Hecke correspondence [**E**]. All work in arithmetic quantum chaos concerns joint eigenfunctions of Δ and of the Hecke operators T_p. An obvious question is the Hecke correspondence invariance properties of the weak * limits of Hecke ρ_j. This question was raised but not settled in [**Z1**] and an answer will be given in Proposition 6.3 for Hecke operators on spheres and in Proposition 6.6 for

arithmetic hyperbolic quotients . Of course, Lindenstrauss [**L**] has long since proved that $\rho_j \to \omega_L$, but the result we present appears to be new.

- (iii) A weak* limit problem where the canonical relation is not a local canonical graph arose in [**Z10**]. To study nodal sets, the Δ-eigenfunctions of a real analytic Riemannian manifold (M, g) were analytically continued $\varphi_j \to \varphi_j^{\mathbb{C}}$ to its Grauert tubes M_τ and then restricted to geodesic arcs $\gamma : \mathbb{R} \to \partial M_\varepsilon$. The pullbacks $\gamma^* |\varphi_j^{\mathbb{C}}|^2$ can be normalized to form a bounded sequence of measures on compact intervals of $\mathbb{R}$. All of their weak * limits are constant multiples of Lebesgue measure. The constants depend on whether the geodesic is closed or not. Related pointwise Weyl laws for $|\varphi_j^{\mathbb{C}}(\zeta)|^2$ on all of ∂M_ε have been proved [**Z8**].
- (iv) Pointwise phase space Weyl laws for matrix elements of coherent state projectors $F_\hbar = \psi_{x,\xi}^{\hbar_j} \otimes \psi_{x,\xi}^{\hbar_j *}$ were obtained in [**PU**]. They are somewhat similar to modulus squares $|\varphi_j^{\mathbb{C}}(\zeta)|^2$ but involve a different FBI transform. It does not seem that the weak* limit problem was studied explicitly before, but the results are rather similar to the restrictions $\gamma^* |\varphi_j^{\mathbb{C}}|^2$.
- (v) Both of the above problems are special cases of weak* limit problems for ρ_j on algebras of Toeplitz operators associated to invariant symplectic cones Σ under the geodesic flow. The results in this setting are parallel to the case of ρ_j as states on the algebra $\Psi^0(M)$ in the sense of [**Z4**]. Different algebras of Fourier integral operators associated to idempotent canonical relations were introduced in [**GuSt**]. In a special case, the the weak* limit problem was studied in [**Z9**] (see also [**ST**]).

There is a long-standing question as to the uniquess of weak* limits (2) when the geodesic flow is sufficiently chaotic. The larger the class of 'test' operators one can use, the more control one has over the limits. In general one would like to study the most general possible microlocal defect measures. We refer to [**Zw**] for general background.

In this article, we concentrate on the case where C is a local canonical graph and the order $r = 0$, and only briefly summarize results on weak * limits of matrix elements in the other cases above. In the canonical graph case, the family $\{\rho_j\}$ of functionals on $I^0(M \times M, C)$ is uniformly bounded and all weak* limits are complex measures on C. More precisely, they are linear functionals of the symbol σ_F of F, which is a $\frac{1}{2}$-density along C (times a Maslov factor, which will be ignored here for simplicity of exposition and because the results do not depend on the Maslov factor). The local Weyl law for Fourier integral operators associated to local canonical graphs was studied in [**Z2**] (see also [**TZ2, JSS**]).

Many of the results for local canonical graphs turn out to be negative: the weak * limits of the diagonal functionals ρ_j are very often zero, since most canonical graphs C are almost nowhere invariant under a given geodesic flow. The graph in the Hecke case is invariant §6, and so the question becomes one of determining when the limits are trivial and when they are not. The local canonical (or isotropic) relation in the case of $\gamma^* |\varphi_j^{\mathbb{C}}(z)|^2$ or for $|\langle \varphi_j, \psi_{x,\xi}^{\hbar} \rangle|^2$ are not invariant but can be time averaged to become invariant, and the weak * limits (when suitably normalized) are often non-trivial.

The results also suggest that off-diagonal elements

$$\rho_{ij}(F) = \langle F\varphi_i, \varphi_j \rangle \tag{8}$$

are often more natural when testing against Fourier integral operators. They satisfy

$$\rho^t_{ij}(F) := \rho_{ij}(U^{-t}FU^t) = e^{it(\lambda_i - \lambda_j)}\rho_j(F), \tag{9}$$

and intuitively correspond to canonical transformations which change the energy level. However, the weak* limits are again trivial when the graph is almost nowhere invariant.

0.1. Results for local canonical graphs. To state our results, we need to introduce some further notation. A Fourier integral operator is an operator F whose Schwartz kernel may be locally represented as a finite sum of oscillatory integrals,

$$K_F(x,y) \sim \int_{\mathbb{R}^N} e^{i\varphi(x,y,\theta)} a(x,y,\theta)d\theta \tag{10}$$

for some homogeneous phase φ and amplitude a. It is well known that F is determined up to compact operators by the canonical relation

$$C = \{(x, \varphi'_x, y, -\varphi'_y) : \varphi'_\xi(x,y,\xi) = 0\} \subset T^*M \times T^*M,$$

together with the principal σ_F of F, a 1/2-density along C.We denote by $I^0(M, \times M, C)$ the class of Fourier integral operators of order zero and canonical relation C. We refer to [**HoI-IV**] for the background. Then we may regard (1) as defining continuous linear functionals on $I^0(M \times M, C)$ with respect to the operator norm. We recall that C is a local canonical graph when both projections in the diagram

$$\begin{array}{ccc} & C \subset T^*M \times T^*M & \\ \pi_X \swarrow & & \searrow \pi_Y \\ T^*M & \iff & T^*M \end{array} \tag{11}$$

are (possibly branched) covering maps. If we equip C with the symplectic volume measure pulled back by π_X from T^*M, then we may consider symbols σ_F as functions on C. Some well-known examples are:

- $F = T_g$ is translation by an isometry of a Riemannian manifold (M,g) possessing an isometry.
- F is a Hecke operator $Tf(x) = \sum_{j=1}^k (f(g_j x) + f(g_j^{-1}x))$ on S^n or on an arithmetic hyperbolic manifold corresponding to a finite set $\{g_1, \dots, g_k\}$ of isometries of the universal cover. In this case C is the graph of the cotangent lift of the Hecke correspondence and is a local canonical graph [**RS**, **LPS**].
- $F_t = U^t = e^{it\sqrt{\Delta}}$ or its self-adjoint part $\cos t\sqrt{\Delta}$.
- $F = W^*W$ where $Wf = \gamma_H B U^t$ where γ_H is restriction to a hypersurface $H \subset M$ and $WF'(B)$ is disjoint from the cotangent directions to H [**Ta**, **GS**, **TZ**, **TZ2**, **DZ**] among many articles.
- F is a semi-classical quantum map in the setting of positive Hermitian holomorphic line bundles over Kähler manifolds [**Z6**].

The first result is:

PROPOSITION 1. *If C is a local canonical graph and $F \in I^0(M \times M, C)$ then the weak limits ρ_∞ of $\langle F\varphi_j, \varphi_j\rangle$ are measures on $SC := C \cap S^*M \times S^*M$, i.e. $|\rho_\infty(F)| \leq C \sup_{SC} ||\sigma_F||_{C^0}$.*

As discussed above (7) we also write the limit functional as a functional on the symbol, and thus have $|\rho_\infty(\sigma_F)| \leq C \sup_{SC} ||\sigma_F||_{C^0}$.

The proof is quite similar to that for $A \in \Psi^0(M)$. But it can be useful to interpret the quantum limits as living on SC rather than on S^*M, as will be seen in the case of Hecke operators.

The weak* limit problem for $\{\rho_j\}$ on $I^0(M \times M, C)$ is not so different from that of $\Psi^0(M)$ since there often exists an elliptic element F_0 of $I^0(M \times M, C)$, i.e. one with nowhere vanishing symbol, and then all of the elements have the form AF_0 or F_0B where $A, B \in \Psi^0(M)$. But the canonical relation of $U^{-t}FU^t$ equals the image

$$C_t := (G^{-t} \times G^t)(C)$$

of C under the map $G^{-t} \times G^t$ of $T^*M \times T^*M$, and only coincides with the canonical relation C of F if C is G^t-invariant. In general,

$$U^{-t}FU^t \in I^0(M \times M, C_t),$$

so that by (6) ρ_j^t induces a functional on $I^0(M \times M, C_t)$.

The following initial result shows that the invariance properties of quantum limits depend on whether the canonical relation is invariant under the geodesic flow.

PROPOSITION 2. *Let ρ_∞^t be a weak* limit of the functionals ρ_j^t on $I^0(M \times M, C)$. Then there exists a family of measures μ_t on C_t such that*

$$\rho_\infty^t(A) = \int_{C_t} (G^t \times G^{-t})^* \sigma_A \quad d\mu_t, \qquad A \in I^0(M \times M, C)$$

with $\mu_t = \mu$ on $C_t \cap C$.

Thus,

COROLLARY 1. *Let ρ_∞ be a limit of the sequence of functionals $\rho_j(F) = \langle F\varphi_j, \varphi_j \rangle$ on $I^0(M, \times M, C)$. Suppose that the canonical relation C is invariant under the geodesic flow $G^{-t} \times G^t$. Then ρ_∞ is a G^t-invariant signed measure of mass ≤ 1 on C.*

Of course, the quantum invariance (6) implies that

$$\int_{C_t} (G^t \times G^{-t})^* \sigma_A \quad d\mu_t = \int_C \sigma_A \quad d\mu.$$

But Proposition 2 does not give any non-trivial invariance conditions on the set where canonical relation C is nowhere invariant under the geodesic flow $G^{-t} \times G^t$. The next result identifies the limit measure as zero along a subsequence of density one. We refer to this as the 'almost-orthogonality' of $F\varphi_j$ and φ_j. In the following, $n = \dim M$ so that $2n = \dim C$.

PROPOSITION 3. *Let $F \in I^0(M \times M, C)$ and assume that for $t > 0$, the set $\mathcal{C}_t = C_t \cap C$ has Minkowski $2n$- measure zero. Then there is a density one subsequence of eigenfunctions so that $\langle F\varphi_{\lambda_j}, \varphi_{\lambda_j} \rangle \to 0$.*

A special case of this almost orthogonality result was one of the the main ingredients in the proof in [**TZ, TZ2, DZ**] of the quantum ergodic restriction theorem along hypersurfaces. See Theorem 10 of [**TZ**] or §8 of [**TZ2**]. Although this statement is reminiscent of quantum ergodicity, it does not use any dynamical properties such as ergodicity of G^t. As Proposition 3 indicates, almost-orthogonality can sometimes be understood in terms of localization on energy surfaces of eigenfunctions. The nowhere commuting condition in that result implies that $F\varphi_\lambda$ localizes on a disjoint set from φ_j and thus the two states are almost orthogonal. In this example there are no sparse exceptional subsequences of eigenfunctions. See also Lemma 3.2 of §3.1. But this is not always the case, for instance localization does not seem to play a role in the Kähler analogue in Proposition 5.

0.2. Reality. Additional invariance properties arise when F is self-adjoint or real due to the fact that the eigenfunctions are real valued. We say that F is real if $cF = Fc$ where c denotes complex conjugation. We note that

$$\langle F^* \varphi_j, \varphi_j \rangle = \overline{\langle F\varphi_j, \varphi_j \rangle} = \langle \overline{F}^* \varphi_j, \varphi_j \rangle = \langle F^t \varphi_j, \varphi_j \rangle.$$

We recall that the transpose of a Lagrangian manifold Λ is defined by

$$\Lambda^t = \{(y, \eta, x, \xi) : (x, \xi, y, \eta) \in \Lambda\},$$

i.e. it is the image of Λ under the involution,

$$\iota(x, \xi, y, \eta) = (y, \eta, x, \xi).$$

The 'conjugate' Lagrangian is defined by

$$\Lambda^* = \{(y, \eta, x, -\xi) : (x, \xi, y, \eta) \in \Lambda\},$$

i.e. it is the image under $c \circ \iota$ where c is the conjugation involution $(x, \xi) \to (x, -\xi)$ in the second variable; the notation is consistent since complex conjugation is the quantization of the map c. We say that a canonical relation C is symmetric if $C^t = C$ and self-adjoint if $C^* = C$. A self-adjoint Fourier integral operator always has a self-adjoint canonical relation and a real Fourier integral operator has a canonical relation invariant under c. When F is self-adjoint we obtain an addtional invariance principle if we consider symbols defined by the functions $\pi_X^* a, \pi_Y^* a$ (cf. (11)):

PROPOSITION 4. *If F is real and self-adjoint then $(\pi_X)_* \mu = (\pi_Y)_* \mu$ for any quantum limit measure μ on C of Proposition* 2.

This additional principle is useful in obtaining relations between limit measures of Hecke operators. See §6.

0.3. Quantum maps in the Kähler setting. There is a natural analogue of Proposition 3 in the Kähler setting. We let (M, ω, L) be a compact polarized Kähler manifold. That is, $L \to M$ is a holomorphic line bundle equipped with a Hermitian metric h whose curvature form Θ_h equals $i\omega$. Thus, $\omega \in H^{1,1}(M, 2\pi\mathbb{Z})$. We also denote the kth tensor power of L by L^k and denote the space of holomorphic sections by $H^0(M, L^k)$; we also put $d_k = \dim H^0(M, L^k)$. We refer to [**Z6, Z7**] for background.

We further χ_1, χ_2 be two quantizable symplectic diffeomorphisms of (M, ω). The definition of quantizable is from [**Z6, Z7**] , which generalizes the implicit standard notion for special cases such as cat maps on a complex one dimensional torus (elliptic curve). Namely, χ_j are symplectic diffeomorphisms which possess lifts as contact transformations of the unit circle bundle $X_h = \partial D_h^*$ where D_h^* is the unit co-disc bundle in the dual line bundle L^* of L with respect to the dual metric h. An exposition of the key notions Kähler quantization can be found [**Gu**].

We let $\{U_{\chi,k}\}_{k=1}^{\infty}$ denote the semi-classical quantization of a quantizable symplectic map χ as a sequence of unitary operators on the Hilbert spaces $H^0(M, L^k)$.

Thus, $F = U_{\chi_2,k}$ denotes the quantum map quantizing χ_2. As discussed in [**Z6, Z7**] the quantizations have the form $U_{\chi,k} := \Pi_{h^k}\sigma_{k,\chi}T_\chi\Pi_{h^k}$ where $\Pi_{h^k} : L^2(M, L^l) \to H^0(M, L^k)$ is the orthogonal projection (Szegö kernel), where T_χ is the translation operator by χ and where $\sigma_{k,\chi}$ is a symbol designed to make $U_{\chi,k}$ unitary. More precisely, T_χ is the translation operator by the lift of χ to X_h.

PROPOSITION 5. *Let $\varphi_{k,j}$ denote the eigensections of $U_{\chi_1,k}$. Suppose that χ_1, χ_2 almost nowhere commute in the sense that the set $\{z \in M : \chi_1\chi_2(z) = \chi_2\chi_1(z)\}$ has measure zero. Then*

$$\frac{1}{d_k}\sum_{j=1}^{d_k} |\langle U_{\chi_2,k}\varphi_{k,j}, \varphi_{k,j}\rangle|^2 \to 0.$$

As a simple example, suppose that A_1, A_2 are two non-commutating elements of the θ subgroup of $SL(2,\mathbb{Z})$ (i.e. are congruent to the identity modulo 2). The associated symplectic maps χ_1, χ_2 of $\mathbb{R}^2/\mathbb{Z}^2$ are then quantizable and almost nowhere commute. So the eigenfunctions of one 'quantum cat map' give rise to zero quantum limits for the other.

An interesting comparison to Proposition 3 is that the semi-classical eigensections $\varphi_{k,j}$ do not appear to have any localization properties on the phase space X_h which account for the almost orthogonality of the matrix elements.

0.4. Pointwise squares as matrix elements. Let $\{\psi_\hbar\}$ be a semi-classical Lagrangian state, for instance a coherent state [**CR**] in the Schrödinger representation, or coherent states induced by a Bergman reproducing kernel, or the sequence of Gaussian beams associated to a closed geodesic [**R, BB**]. In each case, we consider the norm-squares as matrix elements of semi-classical Fourier integral operators,

$$|\langle \psi_\hbar, \varphi_j\rangle|^2 = \langle F_\hbar\varphi_j, \varphi_j\rangle, \quad \text{where } F_\hbar = \psi_\hbar \otimes \psi_\hbar^*. \tag{12}$$

The Schwartz kernel of $F_\hbar(x,y) = \psi_\hbar(x)\overline{\psi_\hbar(y)}$ inherits an oscillatory integral representation from that of $\psi_\hbar$. One relevant normalization is to take $||\psi_\hbar||_{L^2} = 1$.

The underlying Lagrangian or isotropic submanifold of $\{\psi_\hbar\}$ may or may not be invariant under the geodesic flow. For instance, they are not invariant for coherent states $\psi^\hbar_{x,\xi}$ (where the isotropic submanifold is a point), but they are for the sequence of highest weight spherical harmonics Y^k_k on the standard S^2 or for more general Gaussian beams. However by an averaging argument (see §4) one can show that the weak* limits must be non-negative measures on the 'orbit' of the underlying Lagrangian or isotropic submanifold under the geodesic flow. In the case of a local canonical graph this 'flowout' could be dense in $T^*M \times T^*M$ if C is almost nowhere invariant under $G^t \times G^t$ but for an isotropic submanifold the flowout can be a closed submanifold and one can have non-trivial limits.

For the sake of brevity, we only give explain how the recent results in [**Z10**] fit into the picture of matrix elements of Fourier integral operators, which was not the approach used in that article. There are many related examples that we will not consider here.

0.5. Idempotent canonical relations and algebras of Fourier integral operators. There are other settings where the weak* limit problem is of interest. One is where $I^0(M \times M, C)$ is a $*$ algebra of Fourier integral operators. This occurs when the canonical relation is idempotent in the sense that $C^* = C = C^2$. One such situation is the algebra of Fourier integral operators associated to a symplectic cone

$\Sigma \subset T^*X$. In fact C need not be a Lagrangian submanifold. It is sufficient that C be isotropic (see [**W**]). An example is when γ is a closed geodesic of a Riemannian manifold and where $\Sigma = \mathbb{R}_+\dot{\gamma}$. Then C is the diagonal of $\Sigma \times \Sigma$. The averaged coherent state projections

$$(13) \qquad \langle \psi^{\hbar}_{x,\xi} \otimes (\psi^{\hbar}_{x,\xi})^* \rangle_L := \frac{1}{L} \int_0^L U^t \left(\psi^{\hbar}_{x,\xi} \otimes (\psi^{\hbar}_{x,\xi})\right)^* U^{-t} dt$$

are Toeplitz operators in the case $\Sigma = \mathbb{R}_+\dot{\gamma}$.

Idempotent canonical relations also arise as leaf equivalence relations of null foliations of co-isotropic submanifolds $\Sigma \subset T^*M$, also known as flowouts. In the case where the null-foliation is a fiber bundle with compact fiber over a leaf space S (a symplectic manifold), the algebra was denoted $\mathcal{R}_\Sigma$ and was studied in [**GuSt**]. For the sake of brevity we omit further discussion and refer to [**Z9**] and to [**ST**, **GU**] for the study of quantum ergodicity in this setting.

1. Background

We recall that a Fourier integral operator $A : C^\infty(X) \to C^\infty(Y)$ is an operator whose Schwartz kernel may be represented by an oscillatory integral

$$K_A(x, y) = \int_{\mathbb{R}^N} e^{i\varphi(x,y,\theta)} a(x, y, \theta) d\theta$$

where the phase φ is homogeneous of degree one in θ. The critical set of the phase is given by

$$C_\varphi = \{(x, y, \theta) : d_\theta \varphi = 0\}.$$

When the map

$$\iota_\varphi : C_\varphi \to T^*(X, Y), \quad \iota_\varphi(x, y, \theta) = (x, d_x\varphi, y, -d_y\varphi)$$

is an immersion the phase is called non-degenerate. Less restrictive is where the phase is clean, i.e. $\iota_\varphi : C_\varphi \to \Lambda_\varphi$, where Λ_φ is the image of ι_φ, is locally a fibration with fibers of dimension e. From [**HoI-IV**] Definition 21.2.5, the number of linearly independent differentials $d\frac{\partial \varphi}{\partial \theta}$ at a point of C_φ is $N - e$ where e is the excess.

We work in the polyhomogeneous framework of [**HoI-IV**], and asume that classical poly-homogeneous symbols

$$a(x, y, \theta) \sim \sum_{k=0}^{\infty} a_{-k}(x, y, \theta), \; (a_{-k} \text{ positive homogeneous of order } \text{-k in } \theta.$$

All of the results and notions of this note geeneralize to semi-classical Fourier integral operators with semi-classical symbols $a \in S^{0,0}(T^*H \times (0, h_0]$ of the form

$$a_\hbar(s, \sigma) \sim \sum_{k=0}^{\infty} \hbar^k \, a_{-k}(s, \sigma), \; (a_{-k} \in S^{-k}_{1,0}(T^*H)).$$

Since there is no essential difference in the weak* limit results in the two settings, we only consider the poly-homogeneous one.

We a recall that the order of $F : L^2(X) \to L^2(Y)$ in the non-degenerate case is given in terms of a local oscillatory integral formula by $m + \frac{N}{2} - \frac{n}{4}$,, where $n = \dim X + \dim Y$, where m is the order of the amplitude, and N is the number of phase variables in the local Fourier integral representation (see [**HoI-IV**], Proposition 25.1.5); in the general clean case with excess e, the order goes up by $\frac{e}{2}$ ([**HoI-IV**], Proposition 25.1.5'). Further, under clean composition of operators of orders m_1, m_2, the order of the composition is $m_1 + m_2 - \frac{e}{2}$ where e is the so-called excess (the fiber dimension of the composition); see [**HoI-IV**], Theorem 25.2.2.

The symbol $\sigma(\nu)$ of a Lagrangian (Fourier integral) distributions is a section of the bundle $\Omega_{\frac{1}{2}} \otimes \mathcal{M}_{\frac{1}{2}}$ of the bundle of half-densities (tensor the Maslov line bundle). In terms of a Fourier integral representation it is the square root $\sqrt{d_{C_\varphi}}$ of the delta-function on C_φ defined by $\delta(d_\theta \varphi)$, transported to its image in T^*M under ι_φ. If $(\lambda_1, \dots, \lambda_n)$ are any local coordinates on C_φ, extended as smooth functions in neighborhood, then

$$d_{C_\varphi} := \frac{|d\lambda|}{|D(\lambda, \varphi'_\theta)/D(x, \theta)|},$$

where $d\lambda$ is the Lebesgue density.

1.1. Local Weyl law for Fourier integral operators. It was proved in [**Z2**] (see also [**TZ, TZ2, DZ**]) that if average the functionals ρ_j, then the limit is a measure on the unit vectors in the intersection $C \cap \Delta_{T^*M \times T^*M}$ of the canonical relation C with the diagonal in $T^*M \times T^*M$. That is, one has the local Weyl law,

$$\frac{1}{N(\lambda)} \sum_{j: \lambda_j \leq \lambda} \rho_j \to \rho_{LWL}, \tag{14}$$

where the local Weyl law measure is given by

$$\rho_{LWL}(F) = \int_{S(C \cap \Delta_{T^*M \times T^*M})} \sigma_F d\nu,$$

where $d\nu$ is a 'half-density measure' and $S(C \cap \Delta_{T^*M \times T^*M})$ is the set of unit covectors in $C \cap \Delta_{T^*M \times T^*M}$. In the case where C is a local canonical graph, this intersection is the fixed point set of the correspondence χ and we write it as $SFix(\chi)$.

Note that the trace operation concentrated the average of the ρ_j on the diagonal part of C. The individual matrix elements do not have this property, even though $\langle F\varphi, \varphi \rangle = TrF\varphi \otimes \varphi^*$ is a trace.

2. Invariant states on $I^0(M \times M, C)$: Proof of Propositions 1 and 2

2.1. Fourier integral operators associated to local canonical graphs. The proof of Proposition 2 is similar to the case of pseudo-differential operators proved in [**W**].

We first recall

THEOREM 2.1. *([**HoI-IV**], Theorem 25.3.1) If C is a local canonical graph and $A \in I^0(M \times M, C)$, then $A : L^2(M) \to L^(M)$ is bounded, and it is compact if the symbol of A tends to 0 as $|\xi| \to \infty$.*

We then prove

LEMMA 2.2. *If C is a local canonical graph and $A \in I^0(M \times M, C)$ then*

$$\sup_{S^*M} |\sigma_A| = \inf_K ||A + K||. \tag{15}$$

PROOF. The equality (15) is well known for $A \in \Psi^0(M)$. To generalize it to Fourier integral operators associated to local canonical graphs it suffices to use that, in a sufficiently small cone, C is the graph of a canonical transformation. Then as in the proof of [**HoI-IV**], Theorem 25.3.1, $A^*A \in \Psi^0(M)$ and $\sigma_{A^*A} = |\sigma_A|^2$. It follows that

$$\sup_{T^*M} |\sigma_A|^2 = \inf_{K \text{ compact}} ||A^*A + K|| = \inf_{K \text{ compact}} ||A + K||^2.$$

Here we use that for any $u \in L^2$ with $||u|| = 1$, $||(A+K)u||^2 = \langle (A^*A+K_1)u, u\rangle$ for another compact operator K_1 and that $||A^*A+K|| = \sup_{||u||=1} |\langle (A^*A+K_1)u, u\rangle|$ when K is self-adjoint. □

If K is any compact operator on $L^2(M)$ then $\langle K\varphi_j, \varphi_j\rangle \to 0$. Indeed, $\varphi_j \to 0$ weakly in L^2 and so $K\varphi_j \to 0$ in norm. The principal symbol of F determines F up to an element of $I^{-1}(M \times M, C)$ and the operators in this class are compact. This proves Proposition 1.

We now complete the proof of Propositions 1 - 2.

PROOF. For any compact operator K, $\langle K\varphi_j, \varphi_j\rangle \to 0$. Hence, any limit of $\langle A\varphi_k, \varphi_k\rangle$ is equally a limit of $\langle (A+K)\varphi_k, \varphi_k\rangle$. By the norm estimate, the limit is bounded by $\inf_K ||A+K||$ (the infimum taken over compact operators). Hence any weak limit is bounded by a constant times $||\sigma_A||_{L^\infty}$ and is therefore continuous on $C(S^*M)$. It is a positive functional since each ρ_j is and hence any limit is a probability measure.

To prove the invariance of the limit measure, we apply an Egorov type theorem to $U^{-t}FU^t$ for $F \in I^0(M \times M, C)$ and for fixed t. The canonical relation of the composition is the composition

$$\Gamma_t^* \circ C \circ \Gamma_t = C,$$

where

$$\Gamma_t = \{(x, \xi, G^t(x, \xi)) : |\tau| + |\xi| = 0\}.$$

Hence only the symbol σ_F is changed. If we choose a nowhere vanishing half-density on C (e.g. the graph half-density corresponding to the symplectic voluime density on T^*M), then σ_F may be identified with a scalar function and the composite symbol is its pull-back under G^t.

By invariance of the ρ_k, any limit of $\rho_k(A)$ is a limit of $\rho_k(Op(\sigma_A \circ \Phi^t))$ and hence the limit measure is invariant. It is also time-reversal since the eigenfunctions are real-valued, i.e. complex conjugation invariant. □

2.2. Off-diagonal matrix elements. A similar argument applies to off-diagonal matrix elements

$$\rho_{jk}(A) = \langle A\varphi_j, \varphi_k\rangle. \tag{16}$$

The discussion is very similar to that in [**Z5**] in the pseudo-differential case.

PROPOSITION 2.3. *Let ρ_∞ be a limit of the sequence of functionals* (16) *with the gap $\lambda_j - \lambda_k \to \tau$ on $I^0(M, \times M, C)$. Suppose that the canonical relation C is invariant under the geodesic flow $G^{-t} \times G^t$. Then ρ_∞ is a signed G^t-eigennmeasure of mass ≤ 1 on C.*

The only change to the proof of Proposition 2 is in the last step. The functionals ρ_{jk} are no longer invariant but rather satisfy (9). It follows that if $\lambda_j - \lambda_k \to \tau$ then any limit measure is a G^t eigenmeasure on C with eigenvalue $e^{it\tau}$.

3. Almost orthogonality: Proof of Proposition 3

This section is motivated by the proof of the quantum ergodic restriction theorem in [**TZ, TZ2**] (see also [**DZ**]). The criterion for QER in those papers is an almost nowhere commutativity condition between two canonical transformations, or equivalently an almost nowhere invariance problem. It is not clear that the condition for QER in those papers is sharp.

We now prove Proposition 3.

PROOF. We are assuming that for $t > 0$, the set $\mathcal{C}_t = \{(x, \xi) \in S^*M : G^t\chi = \chi G^t\}$ has Liouville measure zero.

It suffices to show that

$$\frac{1}{N(\lambda)} \sum_{\lambda_j \leq \lambda} |\langle F\varphi_{\lambda_j}, \varphi_{\lambda_j}\rangle|^2 = o(1) \text{ as } \lambda \to \infty. \tag{17}$$

We put

$$F(t) = U^{t*}FU^t, \;\; \langle F\rangle_T := \frac{1}{T}\int_{-T}^{T} F(s)ds, \;\text{ where } U^t = e^{it\sqrt{\Delta}}.$$

Then $\langle F(t)\varphi_{\lambda_j}, \varphi_{\lambda_j}\rangle = \langle F\varphi_{\lambda_j}, \varphi_{\lambda_j}\rangle$.

For any operator A we have

$$|\langle A\varphi_\lambda, \varphi_\lambda\rangle|^2 \leq \langle A^*A\varphi_\lambda, \varphi_\lambda\rangle.$$

It follows that

$$|\langle F\varphi_{\lambda_j}, \varphi_{\lambda_j}\rangle|^2 \leq \langle \langle F\rangle_T^* \langle F\rangle_T \varphi_\lambda, \varphi_\lambda\rangle.$$

Hence,

$$\frac{1}{N(\lambda)} \sum_{\lambda_j \leq \lambda} |\langle F\varphi_{\lambda_j}, \varphi_{\lambda_j}\rangle|^2 \leq \frac{1}{N(\lambda)} \sum_{\lambda_j \leq \lambda} \langle \langle F\rangle_T^* \langle F\rangle_T \varphi_\lambda, \varphi_\lambda\rangle. \tag{18}$$

We now let $\lambda \to \infty$ and use the local Weyl law (14) for Fourier integral operators. We have,

$$\langle F\rangle_T^* \langle F\rangle_T = \frac{1}{T^2}\int_{-T}^{T}\int_{-T}^{T} F(s)^*F(t)dsdt.$$

Since we are taking a trace, we can cycle the U^t to get

$$\langle F\rangle_T^* \langle F\rangle_T = \frac{1}{T^2}\int_{-T}^{T}\int_{-T}^{T} U(t-s)^*F^*U(t-s)Fdsdt.$$

We change variables to $u = \frac{t-s}{2}, v = \frac{t+s}{2}$ and simplify to get

$$\langle F\rangle_T^* \langle F\rangle_T = \frac{1}{T}\int_{-T}^{T} U(t)^*F^*U(t)F\rho_T(t)dt.$$

Here, $\rho_T(t)$ is the measure in $[-T, T] \times [-T, T]$ of $\{(s, s') : s - s' = t\}$. For each t the local Weyl law gives

$$\frac{1}{N(\lambda)} \sum_{\lambda_j \leq \lambda} \langle U(t)^*F^*U(t)F\varphi_\lambda, \varphi_\lambda\rangle \to \int_{SFixG^{-t}\chi^{-1}G^t\chi} \sigma_{t,F}, \tag{19}$$

where $\sigma_{t,F}$ is a composite density on the fixed point set. The fixed point set is exactly the set where $\chi G^t = G^t\chi$ and has measure zero for all $t \neq 0$. Hence the integral tends to zero for all $T > 0$.

□

This arguments works too well because the assumption is so strong. A related argument is to just assume that there exists t_0 so that $\chi G^{nt_0} = G^{nt_0}\chi$ only holds on a set of measure zero. It is this argument which was in effect used in [**TZ, TZ2**].

PROPOSITION 3.1. *Let F be a Fourier integral operator associated to a symplectic correspondence χ. Assume that there exists $t_0 \neq 0$ so that for $n = 1, 2, 3\dots$, the set $\mathcal{C}_t = \{(x,\xi) \in S^*M : G^{nt_0}\chi = \chi G^{nt_0}\}$ has Liouville measure zero. Then there is a density one subsequence of eigenfunctions so that $\langle F\varphi_{\lambda_j}, \varphi_{\lambda_j}\rangle \to 0$.*

PROOF. We define

$$\langle F \rangle_M := \frac{1}{2M} \sum_{m=-M}^{M} U^{-mt_0} F U^{mt_0}.$$

Going through the same argument gives the upper bound

$$\frac{1}{N(\lambda)} \sum_{\lambda_j \leq \lambda} |\langle F\varphi_{\lambda_j}, \varphi_{\lambda_j}\rangle|^2 \leq \frac{1}{N(\lambda)} \sum_{\lambda_j \leq \lambda} \langle \langle F\rangle_M^* \langle F\rangle_M \varphi_\lambda, \varphi_\lambda\rangle. \tag{20}$$

We then have

$$\begin{aligned} \langle F\rangle_M^* \langle F\rangle_M &= \tfrac{1}{M^2} U^{t_0(m-n)*} F^* U^{t_0(m-n))} F \\ &= \tfrac{1}{M} \textstyle\sum_{p=-M}^{M} \tfrac{\#p}{M} U^{t_0 p *} F^* U^{t_0 p} F, \end{aligned}$$

where $\#p = \#\{(m,n) \in [-M,M] \times [-M,M] : m - n = p\}$. We then apply the local Weyl law and find that the only term which makes a non-vanishing contribution is $p = 0$. So it is $O(\frac{1}{M})$.

□

3.1. Almost disjoint energy surfaces. The commutator $[\sqrt{\Delta}, F]$ is always of order one if $F \in I^0(M \times M, C)$. The symbol of $\sqrt{\Delta} F$ at (x,ξ,y,η) is the product $|\xi|_x \sigma_F$ while in the other order we have $|\eta|_y \sigma_F$. So they do not cancel unless χ preserves $S^*_g X$, which is not the case when they almost never commute.

LEMMA 3.2. *Suppose that χ and G^t almost never commute. Then $\chi(S_g^* M) \cap S_g^* M$ has Liouville measure zero.*

Let $H(x,\xi) = |\xi|_g$ be the metric norm function. The Hamiltonian flow of $\chi^* H$ is $\chi G^t \chi^{-1}$. The orbit of (x,ξ) under this flow is almost certainly disjoint from the G^t orbit of (x,ξ). Also the Hamiltonian vector field ξ_H of H almost certainly satisfies $\chi_* \xi_H \neq \xi_H$. If $\chi^* H = H$ on an open set, then we take the Hamiltonian vector field of both sides to get $\chi_* \xi_H = \xi_H$. Similarly for any set of positive measure.

We can then give a second proof of Proposition 3.

PROOF. It is well-known that Δ-eigenfunctions concentrate microlocally on the energy surfaces $|\xi| = E$ (see [**Zw**], Theorem 6.4). In the homogeneous setting we may identify all the energy surfaces with S^*M. It follows that $F\varphi_j$ microlocally concentrates on $\chi(S^*M)$. In other words we may construct semi-classical cutoffs $Op_\hbar(b)$ to S^*M and $Op_\hbar(\chi^* b)$ so that $Op_\hbar \varphi_j = \varphi_j + \mathcal{O}(\hbar)$, $Op_\hbar(\chi^* b) F\varphi_j = F\varphi_j + \mathcal{O}(\hbar)$. By Lemma 3.2, the intersection has Liouville measure zero. It follows that

$$\langle F\varphi_j, \varphi_j\rangle = \langle Op_\hbar(\chi^* b) F\varphi_j, Op_\hbar(b)\varphi_j\rangle + \mathcal{O}(\hbar) \leq ||Op_\hbar(\chi^* b) Op_\hbar(b)\varphi_j|| + \mathcal{O}(\hbar).$$

But $Op_\hbar(\chi^* b) Op_\hbar(b) = Op_\hbar(b\chi^* b) + \mathcal{O}(\hbar)$. Then

$$\frac{1}{N(\lambda)} \sum_{\lambda_j \le \lambda} |\langle F\varphi_{\lambda_j}, \varphi_{\lambda_j}\rangle|^2 \le \frac{1}{N(\lambda)} \sum_{\lambda_j \le \lambda} |\langle Op_\hbar(b\chi^* b)\varphi_{\lambda_j}, \varphi_{\lambda_j}\rangle|^2 + \mathcal{O}(\lambda^{-1})$$

$$\le \frac{1}{N(\lambda)} \sum_{\lambda_j \le \lambda} \langle (Op_\hbar(b\chi^* b)^* Op_\hbar(b\chi^* b)\, \varphi_{\lambda_j}, \varphi_{\lambda_j}\rangle|^2 + \mathcal{O}(\lambda^{-1})$$

$$\to \int_{S^* M} b^2 \chi^* b^2 \text{ as } \lambda \to \infty.$$

But by assumption, for any $\varepsilon > 0$ one may construct b so that the support of $b \times \chi^* b$ has volume $\le \varepsilon$. It follows that $\int_{S^*M} b^2 \chi^* b^2 \le \varepsilon$, and therefore $\frac{1}{N(\lambda)} \sum_{\lambda_j \le \lambda} |\langle F\varphi_{\lambda_j}, \varphi_{\lambda_j}\rangle|^2 \to 0$. □

The proof indicates that although a subsequence of density one of the matrix elements tends to zero, not all them need to. It could be the case that a subsequence of eigenfunctions φ_{j_k} concentrates microlocally on a closed geodesic γ and that $\chi(\gamma) = \gamma$. Then even though $\chi(S^*M) \cap S^*M$ has Liouville measure zero, the full sequence of matrix elements $\langle F\varphi_j, \varphi_j\rangle$ need not tend to zero.

Remark: It is natural to ask how sharp the almost nowhere invariance or commutation conditions are in the proof of almost orthogonality. That is, we ask whether the following converse to Proposition 3 is true:

Question: Denote the canonical relation of F by C. Then in the notation of this section, do we have, if for all $F \in I^0(C)$,

$$\frac{1}{N(\lambda)} \sum_{\lambda_j \le \lambda} |\langle F\varphi_{\lambda_j}, \varphi_{\lambda_j}\rangle|^2 \to 0 \implies \text{C is almost nowhere invariant under } g^t?$$

3.2. Almost nowhere commuting quantum maps on Kähler manifolds. We now prove the analogue, Proposition 5, in the Kähler setting. Let χ_1, χ_2 be two quantizable symplectic diffeomorphisms of a compact Kähler manifold (M, ω) which almost nowhere commute in the sense that the set where $\chi_1\chi_2 = \chi_2\chi_1$ has measure zero. We then let U_{χ_1} denote the quantization of χ_1 as a unitary operator on $H^0(M, L^k)$ and we let F denote any quantum map quantizing χ_2.

An interesting comparison to the previous case is that the semi-classical eigensections $\varphi_{k,j}$ do not appear to have any localization properties which account for the almost orthogonality of the matrix elements.

Proof of Proposition 5. We again consider the partial time average

$$\langle F\rangle_M := \frac{1}{2M} \sum_{m=-M}^{M} U^{-m} F U^m.$$

Going through the same argument gives the upper bound

$$\frac{1}{N(\lambda)} \sum_{\lambda_j \le \lambda} |\langle F\varphi_{\lambda_j}, \varphi_{\lambda_j}\rangle|^2 \le \frac{1}{N(\lambda)} \sum_{\lambda_j \le \lambda} \langle \langle F\rangle_M^* \langle F\rangle_M \varphi_\lambda, \varphi_\lambda\rangle. \tag{21}$$

We then have

$$\begin{aligned} \langle F\rangle_M^* \langle F\rangle_M &= \tfrac{1}{M^2} U^{t_0(m-n)*} F^* U^{t_0(m-n))} F \\ &= \tfrac{1}{M} \sum_{p=-M}^{M} \tfrac{\#p}{M} U^{t_0 p*} F^* U^{t_0 p} F, \end{aligned}$$

where $\#p = \#\{(m,n) \in [-M,M] \times [-M,M] : m-n = p\}$. We then apply the local Weyl law and find that the only term which makes a non-vanishing contribution is $p = 0$. So it is $O(\frac{1}{M})$ and therefore the limit equals zero. □

4. Modulus squares as matrix elements

In this section, we tie together the weak * limit problem for Fourier integral operators with some recent results on modulus squares $|\varphi_j^{\mathbb{C}}(z)|^2$ of analytic continuations of eigenfunctions to Grauert tubes and their restrictions to geodesic arcs (cf. [**Z8**, **Z10**]), and to some related pointwise Weyl laws for coherent state projections in [**PU**]. The relevant matrix elements differ from the preceding ones in that the underlying canonical or wave front relations are not local canonical graphs. The results are correspondingly different: the canonical relations are not invariant under the geodesic flow, but Proposition 3 is false for them. When properly normalized the weak* limits can be non-zero. In fact, we will use an averaging argument or a flowout construction to make the canonical relation geodesic flow invariant when (x, ξ) is a periodic point. This is impossible for a local canonical graph which is almost nowhere invariant.

4.1. $|\varphi_j^{\mathbb{C}}(z)|^2$ on a Grauert tube. We first consider pointwise modulus squares of $|\varphi_j^{\mathbb{C}}(z)|^2$ of analytic continuations of eigenfunctions to Grauert tubes in the complexification of M. We refer to [**Z8**, **Z10**] and their references for background on the analytic continuation and the geometry of Grauert tubes.

As discussed at length in [**GS1**, **Z8**], a Grauert tube M_ε is a strictly pseudoconvex domain in the complexification $M_{\mathbb{C}}$ of a real analytic Riemannian mannifold (M, g). Its defining function $\rho(z)$ is the analytic continuation of the squared distance function $r^2(x, y)$ to $x = z, y = \bar{z}$. The Grauert tube function is its square root $\sqrt{\rho}(z)$ (we are ignoring here some constants). A key point is that is the image of a ball bundle $B_\varepsilon^* M$ under the imaginary time exponential map $E : B_\varepsilon^* M \to M_\varepsilon$, $E(x, \xi) = \exp_x i\xi$. The Grauert tube function $\sqrt{\rho}$ corresponds to the norm function $|\xi|_g$ of the metric under E. Moreover E conjugates the geodesic flow to a Hamiltonian flow g^t on M_τ with respect to its adapted Kähler form.

The principal Fourier integral operator in this context is the Poisson kernel $P^\varepsilon(z, y)$ on $\partial M_\varepsilon \times M$, defined as follows: The wave group of (M, g) is the unitary group $U(t) = e^{it\sqrt{\Delta}}$. Its kernel $U(t, x, y)$ solves the 'half-wave equation',

$$\left(\frac{1}{i}\frac{\partial}{\partial t} - \sqrt{\Delta_x}\right) U(t, x, y) = 0, \quad U(0, x, y) = \delta_y(x). \tag{22}$$

The Poisson-wave kernel $P^\tau(z, y)$ is the analytic continuation $U(i\tau, x, y)$ of the wave kernel with respect to time, $t \to i\tau \in \mathbb{R} \times \mathbb{R}_+$ and then in x, i.e.

$$P^\tau(z, y) = U(i\tau, z, y). \tag{23}$$

Thus, the Poisson- kernel has the eigenfunction expansion for $\tau > 0$,

$$U(i\tau, x, y) = \sum_j e^{-\tau\lambda_j} \varphi_{\lambda_j}^{\mathbb{C}}(z) \varphi_{\lambda_j}(y), \quad z \in \partial M_\tau, y \in M. \tag{24}$$

Thus, $P^\tau \varphi_j = e^{-\tau\lambda_j}\varphi_j^{\mathbb{C}}$.

The key fact is that $P^\tau(z,y)$ is a Fourier integral operator with complex phase. It is adapted to the symplectic isomorphism of T^*M with the symplectic cone $\Sigma_\tau \subset T^*\partial M_\tau$ generated by the contact form α_z. This result was stated by Boutet de Monvel in [**Bou**] but only recently have detailed proofs been published (cf. [**Z8**, **St**]). The main ingredient is the analytic continuation of the wave kernel due to J. Hadamard.

If we define

$$\Phi_z^\tau(y) = P^\tau(z,y) \in L^2(M), \tag{25}$$

then

$$e^{-\tau\lambda_j}\varphi_j^{\mathbb{C}}(z) = \langle \Phi_z^\tau, \varphi_j\rangle_{L^2(M)}. \tag{26}$$

It follows that

$$e^{-2\tau\lambda_j}|\varphi_j^{\mathbb{C}}(z)|^2 = |\langle \Phi_z^\tau, \varphi_j\rangle|^2 = \langle \Phi_z^\tau \otimes (\Phi_z^\tau)^*\varphi_j, \varphi_j\rangle, \tag{27}$$

where $\Phi_z^\tau \otimes (\Phi_z^\tau)^*$ is the rank one projector onto $\Phi_z^\tau(y)$. Since $P^\tau(z,y)$ is a homogeneous Fourier integral kernel with complex phase on $\partial M_\tau \times M$, the Schwartz kernel $\Phi_z^\tau(y)\overline{\Phi_z^\tau(y')}$ of $\Phi_z^\tau \otimes (\Phi_z^\tau)^*$ is also a Fourier integral kernel with complex phase on $\partial M_\tau \times M \times M$. If we fix z then $\Phi_z^\tau(y)\overline{\Phi_z^\tau(y')}$ is a Fourier integral kernel with complex phase on $M \times M$. The associated 'canonical relation' is actually an isotropic relation. It is a product relation of the form,

$$\Lambda_z \times \Lambda_z^* \subset T^*M \times T^*M, \tag{28}$$

where

$$\Lambda_z = \{(x,\xi) : E(x,\xi) = z\}.$$

Fourier integral operators associated to isotropic relations were introduced in [**W**].

One of the main problems with the 'states' (27) is that they are not normalized. It is proved in [**Z8**] (Corollary 2) that $e^{-2\varepsilon\lambda_j}|\varphi_j^{\mathbb{C}}(z)|^2 \leq C\lambda^{m-1}$. The upper bound is sharp (it is attained by highest weight spherical harmonics on the standard S^m) but it is generally not attained on a generic analytic Riemannian manifold, nor is it attained at general points z even when it is attained at some point; and even when it is attained at some z, it is only attained by a sparse subsequence.

To deal with these issues, we first observe that the diagonal matrix elements of $\Phi_z^\tau \otimes (\Phi_z^\tau)^*$ are the same as for the partial time averages

$$\langle \Phi_z^\tau \otimes (\Phi_z^\tau)^*\rangle_T := \frac{1}{2T}\int_{-T}^{T} U^t\left(\Phi_z^\tau \otimes (\Phi_z^\tau)\right)^* U^{-t}dt. \tag{29}$$

The full time average is the limit $T \to \infty$ in (29) in the weak operator topology. In general, the Fourier integral properties of $\Phi_z^\tau \otimes (\Phi_z^\tau)^*$ are destroyed by infinite time averaging, or at best they are unclear. But if z is a periodic point for g^t of period L, then $\langle \Phi_z^\tau \otimes (\Phi_z^\tau)^*\rangle_L$ is a Fourier integral operator with complex phase on M which commutes to leading order with $\sqrt{\Delta}$, i.e.

$$\begin{aligned}
[\sqrt{\Delta}, \langle \Phi_z^\tau \otimes (\Phi_z^\tau)^*\rangle_L] &= \tfrac{1}{2L}\int_{-L}^{L} \tfrac{d}{dt}U^t\left(\Phi_z^\tau \otimes (\Phi_z^\tau)\right)^* U^{-t}dt \\
&= \tfrac{1}{2L}\left(U^L\left(\Phi_z^\tau \otimes (\Phi_z^\tau)\right)^* U^{-L} - \left(\Phi_z^\tau \otimes (\Phi_z^\tau)\right)^*\right).
\end{aligned}$$

So for a periodic point, we set T equal to an integer multiple of the periodic and substitute the time average into (26). Geometrically, this corresponds to replacing (28) by its flowout

$$\bigcup_{t\in[0,L]} \Lambda_{g^t(z)} \times \Lambda_{g^t(z)} = \gamma \times \gamma, \tag{30}$$

where γ is the periodic orbit of z. Obviously it is invariant under $g^t \times g^t$.

Secondly, we pull back $\varphi_j^{\mathbb{C}}$ under a parametrization $\gamma^\tau(t)$ of the g^t orbit of z on ∂M_τ. That is, we restrict the complexified eigenfunction to the orbit of z. We then re-normalize $\gamma^{\tau *} e^{-2\tau\lambda_j} |\varphi_j^{\mathbb{C}}(z)|^2$ to have mass one on γ. Here we use the orbit of g^t on ∂M_τ and define

$$U_j^\tau(t) = \frac{\varphi_j^{\mathbb{C}}(\gamma^\tau(t))}{||\gamma^{\tau *}\varphi_j^{\mathbb{C}}||_{L^2([0,L])}}. \tag{31}$$

We then have

Proposition 4.1. [**Z10**] *(Proposition 2) If z is a periodic point, and if U_j does not vanish identitically on the orbit of z, then the unique weak * limit of the positive unit mass measures $\{|U_j|^2\}$ on the orbit γ of z in ∂M_τ is $\frac{1}{L}dt$, the normalized periodic orbit measure on γ. Equivalently, $\{|U_j|^2\} \to dt$ weakly as $j \to \infty$.*

The proof is given in [**Z10**]. One proves an Egorov type theorem in the class of Toeplitz operators. It shows that the weak* limits must be invariant under g^t. Since they are invariant probability measures on γ the only possible limit is $\frac{1}{L}dt$.

In [**Z8**], an asymptotic formula for the average over the spectrum of the matrix elements (27) is given. It involves the stability matrix of the geodesic flow along the orbit of z. However, this data cancels when we take the quotient with $||\gamma^{\tau *}\varphi_j^{\mathbb{C}}||_{L^2([0,L])}$.

One may understand Proposition 4.1 in terms of time averages of the Fourier integral operator $\Phi_z^\tau \otimes \Phi_z^{\tau *}$. The leading order evolution $U^t \Phi_z^\tau \otimes (\Phi_z^\tau) U^{-t}$ is somewhat analogous to the evolution of the standard coherent states

$$\psi_{x,\xi}^\hbar(y) = 2^{-n/4}(2\pi\hbar)^{-\frac{3n}{4}} e^{-i\frac{\xi\cdot x}{2\hbar}} e^{i\frac{\xi\cdot y}{\hbar}} e^{-\frac{(x-y)^2}{2\hbar}} \tag{32}$$

on $\mathbb{R}^n$, although it is closer to that of coherent states in the Bargmann-Fock representation. The evolution of coherent states has been studied extensively in various settings. The model case of evolution of standard coherent states in the Schrödinger or Bargmann-Fock representations under linear Hamilton flows is discussed in detail in [**CR**], and the evolution of coherent states on manifolds are discussed in [**PU**]. The case relevant here is that of the Poisson FBI transform to Grauert tubes and the the Poisson coherent states are discussed in [**Z8, Z10**]. In each case, to leading order the projection evolves as that of a distorted coherent state projection $\Phi^\tau_{g^L(z)} \otimes (\Phi^\tau_{g^L(z)})^*$ 'centered' on the orbit of z. The shape distortion is due to the Jacobi stability matrix (i.e. Dg^t along the orbit). Thus, the time average (29) with $T = L_\gamma$ (the period of the orbit) is a Fourier integral operator with complex phase space with wave front set along $\gamma \times \gamma$ and with symbol constant along the orbit.

It would be interested to find the optimal analogue of Proposition 4.1 when z is not a periodic point, e.g. when it is a regular point for g^t and when g^t is ergodic on ∂M_τ. This case is also studied in [**Z10**]. The problem is that one cannot L^2 normalize on the entire orbit unless one uses a weight, e.g. the characteristic function of an interval. But it is apriori possible that the local L^2 norms are incommensurable along the orbit. This cannot happen over fixed compact sets but the norms of the restrictions could have different orders of magnitudes on parameter intervals for $\gamma : \mathbb{R} \to \partial M_\tau$ which are separated by an amount greater than the Ehrenfest time $C \log \lambda$.

4.2. Coherent states. A variation on the preceding example is to study matrix elements for coherent state projectors, where the coherent states are defined by (32), or more generally (as in [**PU**]),

$$\psi^a_{x,\xi}(y) = \rho(x-y)2^{-n/4}(2\pi\hbar)^{-\frac{3n}{4}} e^{-i\frac{\xi\cdot x}{2\hbar}} e^{i\frac{\xi\cdot y}{\hbar}} \hat{a}(\frac{(x-y)}{\sqrt{\hbar}})$$

for $a \in \mathcal{S}(\mathbb{R})$. The coherent states (or wave packet) transform is defined by $f \to \langle f, \psi^\hbar_{x,\xi}\rangle$. Like (26), it is an FBI transform but it is adapted to the heat kernel rather than the Poisson kernel and has different inversion properties.

As above, $\psi^a_{x,\xi} \otimes \psi^{a*}_{x,\xi}$ is a semi-classical Fourier integral operator with complex phase. If (x, ξ) is a periodic point of period L we again consider,

$$\langle \psi^a_{x,\xi} \otimes \psi^{a*}_{x,\xi}\rangle_L := \frac{1}{2L}\int_{-L}^{L} U^t(\psi^a_{x,\xi} \otimes \psi^{a*}_{x,\xi})U^{-t}dt.$$

As above, $\langle\langle \psi^a_{x,\xi} \otimes \psi^{a*}_{x,\xi}\rangle_L \psi_j, \psi_j\rangle = \langle \psi^a_{x,\xi} \otimes \psi^{a*}_{x,\xi}\psi_j, \psi_j\rangle$. To leading order in $\hbar$ it is the same as

$$\langle \psi^a_{x,\xi} \otimes \psi^{a*}_{x,\xi}\rangle_L \simeq \frac{1}{2L}\int_{-L}^{L} \psi^{a_t}_{G^t(x,\xi)} \otimes \psi^{a_t*}_{G^t(x,\xi)} dt,$$

where a_t is a deformed symbol whose principal part is $a_0 \circ dg^t$ (cf. [**CR**, **PU**]).

As in the setting of analytic continuation to Grauert tubes, we restrict the diagonal matrix elements $|\langle\langle \psi^a_{x,\xi} \otimes \psi^{a*}_{x,\xi}\rangle_L \psi_j, \psi_j\rangle|^2$ to the orbit of (x, ξ) and view them as a sequence of positive measures. When (x, ξ) is a periodic point, we restrict to the pull back to $[0, L]$ and divide by the mass to obtain a sequence of probability measures on $[0, L]$. The weak* limits are then constant multiples of Lebesgue measure as in Proposition 4.1.

5. Proof of Proposition 4

This additional symmetry of Proposition 4 comes from the fact that $I^0(M \times M, C)$ is a right and left module over $\Psi^0(M)$ with respect to composition of operators, and that on the symbol level the left and right compositions commute We thus consider the case where $\rho_j(AF) \sim \rho_j(FA)$ where $A \in \Psi^0(X)$, e.g. if $F^* = F$ as in the cases $\cos t\sqrt{\Delta}$ or T_p, or if φ_j is an eigenfunction of F. We restate Proposition 4 and refer to the diagram (11).

PROPOSITION 5.1. *Suppose that $\rho_j(AF) \sim \rho_j(FA)$ and that the canonical relation C is invariant under the geodesic flow G^t. Then the limit measures are among the G^t-invariant signed measures of mass ≤ 1 on C satisfying $\rho_\infty(\pi_X^* a\sigma) = \rho_\infty(\pi_Y^* a\sigma)$.*

COROLLARY 5.2. *Let $\iota : C \to C$ be the involution $\iota(\xi, \eta) = (\eta, \xi)$. Let ν_∞ be the limit on Ψ^0. Then $\iota^*\rho_\infty = \rho_\infty$. Equivalently,*

$$\pi_{1*}\rho_\infty = \pi_{2*}\rho_\infty = \nu_\infty.$$

In the case where C is the graph of a canonical transformation χ, this says that if we identify ρ_∞ with a measure on the domain, then $\chi^*\rho_\infty = \rho_\infty$. In the general case of a local canonical graph, it does not say that ρ_∞ is invariant under the correspondence but rather than it is invariant under each local branch.

PROOF. If we fix one (elliptic) element $F \in I^0(M \times M, C)$ (i.e. with nowhere vanishing symbol) then $I^0(M \times M, C)$ is spanned by sums of operators AFB with $A, B \in \Psi^0(M)$. Thus ρ_j induces functionals on $\Psi^0(M) \times \Psi^0(M)$ of the form

$$\rho_j(AFB) := \langle AFB\varphi_j, \varphi_j\rangle, \tag{33}$$

which can be normalized to have mass one by

$$\hat{\beta}_j(AFB) := \frac{\langle AFB\varphi_j, \varphi_j\rangle}{\langle F\varphi_j, \varphi_j\rangle}, \quad \text{if } F\varphi_j \neq 0. \tag{34}$$

Thus, we can define right and left functionals

$$\beta_j^L(A) = \rho_j(AF), \quad \beta_j^R(A) = \rho_j(FA) \text{ on } \Psi^0(M)$$

and obtain signed limit measures on S^*M from the weak* limits. Then $\{\beta_j^L\}$ and $\{\beta_j^R\}$ have weak* limits along the same subsequences and the limits are the same. Then

$$\rho_\infty(\pi_X^* a\sigma) \sim \rho_j(AF) = \langle AF\varphi_j, \varphi_j\rangle = \overline{\langle F^*A^*\varphi_j, \varphi_j\rangle} = \overline{\langle FA^*\varphi_j, \varphi_j\rangle} \sim \pi_Y^* a\sigma$$

since $\sigma_{A^*} \sim \overline{\sigma_A}$ and since $\overline{\sigma_F} = \sigma_F$.

□

Remark: When F is associated to a symplectic diffeomorphism χ of $T^*M\backslash 0$, then the left and right functionals are related as follows:

$$\beta_j^R(A) = \rho_j(FA) = \rho_j(FAF^{-1}F) = \beta_j^L(F^{-1}AF). \tag{35}$$

By Egorov's theorem $F^{-1}AF$ is a pseudo-differential operator with symbol $\sigma_A \circ \chi$, and so along any sequence with a weak* limit

$$\beta_\infty^R(\sigma_A) = \beta_\infty^L(\sigma_A \circ \chi). \tag{36}$$

6. Isometries and Hecke operators

In this section we consider the weak* limits in Proposition 1 and Corollary 1 in the case of isometries and sums of isometries known as Hecke operators. By lifting the weak* limit problem to the canonical relation instead of S^*M and using Propositon 5.1, we obtain a new invariance principle. We also make concrete identifications of the canonical relations.

In discussing the canonical relations of Hecke operators we make use of the well-known co-tangent lift $f_\sharp$ of a diffeomorphism $f : X_1 \to X_2$, defined by

$$f_\sharp(x_1, \xi_1) = (x_2, \xi_2), \quad \text{with } \begin{cases} x_2 = f(x_1), \\ \xi_1 = df_{x_1}^* \xi_2 \in T_{x_1}^* X_1, \end{cases} \tag{37}$$

where

$$(df_{x_1})^* : T_{x_2}^* X_2 \to T_{x_1}^* X_1,$$

so that

$$f_\sharp|_{T^*_{x_1}} = (df_{x_1})^{*-1}.$$

Then $f_\sharp$ is a symplectic diffeomorphism.

6.1. Isometries: $F = T_g$. We begin with the simplest example where $F = T_g$ is translation by an isometry $T_g f(x) = f(gx)$ of a Riemannian manifold (M, ds^2). The canonical relation is the graph Γ_g of the lift (37) of T_g to T^*M. Since g is an isometry, T_g commutes with Δ and we consider thier joint eigenfuntions with $T_g\varphi_j = e^{i\theta_j}\varphi_j$ for some $e^{i\theta_j} \in S^1$ (such as spherical harmonics Y^ℓ_m if g is a rotation around the vertical axis). The relevant space of operators $I^0(M \times M, \Gamma_g)$ is spanned by sums of operators AT_gB with $A, B \in \Psi^0(M)$ and so it suffices to consider the functionals

$$\hat{\rho}_j(AT_gB) = \langle AT_gB\varphi_j, \varphi_j\rangle = \langle AT_gBT_g^{-1}T_g\varphi_j, \varphi_j\rangle. \tag{38}$$

Note that $AT_gBT_g^{-1} \in \Psi^0(M)$. As one sees from the case $A = B = I$, a subsequence $\hat{\rho}_{j_k}$ can only have a unique weak* limit if the associated eigenvalues $e^{i\theta_{j_k}}$ have a limit. To define matrix elements with larger subsequential limits, we re-normalize the functionals by

$$\rho_j(AT_gB) := \frac{\langle AT_gB\varphi_j, \varphi_j\rangle}{\langle T_g\varphi_j, \varphi_j\rangle} = \langle AT_gBT_g^{-1}\varphi_j, \varphi_j\rangle. \tag{39}$$

Here, we assume $\langle T_g\varphi_j, \varphi_j\rangle \neq 0$. The reader may prefer the $\hat{\rho}_j$; the methods and results apply equally to them and to (39).

We could regard the weak* limits as measures on S^*M or on $S\Gamma_g$, the graph of the lift of T_g on S^*M, which has the form $\{(\zeta, g\cdot\zeta) : \zeta \in S^*M\}$. To illustrate the lift, we write the quantum limit as

$$\rho_\infty(AT_gB) = \int_{S^*M} a(\zeta)b(g\cdot\zeta)d\nu(\zeta, g\cdot\zeta). \tag{40}$$

PROPOSITION 6.1. *Let $g \in Isom(M, ds^2)$ and let ν_g be a weak* limit measure for the functionals $\rho_j(F) = \frac{\langle F\varphi_j, \varphi_j\rangle}{\langle T_g\varphi_j, \varphi_j\rangle}$ on $I^0(M \times M, C_g)$. Suppose that $T_g\varphi_j = e^{-i\theta_j}\varphi_j$. Then under the identification $S^*M \to C_g, \zeta \to (\zeta, g\cdot\zeta)$, $\nu = \pi^*\bar{\nu}$ is a signed measure of mass ≤ 1 on S^*M which is invariant under both G^t and g.*

PROOF. We have,

$$\int_{SC_g} \sigma_{AT_gB}d\nu = \lim_{k\to\infty} \frac{\langle AT_gB\varphi_{j_k}, \varphi_{j_k}\rangle}{\langle T_g\varphi_{j_k}, \varphi_{j_k}\rangle}.$$

Using $A = I$ and then $B = I$ we get $\pi_*\nu = \bar{\nu}, \rho_*\nu = g_*\pi_*\nu = g_*\bar{\nu} = \nu$. In the notation (40), this says,

$$\int_{S^*M} a(\zeta)d\nu(\zeta, g\cdot\zeta) = \int_{S^*M} a(\zeta)d\omega(\zeta),$$

and

$$\int_{S^*M} b(\zeta)d\omega(\zeta) = \int_{S^*M} b(g\cdot\zeta)d\nu(\zeta, g\cdot\zeta) = \int_{S^*M} b(\zeta)d\nu(g^{-1}\zeta, \zeta),$$

which implies

$$\int_{S^*M} a(\zeta)d\nu(\zeta, g\cdot\zeta) = \int_{S^*M} a(\zeta)d\nu(g^{-1}\zeta, \zeta),$$

i.e. $d\nu(\zeta, g\cdot\zeta) = d\nu(g^{-1}\zeta, \cdot\zeta)$ or

$$\int_{S^*M} a(\zeta)d\omega(\zeta) = \int_{S^*M} a(g\zeta)d\omega(\zeta). \tag{41}$$

□

6.2. Hecke operators on spheres. The now consider the interesting example of self-adjoint Hecke operators on S^n, i.e. sums

$$Tf(x) = \frac{1}{2d}\sum_{j=1}^{d}(f(g_jx) + f(g_j^{-1}x)), \quad g_j \in SO(n+1) \tag{42}$$

of isometries on S^n. Note that T is normalized so that $T1 = 1$ and $||T|| = 1$. They are a helpful guide to Hecke correspondences on hyperbolic quotients in the next section, and have a considerable literature of their own (see [**LPS**] and subsequent articles). The main result of this section, Proposition 6.3, gives a new invariance principle for the quantum limits of joint eigenfunctions of T and Δ. A key point is to view the limit measures as measures on the canonical relation, which we may identify with $\bigcup_{j=1}^{2d} S^*S^n$, rather than on S^*S^n.

A Hecke operator (42) is a discrete Radon transform $T = \rho_*\pi^*$ corresponding to the trivial cover,

$$\begin{array}{ccc} & \bigcup_{j=1}^{2d} S^n & \\ \pi \swarrow & & \searrow \rho \\ S^n & & S^n, \end{array} \tag{43}$$

where $\pi(x,j) = x$ and $\rho(x,j) = g_jx$. The canonical relation is the cotangent lift of the graph

$$\mathcal{G}_{T_{g_j}} = \{(z, g_jz) : x \in S^n)\}$$

of the isometric correspondence

$$C_T(x) = \{g_1x, \ldots, g_{2d}x\}.$$

Thus we have a second diagram

$$\begin{array}{ccc} & \mathcal{G}_{T_g} \subset S^n \times S^n & \\ \pi \swarrow & & \searrow \rho \\ S^n & & S^n. \end{array} \tag{44}$$

The graph of the correspondence is an immersed submanifold of $S^n \times S^n$ with self-intersections when the graphs of the individual isometries g_j intersect. This is clear since $\iota(x,j) = \iota(x',j') \implies x = x'$ and $g_jx = g_{j'}x$ or $g_{j'}^{-1}g_jx = x$. The diagrams (43) and (44) and are related as follows.

LEMMA 6.2. *The map*

$$\iota : \bigcup_{j=1}^{2d} S^n \to \mathcal{G}_{T_g}$$

defined by

$$\iota(\hat{x}) = (\pi(\hat{x}), \rho(\hat{x})), \quad or \ \iota(x, j) = (x, g_j x)$$

is an immersion whose image is the graph.

The canonical relation $C_T \subset T^*(S^n \times S^n)$ of the Hecke operator T is the graph of the sum of cotangent lifts of the invidivual isometries, and the cotangent lift of the map in Proposition 6.2 gives a parametrization

$$\iota : \bigcup_{j=1}^{2d} S^* S^n \to C_T$$

defined by

$$\iota(\hat{x}) = (\pi_\sharp(\hat{x}), \xi, \rho(\hat{x}), \xi), \quad \text{or} \ \iota_\sharp(x, \xi, j) = (x, g_j x, \xi, Dg_j^{*-1}\xi). \tag{45}$$

Thus, we may consider quantum limit measures as measures on $\bigcup_{j=1}^{2d} S^* S^n$, i.e. as a finite set $\{\frac{1}{2}(\nu_j + \overline{\nu_j}\}_{j=1}^{d}$ of real signed measures on $S^* S^n$.

When $\langle T u_j, u_j \rangle \neq 0$, we can define the Wigner functionals by the following normalized matrix elements (cf. (39))

$$\rho_{j,T}(ATB) = \frac{\langle ATBu_j, u_j \rangle}{\langle Tu_j, u_j \rangle}, \quad A, B \in \Psi^0(S^n). \tag{46}$$

We denote by $\hat{\rho}_j$ the other normalization as in (38).

We can also define the Wigner functionals for the individual isometries g_j:

$$\rho_{j,g_k}(ATB) = \frac{\langle AT_{g_k} Bu_j, u_j \rangle}{\langle T_{g_k} u_j, u_j \rangle}, \quad A, B \in \Psi^0(S^n). \tag{47}$$

Note that

$$\rho_{j,g_k^{-1}}(ATB) = \overline{\rho_{j,g_k}(B^* T A^*)}.$$

Exactly as in Proposition 6.1, the weak * limits of the ρ_{j,g_k} for each k are linear functionals of the form

$$\nu_k(ATB) = \int_{S^* S^n} a(\zeta) b(g_k \zeta) d\nu_k(\zeta, g_k \zeta), \tag{48}$$

although they are not necessarily invariant under g_k. It follows that

$$\int_{\hat{\Gamma}_T} a(x) b(y) d\nu_T(x, y)) = \sum_k \int_{S^* S^n} a(\zeta) b(g_k \zeta) d\nu_k(\zeta, g_k \zeta).$$

Lift $\{u_j\}$ to $\{\pi^* u_j\}$ on $\bigcup_{j=1}^{2d} S^n$ and let ν be a quantum limit measure of the sequence. If A, B are microsupported on the kth component, then the quantum limit is ν_k. That is,

$$\int_{\hat{\Gamma}_T} a(x) b(y) d\omega_T(x, y)) = \sum_k \int_{S^* S^n} a_k(\zeta) b_k(g_k \zeta) d\nu_k(\zeta, g_k \zeta).$$

In Proposition 6.1, the measure ν_k is defined on the graph of the lift of T_{g_k} to $S^* M$; it may (and will) be identified with a measure on $S^* M$. Applying Proposition 4, we get

PROPOSITION 6.3. *Suppose that $\{u_{j_k}\}$ is a sequence of joint $\Delta - T$ eigenfunctions for which (1) has a unique weak * limit μ on S^*S^n and (39) has a unique weak* limit ν on $\bigcup_{j=1}^{2d} S^*S^n$ as in Proposition 6.2. Then $\pi_*\nu = \rho_*\nu = \mu$. Moreover, each $\rho_{j,k,g,T}$ in (47) has a weak limit ν_k, and*

$$\mu = \frac{1}{2d}\sum_{j=1}^{2d}\nu_j = \frac{1}{2d}\sum_{j=1}^{2d}(g_j)_*\nu_j = \frac{1}{2d}\sum_{j=1}^{2d}(g_j^{-1})_*\nu_j. \tag{49}$$

The measures ν_j are absolutely continuous with respect to μ and for each g_j, and also $T_{g_j*}\nu_j$ is absolutely continuous with respect to μ. Similarly for all powers T^n.

Remark: If we use $\hat{\rho}_j$ (cf. (38)) instead of (46) then the statement should be: the sequence of eigenvalues $\rho_{j_k}(T)$ of T-eigenvalues has a limit $\rho(T)$ and

$$\mu = \frac{1}{2d}\sum_{j=1}^{2d}\nu_j = \frac{\rho(T)}{2d}\sum_{j=1}^{2d}(g_j)_*\nu_j. \tag{50}$$

PROOF. First, the statement that $\mu = \frac{1}{2d}\sum_{j=1}^{2d}\nu_j$ follows from the fact that the quantum limit ν is calculated on $\bigcup_{j=1}^{2d} S^n$ using the pullbacks of $\{u_{j_k}\}$. If we test against an operator which is also pulled back we get $2d$ times the quantum limit on the base. The statement that $\mu = \frac{1}{2d}\sum_{j=1}^{2d}(g_j)_*\nu_j$ comes the fact that $\{u_{j_k}\}$ is a sequence of T-eigenfunctions.Then the second equality is obvious by taking limits of $\langle TAu_{j_k}, u_{j_k}\rangle$ for operators pulled back from the base.

Since T_{g_k} is unitary,

$$|\langle T_{g_k}A\varphi_j, \varphi_j\rangle|^2 \leq \langle A^*A\varphi_j, \varphi_j\rangle \tag{51}$$

and that implies $\nu_k << \mu$. Indeed, (51) implies that along the relevant subsequence,

$$\lim |\langle T_{g_k}A\varphi_j, \varphi_j\rangle|^2 = |\int_{S^*M}\sigma_A d\nu_k|^2 \leq \int_{S^*M}\sigma_A^2 d\mu. \tag{52}$$

Also, if $A^* = A$,

$$|\langle T_{g_k}A\varphi_j, \varphi_j\rangle|^2 = |\langle T_{g_k}AT_{g_k}^{-1}T_{g_k}\varphi_j, \varphi_j\rangle|^2 \leq \langle (T_{g_k}AT_{g_k}^{-1})^2\varphi_j, \varphi_j\rangle \tag{53}$$

and

$$\lim |\langle T_{g_k}A\varphi_j, \varphi_j\rangle|^2 = |\int_{S^*M}\sigma_A d\nu_k|^2 \leq \int_{S^*M}(T_{g_k}\sigma_A)^2 d\mu, \tag{54}$$

and therefore $d\nu_k << g_{k*}^{-1}d\mu$. Also

$$|\langle AT_{g_k}\varphi_j, \varphi_j\rangle|^2 = |\langle T_{g_k}T_{g_k}^{-1}AT_{g_k}\varphi_j, \varphi_j\rangle|^2 \leq \langle (T_{g_k}^{-1}AT_{g_k})^2\varphi_j, \varphi_j\rangle \tag{55}$$

and

$$\lim |\langle T_{g_k}^{-1}A\varphi_j, \varphi_j\rangle|^2 = |\int_{S^*M}\sigma_A d\nu_k|^2 \leq \int_{S^*M}(T_{g_k}^{-1}\sigma_A)^2 d\mu, \tag{56}$$

and therefore $d\nu_k << g_{k*}d\mu$.

Here we are using (35)-(36) termwise on each term and the fact that φ_j is a T-eigenfunction. That is, ν_j is essentially the left limit and $g_{j*}\nu_j$ is the right limit. Inequality (51) is the inequality for the right functional and (53) is the inequality for the left functional. □

Remark: If instead of ρ_j we use $\hat{\rho}_j$, then in order for the weak limit to exist, we need afortiori that the sequence $\langle Tu_{j_k}, u_{j_k}\rangle$ tends to a limit, i.e. that the associated sequence of eigenvalues tends to a limit. The rest proceeds as above.

COROLLARY 6.4. *In the notation of Proposition* 6.3, *let* Λ_j *be the singular support of* ν_j *and let* Λ *be the singular support of* μ. *Then* $\Lambda = \bigcup_j \Lambda_j = \bigcup_j g_j(\Lambda_j)$.

PROOF. Let Λ_j be the singular support of ν_j. In view of (49), $\Lambda \subset \bigcup_{j=1}^{d} \Lambda_j$. Strict inclusion is apriori possible since cancellations may occur among the terms. On the other hand, since all $\nu_j << \mu$, $\Lambda_j \subset \Lambda$ for all j and therefore $\bigcup_{j=1}^{2d} \Lambda_j \subset \Lambda$. It follows that

$$\Lambda = \bigcup_{j=1}^{2d} \Lambda_j. \tag{57}$$

It also follows from (49) that $\Lambda \subset \bigcup_{j=1}^{2d} g_j\Lambda_j$, since the singular support of $g_{j*}\nu_j$ is $g_j\Lambda_j$. But also by (54), $g_{j*}\nu_j << \mu$, so $g_{j*}\nu_j$ has singular support in Λ for all j, i.e. $g_j\Lambda_j \subset \Lambda$ and therefore

$$\Lambda = \bigcup_{j=1}^{2d} g_j\Lambda_j. \tag{58}$$

□

Remark: If $T\varphi_j = \rho_j(T)\varphi_j$ then also $T^\ell\varphi_j = \rho_j(T)^\ell\varphi_j$. The Proposition and Corollary apply equally to all powers of T.

For generic T, we can obtain some simple restrictions on weak* limits from Corollary 6.4 and Remark 6.2. For simplicity, we assume that $n = 2$, so that each maximal abelian subgroup is one dimensional. We assume that each g_j is a topological generator in a maximal abelian subgroup, i.e. that it is a rotation with an irrational angle. We also assume that the g_j are pairwise independent in the sense that no two generate the same circle of rotations (unless they are inverses).

As a sample application, we show that Λ cannot be a single closed geodesic γ, i.e. $d\mu$ cannot be a constant multiple of the invariant probability measure μ_γ on γ. Otherwise, we would have $\nu_j = c_j\mu_\gamma$ with $|c_j| \leq 1$ for all j. Certainly $c_{j_0} \neq 0$ for some j_0. Corollary 6.4 then forces $g_{j_0}\gamma = \gamma$, i.e. g_{j_0} is a generator of γ. But by the assumption that the g_j are independent, this forces the other $c_j = 0$ (i.e. except for $g_{j_0}, g_{j_0}^{-1}$). By Proposition 6.3 we would then have $\mu_\gamma = \frac{1}{2d}(c_{j_0} + \overline{c_{j_0}})\mu_\gamma = \frac{\rho(T)}{2d}(c_{j_0} + \overline{c_{j_0}})\mu_\gamma$. This leads to several contradictions. First, $|c_{j_0}| \leq 1$ and $d \geq 1$, so it is impossible that $1 = \frac{1}{2d}(c_{j_0} + \overline{c_{j_0}})$. The equation also forces $\rho(T) = 1$. This is a contradiction when T has a spectral gap [**LPS**]. Further contradictions arise if we consider powers of T, since almost all terms in the weak * limit sums must vanish for similar reasons.

6.3. Hecke operators on hyperbolic quotients. We now generalize Proposition 6.3 to arithmetic hyperbolic quotients. The arguments are essentially the same, with $\mathbf{H}$ replacing S^n everywhere. The one difference is that we have an additional discrete group Γ operating which commutes with the Hecke operator so that it acts on the quotient. However, we only need to use it to ensure that the relevant cover of $\mathbf{H}$ is finite sheeted.

Let Γ is a co-compact (or cofinite) discrete subgroup of $G = PSL(2,\mathbb{R})$ and let $\mathbf{X}_\Gamma$ be the corresponding compact (or finite area) hyperbolic surface. An element $g \in G, g \notin \Gamma$ is said to be in the commensurator $Comm(\Gamma)$ if

$$\Gamma'(g) := \Gamma \cap g^{-1}\Gamma g$$

is of finite index in Γ and $g^{-1}\Gamma g$. More precisely,

$$\Gamma = \bigcup_{j=1}^{d} \Gamma'(g)\gamma_j, \quad \text{(disjoint)},$$

or equivalently

$$\Gamma g \Gamma = \bigcup_{j=1}^{d} \Gamma \alpha_j, \quad \text{where } \alpha_j = g\gamma_j.$$

It is also possible to choose α_j so that $\Gamma g \Gamma = \bigcup_{j=1}^{d} \alpha_j \Gamma$. We refer to [**Sh**] for background, or [**Z1**] for notation of this section.

Similar to (43), we then have a diagram of finite (non-Galois) covers:

$$\begin{array}{ccc} & \Gamma'(g)\backslash\mathbf{H} & \\ \pi \swarrow & & \searrow \rho \\ \Gamma\backslash\mathbf{H} & \Longleftrightarrow & g^{-1}\Gamma g\backslash\mathbf{H}. \end{array} \tag{59}$$

Here,

$$\pi(\Gamma'(g)z) = \Gamma z, \;\; \rho(\Gamma'(g)z) = g\Gamma g^{-1}(g\gamma_j z) = g\Gamma z,$$

where in the definition of ρ any of the γ_j could be used. The horizontal map is $z \to g^{-1}z$. The Radon transform $\rho_*\pi^*$ of the diagram defines a Hecke operator $T_g : L^2(\mathbf{X}_\Gamma) \to L^2(\mathbf{X}_\Gamma)$,

$$T_g f(x) = \frac{1}{d}\sum_{j=1}^{d} f(g\gamma_j x).$$

Then $T_{g^{-1}}u(z) = u(g^{-1}z)$ takes Γ-invariant functions to $g^{-1}\Gamma g$-invariant functions. Hecke operators are self-adjoint and commute with the hyperbolic Laplacian and we can consider joint eigenfunctions of Δ and T_g,

$$T_g u_j = \rho_j(T_g)u_j.$$

The main difference to the case of S^n is that the covering (59) is not trivial, i.e. not a disjoint union of d copies of the base. However, below we uniformize so that it does become trivial.

The Hecke operator is a kind of averaging operator over orbits of the Hecke correspondence, which is the multi-valued holomorphic map $C_g(z) = \{\alpha_1 z, \dots, \alpha_d z\}$. Its graph $\mathcal{G}_g = \{(z, \alpha_j z) : z \in \mathbf{X}_\Gamma)\}$ is an algebraic curve in the product $\mathbf{X}_\Gamma \times g^{-1}\Gamma g\backslash\mathbf{H}$. Similar to (44) we have a second diagram

$$\begin{array}{ccc} & \mathcal{G}_g \subset \mathbf{X}_\Gamma \times g^{-1}\Gamma g\backslash\mathbf{H} & \\ \pi_1 \swarrow & & \searrow \rho_1 \\ \Gamma\backslash\mathbf{H} & \Longleftrightarrow & g^{-1}\Gamma g\backslash\mathbf{H}. \end{array} \tag{60}$$

Here, π_1, ρ_1 are the natural projections. The following is a kind of analogue of Lemma 6.2.

LEMMA 6.5. *The map*

$$\iota : \Gamma'(g)\backslash \mathbf{H} \to \Gamma_g$$

defined by

$$\iota(z) = (\pi(z), \rho(z)),$$

is a local diffeomorphic parametrization, and $\pi_1 \iota = \pi, \rho_1 \iota = \rho$.

PROOF. It is obvious that ι is well-defined, takes its values in Γ_g and intertwines the projections. Each of the maps π, ρ is itself a local diffeomorphism and therefore ι also is. The 'fiber' over $(z, \alpha_j z)$ is multiple if and only if there exist $\alpha_j, \alpha_k, j \neq k$ such that $\alpha_j z = \alpha_k z$ and this can only happen for a finite set of z. □

We now uniformize and consider Hecke operators on $\mathbf{H}$ as Γ-periodic versions of the Hecke operators on spheres. We then obtain a picture similar to that of S^n, where

$$\begin{array}{ccc} & \bigcup_{j=1}^d \mathbf{H} \times \{j\} & \\ \pi \swarrow & & \searrow \rho \\ \mathbf{H} & & \mathbf{H}, \end{array} \tag{61}$$

where $\pi(x, j) = x$ and $\rho(x, j) = g\gamma_j x$. The Radon transform $\rho_* \pi^*$ of the diagram defines the Hecke operator $T_g : L^2(\mathbf{H}) \to L^2(\mathbf{H})$. Exactly as in (45) in th case of S^n, the map

$$\iota : \bigcup_{j=1}^d \mathbf{H} \times \{j\} \to \mathcal{G}_{T_g} \subset \mathbf{H} \times \mathbf{H}$$

defined by

$$\iota(\hat{x}) = (\pi(\hat{x}), \rho(\hat{x})), \quad \text{or} \quad \iota(x, j) = (x, g\gamma_j x) \tag{62}$$

is an immersion whose image is the graph Γ_T. The self-intersection points occur when $\iota(x, j) = \iota(x', j') \implies x = x'$ and $g_j x = g_{j'} x$ or $g_{j'}^{-1} g_j x = x$. If we take the quotient by Γ of diagram (44) we get diagram (59). But it is preferable to regard all of the functionals as defined on pseudo-differential operators on $\mathbf{H}$ with compact spatial support. Taking the quotient by Γ amounts to cutting off Γ-invariant pseudo-differential operators to fundamental domains, but nothing is lost (and some generality is gained) by using non Γ-invariant pseudo-differential operators.

6.4. Quantum limits on H. On $\mathbf{H}$ we can define the Wigner functionals for the individual isometries g_j:

$$\rho_{j, g_k}(ATB) = \frac{\langle AT_{g_k} B u_j, u_j \rangle}{\langle T u_j, u_j \rangle}, \quad A, B \in \Psi_c^0(\mathbf{H}). \tag{63}$$

By Proposition 6.1, the limits have the form

$$\int_{S^*\mathbf{H}} a(\zeta) b(g_k \zeta) d\nu_k(\zeta, g_k \zeta).$$

As with S^n it follows that

$$\int_{\hat{\Gamma}_T} a(x)b(y)d\nu_T(x,y)) = \sum_k \int_{S^*\mathbf{H}} a(\zeta)b(g_k\zeta)d\nu_k(\zeta, g_k\zeta).$$

Here, $d\nu_k$ is a measure on the kth copy of $\mathbf{H}$. The Hecke limit measure is the sum

$$\nu = \sum_j \nu_{g_j} \quad \text{on} \quad \bigcup_j S^*\mathbf{H} \times \{j\},$$

with ν_{g_j} living on the jth copy $S^*\mathbf{H} \times \{j\}$.

When $\langle T_p u_j, u_j\rangle \neq 0$, we can define the Wigner functionals by the following normalized matrix elements:

$$\rho_{j,g}(AT_pB) = \frac{\langle AT_pBu_j, u_j\rangle}{\langle T_p u_j, u_j\rangle}, \quad A, B \in \Psi^0_c(\mathbf{H}). \tag{64}$$

Completely analogously to Proposition 6.3, we have:

PROPOSITION 6.6. *Suppose that $\{u_{j_k}\}$ is a sequence of joint $\Delta - T$ eigenfunctions for which* (1) *has a unique weak * limit μ on $S^*\mathbf{H}$ and* (39) *has a unique weak* limit ν on $\bigcup_{j=1}^d S^*\mathbf{H}$. Then $\pi_*\nu = \rho_*\nu = \omega$. Moreover, each $\rho_{j,k,g,T}$ in* (47) *has a weak limit ν_k and*

$$\mu = \frac{1}{d}\sum_{j=1}^d \nu_j = \frac{1}{d}\sum_{j=1}^d (g_j)_*\nu_j. \tag{65}$$

The measures ν_j are absolutely continuous with respect to μ and for each g_j, and also $T_{g_j *}\nu_j$ is absolutely continuous with respect to μ. Similarly for all powers T^n. Moreover, μ is Γ-invariant.

The proof is essentially the same as for Proposition 6.3 and is omitted. The fact that $\mathbf{H}$ is of infinite volume and the pseudo-differential operators are spatially compactly supported does not change the proof in any significant way. The only new statement is that μ is Γ-invariant, which is obvious from the fact that T commutes with Γ.

COROLLARY 6.7. *In the notation of Proposition 6.3, let Λ_j be the singular support of ν_j and let Λ be the singular support of μ. Then $\Lambda = \bigcup_j \Lambda_j = \bigcup_j g_j(\Lambda_j)$.*

Again the proof is the same as for Corollary 6.4.

6.5. Quantum limits on $S^*\Gamma'(g)\backslash\mathbf{H}$. Instead of viewing ν as a measure on $\bigcup_j S^*\mathbf{H} \times \{j\}$ as in Proposition 6.6, we may take the quotient by Γ and view it as a measure on $S^*(\Gamma'(g)\backslash\mathbf{H}) = \Gamma'(g)\backslash G$. Note that the quotient by Γ glues together the d disjoint copies of $\mathbf{H}$ and then takes the quotient of the resulting $\mathbf{H}$ by $\Gamma'(g)$. To explain the Γ-invariance properties we prove

LEMMA 6.8. *$\nu = \sum_j \nu_j$ is a Γ-invariant measure on $S^*\mathbf{H} \simeq G$.*

PROOF. Let A, B be compactly supported pseudo-differential operators on $\mathbf{H}$. Since $T_g L_\gamma = L_\gamma T_g$,

$$\begin{aligned}\langle AT_gBu_j, u_j\rangle &= \langle AT_gBL_\gamma u_j, L_\gamma u_j\rangle \\ &= \langle L_\gamma^* AT_gBL_\gamma u_j, u_j\rangle \\ &= \langle L_\gamma^* AL_\gamma T_g L_\gamma^* BL_\gamma u_j, u_j\rangle.\end{aligned}$$

Taking the limit gives $\gamma_* \nu = \nu$. □

References

[BB] V. M. Babič and V. S. Buldyrev, *Short-wavelength diffraction theory*, Springer Series on Wave Phenomena, vol. 4, Springer-Verlag, Berlin, 1991. Asymptotic methods; Translated from the 1972 Russian original by E. F. Kuester. MR1245488 (94f:78004)

[BL] Jean Bourgain and Elon Lindenstrauss, *Entropy of quantum limits*, Comm. Math. Phys. **233** (2003), no. 1, 153–171, DOI 10.1007/s00220-002-0770-8. MR1957735 (2004c:11076)

[Bou] Louis Boutet de Monvel, *Convergence dans le domaine complexe des séries de fonctions propres* (French, with English summary), C. R. Acad. Sci. Paris Sér. A-B **287** (1978), no. 13, A855–A856. MR551763 (80i:58040)

[CdV] Yves Colin de Verdière, *Quasi-modes sur les variétés Riemanniennes* (French), Invent. Math. **43** (1977), no. 1, 15–52. MR0501196 (58 #18615)

[CR] Monique Combescure and Didier Robert, *Coherent states and applications in mathematical physics*, Theoretical and Mathematical Physics, Springer, Dordrecht, 2012. MR2952171

[Co] Alain Connes, *Geometry from the spectral point of view*, Lett. Math. Phys. **34** (1995), no. 3, 203–238, DOI 10.1007/BF01872777. MR1345552 (96j:46074)

[Co2] A. Connes, *The action functional in noncommutative geometry*, Comm. Math. Phys. **117** (1988), no. 4, 673–683. MR953826 (91b:58246)

[DZ] Semyon Dyatlov and Maciej Zworski, *Quantum ergodicity for restrictions to hypersurfaces*, Nonlinearity **26** (2013), no. 1, 35–52, DOI 10.1088/0951-7715/26/1/35. MR3001760

[E] Martin Eichler, *Modular correspondences and their representations*, J. Indian Math. Soc. (N.S.) **20** (1956), 163–206. MR0085296 (19,18a)

[GL] François Golse and Eric Leichtnam, *Applications of Connes' geodesic flow to trace formulas in noncommutative geometry*, J. Funct. Anal. **160** (1998), no. 2, 408–436, DOI 10.1006/jfan.1998.3305. MR1665292 (2000a:58078)

[GS] Allan Greenleaf and Andreas Seeger, *Fourier integral operators with fold singularities*, J. Reine Angew. Math. **455** (1994), 35–56, DOI 10.1515/crll.1994.455.35. MR1293873 (95h:58130)

[Gu] Victor Guillemin, *Some classical theorems in spectral theory revisited*, Equations (Inst. Adv. Study, Princeton, N.J., 1977/78), Ann. of Math. Stud., vol. 91, Princeton Univ. Press, Princeton, N.J., 1979, pp. 219–259. MR547021 (81b:58045)

[GS1] Victor Guillemin and Matthew Stenzel, *Grauert tubes and the homogeneous Monge-Ampère equation*, J. Differential Geom. **34** (1991), no. 2, 561–570. MR1131444 (93e:32018)

[GuSt] Victor Guillemin and Shlomo Sternberg, *Some problems in integral geometry and some related problems in microlocal analysis*, Amer. J. Math. **101** (1979), no. 4, 915–955, DOI 10.2307/2373923. MR536046 (82b:58087)

[GU] Victor Guillemin and Alejandro Uribe, *Reduction and the trace formula*, J. Differential Geom. **32** (1990), no. 2, 315–347. MR1072909 (92h:58191)

[HZ] Andrew Hassell and Steve Zelditch, *Quantum ergodicity of boundary values of eigenfunctions*, Comm. Math. Phys. **248** (2004), no. 1, 119–168, DOI 10.1007/s00220-004-1070-2. MR2104608 (2005h:35255)

[HoI-IV] L. Hörmander, *Theory of Linear Partial Differential Operators I-IV*, Springer-Verlag, New York (1985).

[JSS] D. Jakobson, Y. Safarov and A. Strohmaier, The semi-classical theory of discontinuous systems and ray-splitting billiards (preprint, 2013).

[KV] Mikhail Karasev and Yuri Vorobjev, *Integral representations over isotropic submanifolds and equations of zero curvature*, Adv. Math. **135** (1998), no. 2, 220–286, DOI 10.1006/aima.1997.1703. MR1620834 (99d:58073)

[K] Yu. A. Kordyukov, *Classical and quantum ergodicity on orbifolds*, Russ. J. Math. Phys. **19** (2012), no. 3, 307–316, DOI 10.1134/S1061920812030041. MR2965547

[L] Elon Lindenstrauss, *Invariant measures and arithmetic quantum unique ergodicity*, Ann. of Math. (2) **163** (2006), no. 1, 165–219, DOI 10.4007/annals.2006.163.165. MR2195133 (2007b:11072)

[LPS] A. Lubotzky, R. Phillips, and P. Sarnak, *Hecke operators and distributing points on the sphere. I*, Comm. Pure Appl. Math. **39** (1986), no. S, suppl., S149–S186, DOI 10.1002/cpa.3160390710. Frontiers of the mathematical sciences: 1985 (New York, 1985). MR861487 (88m:11025a)

[MT] Richard B. Melrose and Michael E. Taylor, *Near peak scattering and the corrected Kirchhoff approximation for a convex obstacle*, Adv. in Math. **55** (1985), no. 3, 242–315, DOI 10.1016/0001-8708(85)90093-3. MR778964 (86m:35095)

[PU] T. Paul and A. Uribe, *On the pointwise behavior of semi-classical measures*, Comm. Math. Phys. **175** (1996), no. 2, 229–258. MR1370095 (97a:58189)

[R] J. V. Ralston, *On the construction of quasimodes associated with stable periodic orbits*, Comm. Math. Phys. **51** (1976), no. 3, 219–242. MR0426057 (54 #14003)

[Rez] A. Reznikov, Norms of Geodesic Restrictions of Eigenfunctions on hyperbolic surfaces and representation theory, (arXiv:math/0403437).

[RS] Zeév Rudnick and Peter Sarnak, *The behaviour of eigenstates of arithmetic hyperbolic manifolds*, Comm. Math. Phys. **161** (1994), no. 1, 195–213. MR1266075 (95m:11052)

[ST] Robert Schrader and Michael E. Taylor, *Semiclassical asymptotics, gauge fields, and quantum chaos*, J. Funct. Anal. **83** (1989), no. 2, 258–316, DOI 10.1016/0022-1236(89)90021-9. MR995750 (90i:58198)

[Sh] Goro Shimura, *Introduction to the arithmetic theory of automorphic functions*, Publications of the Mathematical Society of Japan, vol. 11, Princeton University Press, Princeton, NJ, 1994. Reprint of the 1971 original; Kanô Memorial Lectures, 1. MR1291394 (95e:11048)

[St] M. Stenzel, to appear.

[Ta] Daniel Tataru, *On the regularity of boundary traces for the wave equation*, Ann. Scuola Norm. Sup. Pisa Cl. Sci. (4) **26** (1998), no. 1, 185–206. MR1633000 (99e:35129)

[TZ] John A. Toth and Steve Zelditch, *Quantum ergodic restriction theorems. I: Interior hypersurfaces in domains wth ergodic billiards*, Ann. Henri Poincaré **13** (2012), no. 4, 599–670, DOI 10.1007/s00023-011-0154-8. MR2913617

[TZ2] John A. Toth and Steve Zelditch, *Quantum ergodic restriction theorems: manifolds without boundary*, Geom. Funct. Anal. **23** (2013), no. 2, 715–775, DOI 10.1007/s00039-013-0220-0. MR3053760

[W] Alan Weinstein, *On Maslov's quantization condition*, Fourier integral operators and partial differential equations (Colloq. Internat., Univ. Nice, Nice, 1974), Springer, Berlin, 1975, pp. 341–372. Lecture Notes in Math., Vol. 459. MR0436231 (55 #9178)

[Z1] Steven Zelditch, *A Szegö limit formula for conjugates of ψDOs by Hecke operators*, J. Funct. Anal. **77** (1988), no. 1, 198–210, DOI 10.1016/0022-1236(88)90084-5. MR930398 (89g:58213)

[Z2] Steven Zelditch, *Kuznecov sum formulae and Szegö limit formulae on manifolds*, Comm. Partial Differential Equations **17** (1992), no. 1-2, 221–260, DOI 10.1080/03605309208820840. MR1151262 (93e:58184)

[Z3] Steve Zelditch, *Recent developments in mathematical quantum chaos*, Current developments in mathematics, 2009, Int. Press, Somerville, MA, 2010, pp. 115–204. MR2757360 (2012b:58054)

[Z4] Steven Zelditch, *Quantum ergodicity of C^* dynamical systems*, Comm. Math. Phys. **177** (1996), no. 2, 507–528. MR1384146 (98c:46145)

[Z5] Steven Zelditch, *Quantum transition amplitudes for ergodic and for completely integrable systems*, J. Funct. Anal. **94** (1990), no. 2, 415–436, DOI 10.1016/0022-1236(90)90021-C. MR1081652 (92b:58181)

[Z6] Steven Zelditch, *Index and dynamics of quantized contact transformations* (English, with English and French summaries), Ann. Inst. Fourier (Grenoble) **47** (1997), no. 1, 305–363. MR1437187 (99a:58082)

[Z7] Steve Zelditch, *Quantum maps and automorphisms*, The breadth of symplectic and Poisson geometry, Progr. Math., vol. 232, Birkhäuser Boston, Boston, MA, 2005, pp. 623–654, DOI 10.1007/0-8176-4419-9_22. MR2103022 (2005j:53113)

[Z8] S. Zelditch, Phase space Weyl laws and eigenfunction growth on Grauert tubes (preprint, 2013).

[Z9] Steven Zelditch, *On a "quantum chaos" theorem of R. Schrader and M. Taylor*, J. Funct. Anal. **109** (1992), no. 1, 1–21, DOI 10.1016/0022-1236(92)90009-8. MR1183602 (94d:58155)

[Z10] S. Zelditch, Ergodicity and intersections of nodal sets and geodesics on real analytic surfaces, to appear in Jour. Diff. Geom. (arXiv:1210.0834).

[Zw] Maciej Zworski, *Semiclassical analysis*, Graduate Studies in Mathematics, vol. 138, American Mathematical Society, Providence, RI, 2012. MR2952218

DEPARTMENT OF MATHEMATICS, NORTHWESTERN UNIVERSITY, EVANSTON, ILLINOIS 60208-2370

E-mail address: zelditch@math.northwestern.edu

Selected Published Titles in This Series

615 **Plamen Stefanov, András Vasy, and Maciej Zworski, Editors,** Inverse Problems and Applications, 2014

611 **Robert Fitzgerald Morse, Daniela Nikolova-Popova, and Sarah Witherspoon, Editors,** Group Theory, Combinatorics, and Computing, 2014

608 **Robert S. Doran, Greg Friedman, and Scott Nollet, Editors,** Hodge Theory, Complex Geometry, and Representation Theory, 2014

607 **Kiyoshi Igusa, Alex Martsinkovsky, and Gordana Todorov, Editors,** Expository Lectures on Representation Theory, 2014

606 **Chantal David, Matilde Lalín, and Michelle Manes, Editors,** Women in Numbers 2, 2013

605 **Omid Amini, Matthew Baker, and Xander Faber, Editors,** Tropical and Non-Archimedean Geometry, 2013

604 **José Luis Montaña and Luis M. Pardo, Editors,** Recent Advances in Real Complexity and Computation, 2013

603 **Azita Mayeli, Alex Iosevich, Palle E. T. Jorgensen, and Gestur Ólafsson, Editors,** Commutative and Noncommutative Harmonic Analysis and Applications, 2013

602 **Vyjayanthi Chari, Jacob Greenstein, Kailash C. Misra, K. N. Raghavan, and Sankaran Viswanath, Editors,** Recent Developments in Algebraic and Combinatorial Aspects of Representation Theory, 2013

601 **David Carfi, Michel L. Lapidus, Erin P. J. Pearse, and Machiel van Frankenhuijsen, Editors,** Fractal Geometry and Dynamical Systems in Pure and Applied Mathematics II, 2013

600 **David Carfi, Michel L. Lapidus, Erin P. J. Pearse, and Machiel van Frankenhuijsen, Editors,** Fractal Geometry and Dynamical Systems in Pure and Applied Mathematics I, 2013

599 **Mohammad Ghomi, Junfang Li, John McCuan, Vladimir Oliker, Fernando Schwartz, and Gilbert Weinstein, Editors,** Geometric Analysis, Mathematical Relativity, and Nonlinear Partial Differential Equations, 2013

598 **Eric Todd Quinto, Fulton Gonzalez, and Jens Gerlach Christensen, Editors,** Geometric Analysis and Integral Geometry, 2013

597 **Craig D. Hodgson, William H. Jaco, Martin G. Scharlemann, and Stephan Tillmann, Editors,** Geometry and Topology Down Under, 2013

596 **Khodr Shamseddine, Editor,** Advances in Ultrametric Analysis, 2013

595 **James B. Serrin, Enzo L. Mitidieri, and Vicenţiu D. Rădulescu, Editors,** Recent Trends in Nonlinear Partial Differential Equations II, 2013

594 **James B. Serrin, Enzo L. Mitidieri, and Vicenţiu D. Rădulescu, Editors,** Recent Trends in Nonlinear Partial Differential Equations I, 2013

593 **Anton Dzhamay, Kenichi Maruno, and Virgil U. Pierce, Editors,** Algebraic and Geometric Aspects of Integrable Systems and Random Matrices, 2013

592 **Arkady Berenstein and Vladimir Retakh, Editors,** Noncommutative Birational Geometry, Representations and Combinatorics, 2013

591 **Mark L. Agranovsky, Matania Ben-Artzi, Greg Galloway, Lavi Karp, Vladimir Maz'ya, Simeon Reich, David Shoikhet, Gilbert Weinstein, and Lawrence Zalcman, Editors,** Complex Analysis and Dynamical Systems V, 2013

590 **Ursula Hamenstädt, Alan W. Reid, Rubí Rodríguez, Steffen Rohde, and Michael Wolf, Editors,** In the Tradition of Ahlfors-Bers, VI, 2013

589 **Erwan Brugallé, Mariá Angélica Cueto, Alicia Dickenstein, Eva-Maria Feichtner, and Ilia Itenberg, Editors,** Algebraic and Combinatorial Aspects of Tropical Geometry, 2013

For a complete list of titles in this series, visit the
AMS Bookstore at **www.ams.org/bookstore/conmseries/**.